Biomass for Bioenergy and Biomaterials

Emerging Materials and Technologies

Series Editor:
Boris I. Kharissov

Fire-Resistant Paper: Materials, Technologies, and Applications
Ying-Jie Zhu

Sensors for Stretchable Electronics in Nanotechnology
Kaushik Pal

Polymer-Based Composites: Design, Manufacturing, and Applications
V. Arumugaprabu, R. Deepak Joel Johnson, M. Uthayakumar, and P. Sivaranjana

Nanomaterials in Bionanotechnology: Fundamentals and Applications
Ravindra Pratap Singh and Kshitij RB Singh

Biomaterials and Materials for Medicine: Innovations in Research,
Devices, and Applications
Jingan Li

Advanced Materials and Technologies for Wastewater Treatment
Sreedevi Upadhyayula and Amita Chaudhary

Green Tribology: Emerging Technologies and Applications
*T.V.V.L.N. Rao, Salmiah Binti Kasolang, Xie Guoxin, Jitendra Kumar Katiyar, and
Ahmad Majdi Abdul Rani*

Biotribology: Emerging Technologies and Applications
*T.V.V.L.N. Rao, Salmiah Binti Kasolang, Xie Guoxin, Jitendra Kumar Katiyar,
and Ahmad Majdi Abdul Rani*

Bioengineering and Biomaterials in Ventricular Assist Devices
Eduardo Guy Perpétuo Bock

Semiconducting Black Phosphorus: From 2D Nanomaterial to
Emerging 3D Architecture
Han Zhang, Nasir Mahmood Abbasi, and Bing Wang

Biomass for Bioenergy and Biomaterials
Nidhi Adlakha, Rakesh Bhatnagar, and Syed Shams Yazdani

Energy Storage and Conversion Devices: Supercapacitors, Batteries, and
Hydroelectric Cell
Anurag Guar, A.L. Sharma, and Arya Anil

Biomass for Bioenergy and Biomaterials

Edited by
Nidhi Adlakha, Rakesh Bhatnagar, and
Syed Shams Yazdani

CRC Press
Taylor & Francis Group
Boca Raton London New York

CRC Press is an imprint of the
Taylor & Francis Group, an **informa** business

First edition published 2022
by CRC Press
6000 Broken Sound Parkway NW, Suite 300, Boca Raton, FL 33487-2742

and by CRC Press
2 Park Square, Milton Park, Abingdon, Oxon, OX14 4RN

© 2022 Taylor & Francis Group, LLC

CRC Press is an imprint of Taylor & Francis Group, LLC

ISBN: 978-0-367-74555-4 (hbk)
ISBN: 978-0-367-74558-5 (pbk)
ISBN: 978-1-003-15848-6 (ebk)

DOI: 10.1201/9781003158486

Typeset in Times
by codeMantra

Contents

Preface...vii
Editors..ix
Contributors xiii

Chapter 1 Chemistry of Plant Biomass...1

*Senthil Murugan Arumugam, Shelja Sharma, Sandeep Kumar,
Sangeeta Mahala, Bhawana Devi, and Sasikumar Elumalai*

Chapter 2 Lignin to Platform Chemicals and Biomaterials: Chemical and
Biological Perspectives...31

*Yeddula Nikhileshwar Reddy, Kunal Gogde, Shatabdi Paul, and
Jayeeta Bhaumik*

Chapter 3 LCA and TEA for Biomass Conversion Technology 65

*Shilpa Main, Vishwanath H. Dalvi, Yogendra Shashtri, and
Annamma Anil Odaneth*

Chapter 4 Biomass Pre-Treatment and Liquefaction 93

Tejas M. Ukarde, Annamma Anil Odaneth, and Hitesh S. Pawar

Chapter 5 Role of Systems Biology in Enhancing Efficiency of Biocatalysts....127

Trunil Desai, Ahmad Ahmad, and Shireesh Srivastava

Chapter 6 Enzyme-Based Saccharification...153

*D. Sathish, Shivam Aggarwal, Manasa Nagesh Hegde, and
Nidhi Adlakha*

Chapter 7 Enhancement of Biomass for Deconstruction...................................179

*Lavi Rastogi, Deepika Singh, Rajan Kumar Sah,
Aniket Anant Chaudhari, and Prashant Anupama-Mohan Pawar*

Chapter 8 Lignocellulosic Biorefineries – A Step towards a Carbon-
Neutral Economy...211

*Bhawna Madan, Prachi Varshney, Parmeshwar Patil,
Vivek Rathore, Jaya Rawat, and Bharat Newalkar*

Chapter 9   Targeted Strain Engineering to Produce Bioenergy 241

             S. Bilal Jilani, Ali Samy Abdelaal, and Syed Shams Yazdani

Chapter 10   Saccharide to Biodiesel 267

             Farha Deeba, Kukkala Kiran Kumar, and Naseem A. Gaur

Chapter 11   Second-Generation Bioethanol and Biobutanol – Methods
             and Prospects.. 291

             *Guruprasad K, Anurag Singh, Bhawna Madan,
             and Mohan Yama*

Chapter 12   Biological Production of Diols – Current Perspective 327

             *Koel Saha, Divya Mudgil, Sanjukta Subudhi,
             Aishwarya Srivastava, and Nidhi Adlakha*

Chapter 13   Market Analysis of Biomass for Biofuels and Biomaterials 349

             Brajesh Barse, Navin Tamrakar, and Syed Shams Yazdani

Index... 371

Preface

With experts trying to address issues related to global warming and over-utilization of limited fossil fuel reserve at various forums, biofuels have come on the centre stage of discussion as an important alternative for transportation energy. More specifically, biomass-based products have become the target of intense investigation because of its abundance in the form of agricultural and forestry residues. The energy stored in plant biomass can be made available for human use by breaking down the chemical bonds in which they store energy received from sunlight.

The tremendous potential of biomass can easily be comprehended by its use in food, fuel and other industrial sectors. The conversion and utilization of this most abundant carbon source has ruled the scientific and industrial sectors over the last few decades. The evolving knowledge in the field of biomass to value-added products prompted us to edit the proposed book. Accordingly, we embarked on the project associating colleagues and friends to contribute to the 13 chapters for the book entitled "Biomass for Bioenergy and Biomaterials". Each contributing author has been a key scientific player in unravelling different strategies underlying the conversion of biomass to bioenergy and biomaterials. We want to sincerely thank all the authors for their valuable contributions in this venture.

The biomass transformation mostly involves large-scale bioreactions; however, until this decade, it has been tough to anticipate the techno-economic success of the entire process leading to substantial monetary loss. To this end, a new arena of technical and molecular weapons is now available which can assist in guided strain and process engineering. This book presents a comprehensive understanding of the topics, including biological and chemical hydrolysis of biomass; chemistry of plant biomass; life cycle assessment and techno-economic feasibility of bioprocess; strategies for biomass pretreatment; directed strain engineering for bioenergy and bioproducts; conversion of biomass to biomaterials; lignocellulose-based refinery; and market analysis of the existing processes. This book is exclusive as it provides the reader with a complete end-to-end solution for harvesting recalcitrant biomass for value-added products. This book offers an ideal reference guide for academic researchers and industrial engineers in the fields of natural renewable materials, biorefinery of lignocellulose, biofuels and environmental engineering. It can also be used as a comprehensive reference source for university students in metabolic and environment engineering. Each chapter in this book offers many solutions to technical hurdles that come across while bringing the biomass-based technology to market. A very old saying is that a journey of a thousand miles begins with a single step. We hope that this book offers that first step, which will direct you towards your destination.

MATLAB® is a registered trademark of The MathWorks, Inc. For product information, please contact:

The MathWorks, Inc.
3 Apple Hill Drive
Natick, MA 01760-2098 USA
Tel: 508-647-7000
Fax: 508-647-7001
E-mail: info@mathworks.com
Web: www.mathworks.com

Editors

Nidhi Adlakha, PhD, is Principal Investigator and Assistant Professor at the Regional Centre for Biotechnology in Faridabad, Haryana, India. Her research interests are in omics understanding, genetic manipulation, cellulolytic enzymes and biochemicals. She earned a BSc (Hons) in biochemistry at Delhi University, a master's in biochemistry at the prestigious Post Graduate Institute for Medical Education and Research (PGIMER), Chandigarh, and a PhD in life sciences at the International Centre for Genetic Engineering and Biotechnology.

The current research focus of her group is to gain a molecular understanding of cellulase induction mechanism. With the help of molecular tools, the laboratory aims at developing a robust platform for the production of biomass-degrading enzymes. Her group has identified a set of transcription factors and CAZymes that play a role in the early induction mechanism. Another focus of the group is to employ synthetic biology tools to channel biomass towards 2,3-butanediol production. Her initiative was featured in the *Economic Times*.

Dr. Adlakha is involved in teaching industrial biotechnology to postgraduate students at various DBT Institutes. She is a member of the editorial board of *Frontiers in Bioengineering and Biotechnology* and a reviewer of many renowned peer-reviewed journals.

Her early research led to the development of enzymes for the efficient degradation of biomass. In collaboration with the University of Aberdeen, UK, she was further involved in metabolic flux analysis and metabolic engineering of *Paenibacillus* sp. for biochemical production. She has authored papers in various peer-reviewed journals and holds one patent.

Rakesh Bhatnagar, PhD, earned a PhD in biochemistry at the National Sugar Institute, Kanpur, followed by working in Freiburg University, Germany, University of Caen, France, and NIH and USAMRIID, the USA. He pioneered the teaching of biotechnology at the Jawaharlal Nehru University, the first of its kind in the country. Prof. Bhatnagar is well known for developing recombinant vaccine against anthrax. This technology has been transferred to the industry, and the vaccine is undergoing human clinical trials.

Prof. Bhatnagar currently holds two prestigious positions i.e., along with serving as President (Vice-Chancellor) of Amity University Rajasthan, he is also a Senior Vice-President of Ritanand Balved Education Foundation, New Delhi. Prof. Bhatnagar (a former Vice Chancellor of BHU, Varanasi), in addition to being a J. C. Bose Fellow, is an elected Fellow of all three major Science Academies of

India, namely Indian National Science Academy, New Delhi, Indian Academy of
Science, Bangalore, and National Academy of Science, Allahabad. With greater than
184 research papers to his credit, his publications have received more than 5000 cita-
tions. He has supervised more than 58 PhD students. Prof. Bhatnagar's research has
been ranked 7th (the first six were from Pasteur Institute, Harvard Medical School,
NIH and USAMRIID) in the anthrax research globally. Recognizing his extraordi-
nary contribution in the field of research and innovation, he has been awarded with
many prestigious awards and recognition; few of them are President's Award for Best
Innovation by President of India and ICMR Award for outstanding research work in
the field of immunology by ICMR, GoI. He ranked 7th among the top ten eminent
researchers publishing commendable research papers on anthrax by Open Source
Global Anthrax Research Literature.

Prof. Bhatnagar has been a successful administrator and institution builder. He
has served the Jawaharlal Nehru University at various important positions such as
Chairperson, Director and Dean. Because of his efforts, School of Biotechnology has
been ranked at No. 1 in biotechnology teaching programme in the country consis-
tently. He has created a world-class Biosafety Level III laboratory.

Recently, Prof. Bhatnagar has finished his term of Vice Chancellor of Banaras
Hindu University, and during his term of Vice Chancellorship, he has strived hard to
take the University to a new level of excellence.

Syed Shams Yazdani, PhD, is the Group Leader of the
Microbial Engineering Group and Coordinator of the
DBT-ICGEB Centre for Advanced Bioenergy Research
at the International Centre for Genetic Engineering and
Biotechnology in New Delhi, India. His research interests
include metabolic engineering, synthetic biology, cellulolytic enzymes and biofuels. He earned a BSc (Hons) in
chemistry at Aligarh Muslim University and an MSc and a
PhD in biotechnology at Jawaharlal Nehru University. His
group is currently involved in the development of technolo-
gies for fungal enzymes, C5/C6 sugar fermentation and advanced biofuels production such as butanol, fatty alcohols and alkanes.

Dr. Yazdani's recent achievements include the development of a potent enzyme
preparation for use in 2G-ethanol processes with synthetic biology and genome editing tools in the fungal system. His efforts in genome and protein engineering led
to the highest reported value of drop-in fuel alkanes in bacterial systems. These
technologies are at various stages of development for commercial exploitation in collaboration with industries.

Dr. Yazdani serves in various committees of the Government of India, such as the
Department of Biotechnology (DBT), the Department of Science and Technology
(DST), the Biotechnology Industry Research Assistance Council (BIRAC) and many
academic institutions of high reputation. He was featured in *BioSpectrum* magazine
("Next-Gen Biofuel Engineer"). He represents India in the Technical Expert Group
of Synthetic Biology at the UN Convention of Biological Diversity. He leads several
multilateral international collaborative projects, including Indo-UK, Indo-Australia

and Indo-US. He is a member of the editorial board of *Scientific Reports* (Nature Publishing Group), *Journal of Industrial Microbiology and Biotechnology* (JIMB), *Frontiers in Bioengineering and Biotechnology* and *Indian Journal of Biotechnology*, and he is a reviewer for many renowned peer-reviewed journals.

Dr. Yazdani's early career research led to the development of technologies for the production of recombinant streptokinase and recombinant malaria vaccine candidates, which were transferred to industries for cGMP production, clinical trials and commercialization. He was further involved in a breakthrough discovery during his research at Rice University, Houston, USA, as a DBT overseas fellow, where a novel pathway was identified in *E. coli* to produce bioethanol from glycerine, a waste from biodiesel industry. He is the author of more than 70 publications in international journals, such as *JBC*, *Nature Communications* and *Metabolic Engineering*, and he holds 15 patent applications (with three US patents granted).

Contributors

Ali Samy Abdelaal
DBT-ICGEB Centre for Advanced
    Bioenergy Research
International Centre for Genetic
    Engineering and Biotechnology
New Delhi, India
and
Department of Genetics
Faculty of Agriculture
Damietta University
Damietta, Egypt

Shivam Aggarwal
Synthetic Biology and Bioprocessing
    Lab
Regional Centre for Biotechnology
NCR Biotech Science Cluster
Faridabad, India

Ahmad Ahmad
Systems Biology for Biofuel Group
International Centre for Genetic
    Engineering and Biotechnology
    (ICGEB)
New Delhi, India

Senthil Murugan Arumugam
Chemical Engineering Division
DBT-Center of Innovative and Applied
    Bioprocessing
Mohali, India

Brajesh Barse
DBT-ICGEB Centre for Advanced
    Bioenergy Research
and
Microbial Engineering Group
International Centre for Genetic
    Engineering and Biotechnology
    (ICGEB)
New Delhi, India

Jayeeta Bhaumik
Department of Nanomaterials and
    Application Technology
Center of Innovative and Applied
    Bioprocessing (CIAB)
Department of Biotechnology (DBT)
Government of India, Sector 81
    (Knowledge City)
Nagar, India

Aniket Anant Chaudhari
Laboratory of Plant Cell Wall Biology
Regional Centre for Biotechnology
NCR Biotech Science Cluster
Faridabad, India

Sathish D
Synthetic Biology and Bioprocessing
    Lab
Regional Centre for Biotechnology
NCR Biotech Science Cluster
Faridabad, India

Vishwanath H. Dalvi
Department of Chemical Technology
Institute of Chemical Technology
Mumbai, India

Farha Deeba
Yeast Biofuel Group
International Centre for Genetic
    Engineering and Biotechnology
    (ICGEB)
New Delhi, India

Trunil Desai
Systems Biology for Biofuel Group
International Centre for Genetic
    Engineering and Biotechnology
    (ICGEB)
New Delhi, India

Bhawana Devi
Chemical Engineering Division
DBT-Center of Innovative and Applied
    Bioprocessing
Mohali, India
and
Department of Chemical Sciences
Indian Institute of Science Education
    and Research
Mohali, India

Sasikumar Elumalai
Chemical Engineering Division
DBT-Center of Innovative and Applied
    Bioprocessing
Mohali, India

Naseem A. Gaur
Yeast Biofuel Group
International Centre for Genetic
    Engineering and Biotechnology
    (ICGEB)
New Delhi, India

Kunal Gogde
Department of Nanomaterials and
    Application Technology
Center of Innovative and Applied
    Bioprocessing (CIAB)
Department of Biotechnology (DBT)
Government of India, Sector 81
    (Knowledge City)
Nagar, India

Manasa Nagesh Hegde
Synthetic Biology and Bioprocessing
    Lab
Regional Centre for Biotechnology
NCR Biotech Science Cluster
Faridabad, India

S. Bilal Jilani
Microbial Engineering Group
International Centre for Genetic
    Engineering and Biotechnology
New Delhi, India
and
Institute of Biotechnology
Amity University
Manesar, India

Guruprasad K
Corporate Research and Development
    Centre
Bharat Petroleum Corporation Ltd.
Greater Noida, India

Kukkala Kiran Kumar
Yeast Biofuel Group
International Centre for Genetic
    Engineering and Biotechnology
    (ICGEB)
New Delhi, India

Sandeep Kumar
Chemical Engineering Division
DBT-Center of Innovative and Applied
    Bioprocessing
Mohali, India
and
Dr. SSB University Institute of
    Chemical Engineering and
    Technology
Panjab University
Chandigarh, India

Bhawna Madan
Corporate Research and Development
    Centre
Bharat Petroleum Corporation Ltd.
Greater Noida, India

Sangeeta Mahala
Chemical Engineering Division
DBT-Center of Innovative and Applied
  Bioprocessing
Mohali, India
and
Department of Chemical Sciences
Indian Institute of Science Education
  and Research
Mohali, India

Shilpa Main
DBT-ICT Centre for Energy Biosciences
Institute of Chemical Technology
Mumbai, India

Divya Mudgil
DBT-TERI Centre of Excellence
  in Advanced Biofuels and
  Bio-Commodities
Advanced Biofuels Program
The Energy and Resources Institute
New Delhi, India

Bharat Newalkar
Corporate Research and Development
  Centre
Bharat Petroleum Corporation Ltd.
Greater Noida, India

Annamma Anil Odaneth
DBT-ICT Centre for Energy Biosciences
Institute of Chemical Technology
Mumbai, India

Parmeshwar Patil
Corporate Research and Development
  Centre
Bharat Petroleum Corporation Ltd.
Greater Noida, India

Shatabdi Paul
Department of Nanomaterials and
  Application Technology
Center of Innovative and Applied
  Bioprocessing (CIAB)
Department of Biotechnology (DBT)
Government of India, Sector 81
  (Knowledge City)
Nagar, India
and
Regional Centre for Biotechnology
Department of Biotechnology (DBT)
Government of India
Faridabad, India

Hitesh S. Pawar
DBT-ICT Centre for Energy
  Biosciences
Institute of Chemical Technology
Mumbai, India

Prashant Anupama-Mohan Pawar
Laboratory of Plant Cell Wall Biology
Regional Centre for Biotechnology
NCR Biotech Science Cluster
Faridabad, India

Lavi Rastogi
Laboratory of Plant Cell Wall Biology
Regional Centre for Biotechnology
NCR Biotech Science Cluster
Faridabad, India

Vivek Rathore
Corporate Research and Development
  Centre
Bharat Petroleum Corporation Ltd.
Greater Noida, India

Jaya Rawat
Corporate Research and Development
  Centre
Bharat Petroleum Corporation Ltd.
Greater Noida, India

Yeddula Nikhileshwar Reddy
Department of Nanomaterials and
  Application Technology
Center of Innovative and Applied
  Bioprocessing (CIAB)
Department of Biotechnology (DBT)
Government of India, Sector 81
  (Knowledge City)
Nagar, India
and
Department of Chemical Sciences
Indian Institute of Science Education
  and Research
Mohali, India

Rajan Kumar Sah
Laboratory of Plant Cell Wall Biology
Regional Centre for Biotechnology
NCR Biotech Science Cluster
Faridabad, India

Koel Saha
DBT-TERI Centre of Excellence
  in Advanced Biofuels and
  Bio-Commodities
Advanced Biofuels Program
The Energy and Resources Institute
New Delhi, India

Shelja Sharma
Chemical Engineering Division
DBT-Center of Innovative and Applied
  Bioprocessing
Mohali, India

Yogendra Shashtri
Department of Chemical Engineering
Indian Institute of Technology Bombay
Mumbai, India

Anurag Singh
Corporate Research and Development
  Centre
Bharat Petroleum Corporation Ltd.
Greater Noida, India

Deepika Singh
Laboratory of Plant Cell Wall Biology
Regional Centre for Biotechnology
NCR Biotech Science Cluster
Faridabad, India

Aishwarya Srivastava
Regional Centre for Biotechnology
NCR Biotech Science Cluster
Faridabad, India

Shireesh Srivastava
Systems Biology for Biofuel Group
International Centre for Genetic
  Engineering and Biotechnology
  (ICGEB)
New Delhi, India

Sanjukta Subudhi
DBT-TERI Centre of Excellence
  in Advanced Biofuels and
  Bio-Commodities
Advanced Biofuels Program
The Energy and Resources Institute
New Delhi, India

Navin Tamrakar
Independent Consultant for Market,
  Process, Cost and Price Intelligence
  across Specialty Chemicals and
  Advanced Composites Industry
Hadapsar, India

Tejas M. Ukarde
DBT-ICT Centre for Energy Biosciences
Institute of Chemical Technology
Mumbai, India

Prachi Varshney
Corporate Research and Development
    Centre
Bharat Petroleum Corporation Ltd.
Greater Noida, India

Mohan Yama
Corporate Research and Development
    Centre
Bharat Petroleum Corporation Ltd.
Greater Noida, India

1 Chemistry of Plant Biomass

Senthil Murugan Arumugam and Shelja Sharma
DBT-Center of Innovative and Applied Bioprocessing

Sandeep Kumar
DBT-Center of Innovative and Applied Bioprocessing
Panjab University

Sangeeta Mahala and Bhawana Devi
DBT-Center of Innovative and Applied Bioprocessing
Indian Institute of Science Education and Research

Sasikumar Elumalai
DBT-Center of Innovative and Applied Bioprocessing

CONTENTS

1.1 Introduction 2
1.2 Classification of Plant-Derived Biomass 3
    1.2.1 Woody Biomass 3
    1.2.2 Herbaceous Biomass 3
1.3 Plant Cell Wall Composition and Architecture 4
    1.3.1 Chemistry of Cellulose 6
    1.3.2 Chemistry of Hemicellulose 10
        1.3.2.1 Xylan 10
        1.3.2.2 Mannan 13
        1.3.2.3 Xylogalactan 13
        1.3.2.4 Xyloglucan 14
    1.3.3 Chemistry of Lignin 14
    1.3.4 Chemistry of Starch 19
        1.3.4.1 Amylose 20
        1.3.4.2 Amylopectin 21
1.4 Other Low Molecular Weight Constituents 23
    1.4.1 Extractives 23
    1.4.2 Inorganic Constituents 26
    1.4.3 Fluid Content 26
1.5 Conclusions 27
References 27

DOI: 10.1201/9781003158486-1

1.1 INTRODUCTION

Biomass can be defined as a solid organic substance or material derived from living organisms, including plants and animals. Due to the massive production of plant-derived materials, particularly agro-wastes, wood shavings, and forestry residues, and its attractive chemical composition, a large attention is being paid to the exploitation of them for deriving energy and chemicals (Shankar Tumuluru, Sokhansanj, Hess, Wright, & Boardman, 2011). The characteristic material is profoundly employed in the renewable energy generation as a potential resource (Tursi, 2019). Historically, since the mid-18th century, plant biomass is considered the largest energy producer through thermal processing techniques. In recent years, its application in the biofuel for transportation and electricity generation has been increased, particularly in the developed countries, including the United States. It offers several environmental benefits that could reduce CO_2 emissions, which is comparatively more with fossil fuels. This material contains stored chemical energy derived via photosynthesis in the presence of sunlight and water. Thus, it is represented as a solid material that can be burned directly to recover heat or renewable fuels (both liquid and gaseous) through various thermochemical techniques. The International Energy Report 2019 details that the agro-industrial residues delivered nearly five quadrillion British thermal units (Btu) of thermal energy, which is estimated to be nearly 5% of the total primary energy used in the United States alone (Newell, Raimi, & Aldana, 2019). Overall, it is accepted as a potentially scalable feedstock for the production of sustainable fuels and chemicals and, moreover, is believed to have the ability to displace petroleum-derived products. The classified biomass sources utilized for energy and other products manufacturing are forest wood and its processing wastes, agricultural crops and its residues, and other biological materials, including municipal solid wastes, animal wastes (manure), and human sewage wastes (Muscat, de Olde, de Boer, & Ripoll-Bosch, 2020).

The attractive inherent fractional composition of the plant-based material (generally referred to as lignocellulose) encouraged the researchers to exploit it for the potential production of value-added chemicals through different technological routes (Gusiatin & Pawłowski, 2016). Therefore, it has been thoroughly assessed for the potential ethanol production via biological fermentation after employing the pretreatment with that impression. At the same time, the substrate offers challenges in successfully commercializing the bioconversion technologies due to its heterogeneous characteristics. Fundamentally, the solid biomass is made up of cell walls (primary and secondary), constituted with cellulose, hemicellulose, and lignin biogenic polymers; therefore, it is represented as the abundant micromolecular biocomponent available on the planet. Indeed, its presence in plants makes the major difference between the animal cell and plant cell, and furthermore, the plasma membrane surrounds the latter's cell wall. Basically, the membrane functions to provide tensile strength, giving protection against plant stresses (i.e., osmotic and mechanical). It allows the cells to develop turgor pressure (pressure of the cell contents against the development of the cell wall). Mechanistically, the increased turgor pressure leads to plant wilting; therefore, a plant requires enough water supply. Thus, the cell walls help maintain the plant's stems, leaves, and other structures (Lerouxel, Cavalier, Liepman, & Keegstra, 2006).

1.2 CLASSIFICATION OF PLANT-DERIVED BIOMASS

1.2.1 WOODY BIOMASS

Wood biomass is typically originated from the tree materials and can be classified into forestry residues (leaves and branches), sawmill wastes (sawdust), and wood scraps (construction wastes) (Demirbaş, 2005). However, this material may not be suitable for preparing wood lumber, and thus, it is ascertained as a potential renewable resource belonging to the lignocellulose category (Danish & Ahmad, 2018). Like the vegetative plants, afforestation, regeneration, and sapling are maintenance methods that are usually adopted for continuous production. Typically, the timber process yields the largest portion of woody biomass, accounting for nearly 25%–45% of the harvested wood. However, the characteristic biomass possesses low bulk density and fuel value, and therefore, it increases the transport cost per unit. To reduce its impact on transportation, a general practice of comminution (or chipping) with compaction to make them in bundles is followed to increase bulk density. The advantages of the substrate to be used for energy and chemicals production include that it is a: (a) non-food organic feedstock; therefore, its competition with the agricultural food crops development is relatively low, (b) renewable resource, (c) material that requires lower energy input for growth, and (d) energy source material that, moreover, emits maximum CO_2, which is comparable to the net CO_2 released during its natural degradation and therefore represented to be a carbon-neutral material. Another subcategory of this type of biomass is the dedicated energy crops that include, for example, Eucalyptus spp., willow (*Salix*), poplar and perennial grasses (Miscanthus), which are also explored as a potential feedstock for energy production (Danish & Ahmad, 2018). In total, ~1.4 megatons of dry matter per year (Mt DM/yr) of forestry residues are generated across the world.

1.2.2 HERBACEOUS BIOMASS

This type of biomass material originates from plant sources, but has a non-woody stem and is collected as residues at the end of each harvest season. Agricultural crops (or cereal crops) and grasses are the major classification of this type of renewable resource that mainly includes wheat straw, rice straw, bamboo, etc. (Vogel & Jung, 2001). While comparing with the woody biomass, the characteristic material contains high nutrients and low lignin contents. Indeed, its fractional composition depends on the type of plant tissue, growth location, and soil, which contains a varied level of minerals and nutrients and shows a larger impact on plant growth. The comparative chemical composition of a few woods and herbaceous biomass types is presented in Table 1.1. It is analogous, composed mostly of carbohydrates (cellulose and hemicellulose), lignin, and ash contents. It is roughly estimated that ~88 Mt DM/yr of such residues are produced by agricultural crops. However, this massive production creates severe environmental problems and nuisances as they are not properly managed. Therefore, its effective utilization techniques are sought for the sustainable and clean production of value-added chemicals to benefit both environment and people. There emerged various process methods in its effective conversion via

TABLE 1.1

Fraction Composition of Few Representative Biomass Materials (Saha, 2003)

Biomass Type	Cellulose	Hemicellulose	Lignin	Ash
		Fractional Components (% wt. Dry Basis)		
		Wood Biomass		
Hardwood	40–50	24–40	18–25	2–5
Softwood	45–50	25–35	25–35	2–5
		Agri-Residue Biomass		
Rice straw	35–40	18–25	20–26	8–12
Wheat straw	33–40	20–25	12–20	3–7
Corn stover	35–40	20–25	19–22	3–5
Sugarcane bagasse	40–45	30–35	20–30	3–7
Switchgrass	31–46	15–22	17–21	3–5

physical, chemical, and biological routes; however, they are not economically attractive (Adapa, Tabil, & Schoenau, 2009; Liu et al., 2014). This substrate can be further classified into primary, secondary, and tertiary biomasses, based on recovery type. The primary biomass is actually the agri-crop residues generated or left behind in the field after the crop harvest, for example straw, stubble, stover, leaves, sticks, haulms, branches, roots, twigs, trimmings, pruning, and brushes. The agro-industrial residues, i.e., the material generated at production site during post-harvest processing, are collectively called secondary biomass; those include husks, peels, hull, bagasse, pomace, corncobs, etc. The tertiary biomass is collected as residues after processing the agro-industrial products (Vogel & Jung, 2001).

1.3 PLANT CELL WALL COMPOSITION AND ARCHITECTURE

Fundamentally, plant tissues constitute two types of cell walls: primary and secondary cell walls; they differ in function and composition (as illustrated in Figure 1.1). Moreover, their cellular arrangement governs providing mechanical strength and allows the cells to grow and divide by themselves, thereby influencing their shape and size (Keegstra, Talmadge, Bauer, & Albersheim, 1973; Nakano, Yamaguchi, Endo, Rejab, & Ohtani, 2015). The primary wall and middle lamella contents of the apoplast mainly account for the growth of the tissues in the plant. The primary wall's main functions are to provide structural and mechanical support, maintain and determine the cell shape, control the rate and direction of plant growth, and regulate the diffusion of material through the apoplast. Furthermore, it establishes itself as the main textural element of food derived from plant sources (Amos & Mohnen, 2019). As exposed by the literature, the plant's cell wall is surrounded by a polysaccharide-rich primary wall, due to which the beverages derived from plant are often reported to be containing a significant amount of polysaccharides (Holland,

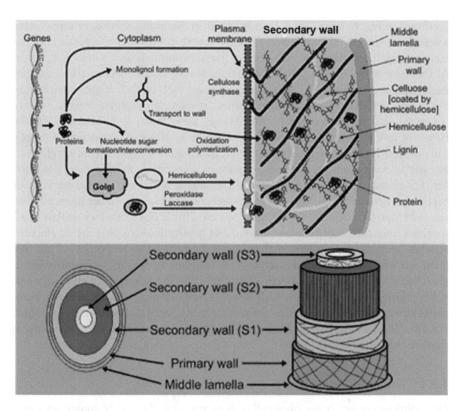

FIGURE 1.1 Overview of development of plant cell wall and its arrangement. (Reprinted with permission from Keegstra et al. (1973).)

Ryden, Edwards, & Grundy, 2020). The cell wall polysaccharides are also used to prepare plant-derived gums, gels, and stabilizers (Cui & Wang, 2009). This exciting architecture of cell wall garnered the plant scientists' interest for its application in the preparation of food and nutritional products. While the analysis of the isolated primary cell walls belonging to the higher plant tissues, those are predominantly composed of polysaccharides. With that, traces of structural glycoproteins (such as hydroxyproline-rich extensins), ionically and covalently bound minerals (e.g., calcium and boron), phenolic esters (i.e., ferulic and coumaric acids), and enzymes are spotted (Hijazi, Velasquez, Jamet, Estevez, & Albenne, 2014). Moreover, cellulose (a polysaccharide composed of 1,4-linked β-D-glucose residues), hemicellulose (a branched polysaccharide composed of 1,4-linked β-D-hexosyl residues as backbone), and pectin (a complex polysaccharide containing 1,4-linked α-D-galacturonic acid) are the classified polysaccharides present in the primary wall of the plants (Rubin, 2008).

The plant's secondary cell exhibits a much thicker and stronger wall, contributing to maximum carbohydrate storage (Figure 1.1). It is constituted by the tracheid, xylem fiber, and sclereid components; those are collectively strengthened by the incorporated aromatic lignin (a micromolecule composed of highly cross-linked phenolics) (as shown in Figure 1.2) (Zhong, Cui, & Ye, 2019). Naturally, secondary walls create a significant impact on human life, since they represent the major constituent of the plant material, and therefore, it is feed to livestock animals as a nutritional source. At the same time, it is established to be apt feedstock for bioprocessing, thereby enabling the reduction in the dependency on fossil fuels by their higher contribution in the bulk generation of renewable biomass resources, which is, in turn, directly converted into fuel or energy (Smith, Wang, York, Peña, & Urbanowicz, 2017). However, numerous technical challenges exist in its effective conversion to energy and product. Thus, both types of plant cell walls contain the cellulose, hemicellulose, and pectin components at different proportions, depending on the type of plant tissue. For instance, the dicot primary cell contains approximately equal amounts of pectin and hemicellulose, whereas the switchgrass contains a higher hemicellulose amount (Marriott, Gómez, & McQueen-Mason, 2016). The secondary wall of those species is constituted of cellulose, lignin, and hemicellulose (either xylan, glucuronoxylan, arabinoxylan, or glucomannan type) at nearly equal proportions. Thus, the plant cell wall's architecture is such that the cellulose microfibrils are embedded in the hemicellulose and lignin system. This cross-linking helps in situ elimination of water molecules out of the cell wall setup, thereby forming the hydrophobic composite structure, rigidly restricting the susceptibility to the hydrolytic enzyme. Thus, plant's secondary wall is established to be the major contributor to its structural characteristics.

1.3.1 CHEMISTRY OF CELLULOSE

The cellulose content in the plant cell wall is represented to be the important structural component. This characteristic polysaccharide contains the repetitive glucose ($C_6H_{12}O_6$) units that are being polymerized at the plasma membrane using the cellulose synthase complex (Rubin, 2008). The $\beta \rightarrow$ (1–4) D-glucose units are arranged in a fashion of 180° rotation between one glucose unit to another, resulting in the synthesis of cellobiose with β-(1→4)-linkages (Figure 1.3) (Heinze, 2015). Thus, the linear chain cellulose is made up of a repeating unit of cellobiose units, with one end having the C4-OH non-reducing group and the other the C1-OH terminating group, and the reducing end has the aldehyde structure (Figure 1.4). However, some technical celluloses obtained through the pulping processes contain extra carbonyl and carboxyl groups (Perrin, Pouyet, Chirat, & Lachenal, 2014). As mentioned, the cell wall has a microfibril-based construction using polymeric constituents. In particular, the arrangement of the secondary cell wall is in the form of layers, viz. outer (S1), middle (S2), and inner (S3) layers (Nakano et al., 2015). These layers differ with the orientation of microfibrils. Layer S2 exhibits a thickest one having steep helices of microfibrils as compared to others. Thus, all of the cell wall layers contain microfibrillar constituents and matrix phases. These phases can be further classified based on the components present in them; for example, the cellulose microfibrils contain a crystalline core with a poor crystalline property on the exterior, whereas the matrix phase of cellulose represents a non-crystalline phase containing pectin, hemicellulose, and other polymers, including lignin. The cellulose crystalline polymorphs can be subdivided into six forms (I, II, III_I, III_{II}, IV_I, and IV_{II}) based on their

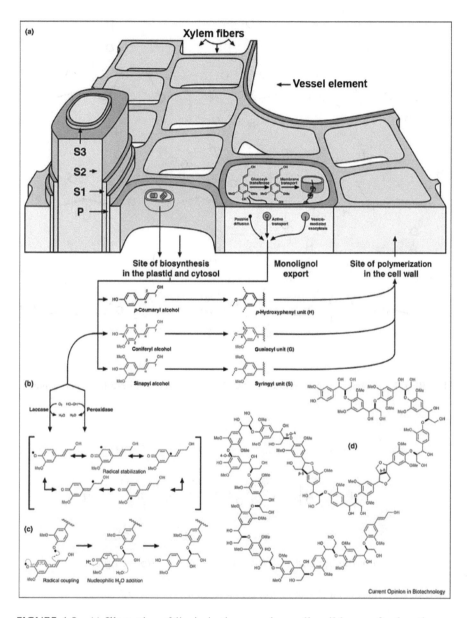

FIGURE 1.2 (a) Illustration of lignin in the secondary cell wall layers of xylem tissues, (b) laccase and peroxidase enzymes involved in the development of lignin polymer, (c) end-wise coupling of a monolignol radical for the lignin polymerization reaction, and (d) lignin polymer model. (Reprinted with permission from Mottiar, Vanholme, Boerjan, Ralph, and Mansfield (2016).)

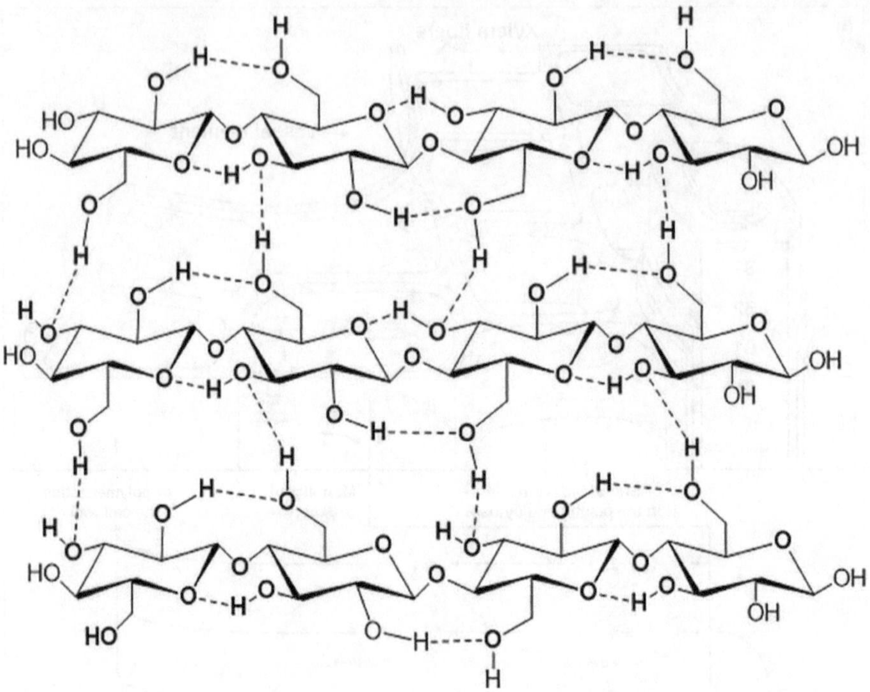

FIGURE 1.3   Polymeric representation of cellulose with inter- and intra-hydrogen bonds.

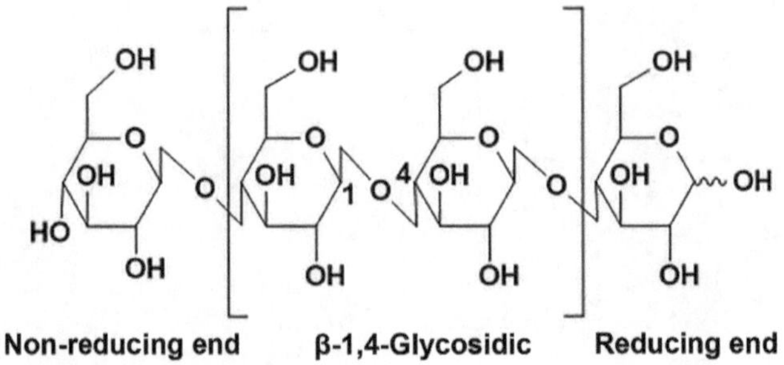

Non-reducing end   β-1,4-Glycosidic   Reducing end

FIGURE 1.4   Illustration of synthesis of cellulose from its monomeric units.

characteristics (Rongpipi, Ye, Gomez, & Gomez, 2019). The I and II cellulose forms are represented to be the general arrangement of cellulose and exist freely in nature. Comparatively, the other forms are less popular and rarely reported. While analyzing the nature of common cellulose forms (I and II), the intramolecular hydrogen bonds are recognized to be responsible for its structural stability and rigidity. The sturdy intra- and inter-chain hydrogen bonding of cellulose microfibrils make the structures of both I and II arrange themselves in parallel and antiparallel directions to the longitudinal axis, respectively (Chami, Khazraji, & Robert, 2013). Additionally, their polymeric characteristics can be determined based on the degree of polymerization (DP), which measures the number of monomeric units present in each of the cellulose chains. Typically, the cellulose chain of the primary cell wall has the DP varying between 2000 and 6000, whereas the secondary cell wall's cellulose measures up to DP 14,000. However, its molecular structure is critical in determining the physico-chemical properties, including hydrophilicity, chirality, degradability, and variability in chemical structure, attributed to the -OH donor group's higher reactivity.

Indeed, the exclusive property of hydrogen bond formation between the networks of cellulose hydroxy groups inspired the researchers to carry out intensive research. Figure 1.2 depicts cellulose's chemical structure consisting of hydroxyl groups of β-1,4-glucan cellulose units at the sites of C2, C3, and C6, while the CH_2OH group of cellulose is positioned relative to the bonds at C4 and C5 sites along with the O5–C5 bonds. The X-ray diffraction analysis of cellulose's crystal structure revealed a monoclinic unit cell, which is made up of two cellulose chains arranged in parallel orientation (Figure 1.5) (Nishiyama, Langan, & Chanzy, 2002). Moreover, the twofold screw axis present within the structure represented the cellulose I and II type crystal structures. Also, it established that the network is arranged in a fashion of intramolecular chain-stiffening hydrogen bonding. The cutting-edge characterization techniques such as NMR revealed that the cellulose $I_β$ crystal structure has different conformations and H-bonds relative to the neighboring chains. Of all forms, a thermodynamically stable form of cellulose is recognized to be cellulose II. However, cellulose I's transformation to cellulose II can be achieved with the reaction using an aqueous NaOH solution.

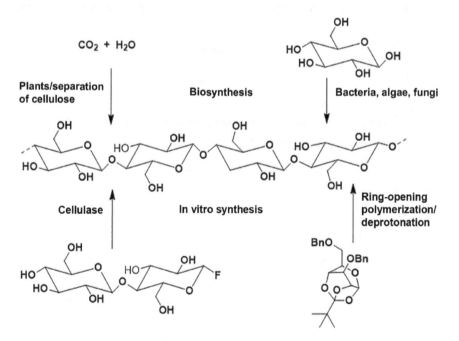

FIGURE 1.5 Illustration of biosynthesis of carbohydrates in plants.

1.3.2 CHEMISTRY OF HEMICELLULOSE

Unlike cellulose, the hemicellulose micromolecule is generally described as a heterogeneous polysaccharide molecule because it is constituted using a group of carbohydrates and lignin molecules formed through different biosynthetic routes (Elumalai & Pan, 2011; Pauly et al., 2013). Similar to cellulose, it functions as a supporting agent in the plant cell wall construction. However, the average DP of hemicellulose is merely 200 (Wyman et al., 2005). While overlooking the structural characteristics, it has a β-(1→4)-linked backbone structure with symmetrical configuration (Figure 1.6). This branched polymer has xylose, glucose, mannose, and galactose sugars in the backbone and, similarly, arabinose, galactose, and 4-O-methyl-D-glucuronic acid in the side chain (Table 1.2). Thus, the heterogeneous macromolecule is a mixture of pentose sugars (such as xylose and arabinose), hexose sugars (glucose, mannose, galactose, rhamnose, and sugar acids (glucuronic and galacturonic acids) (Figure 1.7) (Carvalheiro, Duarte, & Gírio, 2008). However, the proportion of these sugar and substituent units depends on the plant type. It is categorized as a water-insoluble substance, but complete solubilization can be achieved with alkaline solutions (Giummarella & Lawoko, 2017). This property of hemicellulose is appreciated for preparing gum materials. Alternatively, the insoluble characteristics of hemicellulose can be altered through the chemical modification of its structural moieties. Furthermore, this polysaccharide can be broadly classified into xylan, mannan, xylogalactan, and xyloglucan types, based on the type of sugar present in the backbone.

1.3.2.1 Xylan

Xylan is a classified hemicellulose heteropolymer consisting of a β-(1,4)-linked backbone of D-xylose units (Figure 1.8). It is a major type of hemicellulose found

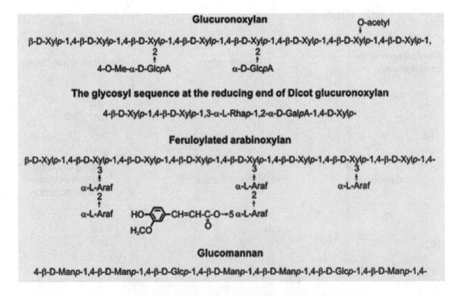

FIGURE 1.6 Illustration of forms of hemicelluloses present in various biomass feedstocks. (Reprinted with permission from Carpita and Gibeaut (1993); Ebringerová (2005).)

TABLE 1.2

Major Hemicellulosic Components Present in Plant Biomass (Carpita & Gibeaut, 1993; Elumalai & Pan, 2011)

Hemicellulose Type	Occurrence	% Availability	Composition			Solubility	DP_n
			Units	Molar Ratios	Linkage		
(Galacto)glucomannan	Softwood	5–8	β-D-Manp β-D-Glup α-D-Galp Acetyl	3 1 1 1	1→4 1→4 1→6	Alkali, water[a]	100
(Galacto)glucomannan	Softwood	10–15	β-D-Manp β-D-Glup α-D-Galp Acetyl	4 1 0.1 1	1→4 1→4 1→6	Alkali, borate	100
(Arabino)glucuronoxylan	Softwood	7–10	β-D-Xylp 4-O-Me-α-D-GlupA α-L-Araf	10 2 1.3	1→4 1→2 1→3	Alkali, DMSO[a], water[a]	100
Arabinogalactan	Larch wood	5–35	α-D-Galp α-L-Araf β-L-Araf β-D-GlupA	6 2/3 1/3 Little	1→3 1→6 1→6 1→3 1→6	Water	200
Glucuronoxylan	Hardwood	15–30	β-D-Xylp 4-O-Me-α-D-GlupA Acetyl	10 1 7	1→4 1→2	Alkali, DMSO[a]	200
Glucomannan	Hardwood	2–5	β-D-Manp β-D-Glup	1-2 1	1→4 1→4	Alkali, borate	200

[a] Represents partial solubility of hemicellulose.

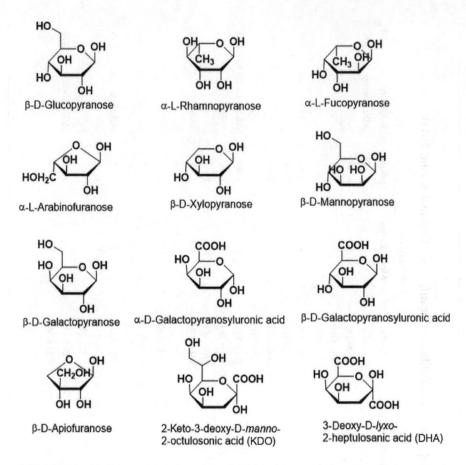

β-D-Glucopyranose   α-L-Rhamnopyranose   α-L-Fucopyranose

α-L-Arabinofuranose   β-D-Xylopyranose   β-D-Mannopyranose

β-D-Galactopyranose   α-D-Galactopyranosyluronic acid   β-D-Galactopyranosyluronic acid

β-D-Apiofuranose   2-Keto-3-deoxy-D-*manno*-2-octulosonic acid (KDO)   3-Deoxy-D-*lyxo*-2-heptulosanic acid (DHA)

FIGURE 1.7 Chemical structures of the individual sugar units commonly present in hemicellulose.

in hardwood and cereals, accounting for 30% of the cell wall components (Sella Kapu & Trajano, 2014). Normally, it accounts for about 30% wt. of the lignocellulosic biomass substrate. Sugar cane, sorghum stalk, corn stalk, and corncob are found to be the major plant sources of this sugar molecule. Recently, it is showed that it could also be obtained from hulls and husks of cereals and seeds. Furthermore, it has three varieties of hemicelluloses, namely homoxylan, glucuronoxylan, and arabinoxylan, classified with respect to the side-chain substituent. The homoxylan-type hemicellulose predominantly contains xylose units in both the backbone and side chain. This type of hemicellulose is recognized to be a unique variety and can be found significantly in seaweeds. The glucuronoxylan is made up of xylose backbone with glucuronic acid in the side chain. It is found abundantly in herbaceous plants. However, the arabinoxylan is made up of xylose backbone with α-L-arabinose in the side chain; therefore, it is also said to be a unique hemicellulose type. Thus, the only dominance of pentose sugars makes the difference in arabinoxylan- and homoxylan-type hemicelluloses. Cereals, including wheat, barley, rice, corn, and

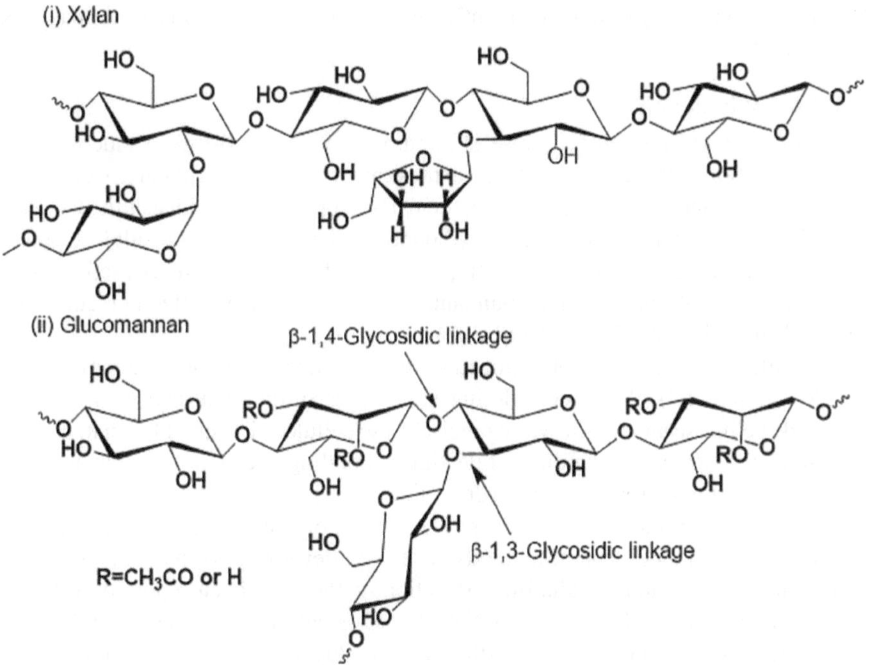

(i) Xylan

(ii) Glucomannan

β-1,4-Glycosidic linkage

R=CH₃CO or H

β-1,3-Glycosidic linkage

FIGURE 1.8 General representation of hemicellulose with backbone and side-chain substitution.

sorghum, are reported to be the major source for this sugar molecule (Spiridon & Popa, 2008).

1.3.2.2 Mannan

Mannan hemicellulose is the other type of heteropolymer having β-(1,4)-D-mannopyranose in the backbone (Figure 1.8). With respect to the side-chain substitution, it can be further classified into three subtypes, such as homomannan, glucomannan, and galactomannan (Carvalheiro et al., 2008). Homomannan refers to the sugar polymer consisting of mannose in the backbone as well as in the side chain. This type of hemicellulose is not commonly found in nature. The glucomannan has a mannose-rich backbone having the side chain containing glucose sugar molecule. It is found in the softwood materials and situated in the secondary cell wall. The last variety (galactomannan) has mannose rich in the backbone and a short side chain consisting of galactose units; it is found abundantly in the plant storage tissues, e.g., guar, Tara, and locust bean. Indeed, the amount of galactose residue present in the hemicellulose significantly influences its solubility and viscosity properties (Spiridon & Popa, 2008).

1.3.2.3 Xylogalactan

Xylogalactan is another classified hemicellulose sugar that contains galactose sugar units in the backbone, which is decorated by the α-D-xylopyranose residues and can be found in the *Prosopis africana* plant's leguminous seeds (Olorunsola, Akpabio, & Ajibola, 2018).

1.3.2.4 Xyloglucan

Likewise, xyloglucan hemicellulose is composed of β-(1→4)-linked glucose units in the backbone; thus, its structural arrangement appears identical to cellulose. However, the backbone is decorated by the α-D-xylopyranose residue at the C6 position. This hemicellulose type is strongly bonded to the microfibrillar cellulose within the cell wall and thus provides a strong resistance for its selective extraction using solvents. This molecule is found abundantly in the tamarind and afzelia leguminous seeds (Schultink, Liu, Zhu, & Pauly, 2014).

Overall, the composition of hemicellulose varies depending on the biomass type (Tables 1.3 and 1.4). While considering its features in the plant cell wall, it exists alongside cellulose and lignin. More precisely, the cellulose is embedded in it, thereby providing the structural rigidity while lignin bonding the entire system together via the formation of lignin–carbohydrate linkages (LCC) (as shown in Figure 1.9) (Tarasov, Leitch, & Fatehi, 2018). Thus, all of these polymers are bound together with the cell wall. Till date, several methods have been developed for its extraction from plant sources, of which the alkaline extraction method has been reported proficient with the involvement of hot aqueous $NaOH/H_2O_2$ solutions, since hemicellulose has poor water solubility property. Nevertheless, xyloglucan hemicellulose solubilizes in hot water; it is widely followed for the preparation of xyloglucan gum via precipitation. Other alternative methods, including microwave treatment, solvent extraction using DMSO and methanol/water mixture, and solvent pressurization (ethanol), are reported to be employed for specific hemicellulose isolation (Tarasov et al., 2018).

1.3.3 CHEMISTRY OF LIGNIN

Lignin is defined as an exceptional polymeric component of biomass because it comprises of heterogeneous aromatic molecules, namely p-coumaryl alcohol, coniferyl alcohol, and sinapyl alcohol; these precursors ultimately yield the corresponding p-hydroxyphenyl (H), guaiacyl (G), and syringyl (S) lignin subunits, respectively (as presented in Figure 1.10). It is polymerized at the surface of the plant's cell wall and considered as the crucial secondary metabolites produced through the phenylalanine and tyrosine metabolic pathways (Liu, Luo, & Zheng, 2018; Mottiar et al., 2016). The role in plants is to provide structural support for their upward growth and enable water transport to a long distance within plant stems. It also extends its duty to provide physical and chemical protection for plants against pathogenic attack (Liu et al., 2018). This feature is often regarded as recalcitrant during the disintegration of biomass for the specific release of fermentable sugar originated from cellulose and/or hemicellulose micromolecular components. Naturally, it is a high molecular weight complex aromatic polymer, representing a composite molecular structure. Moreover, its biosynthesis in plants contributes to a larger extent to their growth, to the development of tissue/organ, and to offering resistance and response against plant stresses (both biotic and abiotic). For its synthesis, the precursors undergo a series of in situ processes, including molecular transport and polymerization, wherein deamination,

TABLE 1.3

Chemical Composition of Some Representative Softwoods (Elumalai & Pan, 2011)

Scientific Name	Common Name	Carbohydrate					Lignin	Tannin	Aliphatic Acids & Proteins	Resin, Gums	Ash
		Cellulose	(Galacto)-glucomannan	Arabino-O-4-methyl-glucuronoxylan	Arabinogalactan	Uronic Anhydride					
Abies alba	White spruce	42.3	12.0	11.5	1.4	3.4	28.9	0.3	0.7	2.0	0.8
Araucaria canadensis	Parana pine	44.3	11.0	4.5	1.8	4.0	29.5	0.3	0.7	4.0	1.4
Pinus sylvestris	Red pine	52.2	11.0	3.1	8.2	5.6	26.3	0.3	0.7	4.0	0.4
Pseudotsuga	Douglas fir	50.4	11.0	7.5	4.7	3.8	27.2	0.4	2.6	2.0	0.2
Thuja plicata	Red cedar	47.5	8.0	11.4	1.2	4.2	32.5	0.7	0.7	1.6	0.3
Tsuga canadensis	Hemlock	44.0	11.0	6.5	0.6	3.3	33.0	0.7	2.2	1.1	0.2

TABLE 1.4
Fractional Composition of Few Representative Hardwood Materials (Elumalai & Pan, 2011)

Scientific Name	Common Name	Carbohydrate					Lignin	Tannin	Aliphatic Acids & Proteins	Resin, Gums	Ash
		Cellulose	(Galacto)-glucomannan	Arabino-O-4-methyl-glucuronoxylan	Arabinogalactan	Uronic Anhydride					
Fagus sylvatica	Beech	36.0	2.7	23.5	1.3	4.8	30.9	0.8	3.2	2.0	0.4
Fraxinus excelsior	Ash	46.0	1.9	16.4	1.7	–	22.0	0.9	0.5	0.6	0.4
Juglans nigra	Walnut	40.8	–	12.6	–	–	29.1	0.6	4.4	5.0	0.8
Swietenia macrophylla	Mahogany	43.9	–	16.0	–	–	28.2	1.9	5.1	1.0	1.1
Populus alba	White poplar	49.0	–	25.6	–	–	23.1	0.6	2.0	2.4	0.2
Quercus robur	Oak	41.1	3.3	22.2	1.6	4.5	19.6	1.2	2.0	0.4	0.3
Ulmus procera	Elm	43.0	3.2	21.8	1.4	3.6	27.3	0.6	1.7	1.7	0.8

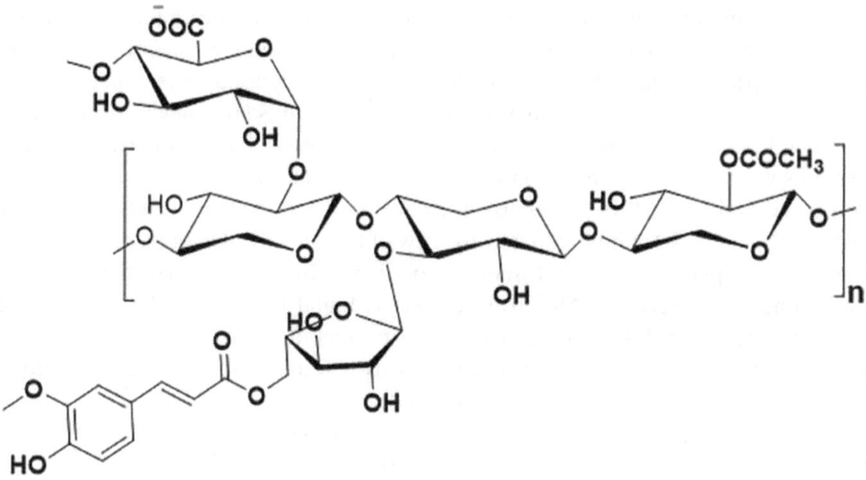

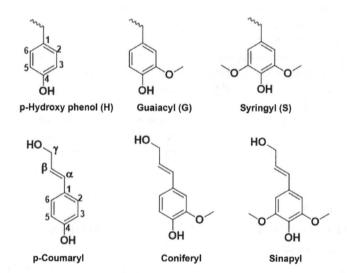

FIGURE 1.9 Illustration of hemicellulose linkage with lignin to form lignin-carbohydrate linkages.

FIGURE 1.10 Chemical structure of lignin precursors and subunits.

methylation, hydroxylation, and reduction reactions are typically undertaken within the cytoplasm. After its synthesis, eventually, it is transported to the ultimate apoplast (Boerjan, Ralph, & Baucher, 2003). Thus, in plants, lignin accounts for ~30% wt.; therefore, it is considered the abundant natural organic carbon available on earth after the plant-derived carbohydrates (Vassilev, Baxter, Andersen, & Vassileva, 2010).

In the presence of peroxidase and laccase enzymes within the secondary cell wall, it forms a polymeric network consisting of different subunit groups, such as alkyl–alkyl,

alkyl–aryl, and aryl–aryl. Thus, all of the subunits differ in the chemical structure with methoxy substitution in the aromatic ring structure at *p*- and/or *m*-position. Besides, hydroxycinnamaldehydes, hydroxystilbenes, tricin flavones, and xenobiotics are also found as rare substituents in the lignin subunits (Bagniewska-Zadworna, Barakat, Łakomy, Smoliński, & Zadworny, 2014). Moreover, the proportion of these subunits vary depending on the plant varieties; for instance, G-unit is predominant in softwood (Jazi et al., 2019). Similarly, the hardwood and the herbaceous plants consist of G-S and G-S-H units, respectively. Few of the ecological factors, such as nutrition, climate, plant growth, and illumination, have also shown a significant impact on its final characteristics. Moreover, carbon is found richer than the neighboring carbohydrate molecules of the major elements present in its structure (carbon, hydrogen, and oxygen) (Table 1.5). For instance, softwood and hardwood lignin's carbon content lies in the range 56%–65%, whereas cellulose has only 44.4%. Therefore, in terms of energy recovery out of these biogenic polymers, lignin is alleged to be the major contributor (Shankar Tumuluru et al., 2011; Vassilev et al., 2010).

In addition, the functional groups, such as phenolic hydroxyl, alcohol hydroxyl, carbonyl, carboxyl, methoxyl, and sulfonic acid, are reported to be its critical structural characteristics for determining its chemical reactivity (Table 1.6). Also, the optical properties, chemical reactivity, dispersion characteristics, and qualitative/quantitative determination are said to be useful in studying the lignin structure. Overall, it is considered a valuable resource for the potential production of energy and aromatic chemicals through the selective cleavage strategy. The subunits are interconnected via phenylpropane β-aryl ether (β-O-4), phenylpropane α-aryl ether (α-O-4), diaryl ether (4-O-5), biphenyl and dibenzodioxocin (5-5), 1,2-diaryl propane (β-1), phenylcoumaran (β-5), and β–β-linked structures (β–β)-linkages (Figure 1.11) (Mottiar et al., 2016). Several model studies have showed that the primary lignin structures are arranged in a random or combinatorial fashion, determined based on the interunit linkage sequences. Another variety of studies have demonstrated that the composite matrix can be altered by the incorporation of certain non-native

TABLE 1.5
Elemental Composition of Average C9 Unit of Milled Wood Lignin (Huang, Fu, & Gan, 2019)

MWL	Elemental Composition of Average C9 Unit
Spruce	$C_9H_{8.83}O_{2.37}(OCH_3)_{0.96}$
Beech	$C_9H_{7.10}O_{2.41}(OCH_3)_{1.36}$
Birch	$C_9H_{9.03}O_{2.77}(OCH_3)_{1.58}$
Wheat straw	$C_9H_{7.39}O_{3.00}(OCH_3)_{1.07}$
Rice straw	$C_9H_{7.44}O_{3.38}(OCH_3)_{1.03}$
Giant reed	$C_9H_{7.81}O_{3.12}(OCH_3)_{1.18}$
Bagasse	$C_9H_{7.34}O_{3.50}(OCH_3)_{1.10}$
Bamboo	$C_9H_{7.33}O_{3.81}(OCH_3)_{1.24}$
Corn stalk	$C_9H_{9.36}O_{4.50}(OCH_3)_{1.23}$

TABLE 1.6
Functional Groups Present in Softwood- and Hardwood-Type Lignins (Huang et al., 2019)

Functional Groups (mol/100 C_9)	Lignin Source			
	Spruce Wood	Birch Wood	Eucalyptus Globules	Eucalyptus Grandis
Methoxy group	92–96	164	164	160
Total hydroxyl groups	–	186	117–121	144
Aliphatic hydroxyl groups	15–20	166	88–91	125
Primary hydroxyl	–	86	68	70
Secondary hydroxyl	–	80	20	55
Benzyl hydroxyl	–	–	16	54
Phenolic hydroxyl	15–30	20	29–30	19
Total carbonyl groups	20	–	24	17
Aldehyde	–	–	9	24
Ketone	–	–	15	8
α–CO	–	–	10	8
Non–conjugated carbonyl	–	–	10	8
-COOH	–	–	4	5
Degree of polycondensation	–	–	18	21

monolignols via artificial lignification. Moreover, the natural lignin and its derivatives possess two fundamental characteristics, i.e., non-crystalline and optically inactive; these have been traditionally considered evidence for the randomness in their structural configuration.

1.3.4 Chemistry of Starch

Starch is another class of carbohydrate polymer present in green plants, particularly in seeds (or grains), leaves, stems, roots, shoots, and vegetable tubers (Seung, 2020). However, its structural morphology varies depending on the plant type. Overall, it appears in a granular form generated through the carbon fixation process via photosynthesis. Naturally, each of the granules contains several million subdivided molecules, such as amylopectin and amylose, differing in the structural configurations (Figure 1.12). On average, the plant starch is composed of 25% amylose and 75% amylopectin by weight (Baba & Arai, 1984). Those molecules differ in the basic solubility nature and, therefore, require a specific technique for isolation. For instance, amylose is a water-soluble substance and can be extracted via a simple hot water technique, whereas amylopectin is insoluble in water and its extraction is achieved through hydrolysis with pullulanase. Structurally, starch is synthesized using the α-D-glucose as repeat units and has the empirical formula $(C_6H_{10}O_5)_n$. Typically, the number of glucose units varies from a hundred to a few thousand in the polymeric network (Baba & Arai, 1984).

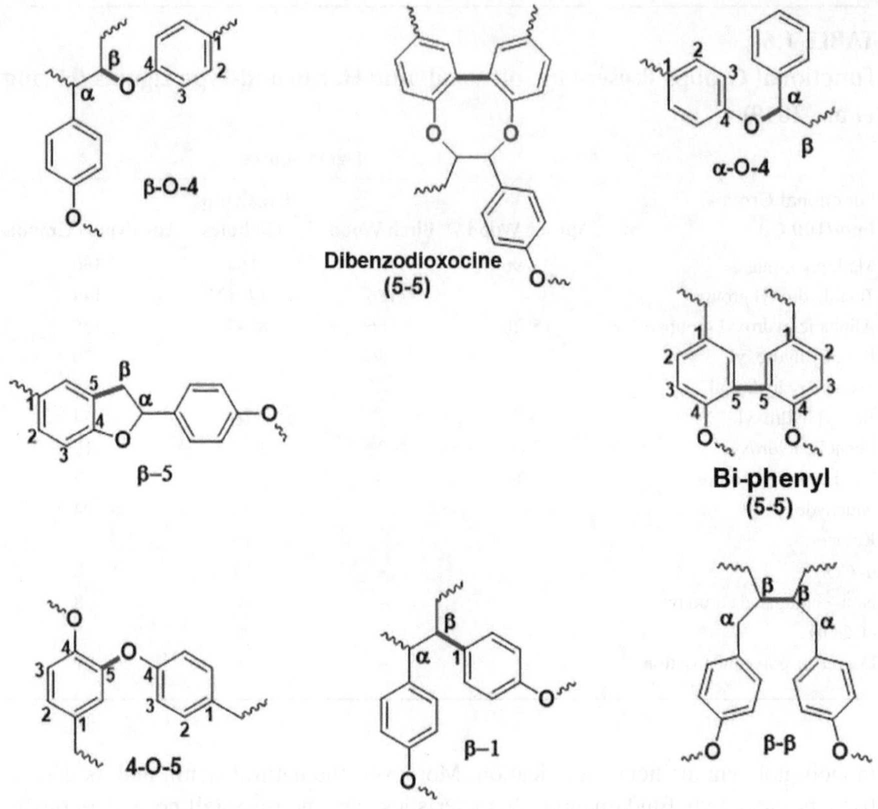

FIGURE 1.11 Representation of major linkages of lignin subunits for the network.

1.3.4.1 Amylose

Amylose is a linear chain molecule consisted of D-glucose units connected via α-(1→4)-linkages (Figure 1.13); its chain length typically varies from 500 to 20,000, and the molecular weight lies between 1.5 and 0.4×10^5. The characteristic molecule exhibits a left-handed-trend α-helical structure of different orientations (at least six anhydroglucose units present in one turn of the helical structure); therefore, it is classified into different structural forms, such as A, B, and V. Although both A and B forms of amylose have a stiff left-handed helical structure with six glucose units per turn, they differ by its style of packing of the helical starch. The A form structure is a single amylose molecule that contains a hydrogen bond forming between O-2 and O-6 atoms of glucose units on its exterior of the helical structure, whereas the V form is developed via co-crystallization of glucose with few organic compounds, including dimethyl sulfoxide, alcohol, iodine, and fatty acids. The formation of hydrogen bonds among the aligned amylose chains promotes retrogradation and results in the release of water molecules, thereby achieving a double-stranded hydrophobic structure. However, the resultant structure offers resistance to the starch-degrading enzymes (amylases) (Seung, 2020; Streb & Zeeman, 2012).

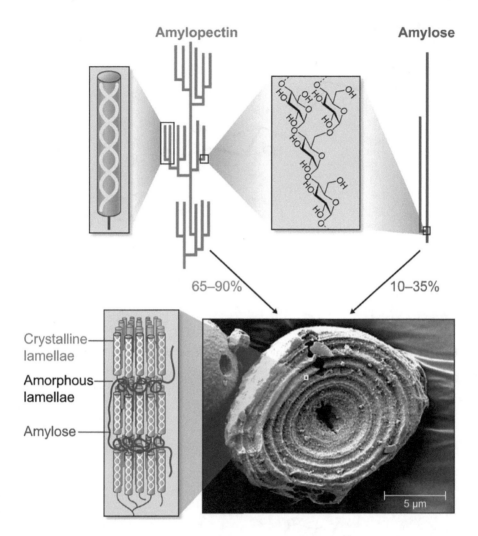

FIGURE 1.12 Schematic illustration of molecular arrangement of starch granule consisting of amylose and amylopectin components. (Reprinted with permission from Refs. Seung (2020); Seung and Smith (2019); Streb and Zeeman (2012).)

1.3.4.2 Amylopectin

Unlike amylose, amylopectin is a highly branched structure (Figure 1.14) possessing the molecular weight between 1×10^7 and 1×10^9. It represents a major component of starch, accounting for nearly 65–85% wt. of the starch granule (Baker, Miles, & Helbert, 2001). This content can be increased in plant starches up to 100% through mutation; this type of starch is commonly known as wax starch (Baba & Arai, 1984). Contrarily, few varieties of mutant plants have showed relatively lower amylopectin in starch due to the effect of mutation. Amylopectin is composed of nearly two million glucose units. The side-chain branches of it are made up of nearly 30 glucose units connected via α-(1→6)-linkage at a frequency of every 20–30 glucose units

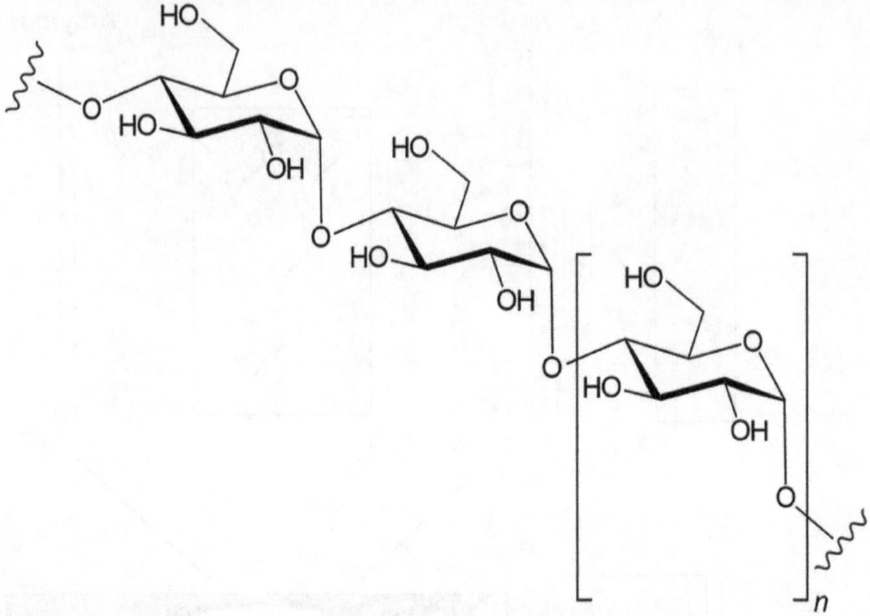

FIGURE 1.13 Generalized linear structure of amylose having glucose subunits.

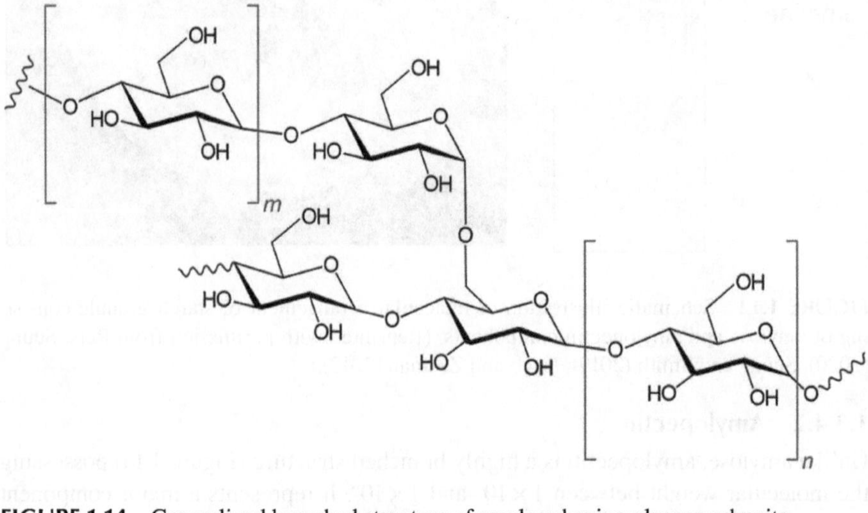

FIGURE 1.14 Generalized branched structure of amylose having glucose subunits.

in the entire chain. Moreover, the individual linear sub-chains of anhydroglucose monomers are joined via α-(1→4) bond, which is terminated in a non-reducing end of glucose units. In total, 4%–5% branching points are aroused at O-6 position of glucose units through the formation of an additional α-(1→6)-linked glucose unit in a single chain. It is further elongated by the α-(1→4)-linked glucan chains on the same pre-existing chains. These give rise to the average chain length of amylopectin between 20 and 30 glucose monomeric units and DP between 10,000 and 1,00,000. However, the natural molecular aggregation attained between the individual amylopectin molecules limits its accurate molecular weight determination obtained from a variety of plant sources. Typically, the side chains are grouped together to form the crystalline zones (or clusters) within the highly branched regions of amylopectin. This enabled subdividing the molecule into A, B, and C types of amylopectin, based on the occurrence of side chains (Martens, Gerrits, Bruininx, & Schols, 2018). The A-type crystalline starch has glucose helixes that are established to be densely packed; in contrast, the B type is a less densely packed structure, thereby leaving room for the formation of water molecules between the molecular branches. The C-type starch is a combination of A- and B-type crystallites. Moreover, in plants, amylopectin deposits in an alternative fashion of amorphous and crystalline shells (also known as growth rings) of 100–400 nm thick during starch biosynthesis. This results in a characteristic water-insoluble starch granule. Furthermore, the shape, size, and distribution of starch granules typically vary depending on the botanical sources along with the surface properties, such as porosity and surface area, to an extent (Seung & Smith, 2019; Streb & Zeeman, 2012).

1.4 OTHER LOW MOLECULAR WEIGHT CONSTITUENTS

Besides all these micromolecules, numerous other organic and inorganic compounds are present in the cell wall of biomasses, including forestry and herbaceous, as extractives (or accessory materials). These components typically account for only a small percentage in the total mass of solid (5–10% wt.). However, they show significant influence on the properties and also processing qualities of lignocellulose toward value addition. Generally, these are broadly classified into organic matter (extractives) and inorganic matter (ash) (Figure 1.15) (Elumalai & Pan, 2011). The organic elemental analysis report of some common biomass types is present in Table 1.7 (Sher, Pans, Sun, Snape, & Liu, 2018).

1.4.1 EXTRACTIVES

A wide range of low molecular weight organic compounds are found in the lignocellulosic biomass and are represented as responsible elements for providing color and protecting the biomass from decay (Kirker, Blodgett, Arango, Lebow, & Clausen, 2013). The extractive contents include a variety of organic matters, such as aromatics (simple/complex phenolics, terpenes), aliphatic acids (essential oils, waxes/fats), alkaloids (glycosides), simple sugars, proteins, gums, resins, mucilages, and saponins (Figure 1.16) (Alper, Tekin, Karagöz, & Ragauskas, 2020; Nascimento, Santana, Maranhão, Oliveira, & Bieber, 2013). Indeed, many of these compounds act

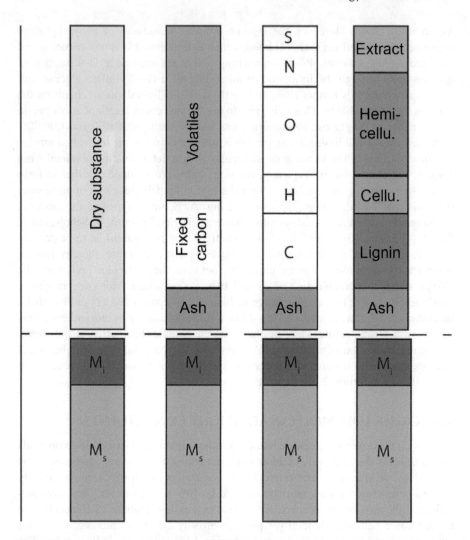

FIGURE 1.15 Overview of fractional composition of plant biomass based on dry basis composition, ultimate, proximate, biochemical (M_i-inherent moisture and M_s-surface moisture) analysis. (Modified from Ref. Vassilev et al. (2010).)

as intermediates in plant's metabolism, e.g., energy reservoir and defense elements against microbial attack. The characteristic extractives exhibit both hydrophobic and hydrophilic natures; therefore, they tend to solubilize in organic solvent(s) or water. Overall, less than 10% of these contents can be found in biomass, for instance, wood species, and moreover, its distribution in a plant varies depending on the species type and location within the same plant. Of the aromatic compounds, tannins ranked the most important one (Arbenz & Averous, 2015). Tannins can be further subdivided into the following three: (a) hydrolyzable tannins, (b) non-hydrolyzable tannins or condensed tannins, and (c) pseudo-tannins. The hydrolyzable tannins are defined as a mixture of poly-galloyl glucose and/or poly-galloyl quinic acid derivatives that

TABLE 1.7
Chemical Composition of Various Biomass Types
(Vassilev et al., 2010)

Biomass	Ultimate Analysis (%, Dry and Ash-Free Basis)				
	C	O	H	N	S
Wood Biomass					
Oak wood	50.6	42.9	6.1	0.3	0.1
Pine bark	53.8	39.9	5.9	0.3	0.07
Poplar	51.6	41.7	6.1	0.6	0.02
Willow	49.8	43.4	6.1	0.6	0.06
Grasses					
Miscanthus	49.2	44.2	6.0	0.4	0.15
Sweet sorghum	49.7	43.7	6.1	0.4	0.09
Switchgrass	49.7	43.4	6.1	0.7	0.11
Straws					
Alfalfa	49.9	40.8	6.3	2.8	0.21
Corn	48.7	44.1	6.4	0.7	0.08
Rape	48.5	44.5	6.4	0.5	0.1
Maize	45.6	43.4	5.4	0.3	0.04
Wheat	46.7	41.2	6.3	0.4	0.1
Rice	41.8	36.6	4.6	0.7	0.08

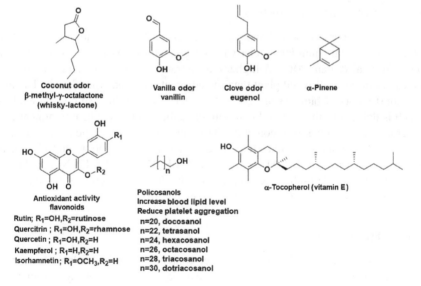

Coconut odor
β-methyl-γ-octalactone
(whisky-lactone)

Vanilla odor
vanillin

Clove odor
eugenol

α-Pinene

Antioxidant activity
flavonoids
Rutin; R_1=OH,R_2=rutinose
Quercitrin ; R_1=OH,R_2=rhamnose
Quercetin ; R_1=OH,R_2=H
Kaempferol ; R_1=H,R_2=H
Isorhamnetin; R_1=OCH$_3$,R_2=H

Policosanols
Increase blood lipid level
Reduce platelet aggregation
n=20, docosanol
n=22, tetrasanol
n=24, hexacosanol
n=26, octacosanol
n=28, triacosanol
n=30, dotriacosanol

α-Tocopherol (vitamin E)

FIGURE 1.16 Structural configuration of biomass extractives as trace constituents (Nascimento et al., 2013).

contain gallic residues from 3 to 12 units per molecule. It can be hydrolyzed by weak acid/base to produce both carbohydrate and phenolic acids. Condensed tannins are also called catechol tannins; they are polymers of flavonoid units connected via C–C bonds, and thus, it is merely susceptible to cleaving through simple hydrolysis. On the contrary, the pseudo-tannins are the representative mixture of low molecular weight compounds being associated with other available organic compounds. Other phenolic substances that include stilbenes and lignans and their respective derivatives are usually present in trace levels (Dai & Mumper, 2010).

Terpenes are another variety of aromatic compounds present in plants, as extractives (Harman-Ware, 2020). Its presence in higher amounts in woods, seeds, leaves, roots, and flowers delivers the perfumery odor. Structurally, it has a characteristic carbon skeleton, consisting of an elementary five-carbon isoprene unit (or 2-methyl-1,3-butadiene). It can be further classified into monoterpenes, sesquiterpenes, diterpenes, and triterpenes, based on the number of isoprene units. A higher level of terpenes can be found in pine tree varieties (Kačík et al., 2012).

Furthermore, saturated/unsaturated higher fatty acids are also found in biomass, as glycerides—esters of glycerol (Santek, Beluhan, & Santek, 2018). The trimesters (or triglycerides) are reported to be dominant among the glycerides. Waxes are categorized as the complex mixture of aliphatic compounds (wax esters) that are composed of majorly of fatty acids, hydrocarbons, and their derivatives. Alcohols also occur in a very widely distributed manner in the plant kingdom, but are available in a composite, and therefore, it is difficult for selective extraction. Ethyl alcohol is found in combination with other substances as esters. The majority of the aliphatic alcohols found in wood biomass are in the form of ester components. In addition, sterols may also present as fatty acid esters or glycosides (Elumalai & Pan, 2011).

1.4.2 Inorganic Constituents

The elemental composition of inorganics present in the plant biomass depends on the environmental conditions under which it grew, particularly the soil characteristics (Elumalai & Pan, 2011). A higher concentration of these matters was found in the cambium layer of wood plant than the adjacent bark and wood layers. The mineral constituents of biomass are generally determined by the ash content analysis, which is defined as the material that remains unburnt after the heat treatment at an extremely high-temperature condition. Till date, more than 50 element types have been found in biomass samples (e.g., wood) by neutron activation analysis and the most common ones are Na, Al, Mg, Si, S, P, Cl, K, Ca, Mn, Fe, Cr, Ni, Cu, Pb, and Zn. In particular, Ca (~80%), K, and Mg are the dominant elements and others are reported as less than 1% of the total (Zhang et al., 2019).

1.4.3 Fluid Content

In plants, cells consist of 85%–95% water content, which is available to carry to their various parts (i.e., from roots to leaves) via liquid transport system or xylem (Zwieniecki, Melcher, & Holbrook, 2001). Sooner or later, this massive water content is evaporated by the transpiration process. This residual water content is responsible for the materials moisture content, normally accounted for 10–60% wt., depending on the plant type. Upon analysis of the fluid matter, it contains a variety of anionic and cationic species. Moreover, the composition and the development of the plant biomass is influenced by the fluid content. For instance, the fast-growing crops contain a higher water level with a wide range of metal elements, including N, P, Na, K, Ca, Cl, and S (Tursi, 2019).

1.5 CONCLUSIONS

This chapter presents a comprehensive review of the inherent constituents of plant-derived biomass and their intra- and inter-associations for gaining fundamental knowledge for their effortless transformation. The intrinsic properties of lignocellulosic biomass, including chemical composition (cellulose, hemicellulose, lignin, and ash), moisture content, bulk density, etc., are highly variable and could influence the economics of the transformation of biomass into value-added products. The inefficiency in its successful bioconversion is due to its complex nature, which reduces the susceptibility to chemicals or enzymes to recover the industrially important molecules, e.g., fermentable sugars, say for fuel ethanol production. Till date, several methods have been developed to reconstruct biomass structure, including physicochemical pretreatment methods involving acid/base, ionic liquids, high-pressure steam, etc. However, those severely damage the biomass to a certain extent and also offer a significant loss of the critical constituents (e.g., cellulose and hemicellulose). Dedicated efforts are still made to overcome the difficulty in disassembling the rigid structure formed in a complex fashion, offering a strong resistant to its decomposition for deriving the bulk chemicals.

REFERENCES

Adapa, Phani Kumar, Lope G Tabil, and Greg J Schoenau. "Compression Characteristics of Selected Ground Agricultural Biomass." *Agricultural Engineering International: CIGR Journal* 11 (2009): 1347–65.

Alper, Koray, Kubilay Tekin, Selhan Karagöz, and Arthur J Ragauskas. "Sustainable Energy and Fuels from Biomass: A Review Focusing on Hydrothermal Biomass Processing." *Sustainable Energy & Fuels* 4, no. 9 (2020): 4390–414.

Amos, Robert A, and Debra Mohnen. "Critical Review of Plant Cell Wall Matrix Polysaccharide Glycosyltransferase Activities Verified by Heterologous Protein Expression." *Frontiers in plant science* 10 (2019): 915.

Arbenz, Alice, and Luc Averous. "Chemical Modification of Tannins to Elaborate Aromatic Biobased Macromolecular Architectures." *Green Chemistry* 17, no. 5 (2015): 2626–46.

Baba, Tadashi, and Yuji Arai. "Structural Characterization of Amylopectin and Intermediate Material in Amylomaize Starch Granules." *Agricultural and Biological Chemistry* 48, no. 7 (1984): 1763–75.

Bagniewska-Zadworna, Agnieszka, Abdelali Barakat, Piotr Łakomy, Dariusz J Smoliński, and Marcin Zadworny. "Lignin and Lignans in Plant Defence: Insight from Expression Profiling of Cinnamyl Alcohol Dehydrogenase Genes during Development and Following Fungal Infection in Populus." *Plant Science* 229 (2014): 111–21.

Baker, Andrew A, Mervyn J Miles, and William Helbert. "Internal Structure of the Starch Granule Revealed by AFM." *Carbohydrate Research* 330, no. 2 (2001): 249–56.

Boerjan, Wout, John Ralph, and Marie Baucher. "Lignin Biosynthesis." *Annual Review of Plant Biology* 54, no. 1 (2003): 519–46.

Carpita, Nicholas C, and David M Gibeaut. "Structural Models of Primary Cell Walls in Flowering Plants: Consistency of Molecular Structure with the Physical Properties of the Walls During Growth." *The Plant Journal* 3, no. 1 (1993): 1–30.

Carvalheiro, Florbela, Luís C Duarte, and Francisco M Gírio. "Hemicellulose Biorefineries: A Review on Biomass Pretreatments." *Journal of Scientific & Industrial Research* 67, no. 11 (2008): 849–64.

Chami Khazraji, Ali, and Sylvain Robert. "Self-Assembly and Intermolecular Forces When Cellulose and Water Interact Using Molecular Modeling." *Journal of Nanomaterials* 2013, (2013): 48–60.

Cui, Steve W, and Qi Wang. "Cell Wall Polysaccharides in Cereals: Chemical Structures and Functional Properties." *Structural Chemistry* 20, no. 2 (2009): 291–97.

Dai, Jin, and Russell J. Mumper. "Plant Phenolics: Extraction, Analysis and Their Antioxidant and Anticancer Properties." [In eng]. *Molecules (Basel, Switzerland)* 15, no. 10 (2010): 7313–52.

Danish, Mohammed, and Tanweer Ahmad. "A Review on Utilization of Wood Biomass as a Sustainable Precursor for Activated Carbon Production and Application." *Renewable and Sustainable Energy Reviews* 87 (2018): 1–21.

Demirbaş, Ayhan. "Estimating of Structural Composition of Wood and Non-Wood Biomass Samples." *Energy Sources* 27, no. 8 (2005): 761–67.

Ebringerová, Anna. "Structural Diversity and Application Potential of Hemicelluloses." *Paper Presented at the Macromolecular Symposia*, 2005.

Elumalai, S, and XJ Pan. "Chemistry and Reactions of Forest Biomass in Biorefining." In *Sustainable Production of Fuels, Chemicals, and Fibers from Forest Biomass*, edited by J. Zhu, X. Zhang and X. Pan, 109–44: ACS Publications, Washington, DC, 2011.

Giummarella, Nicola, and Martin Lawoko. "Structural Insights on Recalcitrance during Hydrothermal Hemicellulose Extraction from Wood." *ACS Sustainable Chemistry & Engineering* 5, no. 6 (2017): 5156–65.

Gusiatin, Zygmunt Mariusz, and Artur Pawłowski. "Biomass for Fuels–Classification and Composition." In *Biomass for Biofuels,* edited by K. Bulkowska, Z. Mariusz Gusiatin, E. Klimiuk, A. Pawlowski, T. Pokoj, 15–36: CRC Press, London, 2016.

Harman-Ware, A.E. "Conversion of Terpenes to Chemicals and Related Products." In *Chemical Catalysts for Biomass Upgrading*, edited by M. Crocker and E. Santillan-Jimenez, 529–68: Wiley-VCH Verlag GmbH, Germany, 2020.

Heinze, Thomas. "Cellulose: Structure and Properties." In *Cellulose Chemistry and Properties: Fibers, Nanocelluloses and Advanced Materials*, edited by Orlando J. Rojas, 1–52: Springer, Switzerland, 2015.

Hijazi, May, Silvia M Velasquez, Elisabeth Jamet, José M Estevez, and Cécile Albenne. "An Update on Post-Translational Modifications of Hydroxyproline-Rich Glycoproteins: Toward a Model Highlighting Their Contribution to Plant Cell Wall Architecture." *Frontiers in Plant Science* 5 (2014): 395.

Holland, Claire, Peter Ryden, Cathrina H Edwards, and Myriam M-L Grundy. "Plant Cell Walls: Impact on Nutrient Bioaccessibility and Digestibility." *Foods* 9, no. 2 (2020): 201.

Huang, Jin, Shiyu Fu, and Lin Gan. "Chapter 2-Structure and Characteristics of Lignin." In *Lignin Chemistry and Applications*, edited by Jin Huang, Shiyu Fu and Lin Gan, 25–50: Elsevier, Netherlands, 2019.

Jazi, Mehdi Erfani, Ganesh Narayanan, Fatemeh Aghabozorgi, Behzad Farajidizaji, Ali Aghaei, Mohammad Ali Kamyabi, Chanaka M Navarathna, and Todd E Mlsna. "Structure, Chemistry and Physicochemistry of Lignin for Material Functionalization." *SN Applied Sciences* 1, no. 9 (2019): 1094.

Kačík, František, Veronika Veľková, Pavel Šmíra, Andrea Nasswettrová, Danica Kačíková, and Ladislav Reinprecht. "Release of Terpenes from Fir Wood During Its Long-Term Use and in Thermal Treatment." [In eng]. *Molecules (Basel, Switzerland)* 17, no. 8 (2012): 9990–99.

Keegstra, Kenneth, Kenneth W Talmadge, WD Bauer, and Peter Albersheim. "The Structure of Plant Cell Walls: III. A Model of the Walls of Suspension-Cultured Sycamore Cells Based on the Interconnections of the Macromolecular Components." [In eng]. *Plant Physiology* 51, no. 1 (Jan 1973): 188–97.

Kirker, GT, AB Blodgett, RA Arango, PK Lebow, and CA Clausen. "The Role of Extractives in Naturally Durable Wood Species." *International Biodeterioration & Biodegradation* 82 (2013/08/01/ 2013): 53–58.

Lerouxel, Olivier, David M Cavalier, Aaron H Liepman, and Kenneth Keegstra. "Biosynthesis of Plant Cell Wall Polysaccharides—A Complex Process. *Current Opinion in Plant Biology* 9, no. 6 (2006): 621–30.

Liu, Qingquan, Le Luo, and Luqing Zheng. "Lignins: Biosynthesis and Biological Functions in Plants." *International Journal of Molecular Sciences* 19, no. 2 (2018): 335.

Liu, Tingting, Brian McConkey, Ted Huffman, Stephen Smith, Bob MacGregor, Denys Yemshanov, and Suren Kulshreshtha. "Potential and Impacts of Renewable Energy Production from Agricultural Biomass in Canada." *Applied Energy* 130 (2014): 222–29.

Marriott, Poppy E, Leonardo D Gómez, and Simon J McQueen-Mason. "Unlocking the Potential of Lignocellulosic Biomass through Plant Science." *New Phytologist* 209, no. 4 (2016): 1366–81.

Martens, Bianca MJ, Walter JJ Gerrits, Erik MAM Bruininx, and Henk A Schols. "Amylopectin Structure and Crystallinity Explains Variation in Digestion Kinetics of Starches across Botanic Sources in an in Vitro Pig Model." *Journal of Animal Science and Biotechnology* 9, no. 1 (2018): 1–13.

Mottiar, Yaseen, Ruben Vanholme, Wout Boerjan, John Ralph, and Shawn D Mansfield. "Designer Lignins: Harnessing the Plasticity of Lignification." *Current Opinion in Biotechnology* 37 (2016): 190–200.

Muscat, A, EM de Olde, IJM de Boer, and R Ripoll-Bosch. "The Battle for Biomass: A Systematic Review of Food-Feed-Fuel Competition." *Global Food Security* 25 (2020/06/01/ 2020): 100330.

Nakano, Yoshimi, Masatoshi Yamaguchi, Hitoshi Endo, Nur Ardiyana Rejab, and Misato Ohtani. "NAC-MYB-Based Transcriptional Regulation of Secondary Cell Wall Biosynthesis in Land Plants." *Frontiers in Plant Science* 6 (2015): 288.

Nascimento, MS, ALBD Santana, CA Maranhão, LS Oliveira, and L Bieber. "Phenolic Extractives and Natural Resistance of Wood." In *Biodegradation-Life of Science*, edited by R. Chamy, 3408–58: Intechopen, London, 2013.

Newell, R, Daniel Raimi, and Gloria Aldana. "Global Energy Outlook 2019: The Next Generation of Energy." *Resources for the Future* 1 (2019): 8–19.

Nishiyama, Yoshiharu, Paul Langan, and Henri Chanzy. "Crystal Structure and Hydrogen-Bonding System in Cellulose Iβ from Synchrotron X-Ray and Neutron Fiber Diffraction." *Journal of the American Chemical Society* 124, no. 31 (2002): 9074–82.

Olorunsola, Emmanuel O, Ekaete I Akpabio, Musiliu O Adedokun, and Dorcas O Ajibola. "Emulsifying Properties of Hemicelluloses." In *Science and Technology Behind Nanoemulsions,* edited by S. Karakuş, 639–59: Intechopen, London, 2018.

Pauly, Markus, Sascha Gille, Lifeng Liu, Nasim Mansoori, Amancio de Souza, Alex Schultink, and Guangyan Xiong. "Hemicellulose Biosynthesis." *Planta* 238, no. 4 (2013): 627–42.

Perrin, Jordan, Frédéric Pouyet, Christine Chirat, and Dominique Lachenal. "Formation of Carbonyl and Carboxyl Groups on Cellulosic Pulps: Effect on Alkali Resistance." *BioResources* 9, no. 4 (2014): 7299–310.

Rongpipi, Sintu, Dan Ye, Enrique D Gomez, and Esther W Gomez. "Progress and Opportunities in the Characterization of Cellulose–An Important Regulator of Cell Wall Growth and Mechanics." *Frontiers in Plant Science* 9 (2019): 1894.

Rubin, Edward M. "Genomics of Cellulosic Biofuels." *Nature* 454, no. 7206 (2008): 841–45.

Saha, Badal C. "Hemicellulose Bioconversion." *Journal of Industrial Microbiology and Biotechnology* 30, no. 5 (2003): 279–91.

Santek, Mirela Ivancic, Suncica Beluhan, and Bozidar Santek. "Production of Microbial Lipids from Lignocellulosic Biomass." In *Advances in Biofuels and Bioenergy,* edited by M. Nageswara-Rao and J. Soneji, 137–64: Intechopen, London, 2018.

Schultink, Alex, Lifeng Liu, Lei Zhu, and Markus Pauly. "Structural Diversity and Function of Xyloglucan Sidechain Substituents." *Plants* 3, no. 4 (2014): 526–42.

Sella Kapu, Nuwan, and Heather L Trajano. "Review of Hemicellulose Hydrolysis in Softwoods and Bamboo." *Biofuels, Bioproducts and Biorefining* 8, no. 6 (2014): 857–70.

Seung, David. "Amylose in Starch: Towards an Understanding of Biosynthesis, Structure and Function." *New Phytologist* 228, no. 5 (2020): 1490–504.

Seung, David, and Alison M Smith. "Starch Granule Initiation and Morphogenesis—Progress in Arabidopsis and Cereals." *Journal of Experimental Botany* 70, no. 3 (2019): 771–84.

Shankar Tumuluru, Jaya, Shahab Sokhansanj, J Richard Hess, Christopher T Wright, and Richard D Boardman. "A Review on Biomass Torrefaction Process and Product Properties for Energy Applications." *Industrial Biotechnology* 7, no. 5 (2011): 384–401.

Sher, Farooq, Miguel A Pans, Chenggong Sun, Colin Snape, and Hao Liu. "Oxy-Fuel Combustion Study of Biomass Fuels in a 20 kWth Fluidized Bed Combustor." *Fuel* 215 (2018): 778–86.

Smith, Peter J, Hsin-Tzu Wang, William S York, Maria J Peña, and Breeanna R Urbanowicz. "Designer Biomass for Next-Generation Biorefineries: Leveraging Recent Insights into Xylan Structure and Biosynthesis." *Biotechnology for Biofuels* 10, no. 1 (2017): 1–14.

Spiridon, Iuliana, and Valentin I Popa. "Hemicelluloses: Major Sources, Properties and Applications." In *Monomers, Polymers and Composites from Renewable Resources*, edited by M. Naceur Belgacem and A. Gandini, 289–304: Elsevier, Netherlands, 2008.

Streb, Sebastian, and Samuel C Zeeman. "Starch Metabolism in Arabidopsis." *The Arabidopsis book/American Society of Plant Biologists* 10, (2012): 1–33.

Tarasov, Dmitry, Mathew Leitch, and Pedram Fatehi. "Lignin–Carbohydrate Complexes: Properties, Applications, Analyses, and Methods of Extraction: A Review." *Biotechnology for Biofuels* 11, no. 1 (2018): 269.

Tursi, Antonio. "A Review on Biomass: Importance, Chemistry, Classification, and Conversion." *Biofuel Research Journal* 6, no. 2 (2019): 962.

Vassilev, Stanislav V, David Baxter, Lars K Andersen, and Christina G Vassileva. "An Overview of the Chemical Composition of Biomass." *Fuel* 89, no. 5 (2010): 913–33.

Vogel, Kenneth P, and Hans-Joachim G Jung. "Genetic Modification of Herbaceous Plants for Feed and Fuel." *Critical Reviews in Plant Sciences* 20, no. 1 (2001): 15–49.

Wyman, Charles E, Stephen R Decker, Michael E Himmel, John W Brady, Catherine E Skopec, and Liisa Viikari. "Hydrolysis of Cellulose and Hemicellulose." *Polysaccharides: Structural Diversity and Functional Versatility* 1 (2005): 1023–62.

Zhang, Ning, Li Wang, Ke Zhang, Terry Walker, Peter Thy, Bryan Jenkins, and Yi Zheng. "Pretreatment of Lignocellulosic Biomass Using Bioleaching to Reduce Inorganic Elements." *Fuel* 246 (2019): 386–93.

Zhong, Ruiqin, Dongtao Cui, and Zheng-Hua Ye. "Secondary Cell Wall Biosynthesis." *New Phytologist* 221, no. 4 (2019): 1703–23.

Zwieniecki, Maciej A, Peter J Melcher, and N Michele Holbrook. "Hydrogel Control of Xylem Hydraulic Resistance in Plants." *Science* 291, no. 5506 (2001): 1059–62.

2 Lignin to Platform Chemicals and Biomaterials
Chemical and Biological Perspectives

Yeddula Nikhileshwar Reddy
Center of Innovative and Applied Bioprocessing (CIAB)
Indian Institute of Science Education and Research

Kunal Gogde
Center of Innovative and Applied Bioprocessing (CIAB)

Shatabdi Paul
Center of Innovative and Applied Bioprocessing (CIAB)
Regional Centre for Biotechnology

Jayeeta Bhaumik
Center of Innovative and Applied Bioprocessing (CIAB)

CONTENTS

2.1 Introduction to Lignocellulosic Biomass 32
2.2 Lignin: Structure and Its Components 33
2.3 Valorization of Lignin: Chemical Perspective.. 35
    2.3.1 Photocatalytic Degradation 37
    2.3.2 Enzymatic Degradation of Lignin 39
        2.3.2.1 Lignin-Degrading (LD) and Lignin-Modifying (LM)
                 Enzymes...44
        2.3.2.2 Lignin Peroxidase (EC 1.11.1.14) 46
        2.3.2.3 Manganese-Dependent Peroxidase (EC 1.11.1.13)...............46
        2.3.2.4 *Laccase* (Lac, Benzenediol: Oxygen Oxidoreductases;
                 EC 1.10.3.2) 46
        2.3.2.5 Superoxide Dismutases.. 47

DOI: 10.1201/9781003158486-2

2.4   Lignin in Biological Agents...47
       2.4.1   Drug Carrier 49
       2.4.2   Microbicidal Agent...50
       2.4.3   Theranostic Agent...50
2.5   Future Perspective 51
References...52

2.1 INTRODUCTION TO LIGNOCELLULOSIC BIOMASS

The plant and animal material present on the earth is termed biomass. Lignocellulosic biomass, derived from plant sources, has been conceded as a crucial source for various biomaterials and platform chemicals. The atmospheric CO_2 and H_2O in the presence of sunlight help in the synthesis of lignocellulosic biomass. It is a complicated model comprising of a multitude of phenolic polymers, polysaccharides, etc., which make up the quintessential part of the woody cell wall (Yousuf, Pirozzi, and Sannino 2019). Structurally, biomass is crafted from lignin, cellulose, and hemicellulose (Figure 2.1). Cellulose, being a glucose-based polysaccharide, is wrapped by another complex carbohydrate hemicellulose. Then, the heteropolyphenolic-based polymer, lignin, provides strength and rigidity to the structure. This structural complexity makes the lignocellulosic biomass recalcitrant to enzymatic degradation too. In the lignocellulosic biomass, cellulose being the major part of the higher plant species contributes about 50% of the total biomass on earth and is the abundantly available polysaccharide with D-glucose units as monomers linked by $\beta(1\rightarrow4)$ bond. Along with cellulose, hemicellulose is made up of various heteropolymer-based polysaccharides consisting of various monomers such as xylose, mannose, galactose, and glucose, constituting about 15%. Hemicellulose is an amorphous polymer with little strength, whilst cellulose is crystalline, strong, and less prone to hydrolysis reactions.

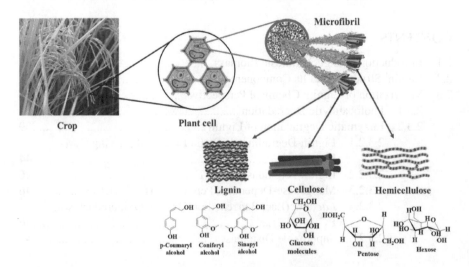

FIGURE 2.1 Chemical components of major fractions of lignocellulosic biomass.

These carbohydrate-based polymers are tightly bound to the aromatic polyphenolic polymer lignin, making it resistant to various chemical and enzymatic hydrolysis. On the contrary, lignin is the largest contributor to the aromatic component and constitutes about 15%–20% of the biomass. This complex spatial mixture is divided into three categories based on the source. The low-value by-product obtained from agro-based industrial sector and forestry is termed waste biomass. The virgin biomass is termed as the naturally occurring biomass from terrestrial plants such as grass, trees, and bushes. However, the usage of energy crops namely elephant grass and switchgrass yields high amounts of lignocellulosic biomass utilized for the manufacturing of second-generation biofuel (Zoghlami and Paës 2019). Apart from the biofuel generation from lignocellulosic biomass, the individual entities of the lignocellulose have several applications ranging from biological applications such as delivery of drug molecules, therapeutic applications to the degradation of the same to platform chemicals. For instance, lignocellulosic biomass has been directly used for the development of nanocellulose (Gupta and Shukla 2020; Nis and Kaya Ozsel 2021). In recent times, as fossil fuels are on the verge of depletion, lignocellulosic biomass remains as the renewable carbon source present on this planet. However, the usage of higher plants is confined to the paper industry, for which cellulose plays a key role where lignin is the effluent. Every year, several tonnes of lignin obtained as the effluent from the cellulose-based industry is diverted into the aquatic bodies, making it one of the major pollutants. This potentiated the researchers to utilize the underutilized renewable source of energy and various chemicals.

2.2 LIGNIN: STRUCTURE AND ITS COMPONENTS

The term lignin was first used by a Swiss Botanist A. P. Candolle in the year 1813, acquired from the Latin word "lignum", denoting wood (Kirk-Othmer(ed) 2007). Lignin is a secondary metabolite produced from the metabolism of tyrosine and phenylalanine in plant cells. It is achieved in three different processes; initially, the monomers essential for lignin structure are biosynthesized in the cytoplasm of the plant cell. Then, the monomers are moved to apoplast for the final polymerization to occur using the enzymes laccases and peroxidases alongside hydrolysis, deamination, reduction, and methylation (Miao and Liu 2010; Bonawitz and Chapple 2010; C. J. Liu, Miao, and Zhang 2011; Ralph et al. 2004). Typically, the polymerization involves three different monomers (monolignols), namely coniferyl alcohol (G component), p-coumaryl alcohol (H unit), and sinapyl alcohol (S component), and various other xenobiotics such as derivatives of stilbenes, hydroxycinnamaldehyde, and flavones, etc., to name a few.(Río et al. 2012; Lan et al. 2015; Río et al. 2017). Due to the presence of a multitude of monomers in the biosynthetic pathway of lignin, it can be understood that the structure is a heteropolymer, polymerized in the cell wall of the plants. This random polymerization is majorly responsible for variable molecular structure in disparate plants. The consequential heteropolyphenolic structure provides rigidity and hydrophobicity to the cell wall, which acts as a defensive system against various microbial infections. Additionally, its metabolism is also involved in providing resistance to various environmental stresses and a barrier against pathogens and pests (Ithal et al. 2007; Schuetz et al. 2014; Moura et al. 2010).

The synthesis of lignin in plant involves the dehydrogenation of monomer units mediated by peroxidases. This generates the corresponding monolignol radicals which fuse to generate the lignin polymer. Fundamentally, the radical being the highly active species, it combines in an unorganized way, resulting in many chiral centres. However, it has been hypothesized that the ultimately obtained polymer of lignin is optically inactive. Concomitantly, the monolignol radicals bind to other cell wall structures leading to other complexes resulting in a three-dimensional structure. Generally, gymnosperms such as pine and Cycas plant species are rich in G monomer units, dicots such as cotton are rich in G and S monomer units, whilst G, S, and H monomers make up monocotyledons such as paddy and wheat.

This polypropanoid-rich lignin majorly consists of C–C linkages and ether linkages formed during polymerization. According to Gibbs, the lignin structure is homogenous due to dehydrated oligomerization of G units in gymnosperms, whilst softwood contains copolymerized G, S, and H monomers. Moreover, the growth factors required for the development of lignin, such as nutrition, climate, etc., affect the quality of lignin. This structural complexity of lignin is still a major challenge because of the composition of various C–C bonds in the structure. This portrays the reason behind the unorganized deconstruction of lignin by chemical methods when the researchers tried to understand the lignin structure. However, some researchers tried to understand the lignin structure and came to some conclusions. In this regard, initially in 1961, Freundberg achieved a lignin-like polymer structure by polymerizing the coniferyl alcohol by dehydrogenation method. This model upon breakdown leads to the identification of 15 basic units. Later on, he could develop a model compound similar to spruce lignin with 18 basic units in the year 1968 (*Constitution and Biosynthesis of Lignin | Karl Freudenberg | Springer* 2021). Then, Alder was the first to propose a tentative structure (Adler 1977). Considering the above-hypothesized structure of lignin, Nimz, in 1974, proposed a fragment model containing 25 phenylpropanoid units, guaiacyl units, and syringyl units (Nimz 1974). Although the presence of repetitive monomers has been identified, the structural organization was still indistinct. In the same year, by the application of various simulation techniques, Glasser and Glasser proposed a lignin model compound with more than 90 phenylpropanoid units. These simulations also revealed various bonds, namely 5-5′, β-O-4, β-β′, β-5, and other bonds, involved in the structural makeup of lignin and could decipher the molecular weight of around 17 kD (Glasser and Glasser 1974). Besides, these structures became controversial among the researchers. After 3 years, Alder proposed the structure of lignin which contains all the monolignol units alongside 16 phenylpropanoid units, which disclosed that the structure is biosynthesized in the cell wall (Adler 1977). Contemporarily, Wayman and Obiaga (1974) proposed a module assembly model for the lignin structure. Later, many research groups tried to understand the structure of lignin by proposing various structures with different monomer units. Then, in 2011, Crestini proposed a linear oligomer chain structure that is in line with the module assembly model, which is in contrast to the regularly proposed cross-linked network-like structure (Crestini et al. 2011). However, the quantification of the monomers has not been achieved to date. This archetype behaviour of lignin makes it recalcitrant for its valorization.

TABLE 2.1
Methods of Delignification and Their Solubility

Pulping Process	Conditions	Solubility	References
Kraft	170°C, NaOH + Na₂S	Alkaline solutions	Schutyser et al. (2018)
Sulphite	140°C, SO₂ + Na/Ca/Mg. NH₄	Alkaline solutions	El Mansouri, Pizzi, and Salvadó (2007)
Alkaline (soda anthraquinone)	150°C–170°C, NaOH	Water	Rodríguez et al. (2010)
Organosolv	150°C–200°C, acetic acid/formic acid, ethanol, water	Organic solvent	Mandlekar et al. (2018)
Hydrolysed lignin	Enzyme	–	Stücker et al. (2016)
Second-generation lignin from biorefinery	Hydrolytic pretreatment	–	Cotana et al. (2014)
Steam-exploded lignin	High temperature/pressure at 180°C–200°C	Organic solvents	He et al. (2020)
Formaldehyde-assisted fractionation	80°C–100°C, formaldehyde, dioxane, HCl	Organic solvents	Van den Bosch et al. (2018)
Reductive catalyst fractionation	Redox catalyst (H2 donor), 180°C–250°C, organic solvent, and H₂O	–	Qiu et al. (2020)

In recent times, the lignin obtained from the paper and pulp industry has been understood on a whole different level. A recently published review of the resources generated from the paper industry provides us deeper insights into the challenges and innovations which have been observed from the effluent generated from the paper industry (Mandeep, Kumar Gupta, and Shukla 2020). During the sulphite treatment of wood, many sulphonic acid groups are introduced into the structure. This is observed in the side chains of the lignin structure, i.e. the α position. This condensation results in α-6 linkages. Moreover, during the kraft process, it has been hypothesized that a nucleophilic substitution reaction can be observed, which is a result of sulphur atom attacking at the β position of the carbon side chain, enhancing the rigidity in the structure leading to α-5, β-1, 4-O-5 bond formation in the structure (Gierer 1980). A comprehensive review of various lignin model types has been beautifully depicted in the book titled "Structure and Characteristics of Lignin" (2019). Moreover, the extracted lignin from various delignification strategies has differential solubility, which is tabulated in Table 2.1. In almost all types of lignin, to mention hardwood, softwood, and other varieties, the basic structure can be depicted as in Figure 2.2. This figure provides a pictorial representation of basic lignin structure with bond dissociation energies and the % of linkages constituting the lignin structure sources.

2.3 VALORIZATION OF LIGNIN: CHEMICAL PERSPECTIVE

Valorization of lignocellulosic biomass as a whole is a challenging task because of its recalcitrant nature (Himmel et al. 2007). Cellulose and hemicellulose components

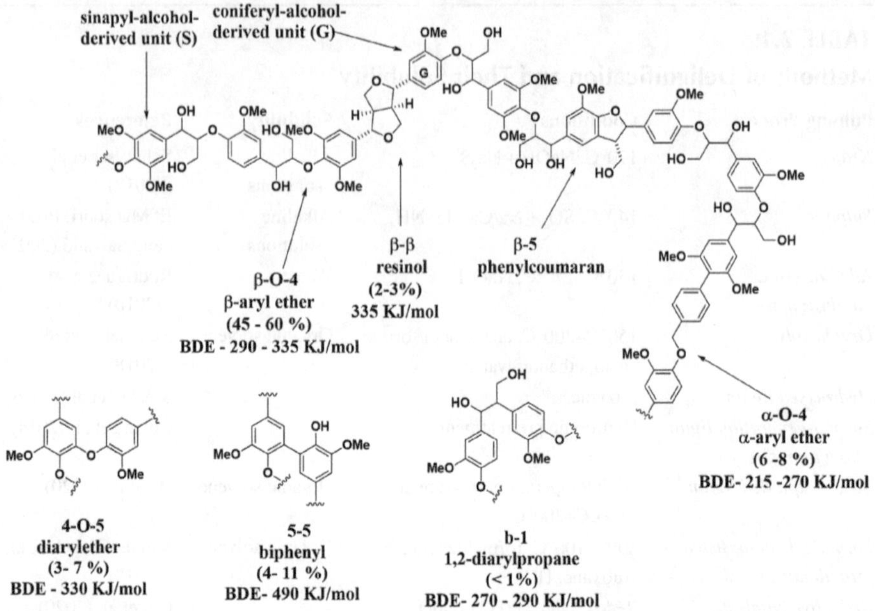

FIGURE 2.2 Lignin structure with linkage abundancy and bond dissociation energy.

of lignocellulosic biomass are consumed in the ethanol manufacturing, paper industry, pulp industry, and others, which primarily involve the delignification process and produce around 100 million tonnes/year, accounting for 700 million USD approximately. It is being envisioned that at a CAGR of 2.2%/year, this is projected to reach more than 900 million USD worldwide (Bajwa et al. 2019). Lignin, being the only aromatic and fuel source, is the underutilized component. Because of the heteropolyphenolic and amorphous nature of lignin, it is a reservoir of aromatic compounds and fuels. Moreover, lignin obtained from rice and wheat straw promises to nurture the bio-based economy, yet these are carbonized or used as fodder. Nevertheless, these also constitute a potentially indispensable source of lignin, from which many value-added products can be made (H. Luo and Abu-Omar 2017). In recent years, despite substantial research to depolymerize lignin, it remains elusive. Lignin being the quintessential polyphenolic polymer, its dry weight accounts for 15%–30%, contrary to the energy ratio of the lignin (40%) in the lignocellulosic biomass(Gundekari, Mitra, and Varkolu 2020). This unique framework and high specific energy make it an exemplary renewable feedstock for platform chemicals and superior-quality fuels.

The technologies, involving the consumption of lignin for a multitude of applications, are far from the desirable level; efforts are being made for the appropriate utilization. At present, 700 lakh tons of lignin is being generated and is anticipated to have a double-digit growth due to the rapid establishment of biorefineries to meet the needs of growing demand. To date, most of the unspent lignin is being mixed into wastewater as black liquor from the paper industry and is serious pollution to the aquatic life besides the loss of aromatic source (Z. Sun et al. 2018; H. Wang et al. 2019).

To have proficient utilization of lignin for the development of aromatic compounds, depolymerization of lignin remains the primary unmet objective. Structurally, the presence of various C–C and C–O–C linkages makes the structure difficult to depolymerize because C–O and C–C bond dissociation energies are higher. This framework also relates to the inadequate solubility in organic solvents used day to day and averts it from physical, biological, and chemical degradations in nature(Shen et al. 2019; Schutyser et al. 2018). H. Wang et al. (2016), when tried to depolymerize lignin, observed that when monomers are obtained, they tend to repolymerize, resulting in more recalcitrant polymer. These intrinsic properties pose intimidating challenges in the selective conversion of lignin to phenol-like compounds. Many strategies have been applied in the past decade. These involve acid or base treatment, oxidative/ reductive depolymerization, and more. However, these techniques involve extremely concise screening of catalyst, solvents, temperature, pH, and other reaction parameters (Xiang et al. 2020). However, progress has been achieved in the field involving novel conversion systems, enzymes, photocatalysis, and pretreatment techniques, to name a few. In this chapter, we will discuss lignin degradation using photocatalytic and enzymatic techniques.

2.3.1 Photocatalytic Degradation

The degradation of the spent mother liquor from the paper industry requires greener and sustainable techniques. Photocatalysis is the most sustainable technique for the degradation of a large number of pollutants, catalysing various reactions, and more. Solar energy is the greener solution for the same as it is a renewable source of energy. Hence, this renewable energy can be efficiently utilized to degrade the renewable source of carbon. Over the last few years, the usage of solar energy has gained considerable interest in the depolymerization of lignin to small molecules or phenolics (Hongji Li et al. 2019). Photocatalysis is a mild, highly efficient, pollution-free, and economical technique. The deployment of photons as an energy source for chemical transformation reactions was first proposed by Ciamician in 1912 (Ciamician 1912). This did not find much interest among the researchers until Fujishima and Honda unearthed the water splitting via photocatalytic route over TiO_2 (Fujishima and Honda 1972). Recently, the photocatalytic process has widely been employed in pollution abatement, chemical synthesis, and fuel production (Nikhileshwar Reddy, Thakur, and Bhaumik 2020; Staveness, Bosque, and Stephenson 2016; X. Liu et al. 2019; Kuehnel and Reisner 2018).

The photocatalytic conversions generally proceed via three different mechanisms, i.e. reductive depolymerization, oxidative depolymerization, and redox-neutral depolymerization, or by providing energy higher than the bandgap energy. The photocatalysts responsible for the transformation reactions can be categorized into two types: homogenous photocatalysts and heterogeneous photocatalysts. In the case of heterogeneous photocatalysts, the photons contain higher energy than the bandgap energy (E_{BG}) of the catalyst and electrons (e^-) move from valence band (VB) to conduction band (CB), which, in turn, generates holes (h^+) on the valence band. These photogenerated charges (e⁻ and h⁺) recombine rapidly or are stabilized/trapped by the internal or surface flaws in the catalyst. If stabilized, the energy is sufficient to initiate or promote a chemical reaction directly by themselves or generate reactive oxygen species (ROS) produced from the catalyst (W. Sun et al. 2018; Kou et al. 2017). These ROS (OH, H_2O, O_2^-) are of great significance even in the biological process too, which will be discussed in the later part of the chapter. These ROS generated in the reaction mixture have the potential to oxidize organic substrate (Nikhileshwar Reddy, Singh Thakur, and Bhaumik 2020). Considering this phenomenon, lignin can be photocatalytically transformed into various phenolic compounds or small molecules, which solely depends on the type of catalyst, the bandgap energy of the catalyst, solvent system, and other reaction parameters.

As mentioned earlier, lignin majorly consists of β-O-4 (50%) linkages; the majority of the photocatalytic process involves the same. This can be achieved in three different ways: reductive cleavage by electrons, oxidizing by holes, or redox-neutral bond disruption by ROS combined with electrons. Nevertheless, the reactions are non-selective; achieving the selectivity of product from lignin to specific platform chemicals under mild conditions is of significant interest among researchers. For example, the conversion of lignosulphonate to vanillin is of high interest in the food, flavour, and fragrance industries. The current methods involve high temperatures and pressures with less appreciable yields (Hibbert and Tomlinson 1937; Bjørsvik and Minisci 1999; Fache, Boutevin, and Caillol 2016). However, the lignin-to-monomer conversion is still not achieved despite the advent of various catalytic approaches and milder reaction processes. Among the few research groups working on the catalytic approaches to depolymerizing lignin, Stephenson's group has initially proposed a redox catalytic approach for benzylic oxidation in lignin model compounds. However, the usage of iridium-based catalysts made the process expensive. This led to the development of cheaper alternatives by the same group, which includes *N*-phenylphenothiazine as a photocatalyst. This organophotocatalyst provided reasonably good yields for the degradation of lignin models to phenol-like compounds (Nguyen, Matsuura, and Stephenson 2014; Bosque et al. 2017; Magallanes et al. 2019).

The other mechanism involved in the photocatalytic approach is oxidation, which can be termed as oxidative depolymerization of lignin. In this mechanism, lignin substrates react with holes produced from the photocatalysts and lead to the formation of radical cations. These radical cations are responsible for the oxidative cleavage of C–O and C–C bonds. The hydroxyl radical is generated from the H_2O, whilst the superoxide anion O_2^- is generated from the electrophilic O_2 by capturing photogenerated electrons. In line with this, Zhang and co-workers succeeded in achieving a binary ionic liquid system for the oxidative cleavage of lignin model compounds. In this reaction, one ionic liquid [PMim][NTf₂] initiates the breakdown of the $C_β$–H bond and the Brønsted acid-based ionic liquid [PrSO₃HMim] [OTf] cleaves the C–O–C bond (Kang et al. 2019). This is also an example of the mild reaction conditions-based organic approach for the photochemical degradation of lignin. Proton-coupled electron transfer is another interesting approach that is involved in the breakdown of the C–C bond. Using this approach, Zhang and his colleagues came up with a redox-neutral depolymerization approach for the disruption of the C–C bonds of β-O-4 linkages in lignin. For this, a photocatalytically active (Ir^{III})-based catalyst was chosen. Upon excitation under blue light, the catalyst excites resulting in (*Ir^{III}) form that draws electrons from the lignin which is intermolecularly bonded with the base. Then, the excited state generated cationic-based radical intermediates which further result in alkoxy-free radical intermediates through the proton-coupled electron process. Then the radical intermediates lacking the alkoxy group undergo Cα–Cβ splitting to selectively yield benzaldehyde and phenyl ether with up to 9% selectivity (Y. Wang et al. 2019). The redox-neutral approach of depolymerization offers the benefit of single-pot reaction conditions, thereby limiting the use of stoichiometric additives.

It is well known that quantum dots (QDs) are excellent semiconductor materials having adiameter smaller than Bohr excitons. They possess exceptional surface and photophysical properties fitting themselves in the field of material sciences, chemistry, biology, and physics. The photocatalytic behaviour of QDs can be tuned by altering the adhered superficial ligands. This approach has been found sustainable when Wu and the team developed a photocatalytic system containing CdS-based quantum dots for the degradation of lignin. Mechanistically, the photon-induced hole was transported from the center of QDs to the lignin-based model compound. This cleaves the Cα–H bond present in the β-O-4 linkage via oxidative dehydrogenation, producing Cα radical and a proton, thereby weakening the dissociation energy of the C–O bond in the β-O-4 linkage. Later, the photoinduced electron via electron tunnelling mechanism moves to the Cα radical from the catalyst to the substrate. This eased the reorganization of the electrons in β-O-4 linkages, thereby facilitating the C–O bond to produce acetophenone and phenoxy anion. As an endpoint, the phenoxy anion is converted into phenol by accepting a proton (Wu et al. 2019).

Moreover, despite the odds in the development of a feasible catalyst for the depolymerization of lignin, many other reaction additives such as acid/base, radical trappers, radical stabilizers, solvents, reactor design, reaction environment, and radical quenchers formed during the reaction play a substantial role in photocatalysis. In the case of reactor design, light source and environment of the reaction such as oxygen environment or inert atmosphere quintessential for the radical generation and stabilization have been reviewed beautifully by Xiang et al. (2020). They provide more insights into a typical photoreaction. In this regard, various photocatalysts for lignin depolymerization are tabulated in Table 2.2. In conclusion, photocatalysis is a mild and sustainable approach which upon significant advancements can provide an economical approach for the valorization of the recalcitrant lignin.

2.3.2 ENZYMATIC DEGRADATION OF LIGNIN

Another catalytic approach for depolymerization of lignin is the usage of enzyme-based techniques. Nature has an answer to every problem. In this regard, enzymes are ubiquitous catalysts which are responsible for carrying out a multitude of processes ranging from synthesis and transformations to degradation, chemically and biologically. Moreover, the lignin upon enzymatic degradation has a prominent influence on its structure (Sheng et al. 2021). Considering this fact, peroxidases and laccases are the major enzyme families which can synthesize lignin and can degrade

TABLE 2.2
Various Photocatalysts Involved in the Depolymerization of Lignin and Lignin Model Compounds

S. No.	Catalyst	Substrate	Product	Conversion (%)	Yield (%)	References
1.	Pb/ZnInS$_4$	(structure, OH)	(1a) phenol (OH), (1b) ketone	94	1a 94, 1b 76	N. Luo et al. (2016)
2.	TiO$_2$	Organosolv black liquor	(2a) HO–, methoxy aldehyde; (2b) OH, methoxy, allyl	94	NA	Prado, Erdocia, and Labidi (2013)
3.	In$_2$S$_3$	(structure, OH, methoxy)	(3a) HO–, methoxy benzaldehyde	NA	NA	Chen et al. (2018)
4.	CdS	(structure, OH)	(4a) phenol (OH), (4b) ketone	99	94	Wu et al. (2018)
5.	ZnInS$_4$	(structure, OH)	(5a), (5b), (5c)	99	5a 83, 5b 90, 5c 6	N. Luo et al. (2017)
6.	Mesoporous graphitic carbon nitride	(structure, OH)	(6a), (6b) CHO, (6c) COOH	96	6a 51, 6b 30, 6c 21	H. Liu et al. (2018)

(Continued)

TABLE 2.2 (Continued)
Various Photocatalysts Involved in the Depolymerization of Lignin and Lignin Model Compounds

S. No.	Catalyst	Substrate	Product	Conversion (%)	Yield (%)	References
7.	Carbazolic copolymers (CzCPs)			NA	**7a** 69–91 **7b** 74–86	J. Luo et al. (2017)
8.	Ligand-controlled CdS quantum dots	Bare lignin	Aromatic monomers	27	NA	Wu et al. (2019)
9.	Ir(ppy)2(bpy)MCFs			99	**9a** 92 **9b** 92	Hao et al. (2018)
10.	CuO₂/CeO₂/ANTs			99	**10a** 98 **10b** 2	Hou et al. (2017)
11.	Vanadium complexes			100	**11a** 85 **11b** 80	Gazi et al. (2015)
12.	[Ir(ppy)2(dtbbpy)]PF₆			NA	**12a** 88 **12b** 89	Nguyen, Matsuura, and Stephenson (2014)

(Continued)

TABLE 2.2 (Continued)
Various Photocatalysts Involved in the Depolymerization of Lignin and Lignin Model Compounds

S. No.	Catalyst	Substrate	Product	Conversion (%)	Yield (%)	References
13.	[Ir[dF(CF$_3$)(ppy)2-(dtbbpy)] PF$_6$/Na$_2$S$_2$O$_8$/Pd(OAc)$_2$		(13a)	96	13a 100	Kärkäs et al. (2016)
14.	[PMim] [NTf$_2$] [PrSO$_3$HMim][OTf]		(14a) (14b)	87.1	14a 78 14b 55	Kang et al. (2019)
15.	N-hydroxyphthalimide (NHPI)		(15a)	100	15a 96	J. Luo and Zhang (2016)
16.	AgSCdS		(16a) (16b) (16c)		16a 16b 16c	Yoo et al. (2020)
17.	[Ir(dF(CF$_3$)bpy)2(5,5'd(CF$_3$)-bpy)]PF$_6$		(M1) (M2) (M3)	NA	M1 0.3, M2 0.3, M3 0.3	S. T. Nguyen, Murray, and Knowles (2020)

(Continued)

TABLE 2.2 (Continued)
Various Photocatalysts Involved in the Depolymerization of Lignin and Lignin Model Compounds

S. No.	Catalyst	Substrate	Product	Conversion (%)	Yield (%)	References
18.	δ-MnO$_2$			94	94	J. Dai et al. (2019)
19.	CuBr$_2$, (NH$_4$)$_2$S$_2$O$_8$			99	96	Cao et al. (2018)
20.	Ni/CdS			95	95	Han et al. (2019)

the same. To find a solution for lignin degradation, many researchers dedicated themselves to describing the complex process of lignin degradation by using enzymes. Henceforth, different enzymatic strategies have been utilized, evolved from microorganisms, especially from bacteria and fungi to utilize this abundantly available lignin (Bugg et al. 2011; Janusz et al. 2017).

For this, three classes of bacteria are majorly involved in lignin degradation. They are *Actinomycetes, γ-proteobacteria*, and *α-proteobacteria* (Bugg et al. 2011). The secretion of the extracellular enzymes from these bacterial species is considered crucial in the decomposition of the lignocellulose (Janusz et al. 2017; Bibb 2005; Sonia, Naceur, and Abdennaceur 2011). Similarly, bacterial enzymes, e.g. *peroxidases, β-esterases, laccases,* and other oxidative enzymes involved in lignin degradation have been reported in recent times (Janusz et al. 2017). Recently, Rashid et al. (2015) have presented that *Sphingobacterium* produces manganese superoxide dismutase, which can produce hydroxyl radical, to oxidize lignin. According to Bugg et al. (2011) and Kumar et al. (2015), recombinant proteins from bacterial strain can easily express and synthesize these enzymes at a large scale due to the small bacterial genome. In this context, *actinobacteria*, various strains of *proteobacteria alongside,* Bacteroides, and archaea have also been reported for the lignin degradation (Priyadarshinee et al. 2016; Tian et al. 2014).

Numerous fungal species have been identified to break down lignin after the first report published by Bumpus et al. (1985) to degrade lignin via oxidation. Later, few more fungal species, namely *Fusarium solani, Penicillium chrysogenum, and F. oxysporum,* have been identified to degrade lignin, with potency lower than that of white-rot fungi (Kirk and Farrell 1987). Fungal species such as white-rot saprophytic fungi, brown-rot fungi, and soft-rot fungi have been identified for the degradation of lignin by the production of various ligninolytic enzymes. All these discussed fungi are capable of degrading the lignocellulosic biomass. However, only white-rot fungi decompose lignin completely to CO_2 and H_2O (Blanchette 1995). *Ustilago maydis, Panaeolus papilionaceus,* and *Coprinopsis friesii* have also been identified to degrade lignin (Heinzkill et al. 1998; Liers et al. 2011; Couturier et al. 2012). In addition to the above-mentioned fungal species in the year 2012, *Alternaria alternata,* an anamorphic fungi, has also been identified to degrade lignin enzymatically (Sigoillot et al. 2012). All these microorganisms secrete enzymes which are tabulated in Table 2.3, and a detailed description of the same is provided in the upcoming part of this chapter.

The enzymes responsible for the degradation of lignin can be categorized into two types, namely lignin-degrading (LD) and lignin-modifying (LM) enzymes. It is observed that LD enzymes have shown an incomplete degradation process, whilst LM enzymes (laccases and peroxidases) can cleave lignin completely (Silva Coelho-Moreira et al. 2013; Desai and Nityanand 2011; Sanchez, Sierra, and J. 2011).

2.3.2.1   Lignin-Degrading (LD) and Lignin-Modifying (LM) Enzymes

Peroxidases (POD) belong to the class II type of LM enzymes and the superfamily of the catalase-peroxidases of plants and animals (Hammel and Cullen 2008; Couturier et al. 2012). All these peroxidases have protoporphyrin IX as the prosthetic group (Pollegioni, Tonin, and Rosini 2015). Further, based on the similarity

TABLE 2.3

Microbial Enzyme-Mediated Degradation of Lignin and Lignin Model Compounds

S. No.	Enzyme	Source	Cleavage	References
1.	Lignin peroxidases	*Bjerkandera sp., Phlebia tremellosa*	Non-specific sites in lignin	Dashtban et al. (2010)
2.	Manganese-dependent peroxidase	*Panus tigrinus*	Non-specific sites in lignin	Lisov, Leontievsky, and Golovleva (2003)
3.	Versatile peroxidases	*Pl. ostreatus*	Lignin model compound	Fernández-Fueyo et al. (2014)
4.	Laccase	*Sinorhizobium meliloti 71 and S. lavendulae 73*	Oxidation of aromatic compounds, e.g. phenolic moieties typically found in lignin	Pawlik et al. (2016) Garcia-Ruiz et al. (2014) Suzuki et al. (2003)
5.	Glyoxal oxidase	*Phanerochaete chrysosporium 74*	Glycol aldehyde from lignin	Kersten (1990) Watanabe et al. (2001)
6.	Aryl-alcohol oxidase	*Agaricales* species 60	Phenolic, non-phenolic aryl-alcohols, primary and secondary alcohols to aldehydes	Hernández-Ortega, Ferreira, and Martínez (2012) Ferreira et al. (2010)
7.	Superoxidases	*Sphingobacterium sp.*	Oxidize Organosolv, kraft lignin, and lignin model substrates	Rashid et al. (2015)
8.	Glutathione-dependent β-etherases	*Sphingobium SYK 6*	Ether cleavage in lignin model compounds	Singh (2004)
9.	Dehydrogenases	*Sphingobium sp.*	β-O-4-aryl ether linkage present in lignin **to** monomers	Reiter et al. (2013)
10.	LigD-Cα-dehydrogenase	*Sphingobium sp.*	Cα-O bond present in lignin	Reiter et al. (2013)
11.	O-Demethylases	*Pseudomonas* and *Acinetobacter*	Methoxy-substituted such as syringate	Abdelaziz et al. (2016)
12.	Aromatic alcohol oxidase	*Geotrichum candidum, Botrytis cinerea*, and *Pleurotus eryngii*	Primary alcohol	Sonoki et al. (2009)
13.	Dioxygenases	*Streptomyces sp.* and *Sphingomonas paucimobilis*	Lignin degradation	Sonoki et al. (2009)
14.	Catalase-peroxidase (Amycol)	*Amycolatopsis sp.*	Degrade phenolic lignin model compounds	Brown et al. (2011)

of amino acid sequences and catalytic behaviour, this superfamily has been divided into three subclasses. In class I, catalase-peroxidases are found in prokaryotes and organelle-localized eukaryotes, which are heme dependent; in Class II are extracellular fungi-based heme peroxidases; and in class III are heme peroxidases obtained from plants (Zámocký, Furtmüller, and Obinger 2009; Lombard et al. 2014). During the same period, Sugano reported that the new heme group of peroxidases known as dye-peroxidases (DyP) isolated from Basidiomycete, *Auricularia auricula-judae, can also cleave* lignin substructure linkages (Sugano 2009; Liers et al. 2010).

2.3.2.2   Lignin Peroxidase (EC 1.11.1.14)

Lignin peroxidases (LiP) are another class of peroxidase, isolated from various fungal species, namely *Phanerochaete chrysosporium* (Tien and Kent Kirk 1983; Paszczyński, Huynh, and Crawford 1986), *Trametes versicolor* (Johansson and Nyman 1993), and white-rot fungi, *namely Bjerkandera sp.* and *Phlebia tremellosa* (Bugg et al. 2011). According to Wong, LiP can actively cleave non-specific sites in the lignin polymer (Wong 2009). Then the similar activity of LiP has also been observed in *Streptomyces viridosporus* and *Acinetobacter calcoaceticus* bacteria (Dashtban et al. 2010). In the same study, the oxidation of phenolic aromatic compounds, lignin and lignin model compounds, was also understood. LiP obtained from *P. chrysosporium* helped in understanding the structural aspects. It was observed that LiP is globular and comprises of eight major and minor α-helices with less amount of β components. This, in turn, is organized into two domains to form active sites in the presence of ferric ion present in the core of the enzyme (Choinowski et al. 1999). Then it is further folded into a three-dimensional structure for stabilization, with glycosylation sites, a Ca^{2+} binding site, and four disulphide bridges with molecular mass of 35–48 kDa (Sigoillot et al. 2012).

2.3.2.3   Manganese-Dependent Peroxidase (EC 1.11.1.13)

Manganese-dependent peroxidase (MnP) belongs to the family of *P. chrysosporium*, which was isolated for the first time, and exists in many isoforms (Glenn and Gold 1985). Later on, these enzymes were also isolated from *Nematoloma frowardii* and *Panus tigrinus* (Hildén et al. 2008; Sutherland et al. 1997). Further studies revealed that the molecular structure of MnP is similar to LiP comprising of two α-helices bridged by heme. The structure is stabilized by disulphide bridges and two calcium ions, which keep the enzyme structure in active form (Sutherland et al. 1997). A review on the characterization of enzymes derived from various white-rot fungi has been nicely portrayed elsewhere (Manavalan, Manavalan, and Heese 2015).

2.3.2.4   *Laccase* (Lac, Benzenediol: Oxygen Oxidoreductases; EC 1.10.3.2)

Laccases and peroxidases are involved in the synthesis and degradation of lignin. Laccases are most widely present in plants, fungi, insects, and bacteria (Manavalan, Manavalan, and Heese 2015). These belong to the most significant part of the ligninolytic enzyme that can act directly on the wood (Riva 2006). Laccase is a copper-containing metalloprotein also named as blue multicopper oxidases and belongs to the polyphenol oxidases group (Baldrian 2006). Laccase efficiently oxidizes the different aromatic and phenolic groups present in lignin in the presence of O_2 acting as an electron acceptor. It has also been reported that this set of enzymes can oxidize inorganic or organic metals such as Mn(II) to Mn(III) and other organometallic mixtures too (Garcia-Ruiz et al. 2014). As mentioned earlier, the location of the enzyme determines the activity of the laccase. The ironical part of nature is that, if the enzyme is present in the intracellular portion of the cell wall, laccases synthesize lignin (Sigoillot et al. 2012). If the same enzyme is derived from fungi either an intracellular or extracellular source, they degrade the β-O-4 phenolic of lignin and β-1 linkages of the lignin (Baldrian 2006; Riva 2006).

2.3.2.5 Superoxide Dismutases

Superoxidases are intracellularly localized enzymes, which protect cells from damage by altering superoxide anions into molecular oxygen and hydrogen peroxide (Rashid et al. 2015). It has been reported that it can effectively oxidize Organosolv- and kraft process-derived lignin, and other lignin model compounds into various aryl groups (Rashid et al. 2015).

Henceforth, it can be understood that the enzymes can construct the lignin as well as deconstruct the lignin. The enzymes responsible for the cleavage of lignin and its model compounds are given in Table 2.3.

Many research groups have done an extensive research to describe the degradation of lignin, which is still a challenging task. In this regard, microorganisms identified till date have been found to secrete different enzymes involved in decomposing this lignin biomaterial as discussed in Table 2.3. We believe that diversifying the fungal and bacterial enzymes according to the niche environment can play a significant role in the cleavage of the specific bond in lignin. However, a rapidly changing climate may have had the tremendous impact on the evolution of fungi.

2.4 LIGNIN IN BIOLOGICAL AGENTS

Many scientists across the globe reported that the agri-biomass-based lignin can be utilized as an excipient and as a pharmacological agent because of its biocompatibility, low cost, environmental friendliness, and hydrophilic properties. There have been numerous studies reported on the biological efficacy of lignin and its capability to transport a drug. Because of the polyphenolic nature, lignin has been tuned for various therapeutic applications (Domínguez-Robles et al. 2020; R. Liu et al. 2020) (Figure 2.3). In the past decade, lignin-based/lignin-derived molecules have been found to have interesting properties which are tabulated in Table 2.4. For example, lignin can reduce cholesterol levels by binding to bile acids in the intestine (Barnard and Heaton 1973). Lignin-derived honokiol (HNK), a small molecule, can be useful in the treatment of insomnia, and it also possesses the antioxidant as well as antimicrobial activity (Yang et al. 2019; Domínguez-Robles et al. 2020).

Upon chemical modification, the therapeutic properties of lignin are enhanced alongside its direct implication in the biomedical field. To note, lignin can be an alternative source of excipient over others in terms of economical, structural, and biocompatibility aspects (Imlimthan et al. 2020).

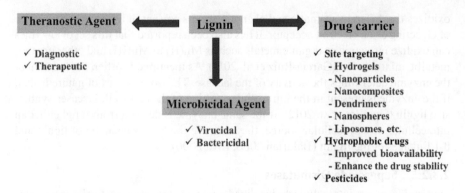

FIGURE 2.3   Various applications of lignin as biological agents.

TABLE 2.4
Biological Activity of Various Lignin Derivatives

S. No.	Biological Activity	Lignin Derivatives	References
1.	Antiviral		
	Inhibiting viral binding and penetration	Lignin–carbohydrate complexes	Lee et al. (2011)
	LA binds to the HIV-1 envelope glycoproteins	Lignosulphonic acid	Karim et al. (2010)
	Inhibiting the replication of herpes simplex virus	Lignosulphonic acid	Karim et al. (2010)
2.	Obesity control		
	Decreasing the oleate-induced apo-B secretion and plasma triglyceride levels	Lignophenols	Norikura et al. (2010)and Sato et al. (2009)
3.	Antidiabetic		
	α-Amylase inhibition and decreasing glucose diffusion	Alkali lignin	Sato et al. (2009)
	Inhibitor of α-glucosidase and decreasing glycaemic levels	Lignosulphonic acid	Hasegawa et al. (2015)
4.	Anticoagulant		
	Inhibition of thrombin and allosteric inhibition of thrombin	Sulphated low molecular weight lignins	Henry and Desai (2014) and Mehta et al. (2016)
5.	Antiemphysema		
	Elastase, oxidation, and inflammation inhibition	Sulphated low molecular weight lignins	Saluja et al. (2013)
6.	Antiproliferative		
	Inhibition of proliferation	Low molecular weight lignin fraction	Sipponen et al. (2018)
7.	Neuroprotective		
	Decrease the effect of ER stress	Lig-8	Ito et al. (2007)

2.4.1 DRUG CARRIER

Lignin is a natural biodegradable polymer obtained from natural resources and is generally non-toxic. In the past decade, polymers has been utilized as a tool to deliver drugs, among which natural polymers play a pivotal role compared to synthetic polymers. In this regard, lignin being a natural phenol-based polymer has proved to be a potential candidate for a multitude of applications such as drug delivery with hydrogels, drug-loaded nanoparticles, bare lignin nanoparticles, nanocomposites, nanocapsules, dendrimers, and nanospheres. In line with this, lignin-based drug carriers can be used for various biomedical and therapeutic applications because of their availability, ease of surface modification, stability, safety, biodegradability, and non-chemical processing. So, this polymer can be an attractive option for drug delivery (Imlimthan et al. 2020; Sriroth and Sunthornvarabhas 2018). For instance, hollow lignin-based nanoparticles have been utilized to load a plant constituent curcumin for various biomedical applications. It has been observed that the bioavailability of the same has been significantly increased. In another case, lignin-based hydrogel has been developed to infuse curcumin for sustained release applications of up to 4 days (Larrañeta et al. 2018).

These lignin-based delivery approaches can be potentiated to deliver a hydrophobic and poorly bioavailable drug. Recently, a study suggested that it can also serve as a pharmaceutical excipient with microcrystalline cellulose (MCC) for tablet manufacturing by direct compression method. This, in turn, affected the release profile of the drug, which was observed in the case of tetracycline. This green and renewable biopolymer can act as a substitute for synthetic polymer (Domínguez-Robles et al. 2019). Anti-infective ointment having lignin-*graft*-polyoxazoline conjugated triazole was also formulated and studied to control persistent inflammation (Mahata et al. 2017). Antibiotics-loaded scaffolds from polycaprolactone-coated chitin-lignin gel (core–shell fibres) were successfully developed, and the release profile of the same was also studied. It was observed that these fibres can provide superior bactericidal effect against common bacteria found on the skin with minimal or no observable cytotoxicity (Abudula et al. 2020).

In another approach, to utilize lignin, lignin-based surfactant in self-nano-emulsifying drug delivery system has been developed and studied for enhancing the bioavailability of the hydrophobic and photosensitive drug resveratrol (RSV) (L. Dai et al. 2018). In another study, complex spherical biocarriers based on lignin–carbohydrate complexes isolated from ginkgo (*Ginkgo biloba* L.) xylem were found to be stable even in an aqueous solution. These are found to be biocompatible and showed promising results for use as a biomaterial when treating the human hepatocyte culture with the same (Zhao et al. 2017). In another study, three different lignin nanoparticles (LNPs): iron (III)-complexed lignin nanoparticles, pure lignin nanoparticles, and Fe_3O_4-infused lignin nanoparticles (Fe_3O_4-LNPs), were formulated for the drug delivery of hydrophobic drugs and cytotoxic agents sorafenib (SFN) and benzazulene (BZL). After encapsulation into the above-mentioned nanoparticles, it was found that the antiproliferative activity of the same has been enhanced alongside the achievement of the delivery of the drug (Figueiredo et al. 2017).

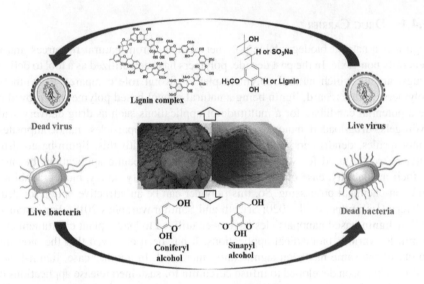

FIGURE 2.4 Lignin and its virucidal and bactericidal effects.

2.4.2 Microbicidal Agent

Lignin has been investigated as an antimicrobial agent. In a variety of plant species, lignin is usually composed of coniferyl alcohol, sinapyl alcohol, and p-coumaryl in varying proportions. Among them, the presence of coniferyl and sinapyl alcohols is responsible for various pharmacological effects (Sriroth and Sunthornvarabhas 2018) (Figure 2.4). The surface modification of cellulose fibres using layer-by-layer deposition of lignosulphonates and chitosan was obtained. Upon antimicrobial analysis, the results demonstrated that the modified cellulose fibres of five-layer thickness unveiled the growth inhibition up to 97% against *E. coli.* (Hui Li and Peng 2015). The lignin-derived zinc oxide nanoparticles were also reported for their antimicrobial and UV-blocking properties (Hui Li and Peng 2015). The antimicrobial properties of lignin-based hydrogels cross-linked using polyethylene glycol of MW 10,000 were evaluated for pathogens *P. mirabilis* and *S. aureus*, responsible for medical device-associated infections, and were found to be highly effective (Larrañeta et al. 2018).

In another study by Lee et al., the isolation of various lignin–carbohydrate complexes (LCs) from the hot water extract of the seeds of *Pimpinella anisum* was found to possess antiviral and immunostimulating substances. These LCs proved to have potential activity against herpes simplex virus types 1 and 2, measles, and human cytomegalovirus. Besides, the LC from *Prunella vulgaris* of MW8500 has been reported for its antiherpes activity. Similarly, pinecone LCs were found to be effective against influenza virus, HIV-1, and HSV (Lee et al. 2011).

2.4.3 Theranostic Agent

Concerning photodynamic therapy, it is a conjugation of three key components: a light source, a photosensitizer, and tissue O_2. A particular wavelength of the light source is required for the excitation of the photosensitizer to produce ROS (Kirar et al. 2021). The

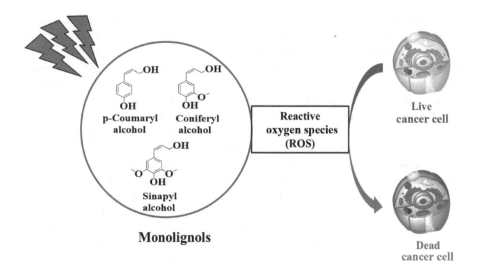

FIGURE 2.5   Photodynamic chemotherapy effect of lignin nanoparticles.

application of this phenomenon, for the generation of ROS against microorganisms, is known as photodynamic antimicrobial chemotherapy (PACT). In recent times, bare lignin upon acetylation can act as an inherent photosensitizer (Marchand et al. 2018). Considering this effect, various research groups tried to accomplish the PACT effect by various means. Porphyrin-loaded lignin nanoparticles were developed and found to be highly effective against the bacteria by producing ROS upon photoexcitation leading to detrimental effect on various Gram-positive strains (Maldonado-Carmona et al. 2020). In another study, a photosensitizer (rose bengal)-based conjugate with lignin-derived metallic and bimetallic (silver- and gold-based) nanocomplexes were developed and these nanoconjugates were doped into polyacrylic acid-based biocompatible and pH-responsive hydrogels. These were deployed to efficiently control the drug delivery through the construction of wound dressings and antimicrobial nanocoatings (Chandna et al. 2020; Kaur et al. 2021). In another study by the same group, a lignin nanosphere-based spray was developed. This showed promising UV-blocking properties and microbicidal properties. This nanospray also exhibited significant photoluminescence properties, which can be deployed in bioimaging, making it an efficient phototheranostic agent (Paul et al. 2021). An image depicting the role of ROS on cancer cells is portrayed in Figure 2.5.

2.5   FUTURE PERSPECTIVE

Lignin, the abundantly available heteroaromatic biopolymer is a potential aromatic source. It is understood that the lignin source plays a major role in structural makeup. This, in turn, makes the researchers elucidate the structure. To date, many lignin model structures have been depicted by various groups. Considering these structures, significant advancements in the development of lignin model compounds have been achieved. This deciphered the mechanisms involved in the synthesis of lignin and degradation of the same. Nevertheless, lignin model compounds played a significant role in the depolymerization techniques such as photocatalysis, pyrolysis, and enzymatic degradation. In parallel, lignin being a biocompatible polymer derived from lignocellulosic biomass has a multitude of applications. The presence of phenolics in the structure of native lignin opened a new door for its inherent antimicrobial property, antioxidant property, and more. In a similar scenario, lignin nanopreparations have found extensive applications as drug carriers and theranostic agents.

Moreover, there are numerous underlying challenges in the utilization and depolymerization of lignin. First, the complete structural elucidation of lignin is necessary to understand the complexity in the structure. Secondly, sustainable approaches to lignin depolymerization via photocatalysis, which faces challenges such as the mechanism of the reaction, selective conversion of native lignin to monomers or relevant phenols, performing visible-light photocatalysis, are some of the critical issues. In the aspect of enzymatic degradation, despite being expensive, deeper insights into the enzyme immobilization techniques for the archetypical conversion of lignin to specific platform chemicals remain elusive. In this perspective, collaborative efforts of other physical, chemical, and biological technologies can render the depolymerization process of lignin efficient. In terms of biological applications of lignin as a drug carrier, microbicidal agent, and theranostic agent, controlled release of encapsulated drugs and enhancement in the activity profiling in terms of site-targeted drug delivery for the chemotherapeutic purpose can further be an admirable goal.

REFERENCES

Abdelaziz, Omar Y., Daniel P. Brink, Jens Prothmann, Krithika Ravi, Mingzhe Sun, Javier García-Hidalgo, Margareta Sandahl, et al. 2016. "Biological Valorization of Low Molecular Weight Lignin." *Biotechnology Advances.* Elsevier Inc. doi:10.1016/j. biotechadv.2016.10.001.

Abudula, Tuerdimaimaiti, Kalamegam Gauthaman, Azadeh Mostafavi, Ahmed Alshahrie, Numan Salah, Pierfrancesco Morganti, Angelo Chianese, Ali Tamayol, and Adnan Memic. 2020. "Sustainable Drug Release from Polycaprolactone Coated Chitin-Lignin Gel Fibrous Scaffolds." *Scientific Reports* 10 (1). Nature Research: 20428. doi:10.1038/s41598-020-76971-w.

Adler, Erich. 1977. "Lignin Chemistry-Past, Present and Future." *Wood Science and Technology* 11 (3). Springer-Verlag: 169–218. doi:10.1007/BF00365615.

Bajwa, D. S., G. Pourhashem, A. H. Ullah, and S. G. Bajwa. 2019. "A Concise Review of Current Lignin Production, Applications, Products and Their Environment Impact." *Industrial Crops and Products* 139 (November). Elsevier B.V.: 111526. doi:10.1016/j. indcrop.2019.111526.

Baldrian, Petr. 2006. "Fungal Laccases-Occurrence and Properties." *FEMS Microbiology Reviews.* doi:10.1111/j.1574-4976.2005.00010.x.

Barnard, D. L., and K. W. Heaton. 1973. "Bile Acids and Vitamin A Absorption in Man: The Effects of Two Bile Acid-Binding Agents, Cholestyramine and Lignin." *Gut* 14 (4). Gut: 316–18. doi:10.1136/gut.14.4.316.

Bibb, Mervyn J. 2005. "Regulation of Secondary Metabolism in Streptomycetes." *Current Opinion in Microbiology* 8 (2). Elsevier Ltd: 208–15. doi:10.1016/j.mib.2005.02.016.

Bjørsvik, Hans René, and Francesco Minisci. 1999. "Fine Chemicals from Lignosulfonates. 1. Synthesis of Vanillin by Oxidation of Lignosulfonates." *Organic Process Research and Development* 3 (5). American Chemical Society: 330–40. doi:10.1021/op9900028.

Blanchette, Robert A. 1995. "Degradation of the Lignocellulose Complex in Wood." *Canadian Journal of Botany* 73 (S1). Canadian Science Publishing: 999–1010. doi:10.1139/b95-350.

Bonawitz, Nicholas D., and Clint Chapple. 2010. "The Genetics of Lignin Biosynthesis: Connecting Genotype to Phenotype." *Annual Review of Genetics.* doi:10.1146/annurev-genet-102209-163508.

Bosch, S. Van den, S. F. Koelewijn, T. Renders, G. Van den Bossche, T. Vangeel, W. Schutyser, and B. F. Sels. 2018. "Catalytic Strategies Towards Lignin-Derived Chemicals." *Topics in Current Chemistry.* Springer International Publishing. doi:10.1007/s41061-018-0214-3.

Bosque, Irene, Gabriel Magallanes, Mathilde Rigoulet, Markus D. Kärkäs, and Corey R.J. Stephenson. 2017. "Redox Catalysis Facilitates Lignin Depolymerization." *ACS Central Science* 3 (6). American Chemical Society: 621–28. doi:10.1021/acscentsci.7b00140.

Brown, Margaret E., Mark C. Walker, Toshiki G. Nakashige, Anthony T. Iavarone, and Michelle C.Y. Chang. 2011. "Discovery and Characterization of Heme Enzymes from Unsequenced Bacteria: Application to Microbial Lignin Degradation." *Journal of the American Chemical Society* 133 (45). American Chemical Society: 18006–09. doi:10.1021/ja203972q.

Bugg, Timothy D.H., Mark Ahmad, Elizabeth M. Hardiman, and Rahman Rahmanpour. 2011. "Pathways for Degradation of Lignin in Bacteria and Fungi." *Natural Product Reports.* The Royal Society of Chemistry. doi:10.1039/c1np00042j.

Bugg, Timothy D.H., Mark Ahmad, Elizabeth M. Hardiman, and Rahul Singh. 2011. "The Emerging Role for Bacteria in Lignin Degradation and Bio-Product Formation." *Current Opinion in Biotechnology.* Elsevier Current Trends. doi:10.1016/j.copbio.2010.10.009.

Bumpus, John A., Ming Tien, David Wright, and Steven D. Aust. 1985. "Oxidation of Persistent Environmental Pollutants by a White Rot Fungus." *Science* 228 (4706). American Association for the Advancement of Science: 1434–36. doi:10.1126/science.3925550.

Cao, Yu, Ning Wang, Xing He, Hong Ru Li, and Liang Nian He. 2018. "Photocatalytic Oxidation and Subsequent Hydrogenolysis of Lignin β-O-4 Models to Aromatics Promoted by In Situ Carbonic Acid." *ACS Sustainable Chemistry and Engineering* 6 (11). American Chemical Society: 15032–39. doi:10.1021/acssuschemeng.8b03498.

Chandna, Sanjam, Neeraj S. Thakur, Ravneet Kaur, and Jayeeta Bhaumik. 2020. "Lignin-Bimetallic Nanoconjugate Doped PH-Responsive Hydrogels for Laser-Assisted Antimicrobial Photodynamic Therapy." *Biomacromolecules* 21 (8). American Chemical Society: 3216–30. doi:10.1021/acs.biomac.0c00695.

Chen, Jun, Wenxia Liu, Zhaoping Song, Huili Wang, and Yimin Xie. 2018. "Photocatalytic Degradation of β-O-4 Lignin Model Compound by In2S3 Nanoparticles under Visible Light Irradiation." *Bioenergy Research* 11 (1). Springer New York LLC: 166–73. doi:10.1007/s12155-017-9886-8.

Choinowski, Thomas, Wolfgang Blodig, Kaspar H. Winterhalter, and Klaus Piontek. 1999. "The Crystal Structure of Lignin Peroxidase at 1.70 Å Resolution Reveals a Hydroxy Group on the C(β) of Tryptophan 171: A Novel Radical Site Formed during the Redox Cycle." *Journal of Molecular Biology* 286 (3). Academic Press: 809–27. doi:10.1006/jmbi.1998.2507.

Ciamician, Giacomo. 1912. "The Photochemistry of the Future." *Science* 36 (926). American Association for the Advancement of Science: 385–94. doi:10.1126/science.36.926.385.

Constitution and Biosynthesis of Lignin | Karl Freudenberg | Springer. 2021. Accessed January 20. http://www.springer.com.sci-hub.ac/gp/book/9783540042747#.

Cotana, Franco, Gianluca Cavalaglio, Andrea Nicolini, Mattia Gelosia, Valentina Coccia, Alessandro Petrozzi, and Lucia Brinchi. 2014. "Lignin as Co-Product of Second Generation Bioethanol Production from Ligno-Cellulosic Biomass." In *Energy Procedia* 45. Elsevier Ltd.: 52–60. doi:10.1016/j.egypro.2014.01.007.

Couturier, Marie, David Navarro, Caroline Olivé, Didier Chevret, Mireille Haon, Anne Favel, Laurence Lesage-Meessen, Bernard Henrissat, Pedro M. Coutinho, and Jean Guy Berrin. 2012. "Post-Genomic Analyses of Fungal Lignocellulosic Biomass Degradation Reveal the Unexpected Potential of the Plant Pathogen Ustilago Maydis." *BMC Genomics* 13 (1). BMC Genomics. doi:10.1186/1471-2164-13-57.

Crestini, Claudia, Federica Melone, Marco Sette, and Raffaele Saladino. 2011. "Milled Wood Lignin: A Linear Oligomer." *Biomacromolecules* 12 (11). American Chemical Society: 3928–35. doi:10.1021/bm200948r.

Dai, Jinhuo, Antonio F. Patti, Gavin N. Styles, Sepa Nanayakkara, Leone Spiccia, Francesco Arena, Cristina Italiano, and Kei Saito. 2019. "Lignin Oxidation by MnO_2 under the Irradiation of Blue Light." *Green Chemistry* 21 (8). Royal Society of Chemistry: 2005–14. doi:10.1039/c8gc03498b.

Dai, Lin, Weiyan Zhu, Rui Liu, and Chuanling Si. 2018. "Lignin-Containing Self-Nanoemulsifying Drug Delivery System for Enhance Stability and Oral Absorption of Trans-Resveratrol." *Particle and Particle Systems Characterization* 35 (4). Wiley-VCH Verlag. doi:10.1002/ppsc.201700447.

Dashtban, Mehdi, Heidi Schraft, Tarannum A. Syed, and Wensheng Qin. 2010. "Fungal Biodegradation and Enzymatic Modification of Lignin." *International Journal of Biochemistry and Molecular Biology*. e-Century Publishing Corporation. www.ijbmb.org.

Desai, S.S., and C. Nityanand. 2011. "Microbial Laccases and Their Applications: A Review." *Asian Journal of Biotechnology* 3 (2). Science Alert: 98–124. doi:10.3923/ajbkr.2011.98.124.

Domínguez-Robles, Juan, Eneko Larrañeta, Mun Leon Fong, Niamh K. Martin, Nicola J. Irwin, Pere Mutjé, Quim Tarrés, and Marc Delgado-Aguilar. 2020. "Lignin/Poly(Butylene Succinate) Composites with Antioxidant and Antibacterial Properties for Potential Biomedical Applications." *International Journal of Biological Macromolecules* 145 (February). Elsevier B.V.: 92–99. doi:10.1016/j.ijbiomac.2019.12.146.

Domínguez-Robles, Juan, Sarah A. Stewart, Andreas Rendl, Zoilo González, Ryan F. Donnelly, and Eneko Larrañeta. 2019. "Lignin and Cellulose Blends as Pharmaceutical Excipient for Tablet Manufacturing Via Direct Compression." *Biomolecules* 9 (9). MDPI AG. doi:10.3390/biom9090423.

Fache, Maxence, Bernard Boutevin, and Sylvain Caillol. 2016. "Vanillin Production from Lignin and Its Use as a Renewable Chemical." *ACS Sustainable Chemistry and Engineering* 4 (1). American Chemical Society: 35–46. doi:10.1021/acssuschemeng.5b01344.

Fernández-Fueyo, Elena, Francisco J. Ruiz-Dueñas, María Jesús Martínez, Antonio Romero, Kenneth E. Hammel, Francisco Javier Medrano, and Angel T. Martínez. 2014. "Ligninolytic Peroxidase Genes in the Oyster Mushroom Genome: Heterologous Expression, Molecular Structure, Catalytic and Stability Properties, and Lignin-Degrading Ability." *Biotechnology for Biofuels* 7 (1). BioMed Central: 2. doi:10.1186/1754-6834-7-2.

Ferreira, Patricia, Aitor Hernández-Ortega, Beatriz Herguedas, Jorge Rencoret, Ana Gutiérrez, María Jesús Martínez, Jesús Jiménez-Barbero, Milagros Medina, and Ángel T. Martínez. 2010. "Kinetic and Chemical Characterization of Aldehyde Oxidation by Fungal Aryl-Alcohol Oxidase." *Biochemical Journal* 425 (3). Biochemical Journal: 585–93. doi:10.1042/BJ20091499.

Figueiredo, Patrícia, Kalle Lintinen, Alexandros Kiriazis, Ville Hynninen, Zehua Liu, Tomás Bauleth-Ramos, Antti Rahikkala, et al. 2017. "In Vitro Evaluation of Biodegradable Lignin-Based Nanoparticles for Drug Delivery and Enhanced Antiproliferation Effect in Cancer Cells." *Biomaterials* 121 (March). Elsevier Ltd: 97–108. doi:10.1016/j.biomaterials.2016.12.034.

Fujishima, Akira, and Kenichi Honda. 1972. "Electrochemical Photolysis of Water at a Semiconductor Electrode." *Nature* 238 (5358). Nature Publishing Group: 37–38. doi:10.1038/238037a0.

Garcia-Ruiz, Eva, Diana M. Mate, David Gonzalez-Perez, Patricia Molina-Espeja, Susana Camarero, Angel T. Martínez, Antonio O. Ballesteros, and Miguel Alcalde. 2014. "Directed Evolution of Ligninolytic Oxidoreductases: From Functional Expression to Stabilization and Beyond." *Cascade Biocatalysis: Integrating Stereoselective and Environmentally Friendly Reactions*. Vol. 9783527335. doi:10.1002/9783527682492.ch1.

Gazi, Sarifuddin, Wilson Kwok Hung Ng, Rakesh Ganguly, Adhitya Mangala Putra Moeljadi, Hajime Hirao, and Han Sen Soo. 2015. "Selective Photocatalytic C-C Bond Cleavage under Ambient Conditions with Earth Abundant Vanadium Complexes." *Chemical Science* 6 (12). Royal Society of Chemistry: 7130–42. doi:10.1039/c5sc02923f.

Gierer, J. 1980. "Chemical Aspects of Kraft Pulping." *Wood Science and Technology* 14 (4). Springer-Verlag: 241–66. doi:10.1007/BF00383453.

Glasser, Wolfgang G., and Heidemarie R. Glasser. 1974. "Simulation of Reactions with Lignin by Computer (Simrel): II. A Model for Softwood Lignin." *Holzforschung* 28 (1). De Gruyter: 5–11. doi:10.1515/hfsg.1974.28.1.5.

Glenn, Jeffrey K., and Michael H. Gold. 1985. "Purification and Characterization of an Extracellular Mn(II)-Dependent Peroxidase from the Lignin-Degrading Basidiomycete, Phanerochaete Chrysosporium." *Archives of Biochemistry and Biophysics* 242 (2). Arch Biochem Biophys: 329–41. doi:10.1016/0003-9861(85)90217-6.

Gundekari, Sreedhar, Joyee Mitra, and Mohan Varkolu. 2020. "Classification, Characterization, and Properties of Edible and Non-Edible Biomass Feedstocks." In *Advanced Functional Solid Catalysts for Biomass Valorization*. Elsevier: 89–120. doi:10.1016/b978-0-12-820236-4.00004-0.

Gupta, Guddu Kumar, and Pratyoosh Shukla. 2020. "Lignocellulosic Biomass for the Synthesis of Nanocellulose and Its Eco-Friendly Advanced Applications." *Frontiers in Chemistry*. Frontiers Media S.A. doi:10.3389/fchem.2020.601256.

Hammel, Kenneth E., and Dan Cullen. 2008. "Role of Fungal Peroxidases in Biological Ligninolysis." *Current Opinion in Plant Biology*. Elsevier Current Trends. doi:10.1016/j.pbi.2008.02.003.

Han, Guanqun, Tao Yan, Wei Zhang, Yi C. Zhang, David Y. Lee, Zhi Cao, and Yujie Sun. 2019. "Highly Selective Photocatalytic Valorization of Lignin Model Compounds Using Ultrathin Metal/CdS." *ACS Catalysis* 9 (12). American Chemical Society: 11341–49. doi:10.1021/acscatal.9b02842.

Hao, Zhongkai, Shuyuan Li, Jiarong Sun, Song Li, and Fang Zhang. 2018. "Efficient Visible-Light-Driven Depolymerization of Oxidized Lignin to Aromatics Catalyzed by an Iridium Complex Immobilized on Mesocellular Silica Foams." *Applied Catalysis B: Environmental* 237 (December). Elsevier B.V.: 366–72. doi:10.1016/j.apcatb.2018.05.072.

Hasegawa, Yasushi, Yukiya Kadota, Chihiro Hasegawa, and Satoshi Kawaminami. 2015. "Lignosulfonic Acid-Induced Inhibition of Intestinal Glucose Absorption." *Journal of Nutritional Science and Vitaminology* 61 (6). Center for Academic Publications Japan: 449–54. doi:10.3177/jnsv.61.449.

He, Qian, Isabelle Ziegler-Devin, Laurent Chrusciel, Sebastien Ngwa Obame, Lu Hong, Xiaoning Lu, and Nicolas Brosse. 2020. "Lignin-First Integrated Steam Explosion Process for Green Wood Adhesive Application." *ACS Sustainable Chemistry and Engineering* 8 (13). American Chemical Society: 5380–92. doi:10.1021/acssuschemeng.0c01065.

Heinzkill, Marion, Lisbeth Bech, Torben Halkier, Palle Schneider, and Timm Anke. 1998. "Characterization of Laccases and Peroxidases from Wood-Rotting Fungi (Family Coprinaceae)." *Applied and Environmental Microbiology* 64 (5). American Society for Microbiology: 1601–06. doi:10.1128/aem.64.5.1601-1606.1998.

Henry, Brian L., and Umesh R. Desai. 2014. "Sulfated Low Molecular Weight Lignins, Allosteric Inhibitors of Coagulation Proteinases Via the Heparin Binding Site, Significantly Alter the Active Site of Thrombin and Factor Xa Compared to Heparin." *Thrombosis Research* 134 (5). Elsevier Ltd: 1123–29. doi:10.1016/j.thromres.2014.08.024.

Hernández-Ortega, Aitor, Patricia Ferreira, and Angel T. Martínez. 2012. "Fungal Aryl-Alcohol Oxidase: A Peroxide-Producing Flavoenzyme Involved in Lignin Degradation." *Applied Microbiology and Biotechnology* 93 (4): 1395–1410. doi:10.1007/s00253-011-3836-8.

Hibbert, Harold, and George H. Tomlinson. 1937. "Manufacture of Vanillin from Waste Sulphite Pulp Liquor." Patent with application number US2069185A.

Hildén, Kristiina S., Ralf Bortfeldt, Martin Hofrichter, Annele Hatakka, and Taina K. Lundell. 2008. "Molecular Characterization of the Basidiomycete Isolate Nematoloma Frowardii B19 and Its Manganese Peroxidase Places the Fungus in the Corticioid Genus Phlebia." *Microbiology* 154 (8). Microbiology (Reading): 2371–79. doi:10.1099/mic.0.2008/018747-0.

Himmel, Michael E., Shi You Ding, David K. Johnson, William S. Adney, Mark R. Nimlos, John W. Brady, and Thomas D. Foust. 2007. "Biomass Recalcitrance: Engineering Plants and Enzymes for Biofuels Production." *Science*. American Association for the Advancement of Science. doi:10.1126/science.1137016.

Hou, Tingting, Nengchao Luo, Hongji Li, Marc Heggen, Jianmin Lu, Yehong Wang, and Feng Wang. 2017. "Yin and Yang Dual Characters of CuOx Clusters for C-C Bond Oxidation Driven by Visible Light." *ACS Catalysis* 7 (6). American Chemical Society: 3850–59. doi:10.1021/acscatal.7b00629.

Imlimthan, Surachet, Alexandra Correia, Patrícia Figueiredo, Kalle Lintinen, Vimalkumar Balasubramanian, Anu J. Airaksinen, Mauri A. Kostiainen, Hélder A. Santos, and Mirkka Sarparanta. 2020. "Systematic in Vitro Biocompatibility Studies of Multimodal Cellulose Nanocrystal and Lignin Nanoparticles." *Journal of Biomedical Materials Research - Part A* 108 (3). John Wiley and Sons Inc.: 770–83. doi:10.1002/jbm.a.36856.

Ithal, Nagabhushana, Justin Recknor, Dan Nettleton, Tom Maier, Thomas J. Baum, and Melissa G. Mitchum. 2007. "Developmental Transcript Profiling of Cyst Nematode Feeding Cells in Soybean Roots." *Molecular Plant-Microbe Interactions* 20 (5). The American Phytopathological Society: 510–25. doi:10.1094/MPMI-20-5-0510.

Ito, Yasushi, Yukihiro Akao, Masamitsu Shimazawa, Norio Seki, Yoshinori Nozawa, and Hideaki Hara. 2007. "Lig-8, a Highly Bioactive Lignophenol Derivative from Bamboo Lignin, Exhibits Multifaceted Neuroprotective Activity." *CNS Drug Reviews* 13 (3). John Wiley & Sons, Ltd: 296–307. doi:10.1111/j.1527-3458.2007.00017.x.

Janusz, Grzegorz, Anna Pawlik, Justyna Sulej, Urszula Świderska-Burek, Anna Jarosz-Wilkolazka, and Andrzej Paszczyński. 2017. "Lignin Degradation: Microorganisms, Enzymes Involved, Genomes Analysis and Evolution." *FEMS Microbiology Reviews*. Oxford University Press. doi:10.1093/femsre/fux049.

Johansson, Tomas, and Per Olof Nyman. 1993. "Isozymes of Lignin Peroxidase and Manganese(II) Peroxidase from the White-Rot Basidiomycete Trametes Versicolor. I. Isolation of Enzyme Forms and Characterization of Physical and Catalytic Properties." *Archives of Biochemistry and Biophysics* 300 (1). Arch Biochem Biophys: 49–56. doi:10.1006/abbi.1993.1007.

Kang, Ying, Xingmei Lu, Guangjin Zhang, Xiaoqian Yao, Jiayu Xin, Shaoqi Yang, Yongqing Yang, Junli Xu, Mi Feng, and Suojiang Zhang. 2019. "Metal-Free Photochemical Degradation of Lignin-Derived Aryl Ethers and Lignin by Autologous Radicals through Ionic Liquid Induction." *ChemSusChem* 12 (17). Wiley-VCH Verlag: 4005–13. doi:10.1002/cssc.201901796.

Karim, Quarraisha Abdool, Salim S. Abdool Karim, Janet A. Frohlich, Anneke C. Grobler, Cheryl Baxter, Leila E. Mansoor, Ayesha B.M. Kharsany, et al. 2010. "Effectiveness and Safety of Tenofovir Gel, an Antiretroviral Microbicide, for the Prevention of HIV Infection in Women." *Science* 329 (5996). American Association for the Advancement of Science: 1168–74. doi:10.1126/science.1193748.

Kärkäs, Markus D., Irene Bosque, Bryan S. Matsuura, and Corey R.J. Stephenson. 2016. "Photocatalytic Oxidation of Lignin Model Systems by Merging Visible-Light Photoredox and Palladium Catalysis." *Organic Letters* 18 (19). American Chemical Society: 5166–69. doi:10.1021/acs.orglett.6b02651.

Kaur, Ravneet, Neeraj Singh Thakur, Sanjam Chandna, and Jayeeta Bhaumik. 2021. "Sustainable Lignin-Based Coatings Doped with Titanium Dioxide Nanocomposites Exhibit Synergistic Microbicidal and UV-Blocking Performance Toward Personal Protective Equipment." *ACS Sustainable Chemistry & Engineering: A-O.* doi: 10.1021/acssuschemeng.1c03637

Kirar, Seema, Neeraj Singh Thakur, Nikhileshwar Reddy Yeddula, Uttam Chand Banerjee, and Jayeeta Bhaumik. 2021. "Insights on the Polypyrrole Based Nanoformulations for Photodynamic Therapy." *Journal of Porphyrins and Phthalocyanines*: A-R. doi:10.1142/s1088424621300032.

Kirk-Othmer(ed). 2007. *Kirk-Othmer Encyclopedia of Chemical Technology*, Index to Volumes 1–26, 5th Edition | Wiley. In. https://www.wiley.com/en-us/Kirk+Othmer+Encyclopedia+of+Chemical+Technology%2C+Index+to+Volumes+1+26%2C+5th+Edition-p-9780471484967.

Kirk, T. K., and R. L. Farrell. 1987. "Enzymatic 'Combustion': The Microbial Degradation of Lignin." *Annual Review of Microbiology* 41 (1). Annual Reviews 4139 El Camino Way, P.O. Box 10139, Palo Alto, CA 94303-0139, USA: 465–505. doi:10.1146/annurev.mi.41.100187.002341.

Kou, Jiahui, Chunhua Lu, Jian Wang, Yukai Chen, Zhongzi Xu, and Rajender S. Varma. 2017. "Selectivity Enhancement in Heterogeneous Photocatalytic Transformations." *Chemical Reviews*. American Chemical Society. doi:10.1021/acs.chemrev.6b00396.

Kuehnel, Moritz F., and Erwin Reisner. 2018. "Solar Hydrogen Generation from Lignocellulose." *Angewandte Chemie - International Edition* 57 (13). Wiley-VCH Verlag: 3290–96. doi:10.1002/anie.201710133.

Kumar, Madan, Jyoti Singh, Manoj Kumar Singh, Anjali Singhal, and Indu Shekhar Thakur. 2015. "Investigating the Degradation Process of Kraft Lignin by β-Proteobacterium, Pandoraea Sp. ISTKB." *Environmental Science and Pollution Research* 22 (20). Springer Verlag: 15690–702. doi:10.1007/s11356-015-4771-5.

Philip J. Kersten. 1990. "Glyoxal oxidase of Phanerochaete chrysosporium: its characterization and activation by lignin peroxidase" Proceedings of the National Academy of Sciences of the United States of America 87 (8). National Academy of Sciences: 2936–40. doi:10.1073/pnas.87.8.2936.

Lan, Wu, Fachuang Lu, Matthew Regner, Yimin Zhu, Jorge Rencoret, Sally A. Ralph, Uzma I. Zakai, Kris Morreel, Wout Boerjan, and John Ralph. 2015. "Tricin, a Flavonoid Monomer in Monocot Lignification." *Plant Physiology* 167 (4). American Society of Plant Biologists: 1284–95. doi:10.1104/pp.114.253757.

Larrañeta, Eneko, Mikel Imízcoz, Jie X. Toh, Nicola J. Irwin, Anastasia Ripolin, Anastasia Perminova, Juan Domínguez-Robles, Alejandro Rodríguez, and Ryan F. Donnelly. 2018. "Synthesis and Characterization of Lignin Hydrogels for Potential Applications as Drug Eluting Antimicrobial Coatings for Medical Materials." *ACS Sustainable Chemistry and Engineering* 6 (7). American Chemical Society: 9037–46. doi:10.1021/acssuschemeng.8b01371.

Lee, Jung Bum, Chihiro Yamagishi, Kyoko Hayashi, and Toshimitsu Hayashi. 2011. "Antiviral and Immunostimulating Effects of Lignin-Carbohydrate-Protein Complexes from Pimpinella Anisum." *Bioscience, Biotechnology and Biochemistry* 75 (3). Biosci Biotechnol Biochem: 459–65. doi:10.1271/bbb.100645.

Li, Hongji, Anon Bunrit, Jianmin Lu, Zhuyan Gao, Nengchao Luo, Huifang Liu, and Feng Wang. 2019. "Photocatalytic Cleavage of Aryl Ether in Modified Lignin to Non-Phenolic Aromatics." *ACS Catalysis* 9 (9). American Chemical Society: 8843–51. doi:10.1021/acscatal.9b02719.

58   Biomass for Bioenergy and Biomaterials

Li, Hui, and Lincai Peng. 2015. "Antimicrobial and Antioxidant Surface Modification of Cellulose
    Fibers Using Layer-by-Layer Deposition of Chitosan and Lignosulfonates." *Carbohydrate
    Polymers* 124 (June). Elsevier Ltd: 35–42. doi:10.1016/j.carbpol.2015.01.071.
Liers, Christiane, Tobias Arnstadt, René Ullrich, and Martin Hofrichter. 2011. "Patterns
    of Lignin Degradation and Oxidative Enzyme Secretion by Different Wood- and
    Litter-Colonizing Basidiomycetes and Ascomycetes Grown on Beech-Wood." *FEMS
    Microbiology Ecology* 78 (1). FEMS Microbiol Ecol: 91–102. doi:10.1111/j.1574-6941.
    2011.01144.x.
Liers, Christiane, Caroline Bobeth, Marek Pecyna, René Ullrich, and Martin Hofrichter.
    2010. "DyP-like Peroxidases of the Jelly Fungus Auricularia Auricula-Judae Oxidize
    Nonphenolic Lignin Model Compounds and High-Redox Potential Dyes." *Applied
    Microbiology and Biotechnology* 85 (6). Appl Microbiol Biotechnol: 1869–79.
    doi:10.1007/s00253-009-2173-7.
Lisov, A. V., A. A. Leontievsky, and L. A. Golovleva. 2003. "Hybrid Mn-Peroxidase from the
    Ligninolytic Fungus Panus Tigrinus 8/18. Isolation, Substrate Specificity, and Catalytic
    Cycle." *Biochemistry (Moscow)* 68 (9). Springer: 1027–35. doi:10.1023/A:1026072815106.
Liu, Chang Jun, Yu Chen Miao, and Ke Wei Zhang. 2011. "Sequestration and Transport of
    Lignin Monomeric Precursors." *Molecules* 16 (1). Molecular Diversity Preservation
    International: 710–27. doi:10.3390/molecules16010710.
Liu, Huifang, Hongji Li, Jianmin Lu, Shu Zeng, Min Wang, Nengchao Luo, Shutao
    Xu, and Feng Wang. 2018. "Photocatalytic Cleavage of C-C Bond in Lignin
    Models under Visible Light on Mesoporous Graphitic Carbon Nitride through π-π
    Stacking Interaction." *ACS Catalysis* 8 (6). American Chemical Society: 4761–71.
    doi:10.1021/acscatal.8b00022.
Liu, Rui, Lin Dai, Chunlin Xu, Kai Wang, Chunyang Zheng, and Chuanling Si. 2020. "Lignin-Based
    Micro- and Nanomaterials and Their Composites in Biomedical Applications."
    ChemSusChem 13 (17). Wiley-VCH Verlag: 4266–83. doi:10.1002/cssc.202000783.
Liu, Xiaoqing, Xiaoguang Duan, Wei Wei, Shaobin Wang, and Bing Jie Ni. 2019.
    "Photocatalytic Conversion of Lignocellulosic Biomass to Valuable Products." *Green
    Chemistry*. Royal Society of Chemistry. doi:10.1039/c9gc01728c.
Lombard, Vincent, Hemalatha Golaconda Ramulu, Elodie Drula, Pedro M. Coutinho, and
    Bernard Henrissat. 2014. "The Carbohydrate-Active Enzymes Database (CAZy) in
    2013." *Nucleic Acids Research* 42 (D1). Nucleic Acids Res. doi:10.1093/nar/gkt1178.
Luo, Hao, and Mahdi M. Abu-Omar. 2017. "Chemicals from Lignin." In *Encyclopedia of
    Sustainable Technologies*. Elsevier: 573–85. doi:10.1016/B978-0-12-409548-9.10235-0.
Luo, Jian, and Jian Zhang. 2016. "Aerobic Oxidation of Olefins and Lignin Model Compounds
    Using Photogenerated Phthalimide-N-Oxyl Radical." *Journal of Organic Chemistry* 81
    (19). American Chemical Society: 9131–37. doi:10.1021/acs.joc.6b01704.
Luo, Jian, Xiang Zhang, Jingzhi Lu, and Jian Zhang. 2017. "Fine Tuning the Redox Potentials
    of Carbazolic Porous Organic Frameworks for Visible-Light Photoredox Catalytic
    Degradation of Lignin β-O-4 Models." *ACS Catalysis* 7 (8). American Chemical
    Society: 5062–70. doi:10.1021/acscatal.7b01010.
Luo, Nengchao, Min Wang, Hongji Li, Jian Zhang, Tingting Hou, Haijun Chen,
    Xiaochen Zhang, Jianmin Lu, and Feng Wang. 2017. "Visible-Light-Driven
    Self-Hydrogen Transfer Hydrogenolysis of Lignin Models and Extracts into
    Phenolic Products." *ACS Catalysis* 7 (7). American Chemical Society: 4571–80.
    doi:10.1021/acscatal.7b01043.
Luo, Nengchao, Min Wang, Hongji Li, Jian Zhang, Huifang Liu, and Feng Wang. 2016.
    "Photocatalytic Oxidation-Hydrogenolysis of Lignin β-O-4 Models via a Dual Light
    Wavelength Switching Strategy." *ACS Catalysis* 6 (11). American Chemical Society:
    7716–21. doi:10.1021/acscatal.6b02212.

Magallanes, Gabriel, Markus D. Kärkäs, Irene Bosque, Sudarat Lee, Stephen Maldonado, and Corey R.J. Stephenson. 2019. "Selective C-O Bond Cleavage of Lignin Systems and Polymers Enabled by Sequential Palladium-Catalyzed Aerobic Oxidation and Visible-Light Photoredox Catalysis." *ACS Catalysis* 9 (3). American Chemical Society: 2252–60. doi:10.1021/acscatal.8b04172.

Mahata, Denial, Malabendu Jana, Arundhuti Jana, Abhishek Mukherjee, Nibendu Mondal, Tilak Saha, Subhajit Sen, et al. 2017. "Lignin-Graft-Polyoxazoline Conjugated Triazole a Novel Anti-Infective Ointment to Control Persistent Inflammation." *Scientific Reports* 7 (1). Nature Publishing Group: 1–16. doi:10.1038/srep46412.

Maldonado-Carmona, Nidia, Guillaume Marchand, Nicolas Villandier, Tan Sothea Ouk, Mariette M. Pereira, Mário J.F. Calvete, Claude Alain Calliste, et al. 2020. "Porphyrin-Loaded Lignin Nanoparticles Against Bacteria: A Photodynamic Antimicrobial Chemotherapy Application." *Frontiers in Microbiology* 11 (November). Frontiers Media S.A.: 2846. doi:10.3389/fmicb.2020.606185.

Manavalan, Tamilvendan, Arulmani Manavalan, and Klaus Heese. 2015. "Characterization of Lignocellulolytic Enzymes from White-Rot Fungi." *Current Microbiology* 70 (4). Springer New York LLC: 485–98. doi:10.1007/s00284-014-0743-0.

Mandeep, Guddu Kumar Gupta, and Pratyoosh Shukla. 2020. "Insights into the Resources Generation from Pulp and Paper Industry Wastes: Challenges, Perspectives and Innovations." *Bioresource Technology*. Elsevier Ltd. doi:10.1016/j.biortech.2019.122496.

Mandlekar, Neeraj, Aurélie Cayla, François Rault, Stéphane Giraud, Fabine Salaün, Giulio Malucelli, and Jin-Ping Guan. 2018. "An Overview on the Use of Lignin and Its Derivatives in Fire Retardant Polymer Systems." In *Lignin - Trends and Applications*. InTech. doi:10.5772/intechopen.72963.

Mansouri, Nour Eddine El, Antonio Pizzi, and Joan Salvadó. 2007. "Lignin-Based Wood Panel Adhesives without Formaldehyde." *Holz Als Roh - Und Werkstoff* 65 (1). Springer: 65–70. doi:10.1007/s00107-006-0130-z.

Marchand, Guillaume, Claude A. Calliste, René M. Williams, Charlotte McLure, Stéphanie Leroy-Lhez, and Nicolas Villandier. 2018. "Acetylated Lignins: A Potential Bio-Sourced Photosensitizer." *ChemistrySelect* 3 (20). Wiley-Blackwell: 5512–16. doi:10.1002/slct.201801039.

Mehta, A. Y., B. M. Mohammed, E. J. Martin, D. F. Brophy, D. Gailani, and U. R. Desai. 2016. "Allosterism-Based Simultaneous, Dual Anticoagulant and Antiplatelet Action: Allosteric Inhibitor Targeting the Glycoprotein Ibα-Binding and Heparin-Binding Site of Thrombin." *Journal of Thrombosis and Haemostasis* 14 (4). Blackwell Publishing Ltd: 828–38. doi:10.1111/jth.13254.

Miao, Yu Chen, and Chang Jun Liu. 2010. "ATP-Binding Cassette-like Transporters Are Involved in the Transport of Lignin Precursors across Plasma and Vacuolar Membranes." *Proceedings of the National Academy of Sciences of the United States of America* 107 (52). National Academy of Sciences: 22728–33. doi:10.1073/pnas.1007747108.

Moura, Jullyana Cristina Magalhães Silva, Cesar Augusto Valencise Bonine, Juliana de Oliveira Fernandes Viana, Marcelo Carnier Dornelas, and Paulo Mazzafera. 2010. "Abiotic and Biotic Stresses and Changes in the Lignin Content and Composition in Plants." *Journal of Integrative Plant Biology*. J Integr Plant Biol. doi:10.1111/j.1744-7909.2010.00892.x.

Nguyen, John D., Bryan S. Matsuura, and Corey R.J. Stephenson. 2014. "A Photochemical Strategy for Lignin Degradation at Room Temperature." *Journal of the American Chemical Society* 136 (4). American Chemical Society: 1218–21. doi:10.1021/ja4113462.

Nguyen, Suong T., Philip R.D. Murray, and Robert R. Knowles. 2020. "Light-Driven Depolymerization of Native Lignin Enabled by Proton-Coupled Electron Transfer." *ACS Catalysis* 10 (1). American Chemical Society: 800–05. doi:10.1021/acscatal.9b04813.

Nikhileshwar Reddy, Yeddula, Neeraj Singh Thakur, and Jayeeta Bhaumik. 2020. "Harnessing the Photocatalytic Potential of Polypyrroles in Water through Nanointervension: Synthesis and Photophysical Evaluation of Biodegradable Polypyrrolic Nanoencapsulates." *ChemNanoMat* 6 (2). Wiley-VCH Verlag: 239–47. doi:10.1002/cnma.201900466.

Nimz, Horst. 1974. "Beech Lignin—Proposal of a Constitutional Scheme." *Angewandte Chemie International Edition in English* 13 (5). John Wiley & Sons, Ltd: 313–21. doi:10.1002/anie.197403131.

Nis, Berna, and Burcak Kaya Ozsel. 2021. "Efficient Direct Conversion of Lignocellulosic Biomass into Biobased Platform Chemicals in Ionic Liquid-Water Medium." *Renewable Energy* 169 (May). Elsevier Ltd: 1051–57. doi:10.1016/j.renene.2021.01.083.

Norikura, Toshio, Yuuka Mukai, Shuzo Fujita, Keigo Mikame, Masamitsu Funaoka, and Shin Sato. 2010. "Lignophenols Decrease Oleate-Induced Apolipoprotein-B Secretion in HepG2 Cells." *Basic and Clinical Pharmacology and Toxicology* 107 (4). Basic Clin Pharmacol Toxicol: 813–17. doi:10.1111/j.1742-7843.2010.00575.x.

Paszczyński, Andrzej, Van Ba Huynh, and Ronald Crawford. 1986. "Comparison of Ligninase-I and Peroxidase-M2 from the White-Rot Fungus Phanerochaete Chrysosporium." *Archives of Biochemistry and Biophysics* 244 (2). Academic Press: 750–65. doi:10.1016/0003-9861(86)90644-2.

Paul, Shatabdi, Neeraj Singh Thakur, Sanjam Chandna, Yeddula Nikhileshwar Reddy, and Jayeeta Bhaumik. 2021. "Development of a Light Activatable Lignin Nanosphere Based Spray Coating for Bioimaging and Antimicrobial Photodynamic Therapy." *Journal of Materials Chemistry B*. Royal Society of Chemistry (RSC). doi:10.1039/d0tb02643c.

Pawlik, Anna, Magdalena Wójcik, Karol Rułka, Karolina Motyl-Gorzel, Monika Osińska-Jaroszuk, Jerzy Wielbo, Monika Marek-Kozaczuk, Anna Skorupska, Jerzy Rogalski, and Grzegorz Janusz. 2016. "Purification and Characterization of Laccase from Sinorhizobium Meliloti and Analysis of the Lacc Gene." *International Journal of Biological Macromolecules* 92 (November). Elsevier B.V.: 138–47. doi:10.1016/j.ijbiomac.2016.07.012.

Pollegioni, Loredano, Fabio Tonin, and Elena Rosini. 2015. "Lignin-Degrading Enzymes." *FEBS Journal*. Blackwell Publishing Ltd. doi:10.1111/febs.13224.

Prado, Raquel, Xabier Erdocia, and Jalel Labidi. 2013. "Effect of the Photocatalytic Activity of TiO_2 on Lignin Depolymerization." *Chemosphere* 91 (9). Elsevier Ltd: 1355–61. doi:10.1016/j.chemosphere.2013.02.008.

Priyadarshinee, Rashmi, Anuj Kumar, Tamal Mandal, and Dalia Dasguptamandal. 2016. "Unleashing the Potential of Ligninolytic Bacterial Contributions towards Pulp and Paper Industry: Key Challenges and New Insights." *Environmental Science and Pollution Research* 23 (23). Springer Verlag: 23349–68. doi:10.1007/s11356-016-7633-x.

Qiu, Shi, Meng Wang, Yunming Fang, and Tianwei Tan. 2020. "Reductive Catalytic Fractionation of Lignocellulose: When Should the Catalyst Meet Depolymerized Lignin Fragments?" *Sustainable Energy and Fuels* 4 (11). Royal Society of Chemistry: 5588–94. doi:10.1039/d0se01118e.

Ralph, John, Knut Lundquist, Gösta Brunow, Fachuang Lu, Hoon Kim, Paul F. Schatz, Jane M. Marita, et al. 2004. "Lignins: Natural Polymers from Oxidative Coupling of 4-Hydroxyphenyl- Propanoids." *Phytochemistry Reviews*. Springer. doi:10.1023/B:PHYT.0000047809.65444.a4.

Rashid, Goran M.M., Charles R. Taylor, Yangqingxue Liu, Xiaoyang Zhang, Dean Rea, Vilmos Fülöp, and Timothy D.H. Bugg. 2015. "Identification of Manganese Superoxide Dismutase from Sphingobacterium Sp. T_2 as a Novel Bacterial Enzyme for Lignin Oxidation." *ACS Chemical Biology* 10 (10). American Chemical Society: 2286–94. doi:10.1021/acschembio.5b00298.

Reiter, Jochen, Harald Strittmatter, Lars O. Wiemann, Doris Schieder, and Volker Sieber. 2013. "Enzymatic Cleavage of Lignin β-O-4 Aryl Ether Bonds Via Net Internal Hydrogen Transfer." *Green Chemistry* 15 (5). Royal Society of Chemistry: 1373–81. doi:10.1039/c3gc40295a.

Río, José Carlos del, Jorge Rencoret, Ana Gutiérrez, Hoon Kim, and Ralph John. 2017. "Hydroxystilbenes Are Monomers in Palm Fruit Endocarp Lignins." *Plant Physiology* 174 (4). American Society of Plant Biologists: 2072–82. doi:10.1104/pp.17.00362.

Río, José Carlos Del, Jorge Rencoret, Pepijn Prinsen, Ángel T. Martínez, John Ralph, and Ana Gutiérrez. 2012. "Structural Characterization of Wheat Straw Lignin as Revealed by Analytical Pyrolysis, 2D-NMR, and Reductive Cleavage Methods." *Journal of Agricultural and Food Chemistry* 60 (23). American Chemical Society: 5922–35. doi:10.1021/jf301002n.

Riva, Sergio. 2006. "Laccases: Blue Enzymes for Green Chemistry." *Trends in Biotechnology.* Elsevier Current Trends. doi:10.1016/j.tibtech.2006.03.006.

Rodríguez, Alejandro, Rafael Sánchez, Ana Requejo, and Ana Ferrer. 2010. "Feasibility of Rice Straw as a Raw Material for the Production of Soda Cellulose Pulp." *Journal of Cleaner Production* 18 (10–11): 1084–91. doi:10.1016/j.jclepro.2010.03.011.

Saluja, Bhawana, Jay N. Thakkar, Hua Li, Umesh R. Desai, and Masahiro Sakagami. 2013. "Novel Low Molecular Weight Lignins as Potential Anti-Emphysema Agents: In Vitro Triple Inhibitory Activity against Elastase, Oxidation and Inflammation." *Pulmonary Pharmacology and Therapeutics* 26 (2). Pulm Pharmacol Ther: 296–304. doi:10.1016/j.pupt.2012.12.009.

Sanchez, Oscar, Rocio Sierra, and Carlos J. Alméciga-Díaz. 2011. "Delignification Process of Agro-Industrial Wastes an Alternative to Obtain Fermentable Carbohydrates for Producing Fuel." In *Alternative Fuel.* InTech. doi:10.5772/22381.

Sato, Shin, Yuuka Mukai, Jyoji Yamate, Toshio Norikura, Yae Morinaga, Keigo Mikame, Masamitsu Funaoka, and Shuzo Fujita. 2009. "Lignin-Derived Lignophenols Attenuate Oxidative and Inflammatory Damage to the Kidney in Streptozotocin-Induced Diabetic Rats." *Free Radical Research* 43 (12). Free Radic Res: 1205–13. doi:10.3109/10715760903247264.

Schuetz, Mathias, Anika Benske, Rebecca A. Smith, Yoichiro Watanabe, Yuki Tobimatsu, John Ralph, Taku Demura, Brian Ellis, and A. Lacey Samuels. 2014. "Laccases Direct Lignification in the Discrete Secondary Cell Wall Domains of Protoxylem." *Plant Physiology* 166 (2). American Society of Plant Biologists: 798–807. doi:10.1104/pp.114.245597.

Schutyser, W., T. Renders, S. Van Den Bosch, S. F. Koelewijn, G. T. Beckham, and B. F. Sels. 2018. "Chemicals from Lignin: An Interplay of Lignocellulose Fractionation, Depolymerisation, and Upgrading." *Chemical Society Reviews.* Royal Society of Chemistry. doi:10.1039/c7cs00566k.

Shen, Xiao Jun, Jia Long Wen, Qing Qing Mei, Xue Chen, Dan Sun, Tong Qi Yuan, and Run Cang Sun. 2019. "Facile Fractionation of Lignocelluloses by Biomass-Derived Deep Eutectic Solvent (DES) Pretreatment for Cellulose Enzymatic Hydrolysis and Lignin Valorization." *Green Chemistry* 21 (2). Royal Society of Chemistry: 275–83. doi:10.1039/c8gc03064b.

Sheng, Yequan, Su Shiung Lam, Yingji Wu, Shengbo Ge, Jinglei Wu, Liping Cai, Zhenhua Huang, Quyet Van Le, Christian Sonne, and Changlei Xia. 2021. "Enzymatic Conversion of Pretreated Lignocellulosic Biomass: A Review on Influence of Structural Changes of Lignin." *Bioresource Technology.* Elsevier Ltd. doi:10.1016/j.biortech.2020.124631.

Sigoillot, Jean Claude, Jean Guy Berrin, Mathieu Bey, Laurence Lesage-Meessen, Anthony Levasseur, Anne Lomascolo, Eric Record, and Eva Uzan-Boukhris. 2012. "Fungal Strategies for Lignin Degradation." In *Advances in Botanical Research*, 61:263–308 Academic Press Inc. doi:10.1016/B978-0-12-416023-1.00008-2.

Silva Coelho-Moreira, Jaqueline da, Giselle Maria, Rafael Castoldi, Simone da Silva Mariano, Fabola Dorneles, Adelar Bracht, and Rosane Marina. 2013. "Involvement of Lignin-Modifying Enzymes in the Degradation of Herbicides." In *Herbicides - Advances in Research.* InTech. doi:10.5772/55848.

Singh, D. 2004. *Environmental Microbiology and Biotechnology* - D. P. Singh - Google Books. https://books.google.com.ec/books?hl=es&lr=&id=NHN25jAMwo8C&oi=fn d&pg=PA97&dq=biocomposting&ots=gAzHLuqBOj&sig=rsB4EC8gBWxrGcA4bh4 OY2Ng6rs&redir_esc=y#v=onepage&q&f=false%0Ahttps://books.google.com/books ?hl=en&lr=&id=NHN25jAMwo8C&oi=fnd&pg=PA10&dq=microbi.

Sipponen, Mika H., Heiko Lange, Mariko Ago, and Claudia Crestini. 2018. "Understanding Lignin Aggregation Processes. A Case Study: Budesonide Entrapment and Stimuli Controlled Release from Lignin Nanoparticles." *ACS Sustainable Chemistry and Engineering* 6 (7). American Chemical Society: 9342–51. doi:10.1021/acssuschemeng.8b01652.

Sonia, Mokni Tlili, Jedidi Naceur, and Hassen Abdennaceur. 2011. "Studies on the Ecology of Actinomycetes in an Agricultural Soil Amended with Organic Residues: I. Identification of the Dominant Groups of Actinomycetales." *World Journal of Microbiology and Biotechnology* 27 (10): 2239–49. doi:10.1007/s11274-011-0687-5.

Sonoki, Tomonori, Eiji Masai, Kanna Sato, Shinya Kajita, and Yoshihiro Katayama. 2009. "Methoxyl Groups of Lignin Are Essential Carbon Donors in C1 Metabolism of Sphingobium Sp. SYK-6." *Journal of Basic Microbiology* 49 (SUPPL. 1). John Wiley & Sons, Ltd: S98–102. doi:10.1002/jobm.200800367.

Sriroth, Klanarong, and Jackapon Sunthornvarabhas. 2018. "Lignin from Sugar Process as Natural Antimicrobial Agent." *Biochemistry & Pharmacology: Open Access* 07 (01). OMICS Publishing Group. doi:10.4172/2167-0501.1000239.

Staveness, Daryl, Irene Bosque, and Corey R.J. Stephenson. 2016. "Free Radical Chemistry Enabled by Visible Light-Induced Electron Transfer." *Accounts of Chemical Research* 49 (10). American Chemical Society: 2295–306. doi:10.1021/acs.accounts.6b00270.

"Structure and Characteristics of Lignin." 2019. In *Lignin Chemistry and Applications*. Elsevier: 25–50. doi:10.1016/b978-0-12-813941-7.00002-3.

Stücker, Alexander, Fokko Schütt, Bodo Saake, and Ralph Lehnen. 2016. "Lignins from Enzymatic Hydrolysis and Alkaline Extraction of Steam Refined Poplar Wood: Utilization in Lignin-Phenol-Formaldehyde Resins." *Industrial Crops and Products* 85 (July). Elsevier B.V.: 300–08. doi:10.1016/j.indcrop.2016.02.062.

Sugano, Y. 2009. "DyP-Type Peroxidases Comprise a Novel Heme Peroxidase Family." *Cellular and Molecular Life Sciences*. Cell Mol Life Sci. doi:10.1007/s00018-008-8651-8.

Sun, Wenting, Sugang Meng, Sujuan Zhang, Xiuzhen Zheng, Xiangju Ye, Xianliang Fu, and Shifu Chen. 2018. "Insight into the Transfer Mechanisms of Photogenerated Carriers for Heterojunction Photocatalysts with the Analogous Positions of Valence Band and Conduction Band: A Case Study of ZnO/TiO$_2$." *Journal of Physical Chemistry C* 122 (27). American Chemical Society: 15409–20. doi:10.1021/acs. jpcc.8b03753.

Sun, Zhuohua, Bálint Fridrich, Alessandra De Santi, Saravanakumar Elangovan, and Katalin Barta. 2018. "Bright Side of Lignin Depolymerization: Toward New Platform Chemicals." *Chemical Reviews*. American Chemical Society. doi:10.1021/acs. chemrev.7b00588.

Sutherland, Greg R.J., Laura Schick Zapanta, Ming Tien, and Steven D. Aust. 1997. "Role of Calcium in Maintaining the Heme Environment of Manganese Peroxidase." *Biochemistry* 36 (12). American Chemical Society : 3654–62. doi:10.1021/bi962195m.

Suzuki, Takashi, Kohki Endo, Masaaki Ito, Hiroshi Tsujibo, Katsushiro Miyamoto, and Yoshihiko Inamori. 2003. "A Thermostable Laccase from Streptomyces Lavendulae REN-7: Purification, Characterization, Nucleotide Sequence, and Expression." *Bioscience, Biotechnology and Biochemistry* 67 (10). Japan Society for Bioscience, Biotechnology, and Agrochemistry: 2167–75. doi:10.1271/bbb.67.2167.

Tian, Jiang Hao, Anne Marie Pourcher, Théodore Bouchez, Eric Gelhaye, and Pascal Peu. 2014. "Occurrence of Lignin Degradation Genotypes and Phenotypes among Prokaryotes." *Applied Microbiology and Biotechnology.* Springer Verlag. doi:10.1007/s00253-014-6142-4.

Tien, Ming, and T. Kent Kirk. 1983. "Lignin-Degrading Enzyme from the Hymenomycete Phanerochaete Chrysasporium Burds." *Science* 221 (4611). Science: 661–63. doi:10.1126/science.221.4611.661.

Wang, Hongliang, Yunqiao Pu, Arthur Ragauskas, and Bin Yang. 2019. "From Lignin to Valuable Products–Strategies, Challenges, and Prospects." *Bioresource Technology.* Elsevier Ltd. doi:10.1016/j.biortech.2018.09.072.

Wang, Hongliang, Libing Zhang, Tiansheng Deng, Hao Ruan, Xianglin Hou, John R. Cort, and Bin Yang. 2016. "ZnCl$_2$ Induced Catalytic Conversion of Softwood Lignin to Aromatics and Hydrocarbons." *Green Chemistry* 18 (9). Royal Society of Chemistry: 2802–10. doi:10.1039/c5gc02967h.

Wang, Yinling, Yue Liu, Jianghua He, and Yuetao Zhang. 2019. "Redox-Neutral Photocatalytic Strategy for Selective C–C Bond Cleavage of Lignin and Lignin Models via PCET Process." *Science Bulletin* 64 (22). Elsevier B.V.: 1658–66. doi:10.1016/j.scib.2019.09.003.

Watanabe, Takashi, Nobuaki Shirai, Hitomi Okada, Yoichi Honda, and Masaaki Kuwahara. 2001. "Production and Chemiluminescent Free Radical Reactions of Glyoxal in Lipid Peroxidation of Linoleic Acid by the Ligninolytic Enzyme, Manganese Peroxidase." *European Journal of Biochemistry* 268 (23). Eur J Biochem: 6114–22. doi:10.1046/j.0014-2956.2001.02557.x.

Wayman, M., and T. I. Obiaga. 1974. "The Modular Structure of Lignin." *Canadian Journal of Chemistry* 52 (11). Canadian Science Publishing: 2102–10. doi:10.1139/v74-304.

Wong, Dominic W.S. 2009. "Structure and Action Mechanism of Ligninolytic Enzymes." *Applied Biochemistry and Biotechnology.* Appl Biochem Biotechnol. doi:10.1007/s12010-008-8279-z.

Wu, Xuejiao, Xueting Fan, Shunji Xie, Jinchi Lin, Jun Cheng, Qinghong Zhang, Liangyi Chen, and Ye Wang. 2018. "Solar Energy-Driven Lignin-First Approach to Full Utilization of Lignocellulosic Biomass under Mild Conditions." *Nature Catalysis* 1 (10). Nature Publishing Group: 772–80. doi:10.1038/s41929-018-0148-8.

Wu, Xuejiao, Shunji Xie, Chenxi Liu, Cheng Zhou, Jinchi Lin, Jincan Kang, Qinghong Zhang, Zhaohui Wang, and Ye Wang. 2019. "Ligand-Controlled Photocatalysis of CdS Quantum Dots for Lignin Valorization under Visible Light." *ACS Catalysis* 9 (9). American Chemical Society: 8443–51. doi:10.1021/acscatal.9b02171.

Xiang, Zhiyu, Wanying Han, Jin Deng, Wanbin Zhu, Ying Zhang, and Hongliang Wang. 2020. "Photocatalytic Conversion of Lignin into Chemicals and Fuels." *ChemSusChem* 13 (17). Wiley-VCH Verlag: 4199–213. doi:10.1002/cssc.202000601.

Yang, Yang, Tianxiao Wang, Juan Guan, Juan Wang, Junyi Chen, Xiaoqin Liu, Jun Qian, et al. 2019. "Oral Delivery of Honokiol Microparticles for Nonrapid Eye Movement Sleep." *Molecular Pharmaceutics* 16 (2). American Chemical Society: 737–43. doi:10.1021/acs.molpharmaceut.8b01016.

Yoo, Hyeonji, Min Woo Lee, Sunggyu Lee, Jehee Lee, Soyoung Cho, Hangil Lee, Hyun Gil Cha, and Hyun Sung Kim. 2020. "Enhancing Photocatalytic β-O-4 Bond Cleavage in Lignin Model Compounds by Silver-Exchanged Cadmium Sulfide." *ACS Catalysis* 10 (15). American Chemical Society: 8465–75. doi:10.1021/acscatal.0c01915.

Yousuf, Abu, Domenico Pirozzi, and Filomena Sannino. 2019. "Fundamentals of Lignocellulosic Biomass." *Lignocellulosic Biomass to Liquid Biofuels.* doi:10.1016/B978-0-12-815936-1.00001-0.

Zámocký, Marcel, Paul G. Furtmüller, and Christian Obinger. 2009. "Two Distinct Groups of Fungal Catalase/Peroxidases." In *Biochemical Society Transactions*, 37. Europe PMC Funders: 772–77. doi:10.1042/BST0370772.

Zhao, Houkuan, Jinling Li, Peng Wang, Shaoqiong Zeng, and Yimin Xie. 2017. "Lignin-Carbohydrate Complexes Based Spherical Biocarriers: Preparation, Characterization, and Biocompatibility." *International Journal of Polymer Science* 2017. Hindawi Limited. doi:10.1155/2017/4915185.

Zoghlami, Aya, and Gabriel Paës. 2019. "Lignocellulosic Biomass: Understanding Recalcitrance and Predicting Hydrolysis." *Frontiers in Chemistry*. Frontiers Media S.A. doi:10.3389/fchem.2019.00874.

3 LCA and TEA for Biomass Conversion Technology

Shilpa Main and Vishwanath H. Dalvi
Institute of Chemical Technology

Yogendra Shashtri
Indian Institute of Technology Bombay

Annamma Anil Odaneth
DBT-ICT Centre for Energy Biosciences,
Institute of Chemical Technology

CONTENTS

3.1 Introduction 65
3.2 Techno Economic Analysis (TEA)..68
    3.2.1 TEA Methodology 68
        3.2.1.1 Technology Maturity..69
        3.2.1.2 Goal and Scope..70
        3.2.1.3 Inventory 71
        3.2.1.4 Assumptions..74
    3.2.2 Case Studies...77
    3.2.3 Challenges and Research Gaps 79
3.3 Life Cycle Assessment...80
    3.3.1 LCA: Methodology 80
        3.3.1.1 Goal and Scope Definition...80
        3.3.1.2 Life Cycle Inventory 83
        3.3.1.3 Impact Assessment...84
        3.3.1.4 Interpretation..85
    3.3.2 Case Studies...85
    3.3.3 Challenges and Research Gaps 87
3.4 Conclusions...89
References..89

3.1 INTRODUCTION

Biomass-based fuels, particularly derived from lignocellulosic (non-food) resources, are expected to a play a major role in our transition towards sustainable energy (Fatma et al., 2018). While biofuels are being scaled up and made techno-economically feasible, it is equally important to ensure that biofuels are sustainable (Lee et al., 2019).

Sustainability has three dimensions, namely economic, environmental, and social, and all dimensions must be met to achieve sustainability. The common definition of sustainability as proposed by the United Nations, as part of the Brundtland Commission Report, is "development that meets the needs of the present without compromising the ability of future generations to meet their own needs" (Brundtland, 1987). This definition clearly recognizes that sustainability goes beyond environmental protection and also identifies the futuristic view. Extending this view to biofuels, the economic and environmental considerations that are encompassed in this definition can be looked upon as the driving forces to guide successful implementation of early stage research and development activities for biofuel conversion processes. The different steps involved in the process need these sustainability studies to be performed at each stage to guide competitive fuel targets and future research investment prospects. Systems-based modelling combined with experimental work has been used to identify current barriers to economic viability and environmental sustainability, and the results of such efforts are critical to guiding future research investment towards competitive fuel targets (Meghana and Shastri, 2020; Gholkar and Shastri, 2020). The various studies reported by the United States Department of Energy (DOE) are an illustration of this concept. This requires a highly interdisciplinary approach and must involve active collaboration between experimental scientists, computational model experts, and industries.

Techno-economic analysis (TEAs) and life cycle analysis (LCAs) are usually aspects of sustainability modelling. Both of them can help make decisions in the pre-commercial stage and optimization in ongoing operations (Xie, 2015) (Figure 3.1). Both involve a foundational model of the process that captures the system's mass and energy flows. For the validity of the results, the fidelity and validation of the sub-process models which constitute the process model are critically essential.

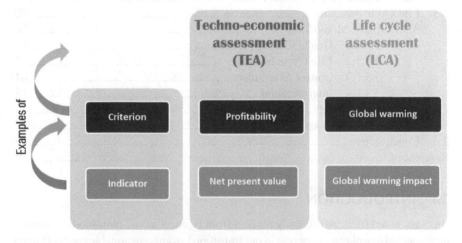

FIGURE 3.1 Life cycle assessment and techno-economic assessment with example criteria and example indicators. (Adapted from Buchner et al., 2018b.)

Feasibility analyses that have failed to provide detailed mass and energy balances are limited to their results being relevant, as it is difficult to validate their stated assumptions. The vast majority of TEAs published in the literature, however, are focused on rigorous models of engineering processes, giving the validity of their results. Nevertheless, for non-optimized model scenarios, the spectrum of recorded TEA results is very high, from \$1.59 per gallon of biofuel to \$33.86 per gallon, and even up to \$104.31 per gallon. This broad range refers to several variables. The clearest distinction is the significant difference in capital cost and productivity of the system. Moreover, economic assumptions and the expected performance, variations in processing technologies, foundational economic methods, performance assumptions, and other modelling decisions are the other discrepancies that lead to a large range in reported results. Some studies model mature "nth-of-a-kind" plants, reflecting more positive predictions about productivity, process performance, facility scales, and the advantages of learning technology. For more tested, and thus lower-risk, schemes, this statement takes advantage of economics.

The carbon sequestration capability of plants is a function of the plant life, which must be accounted for before claiming carbon neutrality of biofuels. More importantly, the conversion of biomass to biofuels itself requires energy and other inputs. These inputs are required at multiple stages such as the growth of the plant, agricultural operations such as harvesting, transportation, and finally in the bio-refinery during the conversion. All these inputs cut into the benefits that one can claim from biofuels, and it is important to ensure that these inputs do not cancel out the benefits of using renewable biofuels. Otherwise, the purpose of promoting biofuels is defeated. Moreover, the energy inputs, as well as other inputs that may be required, lead to carbon emissions. These emissions, thus, affect the net carbon benefit that could be obtained from using biofuels. At the very least, the benefits should outweigh the costs (emissions during production in this case).

It is also important to look beyond only carbon dioxide or greenhouse gas emissions because biofuels require many other inputs that can have environmental impacts other than climate change. Biofuels are very water-intensive. The feedstock for biofuels comes from agriculture, which is very water-intensive. Globally, about 75%–80% of fresh water is consumed in agriculture. As a result, the water footprint of ethanol, one of the key biofuels, has been shown to be between 200 and 700 L/L of ethanol (Murali and Shastri, 2019). Land use is another important aspect, which relates to the famous food vs fuel debate. The concern here is that if biofuel production is commercialized, and too much land is allocated for growing crops for biofuel production, land availability for food crops may reduce. This may drive up the cost of food. As a result, it is acknowledged that biofuels need to be produced from non-food crops or crop residues, mainly lignocellulosic biomass or microalgal biomass. The use of fertilizers and pesticides during the cultivation of biofuel crops can also lead to excessive water pollution problems.

The issues highlighted above indicate that although the impacts due to biofuels may be low during the use phase that is only because they have been moved in space and time. Therefore, a holistic and systematic approach is required to quantify the environmental effects of biofuels and ascertain their sustainability. Life cycle assessment is an approach to the assessment of environmental effects throughout manufacturing or service. Life-cycle analysis is one of the most comprehensive methods for such kind of assessment (Rebitzer et al., 2004). It comes under the category of product-related assessment among the various sustainability assessment techniques. Product-related assessment refers to various material flows associated with the production process and up to the consumption of the product, depending on the boundary condition. The boundary conditions can be typically from the extraction of raw material from the earth's crust (cradle) to either finished products ready for the consumer (gate), or the point where the residues of the products return to the environment (grave).

The goal of this chapter is to discuss the TEA and LCA methodologies. While the methodologies are general, they are discussed in the context of lignocellulose biofuels. After summarizing the key aspects of the methodologies, this chapter reviews recent literature on TEA and LCA of biofuels. The important results are summarized and discussed. This chapter ends with a conclusion regarding the current status and identifies future research needs.

3.2 TECHNO ECONOMIC ANALYSIS (TEA)

Techno-economic analysis (TEA) is a technique for the determination of the technical and economic performance of a process, product, and service. Its primary goal is to determine the production cost of the process as input towards determining the viability of a commercial endeavour based on it. A TEA should be conducted at every stage in the development of technology. During the early stage of technology development, TEA helps to narrow down the choice of alternatives, while in later stages, it allows for a clear assessment of the aspects that require greater focus.

3.2.1 TEA METHODOLOGY

The "T" in "TEA" solely refers to the fact that an economic evaluation is conducted for technology and is based on data obtained from it (Buchner et al., 2018a).

TEA "involves economic impact studies of research, development, technology demonstration, and deployment", which uncover the production cost and market opportunities (Xie, 2015). TEAs typically focus on the production phase, reflecting the perspectives of a producer. The inclusion of further upstream and downstream life cycle stages is possible, for example, to analyse the technical or economic performance of products during the use or disposal phases. TEA is subdivided into the following phases:

a. Technology maturity definition is expressed semi-quantitatively by technology readiness levels (TRLs). This defines the current level of research and development associated with the technology. It also clearly spells out the reason for carrying out the TEA and the decision that hangs on the results.

b. Goal and scope, where analysts define the goal of the study and experts define the aspects to be included and the metrics to be used to conduct the comparisons. The resources available for the exercise are included in the scope and scenario specifications. Resources available, including time, access to databases, access to literature, field studies, etc., need to be specified beforehand.

c. Inventory, where all relevant data are collected.
d. Calculation of indicators, where calculation procedures are specified and results are reported.
e. Interpretation, which is perceived to be a distinct process of the evaluation due to its significance in reporting. Here, analysts evaluate the performance, accuracy, and robustness of the outcomes. Although certain elements of interpretation are conducted during the study, conclusions and guidelines are finalized at the end.

It must be noted that TEA is a highly iterative process whose later phases inform refinements in the earlier phases, specifying and enhancing the assessment for each round. The TEA is summarized into a TEA report, with results from both phases and results (see Figure 3.2).

TEA typically follows the same LCA concept as specified in ISO 14040/14044 (Buchner et al., 2018a).

3.2.1.1 Technology Maturity

As compared to other industries, the time taken for an invention to progress from ideation to commercialization in the chemical industry is comparatively long (up to about 10 years). Reducing the time it takes for an innovation to reach the market has a high opportunity for cost savings or big competitive advantages, which can have a direct impact on a company's innovation plan. This necessitates the need for an accurate method of measuring and conveying the actual state of an invention, which may contribute to a clearer general view of the technology's maturity in terms of research, growth, or implementation. Only after determining the current sophistication of

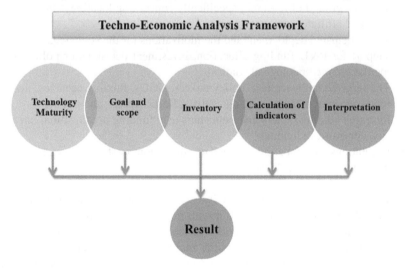

FIGURE 3.2 Techno-economic analysis framework. (Adapted from Zimmermann et al., 2018.)

technology will decisions on the next moves be taken, such as whether to invest in further R&D or to move on to deployment with investments in project management, risk analysis, environmental impact analysis, marketing, and so on. The idea of rating its readiness, known as "technology readiness standards," is a common concept for evaluating technology maturity. TRLs are becoming increasingly popular in academia, business, and policymaking as a method for assessing and communicating the maturity of a technology (Buchner et al., 2019). The definition of TRL is widely used to differentiate stages of research, development, or implementation (Buchner et al., 2018a).

3.2.1.2 Goal and Scope

Setting a goal will set the scope for the study. The goal addresses techno-economic questions, such as the cost or profitability of new technology, product, plant, or project, often for a specific audience.

The first phase describes the intent of the study, including the main questions, context, expected use, limits, and audience of the report. The aim is to classify all other facets of the TEA research. The analysts determine an initial goal at the start of practice, which may be amended or changed at any time during the trial with discretion.

First and foremost, all evaluations must be based on technologically plausible method principles. Before the assessment, a "plausibility check" needs to be conducted by the TEA practitioner (e.g. verifying that the proposed principle does not infringe the first or second laws of thermodynamics, examining mass and energy balances, etc.).

The aims of the TEAs shall be explicitly and unambiguously stated by following the values of the LCA.

- The background of research, specifically concerning location, time, scope, size, and affiliates.
- The anticipated deployment and the motivations for the study (e.g. decision support for R&D funding allocation, investment decisions or policy, and regulation; methodological studies).
- The targeted audience (e.g. policymakers, NGOs, funding agencies, investors, corporate management, journalists, R&D experts, and public).
- Research commissioners and publishers (e.g. funding organization, university, company, individual).
- Limitations on compatibility based on conclusions or strategies (e.g. specific use of product, time, location).

In an attempt to draw on the intent, analysts clarify what aspects of the product they will test and how they will compare it with possible alternatives in the context of the assessment. The significant activities in the scope process explain the intended operation of the product, the subject of the analysis (product system), and the degree to which it is connected to other systems (functional unit, FU), how much it is relative to other systems (reference flow), further defining the system (system elements), and specifying what is included and omitted from the evaluation (system boundary).

The analysts usually derive the parameters for inventory, calculation, and reporting processes based on the systems for comparisons of processes and market penetration for the said technology (Buchner et al., 2019).

3.2.1.3 Inventory

The second stage of TEA is that of inventory. The general approach to developing an inventory model encompasses five interlinked phases: (a) determining data quality specifications, (b) determining related technical processes, (c) gathering technical and (d) economic data, and (e) recording the data obtained (Figure 3.3) (Zimmermann et al., 2018). Based on the given scope and situation, cost estimates and market analysis are performed. These return intermediate outcomes are inputs to the economic impact equation known as "profitability analysis".

Cost evaluation approaches selected from the scope are strongly influenced by the overall objectives of the TEA and the status of the project under TRL design. The input data used for the cost calculation method(s) (Table 3.1) chosen are gathered in the cost inventory and analysed to calculate the reasonable

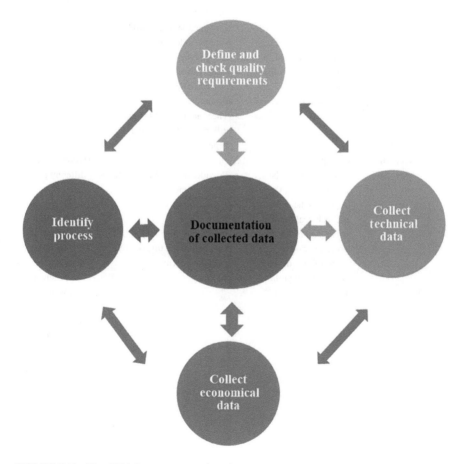

FIGURE 3.3 Five TEA inventory creation phases.

TABLE 3.1

TEA Methods as per Thermal Net Guidelines (Adapted from Lauer, 2008)

S. No.	TEA Methods	Principle	Interpretation and Use	Sensitivity Analysis
1	Static cost–benefit assessment	This is an average one-year cost–benefit comparison that does not consider the interest rate and the inflation rate.	*Advantages*: • Simple and easy to do. • Does not require any tools like computers or calculators. • It is helpful for pre-screening. *Disadvantages*: • Not much reliable.	A sensitivity analysis is not carried out because of the imprecise results of this procedure.
2	Annuity method	It is identical with the static cost–benefit assessment, although the interest rate is included in the annuity calculation. The decrease in profits earned or expenses incurred annually is generally not considered because it is static.	*Advantages*: • If the inflation rate is low and there is no interest rate difference, the annuity method is beneficial and realistic for simple TEA. • Comparable results with a simple, understandable methodology. • Widely used for preliminary project design. *Disadvantage*: • There are no year-to-year differences in costs and benefits.	The sensitivity of the outcomes can be accurately estimated by changing assessment parameters.
3	Net cash flow table	Annual cash flow calculated from the development of the project to the end of the technical life for every year of the project period in terms of the cost paid (in cash) and profit received (in cash).	*Advantages*: • Provides an excellent overview of the revenue/payment schedule over the project period. • It identifies how long it will take before the first positive cash flow occurs.	Hypotheses on different price and cost developments can be readily integrated. The impact of these parameters can be studied in a sensitivity analysis by modifying them in the net cash flow table.

(Continued)

TABLE 3.1 (*Continued*)

TEA Methods as per Thermal Net Guidelines (Adapted from Lauer, 2008)

S. No.	TEA Methods	Principle	Interpretation and Use	Sensitivity Analysis
4	Net present value	This is the most widely used technique for most professional analysts in the TEA. It helps determine whether a project has a positive or negative prospect.	*Advantages*: • The interest rate used for the discount is the minimal rate of return required. • The NPV takes into account the scale of the investment. • NPV calculation is simple (especially with spreadsheets). The NPV uses cash flows instead of net profits (which include non-monetary items such as depreciation). • The NPV uses cash flows instead of net profits (which include non-monetary items such as depreciation). *Disadvantages*: • It could be hard to determine the opportunity cost. This opportunity cost is particularly factored into the initial expenses. As a result, underestimating the original spending will distort the outcome.	The impact of input parameters can be studied in a sensitivity analysis by modifying them in the spreadsheet calculation.
5	Internal rate of return (IRR)	The internal rate of return is the average annual rate of return on the initial investment when all costs and benefits over a given period are factored in. It is calculated on the actual cash value and employs discounted cash flow techniques.	*Advantages*: • The internal rate of return (IRR) is a key metric for assessing the economic quality of a project.	By varying the input parameters of the cash flow model, sensitivity analysis can be readily performed.

cost. To estimate costs, the material and energy balances are calculated. The TEA inventory is augmented by capital investments, indirect operating expenditures, and respective market data (Buchner et al., 2018b). Cost calculation can be immediately preceded by interpretation if it is used as a tool for comparing procedure alternatives that do not have different business ramifications (Buchner et al., 2018a).

TEA accuracy is defined by the accuracy of available information about different unit operations and the economic aspects of the process. All inputs, conversion factors, operational conditions such as time, pressure, temperature, consumables, materials, by-products, and wastes given in each step of processing before the final products are obtained are included in operating parameter data.

Experimental testing, pilot experiments, chemical engineering, process modelling manuals, commercial facilities, and other technical detail sources may be used to collect technical information. Until the final products are collected, it includes all of the input quantities, conversion factors, operating conditions (such as time, temperature, and pressure), consumables, materials, by-products, and wastes created in each step of the method. Mass and energy balance analyses can be performed based on these operational outcomes, and the size and quantity of services and equipment needed can be determined. Data on economic analysis costs may be obtained from procurement and construction firms, materials suppliers, and chemical engineering guides, specifications, databases, experts, and literature. It includes the cost of necessary equipment and the costs of installation and repair, engineering and design charges, labour costs, utility costs, and material costs.

While collecting data from various sources, analysts can harmonize it, which means ensuring that the assumptions are accurate and consistent. The recording of inventory data provides the basis for any assessment, and efficient documentation results in practical analysis, troubleshooting, and interaction at the outset. It is recommended that a model or framework is created with various parts for the evaluation specification, assumptions (separate from base case and scenarios, for technical and economic assumptions), flow data (separate from process elements), measurements, and sensitivity tests, detailed individual estimates, and finally references. The business and comparative assessments would benefit from such technological and economic details, wherever relevant (Buchner et al., 2019).

3.2.1.4 Assumptions

Process Assumptions

A TEA requires certain assumptions to be made regarding the process model. The process model complexity is constrained by the availability of time and information resources. To carry out a general TEA, certain assumptions must be made. The most important assumptions are as follows:

1. *Steady state*: The most common assumption made during TEA is the assumption of the steady state. The process or processes that are being analysed are assumed to have reached a state where the process parameters do not change over time. If cyclic patterns are involved, these patterns are

assumed stable. This allows time to be taken out of the equation when analysing the process.

2. *Stable, reliable equipment operation*: For a TEA, all types of equipment are assumed to operate stably and reliably for the duration of a project. Effects of downtime due to equipment breakdown or maintenance are generally accounted for by the number of operating hours in a year. For reliable, well-run equipment, 8000 h per year are assumed. For less reliable equipment, this number can be lower.

3. *Constant, reliable supply of utilities and raw materials*: A TEA generally does not concern itself with vagaries of the logistics of supply of raw materials or utilities. Again, for reliable supply, 8000 operating hours a year can be considered. For less reliable supply, this number can be lower.

These process assumptions are generally valid for all but the most complex and detailed TEAs. The main reason for their validity is that, in most industrial operations, the process vagaries tend to balance out, especially in reasonably developed, peaceful, and stable economies with reasonably competent and innovative managers and operators. In sites prone to natural phenomena such as earthquakes or flooding, the TEA must consider the cost of earthquake-proof structures or flood barriers/flood pumps, etc.

Financial Assumptions

In addition to assumptions made on the technical side of the analysis, certain assumptions are essential on the economic side of the analysis. The most important assumptions are made in evaluating the capital and operating costs of the process.

Capital costs: Capital costs are notoriously uncertain and can be a strong function of the negotiating power and finesse of the purchasing entity. Most fabricators do not disclose even a ballpark figure for their equipment costs. In addition to equipment costs, there are factors such as the costs of civil work, piping, electric work, controls which are themselves subject to a similar degree of variability and secrecy. Further, the costs of the commodities such as sheet metal, copper, concrete that go into an installation are themselves varying. The most common method of estimating capital costs of equipment is extrapolating from previous equipment whose cost is known. The extrapolation is done using indices such as the Chemical Engineering Index published in the Chemical Engineering Journal. This usually suffices for a preliminary TEA. For a TEA on which a major decision hangs, it is considered safer and more appropriate to develop a relationship with vendors and fabricators to get reliable prices for their products. Budgetary quotes can be invited from various parties for a more definitive equipment cost estimate. Books such as that by Peters and Timmerhaus (ISBN: 978-1259002113) specify costs of various ancillary equipment and services including civil and electrical work, piping, instrumentation and controls, engineering contractor fees (which can be a substantial part of the capital investment), contingency as a multiple of the equipment cost for a given type of installation. These are excellent for a preliminary TEA, but more direct information from various vendors is necessary for a budgetary estimate.

Operating Costs: Raw material and utility costs are somewhat more easily and reliably estimated than the capital costs. They depend directly on the price of various commodities, and very often, for high-value commodities or commodities bought in bulk, transportation is a small part of their on-site costs. The sensitivity of the cost of various raw materials and utilities can be calculated rather straightforwardly from a material and energy balance over the process. Very often, the most likely or the average price over a period of time is chosen for calculating operating costs. If for some reason, the supply of a particular commodity is unreliable at a given site, its procurement cost is set to a higher value. The labour cost is set a multiple of the raw material costs and is generally low (less than 15%) for a chemical plant. However, for budgetary estimates, a detailed analysis of the labour requirement of the plant must be made. This includes not only the cost of the floor operators and workers (who operate in shifts), but also the laboratory, R&D, clerical, accounting, security, and managerial staff including plant managers and supervisors.

Miscellaneous

An assumption made during TEA, especially for products like commodity materials with varying prices, is that no arbitrage occurs. The process is evaluated strictly on its ability to convert raw materials into products cost-effectively and not on whether storage facilities could be used to facilitate the trading of a commodity.

Another assumption during a TEA is that there are no technical, economic, or political disruptions to the functioning of a process.

A further important assumption is related to regulatory oversight of the installation. Compliance can be a costly affair involving lawyers, liaison officers, safety inspectors, financial auditors, etc., and this cost is generally taken as part of the operating costs of a plant.

Interpretations

Interpretation is conducted in accordance with all TEA phases to check the performance, precision, completeness, and reliability of the inventory data (model inputs) and associated intermediate or outcomes (model outputs) regarding the purpose and scope of the study. The main tasks during the interpretation process are the measurement of ambiguity and sensitivity, the evaluation of outcomes, and the development of a multi-criteria decision study.

The interpretation process yields perspectives and constraints that serve as the foundation for decisions and recommendations for future research, development, and implementation. Analysis of uncertainty and sensitivity professionals are now using complexity and sensitivity analysis. To assess the uncertainty and sensitivity of the assessed variables to various inputs, use the following method:

1. Characterization of uncertainty
2. Uncertainty
3. Sensitivity
4. Iterative data quality improvement.

Primarily, the analysts need to define uncertainty. These can exist in the input, model, and scope. Significant input data variability can result from calculation errors, variable probability distributions, or low-accuracy assessments.

Model structure and process uncertainty can result from shortcomings in the model's ability to accurately represent the observed system. Context ambiguity may emerge as a result of methodological choices made during the goal and scope processes (Saltelli, 2002). Second, analysts do an uncertainty analysis on the model input, the model itself, or the context of the model outputs. Through analysing all sources of uncertainty and validating whether the model's output matches the underlying assumptions, the analysis of uncertainty becomes a consistency metric. If model outputs or results are found to be especially vulnerable to such model input values, then additional efforts may be required to evaluate certain values with greater certainty.

3.2.2 CASE STUDIES

Eggeman and colleagues (2005) published a TEA of corn stover pre-treatment technologies. In this paper, they evaluated and compared five different pre-treatment processes (dilute acid, hot water, ammonia fibre explosion (AFEX), ammonia recycle percolation (ARP), and lime) with no pre-treatment and ideal pre-treatment conditions. The Aspen Plus 10 commercial simulator is used to simulate the related processes. Results show that no pre-treatment scenario has a meagre return, which results in a very high average fixed capital per annual gallon of capacity. In all actual pre-treatment scenarios, return and capital requirements for each gallon of capacity were observed to be higher than those for no pre-treatment. All cases, including the optimal pre-treatment cases, were observed to be capital-intensive.

Sassner et al. (2008) investigated the TEA of ethanol production from three feedstock companies: Salix, corn stover, and spruce. In this study, the total cost and energy required for bioethanol production from different feedstocks were compared using a process concept based on SO_2-catalysed steam pre-treatment accompanied by simultaneous saccharification and fermentation (SSF). They simulated a scenario considering SSF, in which 90% of the xylose and arabinose in SSF are converted to ethanol, to investigate the effect of pentose sugar fermentation. In contrast to the base cases, this raised ethanol yields by 32% for Salix, 42% for corn stover, and 8% for spruce. They concluded that for two reasons, the pentose fermentation process necessitated higher heat duties.

Klein-Marcuschamer et al. (2011) studied the cost of enzyme production and how it affected biofuel production economics. They note the general lack of available information detailing the costs of enzymes and their production. This lack of information contributes to the difficulty of studying enzyme production costs. They also note that often in techno-economic studies, enzyme cost contributions to total ethanol production costs are reported per gallon of fuel produced. The per-gallon metric is inherently dependent upon other parts of the ethanol production process besides enzymes (feedstocks, enzyme loading, overall biofuel yield, etc.).

Emanuel (2017) presented the TEA of fed-batch enzymatic hydrolysis using different feedstocks in her thesis. The techno-economics of the batch enzymatic and fed-batch enzymatic processes were compared. Two models were developed: Feedstock Cost Estimation Model (FCEM) and Bioethanol Plant-Gate Price Assessment Model (BPAM), to make the TEA. Sensitivity analyses showed that ethanol processing costs are most sensitive to the costs of corn biomass (raw material prices).

TEA of bioethanol production using lignocellulosic biomass, i.e. corn stover, was performed by Lili Zhao and colleagues (2015). Working with two models, namely BPAM and FCEM, they concluded that the biochemical properties have a strong correlation. The BPAM identifies the variables that impact the rate of cellulose conversion, xylose to ethanol conversion, feedstock cost, and enzyme loading The consumption taxes, value-added tax (VAT), and a feed-in tariff for excess electricity (by-product) are recommended for reduction or removal to facilitate technology.

Technology analysis of cellulosic ethanol production was conducted by Kazi et al. (2010) to understand the ethanol production processes. A matrix was developed after reviewing 35 technologies to capture all aspects related to large-scale processing and implementation. The cost growth analysis was performed using nth plant technology (mature technology). The products of interest were ethanol and butanol. Only ethanol technologies were adopted for the analysis as butanol is at an early stage of technology development. Seven scenarios were considered for the study based on pre-treatment type, on-site enzyme production, parallel fermentation, and pervaporation. The primary assumption for the process design is that it will operate as an nth plant, with reported yields based on experimental data. The inferences from the study were that the enzyme and the feedstock costs are the major concerning factors.

Park et al. evaluated the techno-economic feasibility of a bio-refinery based on Miscanthus using a combination of auto-hydrolysis and pre-treatment followed by enzymatic hydrolysis and further processing of hemicellulose and lignin. The two major by-products studied are hemicellulose and lignin to produce xylitol and polyol, respectively. The impact of value-added generation on the bio-refinery has been studied for energy demand capital, operating costs, and waste generation. A sensitivity analysis was also performed to determine the significant parameters with the greatest impact on economic performance. Using Miscanthus, the study assessed the processing of 1500 metric tonne of biomass per day using the process models developed in Aspen Plus under three different case scenarios. The analysis of capital and operating costs in the financial and general process assumptions indicates that xylitol and polyol production incurs substantial capital investments. The yields and recoveries of the by-products and the recycling of expensive chemicals have strong economic impacts.

Swanson et al. (2010) compared the capital and processing costs of two biomass-to-liquid production plants based on gasification in this report. The aim was to compare two gasification-based refinery scenarios. For the nth plant, high-temperature slagging gasification and low-temperature dry ash gasification are supplemented by Fischer–Tropsch synthesis and hydro-processing to reduce capital expenditure and production costs. In this situation, energy is treated as a process by-product. They used Aspen Plus tools to simulate the process and determine mass and energy balances and equipment costs. They conclude that using current technologies, a biomass-to-liquid plant will produce fuels between $4 and $5 per gallon of gasoline-equivalent. Feedstock costs and the investment return on capital, as well as compressor construction constraints, were considered as the key suspect. The PV of the process was found to have minimal effect by carbon monoxide conversion in the Fischer–Tropsch reactor and feedstock inlet moisture and catalyst lifespan.

Daniel Klein-Marcuschamer et al. (2011) presented the techno-economic model of a bio-refinery, which is based on ionic liquid (IL) pre-treatment technologies. They have stated that the relatively high cost of ILs is the single core issue that stands in the way of commercialization considering its other identified advantages over other pre-treatment technologies. They demonstrated a techno-economic model of a bio-refinery focused on IL pre-treatment technologies. Via extensive sensitivity study, they identified the most important areas in terms of cost reduction and revenue creation, such as lowering IL costs and loading and increasing IL recycling.

Mahsa Dehghanzad et al. (2020) presented the techno-economic model of a bio-refinery, which is based on the whole sweet sorghum plant for biobutanol production. They have compared the economics of six different scenarios such as independent hydrolysis (scenario 1), fermentation (scenario 2), simultaneous saccharification and fermentation (SSF) (scenario 3), co-fermentation of the juice and pre-treated stalk (scenario 4), stalk directly fed to the fermenters without juice extraction (scenario 5), and stalk directly fed to the fermenters without pre-treatment (scenario 6). They used the commercial simulator Aspen Plus to simulate the various processes and Process Economic Analyzer to investigate economic viability. For scenarios 1–6, the butanol output price was 0.62, 0.44, 0.45, 0.61, 0.39, and 0.56 US$/L, respectively. They found that scenario 5 is the most profitable, requiring US$47.75 million in total capital to generate 11,260 tonnes of butanol per year.

3.2.3 Challenges and Research Gaps

The main challenge in doing a TEA is that there are no generalized TEA guidelines available. There is an increasing tendency towards more agile modelling in TEA, which can easily iterate over a wider solution area and helps non-experts create simple models (Vlachokostas et al., 2021). Different groups of experts, such as researchers and corporations, used different approaches to execute TEA. Carrying out TEA using well-known software such as SuperPro Designer® and Aspen Plus necessarily requires significant amounts of technical information that is usually inaccessible during the early assessment of the process. Mothi Bharath Viswanathan et al. developed a spreadsheet-based process modelling and techno-economic framework specifically to aid in combined fermentative-catalytic biorefinery processes. . Early stage models are built on assumptions that are usually used when data are insufficient. In these cases, two methods are used: Bio-PET and Bio-STEAM. Bio-PET has been validated with SuperPro Designer® and published literature, a medium for commercial economic research. Bio-PET holds a crucial niche in the developing bio-based chemical industry. This is because low-cost instruments for performing early stage economic analysis are required for the new bioprocess systems. Such an instrument will offer exciting insights into emerging projects (Lynch 2020).

Inadequate analytical techniques, inefficiencies in the usage of information, simplified models with assumptions, insufficient knowledge, and variation in signal and results are problematic areas of techno-economic research. In the development of biomass processing ventures, a techno-economic model is an invaluable tool.

Techno-economic models are flexible and reliable tools that act as a shared language across technological and financial sectors, enabling more informed project definition and more integrated process decisions.

When a process is scaled up or is scaled up to a commercial stage, techno-economic challenges fall into the frame. Choosing the best-optimized parameters considered for cost economics is the major problem to be kept in mind. Also, equipment used in bioprocessing is included in the estimation. Intensive biomass transport and storage costs, potential market power problems, and local environmental effects are involved in industrial-scale bio-refineries. A solution may be given by alternate means of the use of this cumbersome biomass.

When doing a techno-economic study, one of the most difficult aspects of chemical and biochemical process design is cost estimation. During each stage of the project design, cost estimates are found to be different, such as idea screening, preliminary research, expenditure authorizing, budget management, and construction, as the consistency and quantity of data available during successive phases of the project cycle vary (Cheali et al., 2015).

3.3 LIFE CYCLE ASSESSMENT

In this section, the methodology of LCA is first briefly covered. Thereafter, some of the important studies of LCA of biofuels are reviewed, and the novel insights obtained from those studies are highlighted. Finally, the section discusses some challenges and research needs in the area of LCA of biofuels.

3.3.1 LCA: METHODOLOGY

LCA is still a developing tool with roots in the 1960s when research started on energy requirements. In the 1970s, research in pollution prevention started (Finkbeiner et al., 2006; Finnveden et al., 2009; ISO, 2006a; 2006b). LCA compiles all the emissions and consumption of resources at every stage of the life of the product (Matthews et al., 2020). If the scope of the LCA limits to only a static analysis, it is called attributional LCA. Indirect changes are also considered in what is called the consequential LCA (Rebitzer et al., 2004).

The basic idea in LCA is to consider the complete life cycle of a product, starting from material extraction to waste generation, as shown in Figure 3.4. The resources used as well as emissions and wastes produced at each stage of the life cycle are carefully quantified and summed. This gives the total life cycle impact of the product of interest. It must be noted here that each of the boxes shown in Figure 3.1 could consist of several steps.

A detailed LCA methodology framework has been given by ISO 14040:2006 series, as shown in Figure 3.5 (ISO, 2006a; 2006b). LCA is divided into four stages: goal and scope definition, life cycle inventory, impact assessment, and interpretation. In the context of biofuels, these four stages are described below.

3.3.1.1 Goal and Scope Definition

The goal and scope definition is the first stage of performing an LCA study. Although it is completely qualitative in nature, it sets the foundation of the whole study. The

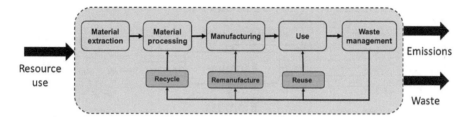

FIGURE 3.4 Concept of life cycle assessment.

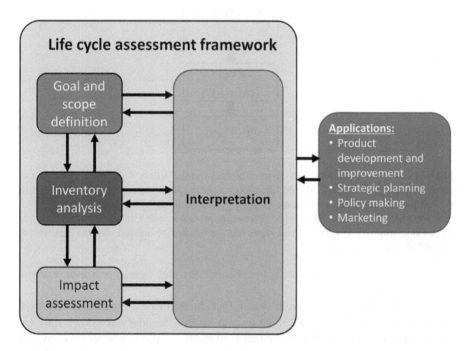

FIGURE 3.5 Life-cycle analysis framework. (Adapted from ISO, 1400.)

goal statement addresses the reason for analysis, intended application, and the audience affected by the system (Matthews et al., 2020). For biofuels, the typical goal is to quantify the total emissions during the production of the biofuel so as to compare it with conventional fossil fuels. This information can be used to make policy recommendations. The goal can also be to identify the major steps responsible for the emissions and address those through process improvement.

ISO requires the addition of 14 important requirements to the scope, two of which are system boundary and functional unit. The system boundary is a very critical aspect of the study and decides its comprehensiveness. The most comprehensive study is performed in a cradle-to-grave manner, as shown in Figure 3.4. Here, as the term suggests, all the steps starting from material extraction and processing to waste disposal are considered. However, depending on the requirement of the study

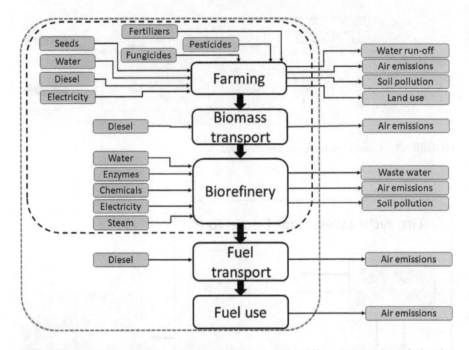

FIGURE 3.6 A typical product system of biofuel and the different system boundaries that can be considered. The black dotted line captures the cradle-to-gate system boundary, while the red dotted line captures the cradle-to-grave system boundary. The blue boxes show inputs that are generally coming from other production processes, but must be considered in the LCA calculations. The yellow boxes show the various environmental impacts that will be quantified by an LCA study.

as well as data availability, other system boundaries such as cradle-to-gate may also be considered.

Figure 3.6 shows two different system boundaries that are commonly employed while studying biofuels. The cradle-to-gate system boundary considers only the stages until the gate of the bio-refinery. In contrast, the cradle-to-grave system boundary considers the fuel transport and fuel use (combustion) stage well. The cradle-to-gate studies are more common in the literature. This is because the use phase emissions are generally known with certainty based on the combustion pro-file. For example, the use phase emissions for ethanol are known based on stoichi-ometry, and the source of the ethanol does not matter. Moreover, for most biofuels, the cradle-to-gate scope has been shown to make the greatest contribution to the overall life cycle impacts.

Once the system boundary is finalized, the detailed product system is developed. The product system is the systematic arrangement of all the steps involved in the complete life cycle. These individual steps are referred to as unit processes in the LCA literature. Figure 3.6 shows the product system for different system boundar-ies. For the sake of illustration, the product system shown here is not comprehensive. In reality, every box in Figure 3.6 will consist of multiple steps. For example, the bio-refinery stage itself will consist of multiple unit processes such as size reduction, pre-treatment, detoxification, hydrolysis, fermentation, separation, and purification. The bio-refinery stage will also include operations such as a boiler for steam generation and wastewater treatment.

The final important part of the goal and scope definition stage is the functional unit (FU). The FU is the basis of quantification of the life cycle impacts and the comparison with other similar products. The FU is decided based on the service provided by the particular product. For LCA of biofuels, primarily two FUs have been used. The first FU commonly used is 1 litre of fuel, such as ethanol. If this FU is used, the life cycle impacts are quantified for the production of 1 litre of fuel. This FU is easier to understand since the fuel is often measured in terms of volume. However, the FU of 1 litre is not correct if the impacts are to be compared with different fuels such as petrol. This is because the energy density of different fuels is different. For example, 1 litre of ethanol cannot be compared with 1 litre of petrol since the energy content of ethanol is less. The service provided by fuel is the energy provision. Therefore, the FU should also be in terms of energy. Consequently, if the objective is to compare LCA results for two or more energy products, a FU such as 1 MJ of energy is more appropriate. In the literature on LCA of biofuels, both FUs have been reported.

3.3.1.2 Life Cycle Inventory

The second stage of LCA is that of life cycle inventory. This stage focuses on performing detailed calculations to quantify the total emissions and resources used. These input–output data are called inventory data. Conceptually, this step is straightforward since it involves performing mass balance for every unit process in the product system and integrating the individual unit process data to get the total life cycle values. However, there are two main operational challenges associated with this stage.

The first challenge is about the availability of the inventory data. For performing inventory calculations, we need production-related data for every unit process that is part of the product system. Generally, the data pertaining to the main process are available easily. For example, the process data of a bio-refinery process are available with the bio-refinery. In fact, it is desirable to use the process data from experimental work from pilot or demonstration-scale studies so that the LCA study will be reliable. However, the product system also includes unit processes that are not within the boundary of the plant. For example, the biochemical processing of biomass to produce fuel involves the step of enzymatic hydrolysis, which requires cellulase enzymes. The enzyme is generally purchased as a cocktail from another vendor and used in the process. The same situation is seen in most other chemical inputs such as acids, alkali, and so on. From a life cycle perspective, the inventory data for all these inputs must be included. However, getting access to these data is not easy. There are three options to get the required inventory data:

- *Primary data collection*: This is the ideal method since it gives accurate data for all the inputs and captures the uniqueness of the specific facility that is providing the material. However, this approach is generally not feasible due to the effort required.

- *Secondary data collection*: This method relies on published data in the scientific literature, technical reports, or industry studies. This is acceptable, but generally, such studies are limited.
- *Databases*: The most popular method to access inventory data is using databases. These databases can be free (e.g. GREET by Argonne National Laboratory and databases by the USDA or NREL in the United States) or can be commercial that need to be purchased (e.g. Ecoinvent®, GaBi®, SimaPro®).

Most of the LCA studies performed for biofuels use a combination of primary data for the main process complemented by databases for inputs such as chemicals, electricity, and fuel.

The second challenge of the inventory step is related to computation. The inventory calculations are performed using a matrix approach, which involves performing matrix inversion operations (Heijungs and Suh, 2002; Matthews et al., 2020). If the product system is very large, this can be computationally challenging and may require the user to modify the system boundary.

Another key issue at this stage is accounting for multiple products from the same product system. If that is the case, then it is not correct to consider all the impacts because of one product, and the impacts need to be divided or allocated. This is the step of impact allocation. Multiple approaches of impact allocation have been proposed. These include mass-based allocation, energy-based allocation, and economic (price-based) allocation. It is not possible to say that one is necessarily better than the other, and LCA analysts often rely on case-specific details before deciding the approach.

Figure 3.5 shows that the goal and scope definition and life cycle inventory stages are interdependent. That is because oftentimes, the availability of data may force the LCA analysts to modify the system boundary.

3.3.1.3   Impact Assessment

The third step in an LCA study is that of impact assessment. The life cycle inventory stage provides data on cumulative emissions and resource use for the product's entire life cycle. However, the data need to be converted to impact categories so that appropriate conclusions can be drawn. This allows the analysts to understand the impact on categories relating to the ecosystem, human health, and resource consumption. For this, impact assessment models have been developed. The impact assessment models convert the inventory data into selected impact categories. The popular impact categories are climate change, freshwater eutrophication, water depletion, ozone layer depletion, and particulate matter formation. However, there are more impact categories relating to human health impacts as well. Different models available for doing impact assessment calculations include ReCiPe, CMP, Eco-indicator-99, and ILCD. The impact assessment stage further includes the option of calculating mid-point or end-point indicators. The impact categories mentioned above are mid-point indicators. However, sometimes the impacts need to be further condensed into fewer categories. This is done by calculating the end-point indicators, which are three, namely ecosystem health, human health, and resource depletion. Most LCA studies of biofuels perform mid-point indicator calculations and focus mainly on the climate change and water depletion impact categories.

3.3.1.4 Interpretation

The fourth step of interpretation is not distinct as such. However, this step indicates that the whole process of performing LCA is iterative in nature. Additionally, the final results that are obtained after impact assessment need to be carefully interpreted. This may involve identifying the emissions hot spot or benchmarking the final impact against an alternative. For biofuels, the benchmarking is generally done with a fossil-based alternative such as petrol or diesel.

3.3.2 CASE STUDIES

Several life-cycle analysis (LCA) studies of lignocellulosic biofuels have been undertaken in the last two decades (Silalertruksa and Gheewala, 2011; Morales et al., 2015). It is not the intention of the authors to review those studies comprehensively here. Borrion et al. (2012) extensively reviewed 53 LCA studies on lignocellulosic biofuels until 2012. Their overall conclusion was that, while most studies predicted that biofuels would have improved (reduced) life cycle impacts, the degree of improvement varied significantly. Singh et al. (2010) also reviewed the literature on LCA and identified the key challenges. Scown et al. (2012) quantified the life cycle of greenhouse gas emissions if the mandated biofuel production in the United States was to be met with ethanol produced from Miscanthus x giganteus. Six different scenarios with different possible ways to allocate the required land were modelled. The overall conclusion was that the GHG intensity of ethanol was about 80%–90% lower than that of gasoline. Wang et al. (2007) developed the GREET (Greenhouse gases, Regulated Emissions, and Energy use in Transportation) model to perform life cycle impact assessment for different fuel and vehicle technologies. The focus of this model is primarily on the transportation sector. The comprehensive model has been used to study biofuels derived from switchgrass by using three different processing options (Wu et al., 2006). They calculated a net GHG impact reduction of 82%–87% compared to the base case and recommended integrating heat and power co-generation via a gas turbine combined cycle. Many more studies have been reported in the literature. However, here we have reviewed studies done in the Indian context. This is important since LCA studies need to be done on a regional basis.

Kadam (2000) used an life-cycle analysisLCA to find out the benefits of using ethanol derived from sugarcane bagasse as a gasoline blend in the Indian context. (The study was done in a mill in Maharashtra.) This is perhaps the first LCA study in the Indian context. The study analysed two scenarios where the excess bagasse in the sugar mill could either be used for burning in an open field or be used to produce ethanol and replace some gasoline through 10% blending. Further, two different processing routes for ethanol production were considered. The system boundary chosen for the study included gasoline production, ethanol production, and the current bagasse disposal method. It should be noted that sugarcane production and the sugar mill were left out of the system boundary. The FU used in the study was 1 dry tonne of bagasse disposed of. Using this FU was justified because the study aimed to find out an effective disposal strategy for excess bagasse. The study revealed that there was a significant difference in the two scenarios as far as the energy derived and the extents of emissions are concerned. Net emissions of hydrocarbons, NO_x, CO, particulates, SO_x, and fossil fuel energy consumption were lower in scenario 2. However, COD levels were higher for the second scenario because of the usage of ammonia in ethanol fermentation.

Mandade et al. (2015) performed an LCA of ethanol production from different feedstocks but with the same conversion process. Cradle-to-gate system boundary was considered, and the FU was 1 litre of ethanol produced. Database for the inventory included journal articles, reports by the Ministry of Agriculture, and also primary data by direct communication with farmers. This is one of the unique features of this study, as the farming data are based on personal communication. Three allocation methods were considered, namely mass-based, energy-based, and economy-based (market price-based). Four parameters have been considered for comparison. They were energy return on investment (EROI), life cycle GHG emissions, life cycle water use, and land use. The study showed that ethanol production from sorghum had the smallest impact among the four parameters, while rice husk, wheat stalk, and cotton stalk also showed potential. Even though bagasse showed good potential, the authors concluded that due to the existing usage of bagasse as fuel (combustion), conversion to ethanol may not be done.

An life-cycle analysis (LCA) on the production of ethanol from sugarcane molasses was done by Soam et al. (2015) in the northern and western parts of India. Two parameters were calculated to compare ethanol production in the two regions – % GHG emission reduction with respect to gasoline and net energy ratio (NER) which is a measure of sustainability or renewability of the biofuel. System boundary chosen included sugarcane farming, transport, sugar production, ethanol production, ethanol transport blending, and combustion in automobiles (cradle-to-grave). The FU used for analysis was 1 tonne of ethanol. The study concluded that sugarcane farming is the most variable factor. When no allocation, ethanol was more polluting than gasoline – 8146.5 kg CO_2-eq in the northern region and 7349 kg CO_2-eq in the western region for the production of 1 tonne of ethanol against the 3602.65 kg CO_2-eq for the production of 1 tonne of gasoline – and the net energy ratio was very low – 0.38 in the northern region and 0.48 in the western region against the NER value of 0.8 for gasoline. However, due to the high GHG emission reduction and NER values, the mass-based allocation and energy-based allocation established the sustainability of ethanol. It was also observed that in the western region, the % GHG emission reduction was lower and the NER was higher as compared to the northern region.

Murali and Shastri (2019) performed a detailed comparison of different processing routes for the production of ethanol from sugarcane bagasse. Their objective was to identify the right processing route using the LCA results. Cradle-to-gate system boundary was considered, and the FU was 1 MJ of energy from ethanol. The process data were collected from the published literature, while utility-related data (e.g. steam and electricity requirement) were estimated using processing engineering calculations. The inventory data related to chemical inputs were taken from the Ecoinvent® database. Their results showed that dilute acid pre-treatment followed by simultaneous saccharification and co-fermentation (SSCF) was the most desirable processing route. It had a lower impact in most impact categories. They also argued that in addition to climate change, other impact categories related to the bio-refinery stage were also important.

Recently, Sreekumar et al. (2020) have conducted a detailed LCA of a novel fractionation-based process developed by the DBT-ICT Bioenergy Centre. It is a feedstock agnostic process that has been proposed to be scaled up for a $100\,m^3$ per day ethanol plant. The unique aspect of this study was that the process data were taken from a demonstration-scale plant, and therefore, had accounted for scale-up issues. The system boundary was cradle-to-gate, and the FU was 1 litre of ethanol. The inventory data for inputs were obtained from the Ecoinvent® database. For a rice straw-based bio-refinery, the results showed that the life cycle climate change impact was $2.82\,kg$ CO_2-eq per litre of ethanol. Of the impact, 86% was due to electricity usage. Consequently, the use of renewable electricity such as hydro could reduce the impact to $0.4\,kg$ of CO_2-eq. This study also highlighted the benefit of having an integrated bio-refinery producing other products such as food-grade CO_2 and methanol, and the benefit of avoided impact such as burning of rice straw. Calculations showed that there could be a net positive benefit due to ethanol production.

Other studies have also been reported in the literature. Achten et al. (2010) focused on Jatropha-based biodiesel production in rural India. They concluded that 82% reduction in non-renewable energy requirement could be achieved by using Jatropha biodiesel, leading to a 55% reduction in GWP. But they also observed an increase in the acidification and eutrophication potential due to Jatropha cultivation. Soam et al. (2016) also compared ethanol production using dilute acid pre-treatment and steam explosion pre-treatment. One tonne of rice straw is considered as the FU. The life cycle GHG emissions ranged between 288 and $292\,kg$ CO_2-eq per tonne of rice straw, and the NER varied between 2.3 and 2.7. Soam et al. (2017) compared different rice straw utilization options using LCA.

3.3.3 CHALLENGES AND RESEARCH GAPS

Although several studies on the life-cycle analysisLCA of biofuels have been reported in the literature, there are still several challenges that need to be addressed (Ahlgren et al., 2015; Shastri, 2017). In this section, some of these challenges are briefly summarized.

- Spatari et al. (2010) identified some limitations of conventional LCA studies for biochemical ethanol production. They claimed that the literature only considered limited pre-treatment options under the assumption of mature technology. They have addressed the problem by considering near-term (c. 2010) and mid-term (c. 2020) technology cases. Importantly, they have also incorporated uncertainties in technology performance through Monte Carlo simulations. The results showed wide variations in the life cycle impacts for different processes. More importantly, their results emphasized that considering the performance of the developing technology instead of mature technology, as normally done, led to substantial differences in the results. This highlighted the value of stochastic analysis.

- Wiloso et al. (2012) observed that a limited number of processing options have been considered. They argued that most studies consider SSF, which is an efficient process. Therefore, these studies underestimated the real impact of processes operating at the commercial scale. This goes back to the performance uncertainty aspect raised by Spatari et al. (2010). There have also been attempts to incorporate the LCA method in process design and optimization steps to come up with a sustainable design of biofuel processes (Gerber et al., 2011; Scott et al., 2013).
- The benefits of biofuels can be case-specific and depend on several assumptions. A surprising diversity in conclusions regarding the true environmental benefits was observed, with some studies arguing that the benefits were possible only under specific circumstances. This was especially true for categories other than greenhouse gas emissions. One of the reasons for this variability is the difference in the system boundaries, allocation methods, impact categories, and data considered in different studies. Borrion et al. (2012) also identified data quality as an issue. Software tools such as Ecoinvent® (http://www.ecoinvent.ch/) have been developed to provide some consistency in data. However, the data provided by these tools must still be carefully customized on a case-specific basis.
- One of the major challenges of implementing life-cycle analysisLCA to biofuels, according to McKone et al. (2011), is the incorporation of spatial heterogeneity in inventories and assessments. This necessitates the regionalization of LCA methodologies as well as inventories.

There are some additional challenges associated with the LCA of biofuels. Typically, attributional LCA based on the process and production data is performed. However, Searchinger et al. (2008) argued that such approaches do not provide the true picture. They stated that the corn-based biofuel production in the United States was leading to changes in the land use pattern elsewhere. Since the change in land use was from forest land to agricultural land, they argued that the resulting carbon footprint is much higher than that reported using a simple attributional approach. The conflicting results are mainly due to the different system boundaries considered by Searchinger et al. (2008), and the point continues to be debated in the literature (Matthews and Tan, 2009). It is, however, now well acknowledged in populous countries such as India that dedicated energy crop cultivation is not sustainable.

There are also conflicting opinions about the use of allocation and the correct method of allocation. In India, the agricultural residue will be used as feedstock. Since agriculture is done mainly for food production, it can be argued that whether any impacts during farming are considered for biofuel LCA. However, the opposing view is that in a scaled-up biofuel sector, the residue will become a commodity generating revenue. Hence, it is appropriate to perform impact allocation. Finally, the challenge regarding the availability of regional inventories continues to persist. Studies done in the Indian context have often used inventory data for other regions due to the lack of data availability. This makes the results less reliable. Therefore, efforts need to be continuously made to compile accurate inventories.

3.4 CONCLUSIONS

This chapter provides a perspective on TEA and LCA in the context of lignocellulosic biofuels. The importance of doing both TEA and LCA is first highlighted by emphasizing the need to ensure the sustainability of the novel processes and technologies. Subsequently, the methodologies are briefly reviewed, and key literature in the area is summarized.

Experts use TEA to figure out how key technical and economic factors affect the overall process and its resultant cost. Process modelling and engineering design are combined with economic evaluation in the TEA. It aids in the evaluation of a process's economic viability and directs science, growth, funding, and policymaking. Many private industry and R&D centres use the stage gate analysis process for the development phase, and TEA blends well with it. LCA has emerged as a powerful tool for rigorous sustainability assessment of biofuels and is being increasingly used. Most of the initial studies were done in the North American and European context, but studies from the Asian region are also being increasingly reported. The studies have shown that biofuels have definite benefits from a climate change perspective. However, a high water footprint has emerged as an area of concern, particularly for regions such as India, where water is already a problem. A literature review has also indicated that there is a lot of scopes to perform region-specific LCA studies so that more reliable conclusions can be drawn. Moreover, issues such as indirect land use change, impact allocation, and development of regional inventory are still not completely resolved. The LCA can also be complemented by a social LCA which accounts for social benefits such as job creation. This is important because biofuels are often promoted not only for their environmental benefits but also as an opportunity to uplift the rural and agrarian economy. The impact of biofuels on the food–energy–water nexus also needs to be addressed.

Finally, it is important to not conduct TEA and LCA studies as a stand-alone exercise after the process has been developed. Early TEA and LCA studies can provide valuable feedback to process development teams so that undesirable options can be ruled out at an early stage. Greater efforts need to be made to ensure that this vision is implemented and becomes a common practice.

REFERENCES

Achten, W. M. J., Almeida, J., Fobelets, V., et al. 2010. Life cycle assessment of Jatropha biodiesel as transportation fuel in rural India. *Applied Energy* 87, 3652–3660

Ahlgren, S., Björklund, A., Ekman, A., et al. 2015. Review of methodological choices in LCA of biorefinery systems—key issues and recommendations. *Biofuels Bioproducts and Biorefining* 9, 606–619.

Borrion, A. L., McManus, M. C., and Hammond, G. P. 2012. Environmental life cycle assessment of lignocellulosic conversion to ethanol: A review. *Renewable and Sustainable Energy Reviews* 16(7), 4638–4650.

Brundtland, G. 1987. Report of the World Commission on Environment and Development. *Our Common Future*. Oxford: Oxford University Press.

Buchner, G. A., Schomäcker, R., Meys, R., and Bardow, A. 2018b. Guiding innovation with integrated life-cycle assessment (LCA) and techno-economic assessment (TEA)-the case of CO_2-containing polyurethane elastomers. *Climate-KIC Technical Report* 12, 2018.

Buchner, G. A., Stepputat, K. J., Zimmermann, A. W., and Schomäcker, R. 2019. Specifying technology readiness levels for the chemical industry. *Industrial & Engineering Chemistry Research* 58(17), 6957–6969.

Buchner, G. A., Zimmermann, A. W., Hohgräve, A. E., and Schomäcker, R. 2018a. Techno-economic assessment framework for the chemical industry—based on technology readiness levels. *Industrial & Engineering Chemistry Research* 57(25), 8502–8517.

Cheali, P., Gernaey, K. V., and Sin, G. 2015. Uncertainties in early-stage capital cost estimation of process design–a case study on biorefinery design. *Frontiers in Energy Research* 3, 3.

Dehghanzad, M., Shafiei, M., and Karimi, K. 2020. Whole sweet sorghum plant as a promising feedstock for biobutanol production via biorefinery approaches: Techno-economic analysis. *Renewable Energy* 158, 332–342.

Eggeman, T., and Elander, R. T. 2005. Process and economic analysis of pre-treatment technologies. *Bioresource Technology* 96(18), 2019–2025.

Emanuel, E. (2017). Techno-economic implications of fed-batch enzymatic hydrolysis. MSc diss., Univ. of Nebraska-Lincoln.

Fatma, S., Amir, H., Muhammad, N., et al. 2018. Lignocellulosic biomass: A sustainable bioenergy source for future. *Protein & Peptide Letters* 25, 148–163.

Finkbeiner, M., Inaba, A., Tan, R., Christiansen, K., and Klüppel, H. J. 2006. The new international standards for life cycle assessment: ISO 14040 and ISO 14044. *The International Journal of Life Cycle Assessment* 11(2), 80–85.

Finnveden, G., Hauschild, M. Z., Ekvall, T., et al. 2009. Recent developments in life cycle assessment. *Journal of Environmental Management* 91(1), 1–21.

Gerber, L., Gassner, M., and Maréchal, F. 2011. Systematic integration of LCA in process systems design: Application to combined fuel and electricity production from lignocellulosic biomass. *Computers & Chemical Engineering* 35(7), 1265–1280.

Gholkar, P., and Shastri, Y. 2020. Renewable hydrogen and methane production from microalgae: Techno-economic and life cycle assessment study. *Journal of Cleaner Production* 279, 123726.

Heijungs, R., and Suh, S. 2002. The computational structure of life cycle assessment. Volume 11, *Eco-efficiency in Industry and Science*. Tukker A (Series Ed.), Springer Science+Business Media, B.V.

ISO. (2006a). *ISO 14040 International Standard. Environmental Management – Life Cycle Assessment – Principles and Framework*. Geneva: International Organisation for Standardization.

ISO. (2006b). *ISO 14040 International Standard. Environmental Management – Life Cycle Assessment – Requirements and Guidelines*. Geneva: International Organisation for Standardization.

Kadam, K. L. 2002. Environmental benefits on a life cycle basis of using bagasse-derived ethanol as a gasoline oxygenate in India. *Energy Policy* 30, 371–384.

Kazi, F. K., Fortman, J., Annexe, R., Kothandaraman, G., and Hsu, D. 2010. Techno-economic analysis of biochemical scenarios for production of cellulosic ethanol. National Renewable Energy Lab. (NREL), Golden, CO (United States), Technical Report NREL/TP-6A2-46588.

Klein-Marcuschamer, D., Simmons, B. A., and Blanch, H. W. 2011. Techno-economic analysis of a lignocellulosic ethanol biorefinery with ionic liquid pre-treatment. *Biofuels, Bioproducts and Biorefining* 5(5), 562–569.

Lauer, M. (2008). Methodology guideline on techno-economic assessment (TEA). In *Workshop WP3B Economics, Methodology Guideline.*

Lee, S. Y., Revathy, S., Wayne, C., et al. 2019. Waste to bioenergy: A review on the recent conversion technologies. *BMC Energy* 1(1), 1–22.

Lynch, M. D. 2020. The Bioprocess TEA Calculator: An online techno-economic analysis tool to evaluate the commercial competitiveness of potential bioprocesses. *bioRxiv.*

Mandade, P., Bakshi, B. R., and Yadav, G. D. 2015. Ethanol from Indian agro-industrial lignocellulosic biomass: A life cycle evaluation of energy, greenhouse gases, land and water. *International Journal of Life Cycle Assessment* 20, 1649–1658.

Matthews, J. A., and Tan, H. 2009. Biofuels and indirect land-use change effects: The debate continues. *Biofuels, Bioproducts & Biorefining* 3, 305–317.

Matthews, S. H., Hendrickson, C. T., and Matthews, D. H. 2020. Life cycle assessment: Quantitative approaches for decisions that matter. Online available at https://www.lca-textbook.com/.

McKone, T. E., Nazaroff, W. W., Berck, P., et al. 2011. Grand challenges for life-cycle assessment of biofuels. *Environmental Science & Technology* 45(5), 1751–1756.

Meghana, M., and Shastri, Y. 2020. Sustainable valorization of sugar industry waste: Status, opportunities, and challenges. *Bioresource Technology* 303(January), 122929.

Morales, M., Quintero, J., Conejeros, R., and Aroca, G. 2015. Life cycle assessment of lignocellulosic bioethanol: Environmental impacts and energy balance. *Renewable & Sustainable Energy Reviews* 42, 1349–1361.

Murali, G., and Shastri, Y. 2019. Life-cycle assessment-based comparison of different lignocellulosic ethanol production routes. *Biofuels* DOI: 10.1080/17597269.2019.1670465

Rebitzer, G., Ekvall, T., Frischknecht, R., et al. 2004. Life cycle assessment: Part 1: Framework, goal and scope definition, inventory analysis, and applications. *Environment International* 30(5), 701–720.

Saltelli, A., 2002. Sensitivity analysis for importance assessment. *Risk Analysis* 22(3), 579–590.

Scott, F., Quintero, J., Morales, M., Conejeros, R., Cardona, C., and Aroca, G. 2013. Process design and sustainability in the production of bioethanol from lignocellulosic materials. *Electronic Journal of Biotechnology* 16(3), 13–13.

Scown, C. D., Nazaroff, W. W., Mishra, U., et al. 2012. Corrigendum: Life-cycle greenhouse gas implications of US national scenarios for cellulosic ethanol production. *Environmental Research Letters* 7(1), 9502.

Searchinger, T., Heimlich, R., Houghton, R. A., et al. 2008. Use of US croplands for biofuels increases greenhouse gases through emissions from land-use change. *Science* 319(-5867), 1238–1240.

Shastri, Y. 2017. Renewable energy, Bioenergy. *Current Opinion in Chemical Engineering* 17, 42–47.

Silalertruksa, T., and Gheewala, S. H. 2011. Long-term bioethanol system and its implications on GHG emissions: A case study of Thailand. *Environmental Science and Technology* 45, 4920–4928.

Singh, A., Pant, D., Korres, N. E., Nizami, A., Prasad, S., and Murphy, J. D. 2010. Key issues in life cycle assessment of ethanol production from lignocellulosic biomass: Challenges and perspectives. *Bioresource Technology* 101(13), 5003–5012.

Soam, S., Borjesson, P., Sharma, P. K., Gupta, R. P., Tuli, D. K., and Kumar, R. 2017. Life cycle assessment of rice straw utilization practices in India. *Bioresource Technology* 228, 89–98.

Soam, S., Kapoor, M., Kumar, R., Borjesson, P., Gupta, R. P., and Tuli, D. K. 2016. Global warming potential and energy analysis of second-generation ethanol production from rice straw in India. *Applied Energy* 184, 353–364.

Soam, S., Kumar, R., Gupta, R. P., Sharma, P. K., Tuli, D. K., & Das, B. 2015. Life cycle assessment of fuel ethanol from sugarcane molasses in northern and western India and its impact on Indian biofuel programme. *Energy* 83, 307–315.

Spatari, S., Bagley, D. M., and MacLean, H. L. 2010. Life cycle evaluation of emerging lignocellulosic ethanol conversion technologies. *Bioresource Technology* 101(2), 654–667.

Sreekumar, A., Shastri, Y., Wadekar, P., Patil, M., and Lali, A. 2020. Life cycle assessment of ethanol production in a rice-straw-based biorefinery in India. *Clean Technologies and Environmental Policy* 22, 409–422.

Swanson, R. M., Platon, A., Satrio, J. A., Brown, R. C., and Hsu, D. D. 2010. Techno-economic analysis of biofuels production based on gasification (No. NREL/TP-6A20-46587). National Renewable Energy Lab.(NREL), Golden, CO (United States).

Vlachokostas, C., Achillas, C., Diamantis, V., Michailidou, A. V., Baginetas, K., and Aidonis, D. (2021). Supporting decision making to achieve circularity via a biodegradable waste-to-bioenergy and compost facility. *Journal of Environmental Management*, 285, 112215.

Wang, M., Wu, Y., and Elgowainy, A. 2007. *Operating Manual for GREET: Version 17*. Argonne, IL, USA: Center for Transportation Research Energy Systems Division, Argonne National Laboratory.

Wiloso, E. I., Heijungs, R., and de Snoo G. R. 2012. LCA of second-generation bioethanol: A review and some issues to be resolved for good LCA practice. *Renewable and Sustainable Energy Reviews* 16(7), 5295–5308.

Wu, M., Wu, Y., and Wang, M. 2006. Energy and emission benefits of alternative transportation liquid fuels derived from Switchgrass: A fuel life cycle assessment. *Biotechnology Progress* 22(4), 1012–1024.

Xie, K. 2015. Life cycle assessment (LCA) and techno-economic analysis (TEA) of various biosystems. PhD diss., IOWA State Univ.

Zhao, L., Zhang, X., Xu, J., et al. 2015. Techno-economic analysis of bioethanol production from lignocellulosic biomass in China: Dilute-acid pre-treatment and enzymatic hydrolysis of corn stover. *Energies* 8(5), 4096–4117.

Zimmermann, A., Müller, L. J., Marxen, A., et al. 2018. Techno-economic assessment & life cycle assessment guidelines for CO_2 utilization.

Sassner, P Galbe, M. and Zacchi, G., 2008. Techno-economic evaluation of bioethanol production from three different lignocellulosic materials. *Biomass and bioenergy*, 32(5), pp.422-430.

4 Biomass Pre-Treatment and Liquefaction

Tejas M. Ukarde, Annamma Anil Odaneth, and Hitesh S. Pawar
Institute of Chemical Technology

CONTENTS

4.1 Introduction 94
4.2 Structure and Composition of LB..95
    4.2.1 Cellulose 96
    4.2.2 Hemicellulose 96
    4.2.3 Lignin...97
4.3 Different Pre-Treatment Strategies 97
    4.3.1 Physical Pre-Treatments..98
    4.3.2 Biological Pre-Treatments...98
    4.3.3 Chemical Pre-Treatments 99
        4.3.3.1 Acid Pre-Treatment 99
        4.3.3.2 Alkali Pre-Treatment 100
        4.3.3.3 Organosolv Pre-Treatment 100
        4.3.3.4 Ionic Liquid Pre-Treatment 101
    4.3.4 Physicochemical Pre-Treatments...101
        4.3.4.1 Steam Explosion 102
        4.3.4.2 Liquid Hot Water Pre-Treatment...102
        4.3.4.3 Ammonia Fibre Explosion 102
        4.3.4.4 Microwave Irradiation 103
        4.3.4.5 Ultrasound Irradiation 103
        4.3.4.6 Wet Oxidation...103
4.4 Hydrothermal Liquefaction (HTL)...104
    4.4.1 Properties of Sub- and Supercritical Water 104
    4.4.2 Reaction Mechanism of HTL 105
        4.4.2.1 Reaction Mechanism for Degradation of Carbohydrates....105
        4.4.2.2 Reaction Mechanism for Degradation of Lignin 108
    4.4.3 Effect of Operating Parameters on HTL of Biomass 109
        4.4.3.1 Effect of Feedstock Type and Particle Size 109
        4.4.3.2 Effect of Temperature 110
        4.4.3.3 Effect of Heating Rate and Thermal Gradient...................110
        4.4.3.4 Effect of Residence Time..111
        4.4.3.5 Effect of Biomass/Water Mass Ratio..................................111

DOI: 10.1201/9781003158486-4

        4.4.3.6   Effect of Pressure.. 112
        4.4.3.7   Effect of Catalyst 112
4.5   Overview of Industrial Applications of HTL 113
4.6   Challenges and Future Perspective.. 114
4.7   Conclusions.. 115
References.. 116

4.1   INTRODUCTION

Presently, the energy requirements of mankind are chiefly met by the limited non-renewable fossil fuels such as coal, oil, and natural gas that may ultimately dry up in near future [1]. Moreover, the extensive use of fossil fuels is one of the key causes of exhaustive greenhouse gas (GHG) emissions resulting in global warming and negative climate changes [2].

In the last two decades, researchers have been attracted to alternative renewable energy sources such as solar, wind, nuclear, hydroelectric, and lignocellulosic biomass (LB) for replacing fossil fuels [3]. Among the existing renewable energy sources, biomass has the potential for becoming the largest contributor to renewable energy, i.e. 10.2%. The energy confined by biomass per year is equal to about 1.08×10^{11} tons of crude oil, while in the case of carbon emission, biomass produces only 17–27 g/kWh in comparison with coal (955 g/kWh), oil (818 g/kWh), and natural gas (446 g/kWh) based on electricity generation [4]. For agro-based economies, LB is the only renewable, abundant, and economic source for the generation of chemicals, fuels, and energy, thereby mimicking the petroleum refinery. The profitability of the petroleum refineries is always dependent on the costs of the feedstocks and the different products that can be made from the raw crude oil. Similarly, the efficiency and the profitability of any renewable source of energy lie in the operational costs of the processing and depend on the value of the final products that are made from the raw material.

The biorefinery is an emerging concept, wherein a broad range of value-added chemicals, green polymeric materials, and energy are targeted from LB via a thermal, biological, or catalytic route with reduced or zero carbon footprint [5]. Commonly, biorefinery processes are divided into three major categories on the basis of mode of action used for biomass conversion: (a) thermal, (b) thermochemical, and (c) biochemical conversion depending upon the type of biofuel expected. While thermal processes require a huge capital cost and an enormous amount of energy, biochemical processes require long processing time and also suffer from low efficiencies [6]. Thus, the thermochemical conversion of biomass into biofuel is a potential substitute for the generation of sustainable and renewable energy via controlled heating and oxidation of biomass.

In addition, the present technologies used for biomass liquefaction can be categorized into indirect and direct liquefaction on the basis of the path followed for liquefaction of biomass. The indirect liquefaction signifies Fischer–Tropsch (F-T) process, in which syngas from biomass gasification is used as the raw material for the production of liquid fuel, including methanol, ethanol, and dimethyl ether which can be used for a variety of applications. However, designing of a catalytic reactor for small-scale biomass conversion and catalysts for the production of specific chemicals conferring to molar ratio of H_2 and CO are required. Thus, the production of syngas from biomass gasification has not been economically feasible till date [7]. The direct liquefaction refers to the production of bio-oil from biomass; the main technologies available are hydrolysis followed by fermentation and thermodynamic liquefaction. The thermodynamic liquefaction could be categorized into fast pyrolysis and hydrothermal liquefaction (HTL). Of them, fast pyrolysis requires a high capital cost and energy but provides low efficiency [6]. Hydrothermal liquefaction of biomass is the thermochemical conversion of biomass into liquid fuels by processing in a hot, pressurized water atmosphere for adequate time at low temperature and heating rates with better yield and quality of liquefied products which can be applied for a spectrum of applications such as the production of fuel, energy, and value-added chemicals after upgradation. Thus, hydrothermal liquefaction is preferred over fast pyrolysis [8]. Also, the HTL of biomass can contribute to providing a dual solution for waste management and fossil fuel depletion.

The present chapter emphasizes the various types of pre-treatment methods such as physical, biological, chemical, and physicochemical that have been used for the fractionation of LB into biofuels. This chapter provides a detailed insight into the structure and components of LB with the existing state-of-the-art liquefaction technologies for the conversion of biomass into bio-oil. Also, the influence of operating parameters such as feedstock type, temperature, residence time, pressure, solvent, and catalyst on HTL product distribution is described. This chapter includes conceivable reaction mechanisms for the HTL of various components of LB into different value-added products such as formic acid, lactic acid, acetic acid, 5-HMF, and levulinic acid. The technical and environmental challenges related to HTL of biomass are discussed with possible future solutions. It provides the vision to develop a robust, efficient, economic, and sustainable biomass HTL process that can help human society move towards a bio-based economy.

4.2   STRUCTURE AND COMPOSITION OF LB

Lignocellulosic biomass (LB) is a common phrase used for all plant-based materials with the composition of cellulose, hemicellulose, lignin, proteins, and other phenolic components. The cellulose and lignin are the significant fractions of LB, and they depend on feedstock variety, age, geographical location, growth conditions, etc. Typically, LB can be categorized into (a) agriculture and forestry residues, (b) herbaceous crops, (c) aquatic biomass, and (d) waste biomass which includes animal waste, sewage waste, and municipal solid waste. Almost all types of biomass mentioned above are lignocellulosic or plant-based, which typically comprise 30–35 wt.% of cellulose $[C_6H_{10}O_5]_n$, 15–35 wt.% of hemicellulose $[C_5H_8O_4]_m$, and 20–35 wt.% of lignin $[C_9H_{10}O_3(OCH_3)0.9-1.7]_x$, with minimal amounts of other compounds such as proteins, lipids, and ash [9]. The general structure of LB present in biomass and its chemistry is shown in Figure 4.1.

Forestry, agriculture, and wastes are heterogeneous by size, composition, structure, and properties. Starches, lipids, and proteins may also be present among the materials. Lower-grade biomass such as municipal solid waste (MSW) may have

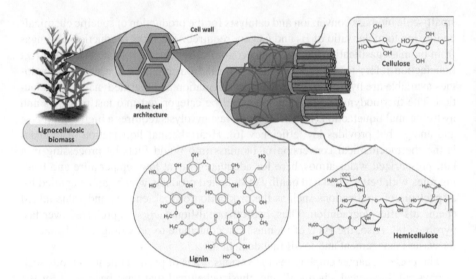

FIGURE 4.1   General structure of LB present in biomass and its chemistry.

relatively high nitrogen contents or ash fractions. This variable composition makes it very difficult to optimize operational parameters at industrial system for converting the feedstock into a final energy carrier (e.g. heat, electricity, gaseous or liquid biofuels). Cellulose, hemicellulose, and lignin contents in common lignocellulosic feedstocks are shown in Table 4.1.

4.2.1 CELLULOSE

Cellulose is a polysaccharide of D-glucose, the main constituent of the plant cell wall which helps plants maintain their structure and form [10]. The cellulose fibrils are stabilized by hydrogen bonding among different strands, which also determines the crystalline or amorphous nature of cellulose [11]. Compact packing of cellulose fibrils in the lignocellulosic matrix is responsible for cellulose's resistivity towards enzyme hydrolysis. This distinctive and complex structure makes cellulose resistant to both biological and chemical treatments. Cellulose is available in waste streams in the form of lignocelluloses or partly purified, e.g. papers, or pure cellulose such as cotton, or mixed with other materials, e.g. citrus wastes.

4.2.2 HEMICELLULOSE

Hemicellulose is another form of polysaccharide present in LB; it is a heterogeneous biopolymer of pentoses (xylose, arabinose) and hexoses (mannose, glucose, and galactose) with acetyl groups held together by β-(1,4)- and β-(1,3)-glycosidic bond, representing about 15%–35% of the total weight of biomass [12]. The acetylated form of sugars in hemicelluloses constrains cellulose accessibility by disturbing hydrophobicity and enzyme recognition [13].

TABLE 4.1

Cellulose, Hemicellulose, and Lignin Content in Common Lignocellulosic Feedstocks (wt.%)

S. No.	Lignocellulosic Feedstock	Cellulose (%)	Hemicellulose (%)	Lignin (%)
1	Sugar cane bagasse	42	25	20
2	Sweet sorghum	45	27	21
3	Hardwood	40–55	24–40	18–25
4	Softwood	45–50	25–35	25–35
5	Corn cob	45	35	15
6	Corn stover	38	26	19
7	Rice straw	32	24	18
8	Nut shells	25–30	25–30	30–40
9	Newspaper	40–55	25–40	18–30
10	Grasses	25–40	25–50	10–30
11	Wheat straw	29–35	26–32	16–21
12	Banana waste	13.2	14.8	14
13	Bagasse	54.87	16.52	23.33
14	Sponge gourd fibre	66.59	17.44	15.46
15	Agricultural residue	5–15	37–50	25–50
16	Leaves	15–20	80–85	0
17	Cotton seeds	80–95	5–20	0
18	Paper	85–99	0	0.15
19	Waste papers from chemical pulps	6–10	50–70	12–20

4.2.3 LIGNIN

The third predominant polymer in LB is the amorphous heteropolymer lignin composed of phenyl propane units such as p-coumaryl, coniferyl, and sinapyl alcohols which are held organized by different linkages, corresponding to 20%–35% of biomass weight [14]. Delignification of biomass results in biomass swelling, disruption of lignin, and elevation in surface area which are responsible for the increase in the cellulose accessibility for cellulolytic enzymes [15].

4.3 DIFFERENT PRE-TREATMENT STRATEGIES

The recalcitrant nature of LB towards enzymatic and microbial decomposition owing to crystallized cellulose and hemicellulose polymer matrix encrusted within highly polymerized phenolic lignin is the main barrier in the industrial valorization of LB for the production of bioenergy. Also, factors such as crystallinity of cellulose, degree of delignification, heterogeneity, and the complexity of fundamental constituents of cell wall can disturb the recalcitrance of LB [16].

Pre-treatment is one of the efficient tools to reduce the recalcitrance of LB via a combination of structural and chemical modifications within LB [17]. To get the components of LB in utilizable form, separation of LB components via pre-treatment is important. Different pre-treatment methods have previously been reported, such as physical, biological, chemical, physicochemical processes and their combinations, to speed up the hydrolysis of LB. Pre-treatment methods used alone are generally more efficient than the above-mentioned combinatory pre-treatment methods (by combining two or more). The ideal pre-treatment methods should have the characteristics such as (a) being applicable to a wide variety and loading of LB, (b) low capital and operational cost, (c) being energy-intensive, and (d) being able to generate no or limited amount of sugar and lignin degradation products that could prevent the growth of fermentative micro-organisms or the activity of hydrolytic enzymes [18].

4.3.1 PHYSICAL PRE-TREATMENTS

Physical pre-treatment of LB includes size reduction, chipping, grinding, shredding, milling, etc., which are used to enhance the digestibility of LB [19]. Size reduction has frequently been used in many hydrolysis methods, but very little information is accessible about its impact on substrate characteristics and energy consumed. Harvesting and pre-conditioning of LB result in biomass particles with varying sizes ranging from 10 to 50 mm. Chipping further cuts the LB into logs of size ranging from 10 to 30 mm; however, grinding and milling can decrease the LB to particles of size ranging from 0.2 to 2 mm [20]. These pre-treatment methods enhance the available surface area and lessen the crystallinity of cellulose and degree of polymerization, resulting in reduced recalcitrance of LB. The milling is preferred over all the above physical pre-treatment methods, in which shear forces produced during milling are responsible for the efficient reduction in particle size and cellulose crystallinity. There are different types of milling, which include vibratory ball milling, disk milling, hammer milling, knife milling, attrition milling, centrifugal milling, colloid milling, pin milling, and extruders. In the case of milling, factors such as biomass type, time, and the kind of milling used have a substantial impact on the available surface area and crystallinity of cellulose. It has been reported that milling of LB increases biogas, bioethanol, and biohydrogen yields [21]. However, milling requires a very high amount of energy and it is not economically viable to this date. Also, there are other different types of advanced pre-treatment methods that have been reported, such as the use of gamma radiation to cleave beta-(1,4)-glycosidic bond, although economic feasibility, environmental, and safety are the major bottlenecks [22].

4.3.2 BIOLOGICAL PRE-TREATMENTS

Biological pre-treatments use enzymes produced by microbes as a catalyst for the degradation of cellulose, hemicellulose, and lignin. In the case of biological pre-treatments, it can be utilized in both stages of hydrolysis. In the first stage of hydrolysis, it can be used to degrade lignin associated with the LB just like other pre-treatment processes. In the next stage of hydrolysis, enzymes such as cellulases can be used to convert cellulose into oligomers and sugar monomers, which is named as enzymatic saccharification. Sun and Cheng reported that white-rot fungi-mediated biological pre-treatment is the most efficient biological pre-treatment of biomass. In the case of brown-, white-, and soft-rot fungi used to mediate biological pre-treatment, brown-rot fungi are known for the hydrolysis of cellulose, while white- and soft-rot fungi are able to hydrolyse cellulose as well as lignin via the production of enzymes such as lignin peroxidase, polyphenol oxidase, manganese-dependent peroxidase, and laccases that lead to the degradation of lignin. Selective delignification of wood and wheat straw via selected white-rot fungi has been reported. Also, Magnusson et al. compared biohydrogen production from barley hulls contaminated by Fusarium head blight and normal barley hulls by using *Clostridium thermocellum* [23]. However, biological pre-treatments require huge space for maintaining optimized growth conditions for a very long residence time of about 10–14 days, making biological pre-treatments less attractive at an industrial scale. Effective biological pre-treatment demands numerous chemical mediators and enzymes to overcome various physical and biochemical hurdles to hydrolysis. In contrast, a synergistic mixture of enzymes can expand the small pores and increase the accessibility through the cell wall [24]. Itoh et al. reported that biological pre-treatment followed by bio-organosolv pre-treatment saved about 15% of the electricity needed for beech wood ethanolysis [25]. Conclusively, biological pre-treatment could be utilized efficiently combined with other pre-treatment methods or its own for low-lignin-containing biomass.

4.3.3 CHEMICAL PRE-TREATMENTS

Chemical pre-treatment methods have frequently been utilized than physical and biological pre-treatment methods due to the efficiency and enhancement of biodegradation of complex materials. Chemicals such as acids, alkalis, organic solvents, and ionic liquids have been used in chemical pre-treatment of LB with a significant impact on LB structure [12].

4.3.3.1 Acid Pre-Treatment

In the case of acid pre-treatment, majorly acetic acid (CH_3COOH), hydrochloric acid (HCl), and sulphuric acid (H_2SO_4) have been employed, which can improve the enzymatic hydrolysis of LB [26]. Acid pre-treatment breaks the Van der Waals forces, hydrogen bonds, and covalent bonds within the native complex structure of LB, which leads to the effective dissolution of hemicellulose and reduction of cellulose. Hydrolysis of hemicelluloses is the chief reaction that occurs during acid pre-treatment and produces sugars such as xylose, galactose, mannose, and glucose. At moderately and highly acidic conditions, these hemicellulose hydrolysis products can be converted into furfural and 5-HMF via dehydration reaction. Thus, dilute acid (<4 wt.%) pre-treatment becomes the majorly utilized pre-treatment which can achieve efficient cellulose hydrolysis at a faster rate. It can hardly dissolve the lignin, but enhances the susceptibility of cellulose to enzymatic hydrolysis. 2% H_2SO_4- and 2% HCl-pre-treated straw resulted in the highest methane yields, i.e. 175.6 and 163.4 mL/g among all the acid pre-treatments [14]. However, dilute and concentrated acids are hazardous, toxic, and corrosive and need to be recovered to make the pre-treatment method inexpensive. Also, acid pre-treatment may require surplus alkali to neutralize the acid hydrolysate.

4.3.3.2 Alkali Pre-Treatment

Alkali pre-treatment involves using bases such as sodium hydroxide (NaOH), potassium hydroxide (KOH), lime, hydrazine, and anhydrous ammonia. The treatment causes the swelling of biomass, an increase in available internal surface area, and a drop in polymerization and cellulose crystallinity. Alkali pre-treatment breaks down the lignin and disrupts the linkages between polysaccharide and lignin, making carbohydrates more accessible for further enzyme hydrolysis. It can also remove acetyl groups and other uronic acid groups associated with hemicellulose and diminish the enzyme availability towards the cellulose surface. Harmsen et al. stated that NaOH, KOH, and $Ca(OH)_2$ are the frequently used compounds in alkali pre-treatment, which requires lengthy processing time at mild conditions [27]. Zheng et al. reported a 73.4% higher methane yield from NaOH-pre-treated corn stover than from untreated corn stover, but toxicity and difficulty in recyclability of NaOH limit the usage of NaOH at a large scale [28]. Similarly, Li et al. reviewed the influence of KOH pre-treatment of corn stover on methane yield and showed a 95.6% improved yield compared to that of untreated corn stover; however, the high cost of KOH and toxicity towards microbes are the drawbacks associated with the usage of KOH for the pre-treatment of biomass [29]. $Ca(OH)_2$ is also one of the potential alkalis for biomass pre-treatment due to its low cost, environmental safety, and comfortable recovery. The usage of $Ca(OH)_2$ in alkali pre-treatment of corn stover has been reported to give a 39.7% higher methane yield than untreated corn stover; however, the dissolution of $Ca(OH)_2$ is the main disadvantage. A combination of KOH and $Ca(OH)_2$ has also been attempted for alkali pre-treatment of corn stover, resulting in a 79.9% higher methane yield than untreated corn stover [29].

4.3.3.3 Organosolv Pre-Treatment

A range of organic or aqueous–organic solvent mixtures with or without catalysts such as HCl or H_2SO_4 have been used for the delignification of biomass. Organic acids such as oxalic acid, acetylsalicylic acid, and salicylic acid can be used as a catalyst that solubilizes the hemicellulose, and lignin can be extracted using the organic solvents or their aqueous solutions. Generally, organosolv pre-treatments are carried out at high temperatures (100°C–250°C) by using low-boiling-point solvents (methanol and ethanol) as well as high-boiling-point solvents (ethylene glycol, glycerol, and tetrahydrofurfuryl alcohol) and other kinds of organic compounds such as ethers, ketones, phenols, and dimethyl sulphoxide [30]. Zhou et al. stated that organosolv pre-treatment can effectively remove the lignin fraction by complete solubilization of hemicellulose via hydrolysis of inter- and intramolecular linkages in lignin and glycosidic linkages in hemicellulose in LB depending upon process conditions [31]. The kinetics of delignification depends on the solvent and biomass used during the pre-treatment. Curvelo and Pereira reported that wood delignification in aqueous methanol happened in three phases (initial, principal, and residual), while delignification of sugarcane bagasse progressed in two phases only (principal and residual) [32]. Organosolv pre-treatment methods are very selective in yielding separate fractions of dry lignin, solubilized hemicellulose, and pure cellulose fraction. Lignin separated from LB by using organosolv pre-treatment is highly pure, of low molecular weight, and sulphur-free and can be utilized as fuel to power pre-treatment plant or can further be purified to high-quality lignin to substitute the polymeric materials such as phenolic powder resin, polyurethane, polyisocyanate foams, and epoxy resin [33]. Thus, by combining organosolv with other pre-treatment methods, efficient, clean, and effective biomass fractionation can be established. Hongzhang and Liying reported a clean wheat fractionation process via steam explosion pre-treatment followed by ethanol delignification [34]. Similarly, Rughani and McGinnis reported the integration of rapid-steam hydrolysis and organosolv pre-treatment for the fractionation of mixed southern hardwood [35]. Solvents need to be recovered to reduce micro-organism growth, enzyme hydrolysis, and anaerobic digestion and check the process costs. However, the cost of chemicals and catalysts can be a major drawback of organosolv pre-treatment and the by-products generated during the various acid-catalysed side reactions such as 5-HMF can be a potential inhibitor for fermentation micro-organisms [15].

4.3.3.4 Ionic Liquid Pre-Treatment

Ionic liquids are molten organic salts with very low vapour pressure, which provide several unique features and dissolve lignin and carbohydrates. Ionic liquids gain more attention owing to the chance of controlling physicochemical properties. Dissolution of LB in ILs is governed by the nature of both the constituents and physical factors of ILs [36]. The ILs can dissolve various substrates such as softwood and hardwood. A probable dissolution mechanism proposes that ILs compete with LB for hydrogen bonding. The usage of ILs in the pre-treatment of LB has resulted in the removal of lignin and hemicellulose and reduced cellulose crystallinity. Cellulose after IL pre-treatment is much less crystalline than the untreated one and more susceptible to enzymatic hydrolysis. Most of the ILs used for LB fractionation are ILs with imidazolium cation. Of the ILs with imidazolium cation, ([EMIM]$^+$[Ac]$^-$) IL is one of the popular ILs for biomass pre-treatment. Li et al. compared the dilute acid pre-treatment and 1-ethyl-3-methylimidazolium acetate ([EMIM]$^+$[Ac]$^-$) IL-mediated pre-treatment of switchgrass, which showed that the IL-mediated pre-treatment resulted in more reduced cellulose crystallinity and lignin content, and increased surface area [37]. Similarly, Silva et al. investigated six different ILs for the dissolution of sugarcane bagasse for efficient enzymatic saccharification [38]. Ninomiya et al. explored the use of cholinium carboxylate ionic liquids for the fractionation of different types of biomass substrates such as bagasse, bamboo powder, and kenaf powder [39]. Different types of ILs are being checked for enhancing the environmental and economic feasibility of the overall process as well as digestibility and fractionation of LB [40]. However, a detailed investigation of the mechanism and process of IL-mediated dissolution and fractionation of LB is still under process. Multi-step IL-based pre-treatment processes trailed by enzyme hydrolysis after cellulose recovery and removal of IL make the process economically non-favourable [41]. Thus, researchers are looking for enzyme-compatible ILs which can enzymatically hydrolyse the fractionated cellulose for one-pot bioethanol production.

4.3.4 Physicochemical Pre-Treatments

This type of pre-treatments uses a hybrid methodology for affecting physical parameters such as cellulose crystallinity, surface area, pore size, and volume, as well as chemical parameters such as digestibility and intermolecular bonding within the LB [42]. These pre-treatment methods have used temperature and/or pressure with chemical processes. Physicochemical pre-treatments are very effective in breaking down the LB, including steam explosion, liquid hot water pre-treatment, ammonia fibre explosion, microwave and ultrasound irradiation, and wet oxidation.

4.3.4.1 Steam Explosion

Steam explosion is a hydrothermal process in which the substrate is exposed to high pressure (0.7–48 bar) and temperature (160°C–260°C). The system is first pressurized gradually and is depressurized rapidly to disrupt intermolecular bonding such as hydrogen bonding and Van der Waals forces [43]. Steam explosion can increase the particle surface area, altering pore size and volume with a reduction in bulk density. The output of this technique can be regulated by tuning the compression rate and decompression [44]. Lower temperatures and longer times (190°C, 10 min.) are more feasible than higher temperatures and shorter times (270°C, 1 min.) as low temperatures avoid the degradation products of sugars which can inhibit fermentation and reduce the yield of biofuels [45]. Datar et al. reported the generation of biogas with high hydrogen yield from hemicellulose by using steam explosion pre-treatment [46]. Also, steam explosion can be made more efficient by using catalysts such as H_2SO_4, CO_2, or SO_2 [47]. Steam explosion is an attractive technique due to low energy input, no recycling and environmental cost, and low usage of chemicals. It is very efficient for the pre-treatment of hardwoods and agricultural residues. Uncatalysed steam explosions have been commercialized in Masonite process to produce fibre board and other products [48].

4.3.4.2 Liquid Hot Water Pre-Treatment

Liquid hot water pre-treatment is very similar to steam explosion methods, wherein LB was treated with water at a high temperature ranging from 160°C to 250°C and at a relatively lower pressure of about 5 bar [43]. Liquid hot water pre-treatment can easily hydrolyse hemicellulose and remove lignin fraction and also avoid the generation of fermentation inhibitors. Antal reported that hot water cleaves hemiacetal linkages and facilitates ethereal linkage breakdown in biomass [49]. The usage of acid catalysts such as dilute acid pre-treatment can increase the efficiency; however, the catalyst can degrade sugars into unwanted products. Thus, Weil et al. maintained the pH between 5 and 7 by using KOH for the minimization of monosaccharides production from yellow poplar wood sawdust [50]. The little cost of solvent is one of the critical advantages of liquid for hot water pre-treatment; however, the solubilized products' concentration is lesser than a steam explosion. This technique has been verified at laboratory scale, while pH-controlled conditions have been used to pre-treat corn fibre at a large scale [48].

4.3.4.3 Ammonia Fibre Explosion

Ammonia fibre explosion technique uses liquid ammonia (alkaline) in 1:1 or 1:2 ratio with dry wt. of biomass for biomass pre-treatment. Ammonia fibre explosion pre-treatment requires a relatively low temperature (60°C–90°C), short processing time (30–60 min.), and pressure above 3 MPa [51]. This pre-treatment causes swelling and alteration in the crystallinity of cellulose with the removal of lignin. Pre-treated biomass using this technique is easily hydrolysable and provides yields near to theoretical enzymatic hydrolysis at relatively low enzyme loading [52]. The advantages of this method are the low cost of ammonia, easy recyclability of ammonia, moderate temperature, short residence time, and high selectivity for lignin reaction [53]. The mild process conditions of ammonia fibre explosion can minimize sugar degradation products and fermentation inhibitors. Ammonia fibre explosion is useful for the pre-treatment of herbaceous plants, agricultural residues, and MSW. However, it has been reported that the ammonia fibre explosion method is less efficient in the case of pre-treatment of high-lignin-containing lignocellulosic biomass such as newspapers (18%–30% lignin) and softwoods [54].

4.3.4.4 Microwave Irradiation

Lignocellulosic biomass can be heated at the molecular level by dipole rotation on the exposure of the biomass to microwaves. The irradiation of microwave to biomass allows a change in the dipole moment of moisture content, which causes swelling of biomass, depolymerization of lignin, and a decrease in cellulose crystallinity [55]. Generally, polar solvents such as H_2O, alcohols, acetonitrile, and acetone have a more remarkable ability to heat as they can absorb more radiations. In contrast, non-polar solvents such as alkanes, toluene, and dichloromethane have low heat ability as they can absorb fewer radiations. There are many reports available on the microwave-assisted pre-treatment of biomass between 130°C and 200°C, over alterable residence time (3–30 min.) and power input (200–800 W) [56]. This process is efficient from energy and residence time perspectives.

4.3.4.5 Ultrasound Irradiation

Ultrasound irradiation is an emerging pre-treatment methodology for biomass pre-treatment. During ultrasound irradiation, the formation of microjets takes place from bubbles generated via ultrasound irradiation, which can travel at a very high speed through LB. Ultrasound irradiation can decrease the cellulose crystallinity, elevate the available cellulose surface area, and loosen the intermolecular linkages between lignin and cellulosic fraction of biomass [57]. There are two ways of using ultrasound irradiation: The first is by using an ultrasound probe that can generate waves and irradiate at a specific frequency in pulse mode. At the same time, the sample can also be located in a warm ultrasonic bath [58]. The microjets generated via ultrasound-assisted pre-treatment of LB can also decrease the average particle size, facilitating greater surface area, and result in higher glucose yield from enzymatic hydrolysis [59]. However, many scientists have reported that the usage of ultrasound for biomass pre-treatment has negative energy efficiencies that make this method commercially unfeasible.

4.3.4.6 Wet Oxidation

The usage of oxidizing agents in an aqueous medium for breaking down LB into a constituent fraction is a beneficial method [60]. Oxidizing agents are responsible for elevated solubilization of hemicellulose and lignin breakdown. In wet oxidation of LB, oxidizing agents such as pressurized gases, air, oxygen, and liquid-based peroxides are used. Pressurized oxygen as an oxidizing agent requires harsh conditions such as temperature in the range from 120°C to 350°C with 0.5–4 h residence time, while peroxide-based oxidizing agent effectively pre-treats LB at milder conditions, at a temperature of 30°C with 8 h residence time. Schmidt et al. stated that the solubilization of hemicellulose depends on the temperature of treatment for wet oxidation. Also, the usage of bases as additives in wet oxidation has been more efficient for LB pre-treatment [61].

4.4 HYDROTHERMAL LIQUEFACTION (HTL)

Presently, the new strategy has extensively been explored for efficient conversion of LB into biofuels without any LB pre-treatment, i.e. liquefaction of LB into aqueous compounds, bio-oil, and bio-char, which can ultimately be converted into value-added chemicals, fuels, and carbon materials, respectively. The generalized scheme for biomass HTL is shown in Figure 4.2. Of the reported thermochemical methods, HTL is efficient and advantageous from an economic and energy perspective. HTL is a technique used for the production of relatively pure biofuel from LB at stringent reaction conditions, i.e. high temperature (250°C–400°C) and pressure (5–25 MPa) in the presence of hot compressed water or alcohols, acetone, and the mixture of solvents such as alcohol–water, phenol–water, and dioxane–water as a solvent system with or without catalysts.

4.4.1 PROPERTIES OF SUB- AND SUPERCRITICAL WATER

HTL has attracted the attention of worldwide scientists owing to characteristic properties of sub- and supercritical water, such as high ion product (Kw), low dielectric constant, low viscosity, and additional mass transfer, which are responsible for driving reaction towards the product side in the absence of catalyst [62]. A comparison of

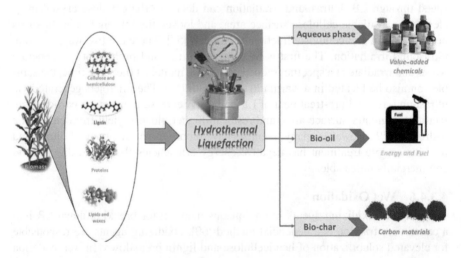

FIGURE 4.2   General process for the HTL of biomass.

TABLE 4.2
Comparison of Properties of Sub- and Supercritical Water

S. No.	Property	Normal Water	Sub-Critical Water		Supercritical Water	
1	Temperature (°C)	25	250	350	400	450
2	Pressure (MPa)	0.1	5	25	25	50
3	Density (g/cm)	1	0.8	0.6	0.17	0.58
4	Dielectric constant (F/m)	78.5	27.1	14.07	5.9	10.5
5	Ionic product	14.0	11.2	12	19.4	11.9
6	Heat capacity (kJ/kg/K)	4.22	4.86	10.1	13.0	6.8
7	Dynamic viscosity (mPa s)	0.89	0.11	0.064	0.03	0.07

characteristic properties of sub- and supercritical water with water at room temperature is shown in Table 4.2 [63].

HTL can be operated at supercritical and sub-critical conditions. Below the critical point of water ($T_c = 373°C$, $P_c = 22.1\,MPa$, and $r_c = 320\,kg\ m^{-3}$), the vapour pressure curve separates liquid and vapour phases, while above the critical point of water, with an increase in temperature, the dielectric constant decreases with a reduction in electronegativity of oxygen which makes water just like any other aqueous solution with H^+ and OH^- ions. The absence of phase boundaries and complete miscibility of supercritical water and gases results in a fast and complete reaction, which makes it an excellent solvent. In the case of sub- and supercritical water-based reactions, water behaves as a reactant and catalyst at a time [64]. The characteristic physicochemical properties of hot compressed water at sub- and supercritical conditions open up new horizons for the usage of sub- and supercritical water as a solvent for biomass liquefaction.

4.4.2 REACTION MECHANISM OF HTL

The exact mechanism of HTL remains ambiguous. However, the reaction mechanism of HTL comprises of three main steps such as (a) depolymerization of biomass components such as cellulose, hemicellulose, and lignin, (b) decomposition of monomers of biomass constituents via cleavage, dehydration, decarboxylation, deamination, etc., followed by (c) recombination, re-polymerization, and condensation of reactive fragments [62]. During HTL, main constituents of biomass such as cellulose, hemicellulose, and lignin undergo hydrolysis to give monomers followed by degradation to produce bio-oil. The plausible reaction pathway for the degradation of each component of biomass is shown in Figure 4.3.

4.4.2.1 Reaction Mechanism for Degradation of Carbohydrates

Cellulose and hemicellulose are the plentiful carbohydrates present in biomass. Different carbohydrates have different hydrolysis rates. From cellulose and hemicellulose, hemicellulose undergoes hydrolysis reaction owing to the crystalline nature of cellulose. In case of degradation of cellulose and hemicellulose, cellulose and

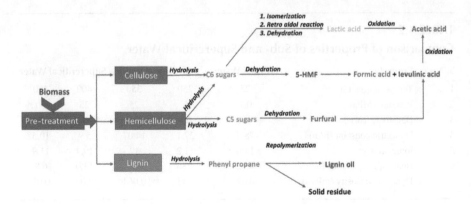

FIGURE 4.3   Plausible reaction pathway for degradation of each biomass component.

hemicellulose firstly undergo hydrolysis to give monomers such as glucose, fructose, xylose, and arabinose, which get degraded into products such as formic acid, lactic acid, acetic acid, 5-HMF, furfural, and levulinic acid [65,66]. Various degradation products of cellulose and hemicellulose are shown below.

A. *Formic acid*: Formic acid is the crucial product formed during the HTL of biomass. The manufacturing of formic acid via HTL of carbohydrates is attracting the attention of worldwide researchers due to its ability to power fuel cells, and it is an intermediate of the hydrogen production process from biomass [67]. During HTL, formic acid is produced via dehydration of glucose or fructose or erythrose followed by acid-catalysed hydrolysis of 5-HMF. It has been stated that alkali addition is essential to inhibit the decay of formic acid and oxidation of organic compounds. While an excess of H_2O_2 is necessary to avoid the formation of acetic acid which can be generated from the oxidation of 5-hydroxy-2-furaldehyde which is the product of glucose dehydration, Yun et al. stated that the hydrothermal treatment of cellulose provides 27% formic acid yield with 50% cellulose conversion at 250°C under alkaline conditions in the presence of H_2O_2 as an oxygen supplier [68]. Similarly, Jin et al. explored the hydrothermal processing of carbohydrates at 250°C in alkaline conditions under H_2O_2 as an oxygen supplier with the highest formic acid yield, i.e. 75% within 2 min. residence time [67]. Similarly, Srokol et al. stated that the rate of formation of formic acid increases with an increase in ionic strength in the presence of sub-critical water by using HCl and NaOH [69].

B. *Lactic acid*: Lactic acid is an essential chemical product which has been used in many sectors like mild acid flavour, pH regulation, food preservative, production of biodegradable plastic, etc. [70]. Thus, many investigators have investigated the lactic acid production from microcrystalline cellulose, glucose, and fructose under sub- and supercritical water conditions [71]. Generally, lactic acid is manufactured through the deconstruction of all hexose monomers. Previous reports have showed that a minute quantity of lactic acid was obtained in the absence of catalyst at 300°C. Jin et al. stated the role of base catalyst in the production of lactic acid under sub-critical water conditions [67]. Lactic acid yield has been elevated by using many alkaline catalysts such as $ZnSO_4$, $Ca(OH)_2$, and NaOH [72]. However, alkaline catalyst leads to corrosion of reactor under hydrothermal conditions. Transition metal ions significantly affect lactic acid yield and selectivity. Earlier, Kong et al. investigated the effect of metal ions such as Zn(II), Ni(-II), Co(II), and Cr(II) on lactic acid production under sub-critical water conditions (300°C, 120 s) [73].

C. *Acetic acid*: Acetic acid is utilized in a variety of arenas such as antibiotics, antiseptics, disinfectants, foods, agriculture, cosmetics, chemical reagent, fungicide, herbicide, pH adjuster, counterirritant, and solvent. And hydrothermal liquefaction of biomass can provide a sustainable and efficient process for the production of acetic acid. Generally, acetic acid is produced by the disintegration of 1,6-anhydro-b-D-glucopyranose or by erythrose decay. It has been stated that acetic acid can be acquired by wet oxidation of organic waste [74]. Also, few reports are available to generate acetic acid from wet oxidation of cellulosic biomass such as rice hulls, potato starch, filter paper powder, and glucose in the presence of H_2O_2 as an oxidizing agent [75]. However, it has been observed that only 11%–13% of carbon mass gets converted into acetic acid [76]. Thus, Jin et al. suggested a two-step process to enhance the yield of acetic acid from carbohydrates, which consists of hydrothermal reaction followed by oxidation [77]. The first step involves an improvement in the yield of 5-hydroxymethyl-2-furaldehyde, 2-furadehyde, and lactic acid which is readily converted into a huge volume of acetic acid in the presence of an oxygen source [78].

D. *5-HMF*: 5-HMF has been known as one of the platform chemicals of the future chemical industry owing to its versatility and being a building block to produce fuels, fuel additives, and polymeric materials that can help humans transform fossil fuel-based economy to bio-based economy. Thus, the production of 5-HMF via HTL of LB can be recognized as one of the sustainable and efficient means. Generally, 5-HMF can be produced from biomass, carbohydrates, glucose, and fructose via dehydration reaction under acidic conditions [79]. Kuster et al. stated that ketohexoses resulted in a superior yield of 5-HMF than aldohexoses [80]. Several mineral acids such as H_3PO_4, HCl, and H_2SO_4 and organic acids such as maleic acid, oxalic acid, and citric acid have been used as catalysts to enhance the yield of 5-HMF [81]. Besides, various heterogeneous catalysts such as zirconium phosphate and zirconium oxide have been reported to produce 5-HMF from carbohydrates and sugars [82]. However, poor selectivity and high purification cost are the major issues in sustainable utilization of 5-HMF to produce fuel and other commodity chemicals. Thus, recently 5-HMF production practices have headed for the usage of biphasic media for the simultaneous extraction of 5-HMF into the organic phase [83]. Also, Caruso et al. testified the synthesis of 5-HMF via a novel electrochemical method from sucrose and fructose with more than 90% yield at room temperature in the presence of dimethyl sulphoxide (DMSO) [84].

E. *Levulinic acid*: Levulinic acid is also identified as 4-oxopentanoic acid or gamma-ketovaleric acid or 3-acetyl propionic acid with the molecular formula $C_5H_8O_3$, and it has a feasible association between biomass and petroleum processing. Multiple derivatives of levulinic acid, such as gamma-valerolactone, ethyl levulinate, and methyltetrahydrofuran, have been recommended for fuel applications [85]. Besides, chemicals derived from levulinic acid have multidimensional applications in various fields as solvents, resins, chemical intermediates, polymers, electronics, batteries, plasticizers, rubber, cosmetics, drug delivery carrier, textiles, and pharmaceuticals [86]. Thus, hydrothermal liquefaction of biomass attracted the attention of worldwide researchers for the sustainable production of levulinic acid. The process of production of levulinic acid from biomass involves the following steps: (a) hydrolysis of carbohydrates, (b) isomerization of glucose to fructose, (c) dehydration of fructose to 5-HMF, and (d) rehydration of 5-HMF into levulinic acid [87]. Previously, several studies have been reported to produce levulinic acid from biomass. Mineral acids such as HCl, H_2SO_4, and methanesulphonic acid have been used as a catalyst to generate levulinic acid from biomass [88]. Also, many heterogeneous catalysts such as $S_2O_8^{2-}/ZrO_2\text{-}SiO_2\text{-}Sm_2O_3$, GaHPMo, acid-modified zeolite, modified zeolite Y, hybrid of HY zeolite, $CrCl_3$, and $AlCl_3$ have been reported to produce levulinic acid from biomass [89, 90]. Similarly, several ionic acids such as $[C_3SO_3Hmim]^+[HSO_4]^-$, $[C_4(Mim)_2]^+[(2HSO_4)(H_2SO_4)_4]$, $[BMimSO_3H]^+[HSO_4]^-$, $[SMim]^+[FeCl_4]^-$, $InCl_3\text{-}[HMim]^+[HSO_4]^-$, and $[C_4Mim]^+[HSO_4]^-$ have been used for the hydrothermal liquefaction of biomass into levulinic acid [91,92].

In addition to these, fragmentation products such as glyceraldehyde, dihydroxyacetone, pyruvaldehyde, erythrose, glycolaldehyde, hydroxyacetone; dehydration products such as 6-anhydroglucose, furfural, 1,2,4-benzenetriol; condensation products involving soluble and insoluble polymeric products, i.e. humins; and carbonized products have been formed during the hydrothermal liquefaction of biomass [93].

4.4.2.2 Reaction Mechanism for Degradation of Lignin

Lignin is an abundant constituent of LB after cellulose and hemicellulose. Lignin is an extremely branched macromolecule that comprises of several oxygen and carbon linkages between alkylated methoxyphenol rings [94]. During hydrothermal conditions, numerous phenols and methoxyphenols are produced as a result of competitive reactions such as hydrolysis of ether linkages, cleavage of C–C bond, demethoxylation, alkylation, and condensation within lignin in the presence of an alkaline environment. To understand in detail the reaction pathways, many researchers have used model phenolic compounds such as catechol, guaiacol, vanillic acid, diphenyl ether, and benzyl phenyl ether. Similarly, wood can be the model substrate for lignin, which has been used to study hydrothermal degradation of lignin with the production of catechol, phenols, and cresols [95]. Zhang et al. studied the hydrothermal processing of kraft pine lignin and organosolv lignin at 374°C, 22 MPa for 10 mins., which resulted in 58%–79% liquid and 15%–20% solid residues [96]. Similarly, Liu et al. reported the base-catalysed hydrothermal treatment of walnut shells at 200°C–300°C with 2-methoxyphenol, 3,4-dimethoxyphenol, and 1,2-benzenediol as major products [97]. Thus, from previous studies, it can be concluded that the degradation of lignin includes hydrolysis of lignin followed by the hydrolysis of methoxy groups.

4.4.3 EFFECT OF OPERATING PARAMETERS ON HTL OF BIOMASS

Reduction in gaseous and residue char and maximization of the yield of bio-oil are the ultimate goals of biomass HTL. The progress of HTL of biomass depends on temperature, pressure, solvent properties, and substrate [98,99]. Among them, temperature and pressure directly affect the reaction by altering activation energy and reaction equilibria [100]. Besides, various parameters that affect the product distribution during HTL, such as heating rate, residence time, slurry concentration, and catalyst, are described in detail below. The schematic representation of the effect of operating parameters on HTL is shown in Figure 4.4.

4.4.3.1 Effect of Feedstock Type and Particle Size

Feedstock type is an important factor that affects the HTL of biomass. Biomass is composed of various components that react differently at identical hydrothermal conditions, and the composition of biomass alters with the feedstock type. Thus, feedstock type significantly affects the performance of HTL [101].

Several sorts of biomass that exist in the world have different chemical compositions with cellulose, hemicellulose, and lignin as major components. It has been reported that several biomass feedstocks resulted in bio-oil with variable constitutions due to the alteration in biomass composition. Karagoz et al. studied the HTL of different biomass

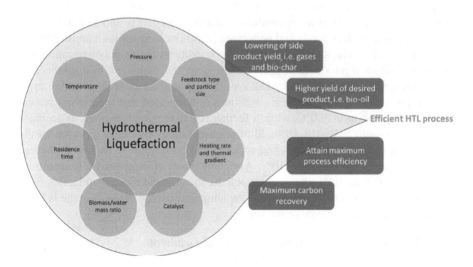

FIGURE 4.4 Diagrammatic illustration of the influence of operating parameters on HTL.

feedstocks such as cellulose, lignin, sawdust, and rice husk at 280°C for 15 min., which resulted in different bio-oil yields [102]. Similarly, Feng et al. investigated the HTL of barks of white pine, white spruce, and white birch at 300°C under 2 MPa for 15 min., which liquefied into 36%, 58%, and 67% bio-oil yields [103]. Generally, biomass with elevated carbohydrate content resulted in better bio-oil yield than biomass with higher lignin content. Reasons for these results are (a) stability and complex nature of lignin which is difficult to liquefy and (b) simpler structure of cellulose which is easy to degrade as compared to lignin. It is reported that the degradation of lignin is difficult; hence, it ends up in residue and its content can alter the yield of bio-oil [104]. Dermibas et al. identified the linear equations for the yield of bio-oil and residue char in terms of lignin content (wt.%) as shown in Eqs. 4.1 and 4.2 below [105].

$$\mathrm{Bio-oil\ yield\ }(\%) = 42.548 - 0.388 \times \mathrm{lignin\ content\ }(\mathrm{wt.\%}) \qquad (4.1)$$

$$\mathrm{Residue\ char\ yield\ }(\%) = 1.979 + 0.868 \times \mathrm{lignin\ content\ }(\mathrm{wt.\%}) \qquad (4.2)$$

In the case of sub-critical water conditions, sub-critical water acts as a heat transfer medium and an extractant and overpowers the heat transfer restrictions in HTL, which makes particle size a minor influencing factor. Thus, the particle size has a minor effect on the biomass HTL [65]. Zhang et al. reviewed the influence of particle size of grass perennials on HTL, which showed that the reduction in particle size does not affect the yield of bio-oil [106].

4.4.3.2 Effect of Temperature

The temperature has a substantial influence on the HTL of biomass. The rivalry within hydrolysis, fragmentation, and re-polymerization reactions defines the function of temperature in HTL of biomass. Also, escalation in thermal energy not only enhances the rate of reactions, but also alters the reaction pathways. A lower temperature supports ionic reactions, while a higher temperature promotes radical reactions [107]. At sub-critical water conditions, temperature rise results in a higher rate of cleavage of chemical bonds and depolymerization of biomass with an increase in bio-oil yield. However, at supercritical water conditions the re-polymerization and re-decomposition of intermediate products are promoted, which results in an increase in char yield and a drop in bio-oil yield. Zhu et al. examined the HTL of barley straw under supercritical and sub-critical conditions and reported that the highest bio-oil yield (34.9 wt.%) was at 300°C, while a further rise in temperature tends to decrease the yield of bio-oil (19.9 wt.%) [108]. Similar observations have been reported for the HTL of jack pine powder at sub- and supercritical conditions by Xu and Etchevery under a hydrogen environment [109]. Also, Sugano et al. suggested a temperature range from 300°C to 315°C that was suitable for the efficient bio-oil production from the HTL of biomass [110]. Conclusively, the temperature requirements of the HTL of biomass change with the type of feedstock.

4.4.3.3 Effect of Heating Rate and Thermal Gradient

The heating rate is also one of the vital influencing parameters that decide the order and extent of the chemical reactions in the whole HTL process. Also, the high actual localized temperature at the reactor wall results in the generation of char at the sidewall due to unwanted reactions [111]. The influence of heating rate on product distribution in the HTL of biomass is poorer than in pyrolysis owing to enhanced dissolution of biomass components and equilibration of intermediate species in boiling pressurized water. In the case of HTL of *Nannochloropsis* and *Chlorella* at 350°C for 60 min., bio-oil yield decreases from 37.2% to 35.8% as the heating rate increases from 10°C/min. to 25°C/min. The slow heating rate results in char formation, but the effect is not that significant [112,113]. Similarly, Akhtar et al. observed an insignificant influence of heating rate on the yield of bio-oil [4]. However, in case of HTL of grassland perennials, Zhang et al. reported a 13% increase in bio-oil yield by rising heating rate from 5°C/min. to 140°C/min [106]. Conclusively, a slow heating rate results in the generation of char owing to the unwanted side reactions, while a high heating rate leads to production of higher amount of gaseous products.

4.4.3.4 Effect of Residence Time

)During the HTL of biomass, residence time has a substantial influence on the of bio-oil yield [114]. Also, it may decide the product composition, bio-oil yield, and biomass conversion. Many scientists have reviewed the influence of residence time on bio-oil yield. Generally, a short residence time results in incomplete conversion, while a long residence time leads to re-polymerization of intermediates, leading to a decrease in the yield of bio-oil yield. Karagoz et al. testified that an increase in residence time at a lower temperature (150°C) results in a higher yield of liquid oil. Additionally, Karagoz et al. observed a difference in the product composition for short and long residence times at 180°C and 250°C [115]. Boocock et al. suggested that a longer reaction time leads to the decomposition of pre-asphaltenes and asphaltenes into lighter products, i.e. light oils or gases [116]. Yu et al. reported a 39.4% bio-oil yield at 280°C for 120 min. residence time, while Alba et al. obtained a 49.4% bio-oil yield at 375°C for 5 min. residence time [114]. Thus, a high temperature requires a short residence time to maximize the bio-oil yield.

4.4.3.5 Effect of Biomass/Water Mass Ratio

The biomass-to-water mass ratio of is one of the crucial factors that significantly affect the bio-oil yield from the HTL of biomass. The occurrence of water in the HTL of biomass enhances the dissolution of intermediates and re-polymerization of small molecules [117]. Besides, water helps to stabilize the free radical and to improve the bio-oil yield. Singh et al. testified the improvement in bio-oil yield and reduction in char yield from 6% to 18% with a drop in mass ratio of biomass to water from 1:3 to 1:12 in the case of HTL of hyacinth [118]. Also, Wang et al. observed the same results due to solvated biomass components [119]. Higher water amounts prevent the formation of char by avoiding cross-linking of aromatic compounds and hydrocarbons, which ultimately increases the bio-oil yield [120].

Contrarily, Liu et al. examined the influence of biomass-to-water ratio on the HTL of cornstalk, in which the bio-oil yield decreases with a rise in cornstalk-to-water ratio from 1:10 to 1:6 [121]. Similarly, Boocock et al. observed a low yield of bio-oil at a very small biomass-to-water ratio [116]. This is because the decrease in biomass-to-water ratio beyond the limit initiates the competition between hydrolysis and re-polymerization. Thus, the optimization of biomass/water ratio is very important to get maximum bio-oil yield.

4.4.3.6 Effect of Pressure

Pressure is one of the critical variables in the HTL of biomass that affect the bio-oil yield. Pressure eases the maintenance of a monophasic environment during the HTL of biomass under both sub-critical and supercritical water conditions, which avoids the large enthalpy inputs required for phase change [122]. Besides, pressure can control the hydrolysis rate and rate of biomass dissolution by enhancing decomposition and extraction of biomass. However, pressure has a negligible influence on the bio-oil yield at supercritical water conditions [123]. Kabyemela et al. stated that the rate of glucose hydrolysis remains unchanged with a decrease in the pyrolysis rate when pressure increases from 30 to 40 MPa [124]. An increased local density of water and cage effect at very high pressures inhibit the cleavage of C–C bonds of biomass. Conclusively, a high pressure owing to high temperatures can reduce the bio-oil yield with increasing char yield, while low pressure leads to incomplete reaction. Thus, the optimization of pressure is important in the case of HTL of biomass.

4.4.3.7 Effect of Catalyst

In the case of HTL of biomass, the catalyst can improve the process efficiency by decreasing the required temperature and pressure for producing maximum bio-oil. Both homogeneous and heterogeneous catalysts have been investigated for the HTL of biomass. Product distribution and selectivity of bio-oil production can be regulated by the usage of a suitable catalyst [65,117,125].

Mineral acids such as H_3PO_4, HCl, and H_2SO_4; organic acids such as formic and acetic acids; and alkalis such as Na_2CO_3, K_2CO_3, NaOH, and KOH have been reported as catalysts for biomass HTL [106,117,118]. However, the HTL of biomass using a weak organic acid as a catalyst leads to the production of bio-oil with high oxygen content, while strong mineral acids have strong corrosiveness, which hinders their commercialization. Thus, alkali salts are considered superior to acid catalysts in the case of homogeneous catalysis because alkali catalysts are responsible for the increase in pH of the liquid phase, which inhibits the dehydration and promotes decarboxylation of biomolecules. Besides, alkalis facilitate water–gas shift reaction to improve the H_2 yield, which acts as a reducing agent. Minowa et al. stated that the usage of catalyst, i.e. Na_2CO_3, led to the improvement of bio-oil yield with a reduction in the quantity of char generated in the case of HTL of cellulose [126]. Similarly, Singh et al. investigated the HTL of water hyacinth using K_2CO_3 and KOH, which resulted in an enhancement in the production of bio-oil efficiency [65].

Heterogeneous catalysts used for the HTL of biomass include metals, metal oxides, and doped metal oxides and are as follows: Pd, Pt, Ru, Co, Mo, Ni, Pt, and Ni supported on SiO_2, Al_2O_3, and zeolite [126]. Heterogeneous catalysts affect the bio-oil yield, product composition, and heating value significantly [127]. Biller et al. studied the HTL of biomass with Co/Mo, Ni/Al, and Pt/Al catalysts, which resulted in an improved bio-oil yield and heating value [113]. However, the dissolution of biomass, poisoning, sintering, and intraparticle diffusion are the main disadvantages of heterogeneous catalysts for the HTL of biomass. Besides, the requirement of reducing gas is one of the major drawbacks of a heterogeneous catalyst. Thus, the selection of catalysts is important to get maximum bio-oil yield.

4.5 OVERVIEW OF INDUSTRIAL APPLICATIONS OF HTL

Considering the depletion of fossil fuel, there is a burning necessity for a sustainable replacement for fossil fuel. And the HTL of biomass is one of the promising substitutes to fossil fuel. Thus, a lot of researches have been done and are still underway throughout the world for the commercialization of the HTL of biomass. Several processes have been reported for the HTL of biomass, such as PERC process, LBL process, HTU process, BFH process, DoS process, B/M process – Mueborit, and LTC process, and are shown in Table 4.3 [128].

a. PERC process has been developed by the Pittsburgh Energy Research Centre (PERC). PERC process converts wood chips into oil, in which the mixture of wood chips is pumped into a tube reactor for 10–30 min. at 330°C–370°C under 200 bar pressure. The process produces 45%–55% of bio-oil yield, and the recycled oil has been used as a hydrogen supply [129]. Further development in this process is halted due to low economic feasibility.

b. Lawrence Berkeley National Laboratory established the LBL process, in which firstly the biomass get hydrolysed with sulphuric acid followed by neutralization with Na_2CO_3. The resultant mixture was pumped through a tube reactor at 330°C–360°C under 100–240 bar pressure for the production of bitumen-like product having a calorific value of 34 MJ/kg [130]. Further development in this process is abandoned due to economic reasons.

c. BFH (Bundesforschungsanstalt für Forst- und Holzwirtschaft, Germany) process is a semi-continuous process for the liquefaction of lignocellulosic biomass, which involves three linked autoclaves for fast heating of the aqueous solvent and fast cooling of the reaction mixture. The process based on catalytic hydrogenolysis using hydrogen, catalyst, and oil gave about 36% oil as liquid tar, 50% carrier oil, 5% coke, and remaining aqueous phase [128].

TABLE 4.3
Overview of HTL Processes of Biomass [128]

S. No.	Process Name	Temperature (°C)	Pressure (bar)	Catalyst	References
1	PERC process	330–370	200	Yes	[129]
2	LBL process	330–360	170–240	Yes	[130]
3	HTU process	265–350	180	No	[131]
4	BFH process	380	100	No	[128]
5	DoS process	350–550	80	No	[132]
6	B/M process – Mueborit	<220	6	No	[128]
7	LTC process	400	1	Yes	[133]

d. Amsterdam-based Shell Research Laboratory has developed a process that converts biomass into bio-oil, named as hydrothermal upgrading process, i.e. HTU process, in which suspended biomass is pumped into the reactor and heated at 300°C–350°C under 120–180 bar pressure for 5–20 min. The product consists of 45% bio-crude oil with 30–35 MJ/kg calorific value, 25% gas, and the remaining aqueous phase [131].

e. DoS process was established by Hochschule für Angewandte Wissenschaften Hamburg, Germany. It is a one-step liquefaction process in which lignocellulosic biomass is heated up to 350°C–500°C under 80 bar pressure. The process is based on fast pyrolysis followed by solvolysis into liquid oil. The overall thermal efficiency of the process is about 70% based on the heating value of the feedstock [132].

f. Umwelttechnik Stefan Bothur has developed the solvolysis-based B/M process – Mueborit. In B/M process, lignocellulosic biomass is solvated at 200°C under 6 bar pressure in a discontinuous tank reactor in the presence of melt of potassium carbonate hydrate, which resulted in about 40% dark brown bio-oil having a calorific value of 35–37 MJ/kg [128].

g. The low-temperature conversion (LTC) process has been established at Giessen–Friedberg University of Applied Sciences by Stadlbauer as a catalytic process, which works in the absence of oxygen at atmospheric pressure at 350°C–400°C temperature. The process has successfully converted sewage sludge, bone, animal fat, tar, fat residue, etc., into bio-oil. The resultant oil has comparable physicochemical properties as diesel fuel [133].

Several processes are available for the liquefaction of LB into bio-oil which can be a potential substitute for fossil fuel. But, the attainment of economic and energetic feasibility is still a challenging task for worldwide researchers.

4.6 CHALLENGES AND FUTURE PERSPECTIVE

Bio-oil produced from the HTL of biomass has the potential to replace fossil fuel, which helps humankind to shift towards a sustainable circular bio-based economy. The study of HTL of biomass is in the development stage, and commercialization is a chief challenge. The major problems associated with the HTL of biomass are as follows: (a) requirement of high temperature and pressure, (b) requirement of a reducing gas, (c) formation of coke, tar, and solid residue responsible for catalyst deactivation, (d) bio-oil with high oxygen content, (e) high nitrogen contents inhibiting the application of bio-oil which thus requires upgradation of bio-oil, and (f) the complexity of lignin, which being one of the important components of biomass decreases the HTL efficiency. In addition to this, there are several challenges associated with the commercialization HTL of biomass such as (a) requirement of a large and consistent supply of biomass, (b) variability of biomass composition, (c) fuel quality, (d) cost competitiveness with other technologies, (e) requirement of the efficient supply chain for collection, storage, and transportation of feedstock, liquefaction products, and intermediates, and (f) change in governmental policies [134]. The challenges associated with the HTL of biomass are shown in Figure 4.5.

FIGURE 4.5   Schematic representation of challenges associated with the HTL of biomass.

To overcome the challenges associated with the HTL of biomass for maximum energy utilization efficiency, a deep understanding of feedstocks, their processing, and catalysis is most essential. Also, biomass liquefaction needs stable support policies from the government which can make a huge impact on the economy of our country.

4.7   CONCLUSIONS

Biomass is an alternative renewable source having the potential to replace fossil fuels from the day-to-day life of humankind. Thus, the production of biofuels, energy, power, etc., from biomass (biorefinery) has attracted the attention of researchers worldwide. Several conventional processes are available in the market, such as pyrolysis, gasification, and liquefaction generation of energy and fuel products from biomass. But the reported processes are associated with several disadvantages, challenges, and barriers. Thus, of the reported options HTL can be seen as one of the best choices to produce second-generation fuel in the form of bio-oil through liquefaction of not only biomass but also any organic feedstock such as municipal solid wastes, plastics, industrial waste liquids, sewage sludge, and animal wastes. This chapter briefly explored the importance of biomass HTL for bio-oil production as a sustainable alternative to fossil fuel. Mainly, the lignocellulosic residues are focused on being utilized for the production of bio-oil through biomass HTL. The HTL of biomass requires a crucial understanding of feedstocks; thus, detailed structural and compositional analyses of biomass are important. Various pre-treatment strategies can also be employed to reduce the recalcitrant nature of LB and to improve the process output as well as process economics. Various plausible reaction routes of HTL has been proposed to obtain the chief products and to enhance knowledge about the process. The HTL process can also be dependent on the operating parameters, which can help the new researchers for improving process output and overall economics. However, presently available HTL processes are associated with several drawbacks such as the requirement of high temperature and pressure, and a reducing gas; generation of coke, tar, and solid residue responsible for catalyst deactivation; and generation of bio-oil having high oxygen and nitrogen contents limiting the application of bio-oil, thus requiring upgradation of bio-oil via hydrogenation, denitrification, deoxygenation, etc., for widening the scope of bio-oil. In addition, the development and advancement of a robust, feasible, and recyclable catalyst according to the feedstock used can be a potential solution for the improvement in HTL efficiency. Also, there are several challenges associated with the commercialization HTL of biomass, such as (a) requirement of a large and consistent supply of biomass, (b) variability of biomass composition, (c) fuel quality, (d) cost competitiveness with other technologies, (e) requirement of the efficient supply chain for collection, storage, and transportation of feedstock, liquefaction products, and intermediates, and (f) change in governmental policies. To conquer the challenges associated with biomass HTL, maximum energy competence and a deep understanding of feedstocks, their processing, and catalysis are most essential. Also, biomass liquefaction processes such as HTL desire long-standing government support policies for commercialization at a large scale that can contribute to shift fossil fuel-based economy towards circular bio-based economy.

REFERENCES

1. Zhao, Na, and Bao Xia Li. "The Effect of Sodium Chloride on the Pyrolysis of Rice Husk." *Applied Energy*, 2016. doi:10.1016/j.apenergy.2016.06.082.
2. Gerssen-Gondelach, S. J., D. Saygin, B. Wicke, M. K. Patel, and A. P.C. Faaij. "Competing Uses of Biomass: Assessment and Comparison of the Performance of Bio-Based Heat, Power, Fuels and Materials." *Renewable and Sustainable Energy Reviews*, 2014. doi:10.1016/j.rser.2014.07.197.
3. Balat, Havva, and Elif Kirtay. "Hydrogen from Biomass - Present Scenario and Future Prospects." *International Journal of Hydrogen Energy*, 2010. doi:10.1016/j.ijhydene.2010.04.137.
4. Akhtar, Javaid, and Nor Aishah Saidina Amin. "A Review on Process Conditions for Optimum Bio-Oil Yield in Hydrothermal Liquefaction of Biomass." *Renewable and Sustainable Energy Reviews*, 2011. doi:10.1016/j.rser.2010.11.054.

5. Salan, T. "A Brief Review of the Thermochemical Platform as a Promising Way to Produce Sustainable Liquid Biofuels in Biorefinery Concept." *Mapping the Development of UK Biorefinery Complexes (NFC 07/008)*, 2017.

6. Gumisiriza, Robert, Joseph Funa Hawumba, Mackay Okure, and Oliver Hensel. "Biomass Waste-to-Energy Valorisation Technologies: A Review Case for Banana Processing in Uganda." *Biotechnology for Biofuels*, 2017. doi:10.1186/s13068-016-0689-5.

7. Zhang, Shiqiu, Xue Yang, Haiqing Zhang, Chunli Chu, Kui Zheng, Meiting Ju, and Le Liu. "Liquefaction of Biomass and Upgrading of Bio-Oil: A Review." *Molecules*, 2019. doi:10.3390/molecules24122250.

8. Basar, Ibrahim Alper, Huan Liu, Helene Carrere, Eric Trably, and Cigdem Eskicioglu. "A Review on Key Design and Operational Parameters to Optimize and Develop Hydrothermal Liquefaction of Biomass for Biorefinery Applications." *Green Chemistry*, 2021. doi:10.1039/d0gc04092d.

9. Akhtar, Nadeem, Kanika Gupta, Dinesh Goyal, and Arun Goyal. "Recent Advances in Pretreatment Technologies for Efficient Hydrolysis of Lignocellulosic Biomass." *Environmental Progress and Sustainable Energy*, 2016. doi:10.1002/ep.12257.

10. Sharma, Hem Kanta, Chunbao Xu, and Wensheng Qin. "Biological Pretreatment of Lignocellulosic Biomass for Biofuels and Bioproducts: An Overview." *Waste and Biomass Valorization*, 2019. doi:10.1007/s12649-017-0059-y.

11. McKendry, Peter. "Energy Production from Biomass (Part 1): Overview of Biomass." *Bioresource Technology*, 2002. doi:10.1016/S0960-8524(01)00118-3.

12. Fengel, Dietrich, and Gerd Wegener. *Wood: Chemistry, Ultrastructure, Reactions. Wood: Chemistry, Ultrastructure, Reactions*, 2011. doi:10.1515/9783110839654.

13. Zhao, Xuebing, Lihua Zhang, and Dehua Liu. "Biomass Recalcitrance. Part II: Fundamentals of Different Pre-Treatments to Increase the Enzymatic Digestibility of Lignocellulose." *Biofuels, Bioproducts and Biorefining*, 2012. doi:10.1002/bbb.1350.

14. Hendriks, A. T. W. M., and G. Zeeman. "Pretreatments to Enhance the Digestibility of Lignocellulosic Biomass." *Bioresource Technology*, 2009. doi:10.1016/j.biortech.2008.05.027.

15. Agbor, Valery B., Nazim Cicek, Richard Sparling, Alex Berlin, and David B. Levin. "Biomass Pretreatment: Fundamentals toward Application." *Biotechnology Advances*, 2011. doi:10.1016/j.biotechadv.2011.05.005.

16. Guerriero, Gea, Jean Francois Hausman, Joseph Strauss, Haluk Ertan, and Khawar Sohail Siddiqui. "Lignocellulosic Biomass: Biosynthesis, Degradation, and Industrial Utilization." *Engineering in Life Sciences*, 2016. doi:10.1002/elsc.201400196.

17. Singh, Seema, Gang Cheng, Noppadon Sathitsuksanoh, Dong Wu, Patanjali Varanasi, Anthe George, Venkatesh Balan, et al. "Comparison of Different Biomass Pretreatment Techniques and Their Impact on Chemistry and Structure." *Frontiers in Energy Research*, 2015. doi:10.3389/fenrg.2014.00062.

18. Chandra, R. P., R. Bura, W. E. Mabee, A. Berlin, X. Pan, and J. N. Saddler. "Substrate Pretreatment: The Key to Effective Enzymatic Hydrolysis of Lignocellulosics?" *Advances in Biochemical Engineering Biotechnology*, 2007. doi:10.1007/10_2007_064.

19. Palmowski, L. M., and J. A. Müller. "Influence of the Size Reduction of Organic Waste on Their Anaerobic Digestion." *Water Science and Technology*, 2000. doi:10.2166/wst.2000.0067.

20. Chang, Vincent S., Barry Burr, and Mark T. Holtzapple. "Lime Pretreatment of Switchgrass." *Applied Biochemistry and Biotechnology - Part A Enzyme Engineering and Biotechnology*, 1997. doi:10.1007/BF02920408.

21. Betiku, E., O. A. Adetunji, T. V. Ojumu, and B. O. Solomon. "A Comparative Study of the Hydrolysis of Gamma Irradiated Lignocelluloses." *Brazilian Journal of Chemical Engineering*, 2009. doi:10.1590/S0104-66322009000200002.

22. Mohanram, Saritha, Dolamani Amat, Jairam Choudhary, Anju Arora, and Lata Nain. "Novel Perspectives for Evolving Enzyme Cocktails for Lignocellulose Hydrolysis in Biorefineries." *Sustainable Chemical Processes*, 2013. doi:10.1186/2043-7129-1-15.

23. Magnusson, Lauren, Rumana Islam, Richard Sparling, David Levin, and Nazim Cicek. "Direct Hydrogen Production from Cellulosic Waste Materials with a Single-Step Dark Fermentation Process." *International Journal of Hydrogen Energy*, 2008. doi:10.1016/j. ijhydene.2008.06.018.

24. Jeremic, Dragica, Robyn E. Goacher, Ruoyu Yan, Chithra Karunakaran, and Emma R. Master. "Direct and Up-Close Views of Plant Cell Walls Show a Leading Role for Lignin-Modifying Enzymes on Ensuing Xylanases." *Biotechnology for Biofuels*, 2014. doi:10.1186/s13068-014-0176-9.

25. Itoh, Hiromichi, Masanori Wada, Yoichi Honda, Masaaki Kuwahara, and Takashi Watanabe. "Bioorganosolve Pretreatments for Simultaneous Saccharification and Fermentation of Beech Wood by Ethanolysis and White Rot Fungi." *Journal of Biotechnology*, 2003. doi:10.1016/S0168-1656(03)00123-8.

26. Pakarinen, O. M., P. L.N. Kaparaju, and J. A. Rintala. "Hydrogen and Methane Yields of Untreated, Water-Extracted and Acid (HCl) Treated Maize in One- and Two-Stage Batch Assays." *International Journal of Hydrogen Energy*, 2011. doi:10.1016/j.ijhydene.2011.08.028.

27. Harmsen, P., L. Bermudez, and R. Bakker. "Literature Review of Physical and Chemical Pretreatment Processes for Lignocellulosic Biomass." *Biomass*, Report/ Wageningen UR, Food & Biobased Research 1184, ISBN 9085857570, 9789085857570.

28. Zheng, Mingxia, Xiujin Li, Laiqing Li, Xiaojin Yang, and Yanfeng He. "Enhancing Anaerobic Biogasification of Corn Stover through Wet State NaOH Pretreatment." *Bioresource Technology*, 2009. doi:10.1016/j.biortech.2009.05.045.

29. Li, Lin, Chang Chen, Ruihong Zhang, Yanfeng He, Wen Wang, and Guangqing Liu. "Pretreatment of Corn Stover for Methane Production with the Combination of Potassium Hydroxide and Calcium Hydroxide." *Energy and Fuels*, 2015. doi:10.1021/ acs.energyfuels.5b01170.

30. Thring, Ronald, Esteban Chornet, and Ralph Overend. "Recovery of a Solvolytic Lignin: Effects of Spent Liquor/Acid Volume Ratio, Acid Concentrated and Temperature." *Biomass*, 1990. doi:10.1016/0144-4565(90)90038-L.

31. Zhao, Xuebing, Keke Cheng, and Dehua Liu. "Organosolv Pretreatment of Lignocellulosic Biomass for Enzymatic Hydrolysis." *Applied Microbiology and Biotechnology*, 2009. doi:10.1007/s00253-009-1883-1.

32. Curvelo, A. A. S., and R. Pereira. "Kinetics of ethanol-water of sugar cane bagasse." *The 8th International Symposium on Wood and Pulping Chemistry Proc*, Helsinki, Finland, 1995; 2: 473–478.

33. Zhang, Y. H. Percival. "Reviving the Carbohydrate Economy via Multi-Product Lignocellulose Biorefineries." *Journal of Industrial Microbiology and Biotechnology*, 2008. doi:10.1007/s10295-007-0293-6.

34. Hongzhang, Chen, and Liu Liying. "Unpolluted Fractionation of Wheat Straw by Steam Explosion and Ethanol Extraction." *Bioresource Technology*, 2007. doi:10.1016/j. biortech.2006.02.029.

35. Rughani, Jagdish, and Gary D. McGinnis. "Combined Rapid-steam Hydrolysis and Organosolv Pretreatment of Mixed Southern Hardwoods." *Biotechnology and Bioengineering*, 1989. doi:10.1002/bit.260330604.

36. Badgujar, Kirtikumar C., and Bhalchandra M. Bhanage. "Factors Governing Dissolution Process of Lignocellulosic Biomass in Ionic Liquid: Current Status, Overview and Challenges." *Bioresource Technology*, 2015. doi:10.1016/j.biortech.2014.09.138.

37. Li, Chenlin, Bernhard Knierim, Chithra Manisseri, Rohit Arora, Henrik V. Scheller, Manfred Auer, Kenneth P. Vogel, Blake A. Simmons, and Seema Singh. "Comparison of Dilute Acid and Ionic Liquid Pretreatment of Switchgrass: Biomass Recalcitrance,

Delignification and Enzymatic Saccharification." *Bioresource Technology*, 2010. doi:10.1016/j.biortech.2009.10.066.

38. Sant'Ana da Silva, Ayla, Seung-Hwan Lee, Takashi Endo, and Elba Bon. "Major Improvement in the Rate and Yield of Enzymatic Saccharification of Sugarcane Bagasse via Pre-Treatment with the Ionic Liquid 1-Ethyl-3-Methylimidazolium Acetate ([Emim][Ac])." *Bioresource Technology*, 2011. doi:10.1016/j.biortech.2011.08.085.

39. Ninomiya, Kazuaki, Hiroshi Soda, Chiaki Ogino, Kenji Takahashi, and Nobuaki Shimizu. "Effect of Ionic Liquid Weight Ratio on Pretreatment of Bamboo Powder Prior to Enzymatic Saccharification." *Bioresource Technology*, 2013. doi:10.1016/j.biortech.2012.10.097.

40. Zavrel, Michael, Daniela Bross, Matthias Funke, Jochen Büchs, and Antje C. Spiess. "High-Throughput Screening for Ionic Liquids Dissolving (Ligno-)Cellulose." *Bioresource Technology*, 2009. doi:10.1016/j.biortech.2008.11.052.

41. Zhu, Zhisheng, Mingjun Zhu, and Zhenqiang Wu. "Pretreatment of Sugarcane Bagasse with $NH_4OH-H_2O_2$ and Ionic Liquid for Efficient Hydrolysis and Bioethanol Production." *Bioresource Technology*, 2012. doi:10.1016/j.biortech.2012.05.111.

42. Zhao, Wei, Ruijin Yang, Yiqi Zhang, and Li Wu. "Sustainable and Practical Utilization of Feather Keratin by an Innovative Physicochemical Pretreatment: High Density Steam Flash-Explosion." *Green Chemistry*, 2012. doi:10.1039/c2gc36243k.

43. Haghighi Mood, Sohrab, Amir Hossein Golfeshan, Meisam Tabatabaei, Gholamreza Salehi Jouzani, Gholam Hassan Najafi, Mehdi Gholami, and Mehdi Ardjmand. "Lignocellulosic Biomass to Bioethanol, a Comprehensive Review with a Focus on Pretreatment." *Renewable and Sustainable Energy Reviews*, 2013. doi:10.1016/j.rser.2013.06.033.

44. Yu, Zhengdao, Bailiang Zhang, Fuqiang Yu, Guizhuan Xu, and Andong Song. "A Real Explosion: The Requirement of Steam Explosion Pretreatment." *Bioresource Technology*, 2012. doi:10.1016/j.biortech.2012.06.055.

45. Wright, J. D. "Ethanol from Biomass by Enzymatic Hydrolysis." *Chemical Engineering Progress*, 1998, 84: 8.

46. Datar, Rohit, Jie Huang, Pin Ching Maness, Ali Mohagheghi, Stefan Czernik, and Esteban Chornet. "Hydrogen Production from the Fermentation of Corn Stover Biomass Pretreated with a Steam-Explosion Process." *International Journal of Hydrogen Energy*, 2007. doi:10.1016/j.ijhydene.2006.09.027.

47. Mosier, Nathan, Charles Wyman, Bruce Dale, Richard Elander, Y. Y. Lee, Mark Holtzapple, and Michael Ladisch. "Features of Promising Technologies for Pretreatment of Lignocellulosic Biomass." *Bioresource Technology*, 2005. doi:10.1016/j.biortech.2004.06.025.

48. Mosier, Nathan S., Richard Hendrickson, Mark Brewer, Nancy Ho, Miroslav Sedlak, Richard Dreshel, Gary Welch, Bruce S. Dien, Andy Aden, and Michael R. Ladisch. "Industrial Scale-up of PH-Controlled Liquid Hot Water Pretreatment of Corn Fiber for Fuel Ethanol Production." *Applied Biochemistry and Biotechnology - Part A Enzyme Engineering and Biotechnology*, 2005. doi:10.1385/abab:125: 2: 077.

49. Antal, Jr. M. J. "Water: A Traditional Solvent Pregnant with New Application." In: White, Jr H. J., editor. *Proceedings of the 12th International Conference on the Properties of Water and Steam*, New York: Begell House, 1996: 23–32.

50. Weil, Joseph, Mark Brewer, Richard Hendrickson, Ayda Sarikaya, and Michael R. Ladisch. "Continuous PH Monitoring during Pretreatment of Yellow Poplar Wood Sawdust Pressure Cooking in Water." In *Applied Biochemistry and Biotechnology - Part A Enzyme Engineering and Biotechnology*, 1998. doi:10.1007/BF02920127.

51. Foster, Brian L., Bruce E. Dale, and Joy B. Doran-Peterson. "Enzymatic Hydrolysis of Ammonia-Treated Sugar Beet Pulp." In *Applied Biochemistry and Biotechnology - Part A Enzyme Engineering and Biotechnology*, 2001. doi:10.1385/ABAB:91-93:1-9:269.

52. Kim, Sung Bae, and Y. Y. Lee. "Diffusion of Sulfuric Acid within Lignocellulosic Biomass Particles and Its Impact on Dilute-Acid Pretreatment." *Bioresource Technology*, 2002. doi:10.1016/S0960-8524(01)00197-3.
53. Kim, Tae Hyun, Jun Seok Kim, Changshin Sunwoo, and Y. Y. Lee. "Pretreatment of Corn Stover by Aqueous Ammonia." *Bioresource Technology*, 2003. doi:10.1016/S0960-8524(03)00097-X.
54. McMillan, J. D. "Pretreatment of Lignocellulosic Biomass." *Enzymatic Conversion of Biomass Fuels Production*, 1994; 566: 292–324.
55. Kostas, Emily T., Daniel Beneroso, and John P. Robinson. "The Application of Microwave Heating in Bioenergy: A Review on the Microwave Pre-Treatment and Upgrading Technologies for Biomass." *Renewable and Sustainable Energy Reviews*, 2017. doi:10.1016/j.rser.2017.03.135.
56. Hu, Jun, Bingxing Jiang, Jing Wang, Yiheng Qiao, Tianyi Zuo, Yahui Sun, and Xiaoxiang Jiang. "Physicochemical Characteristics and Pyrolysis Performance of Corn Stalk Torrefied in Aqueous Ammonia by Microwave Heating." *Bioresource Technology*, 2019. doi:10.1016/j.biortech.2018.11.076.
57. Zheng, Yi, Jia Zhao, Fuqing Xu, and Yebo Li. "Pretreatment of Lignocellulosic Biomass for Enhanced Biogas Production." *Progress in Energy and Combustion Science*, 2014. doi:10.1016/j.pecs.2014.01.001.
58. Wang, Zhenyu, Lijie Qu, Jing Qian, Zhengbin He, and Songlin Yi. "Effects of the Ultrasound-Assisted Pretreatments Using Borax and Sodium Hydroxide on the Physicochemical Properties of Chinese Fir." *Ultrasonics Sonochemistry*, 2019. doi:10.1016/j.ultsonch.2018.09.017.
59. Sasmal, Soumya, Vaibhav V. Goud, and Kaustubha Mohanty. "Ultrasound Assisted Lime Pretreatment of Lignocellulosic Biomass toward Bioethanol Production." In *Energy and Fuels*, 2012. doi:10.1021/ef300669w.
60. Ravindran, Rajeev, and Amit Kumar Jaiswal. "A Comprehensive Review on Pre-Treatment Strategy for Lignocellulosic Food Industry Waste: Challenges and Opportunities." *Bioresource Technology*, 2016. doi:10.1016/j.biortech.2015.07.106.
61. Schmidt, Anette Skammelsen, and Anne Belinda Thomsen. "Optimization of Wet Oxidation Pretreatment of Wheat Straw." *Bioresource Technology*, 1998. doi:10.1016/S0960-8524(97)00164-8.
62. Peterson, Andrew A., Frédéric Vogel, Russell P. Lachance, Morgan Fröling, Michael J. Antal, and Jefferson W. Tester. "Thermochemical Biofuel Production in Hydrothermal Media: A Review of Sub- and Supercritical Water Technologies." *Energy and Environmental Science*, 2008. doi:10.1039/b810100k.
63. Kruse, A., and E. Dinjus. "Hot Compressed Water as Reaction Medium and Reactant. Properties and Synthesis Reactions." *Journal of Supercritical Fluids*, 2007. doi:10.1016/j.supflu.2006.03.016.
64. Uematsu, M., and E. U. Frank. "Static Dielectric Constant of Water and Steam." *Journal of Physical and Chemical Reference Data*, 1980. doi:10.1063/1.555632.
65. Gollakota, A. R. K., Nanda Kishore, and Sai Gu. "A Review on Hydrothermal Liquefaction of Biomass." *Renewable and Sustainable Energy Reviews*, 2018. doi:10.1016/j.rser.2017.05.178.
66. Beims, Ramon Filipe, Yulin Hu, Hengfu Shui, and Chunbao (Charles) Xu. "Hydrothermal Liquefaction of Biomass to Fuels and Value-Added Chemicals: Products Applications and Challenges to Develop Large-Scale Operations." *Biomass and Bioenergy*, 2020. doi:10.1016/j.biombioe.2020.105510.
67. Jin, Fangming, Jun Yun, Guangming Li, Ashushi Kishita, Kazuyuki Tohji, and Heiji Enomoto. "Hydrothermal Conversion of Carbohydrate Biomass into Formic Acid at Mild Temperatures." *Green Chemistry*, 2008. doi:10.1039/b802076k.

68. Yun, J., F. Jin, A. Kishita, K. Tohji, and H. Enomoto. "Formic Acid Production from Carbohydrates Biomass by Hydrothermal Reaction." In *Journal of Physics: Conference Series*, 2010. doi:10.1088/1742-6596/215/1/012126.
69. Srokol, Zbigniew, Anne Gaëlle Bouche, Anton Van Estrik, Rob C.J. Strik, Thomas Maschmeyer, and Joop A. Peters. "Hydrothermal Upgrading of Biomass to Biofuel; Studies on Some Monosaccharide Model Compounds." *Carbohydrate Research*, 2004. doi:10.1016/j.carres.2004.04.018.
70. Matsumura, Yukihiko, Mitsuru Sasaki, Kazuhide Okuda, Seiichi Takami, Satoshi Ohara, Mitsuo Umetsu, and Tadafumi Adschiri. "Supercritical Water Treatment of Biomass for Energy and Material Recovery." *Combustion Science and Technology*, 2006. doi:10.1080/00102200500290815.
71. Shanableh, A. "Production of Useful Organic Matter from Sludge Using Hydrothermal Treatment." *Water Research*, 2000. doi:10.1016/S0043-1354(99)00222-5.
72. Yan, Xiuyi, Fangming Jin, Kazuyuki Tohji, Takehiko Moriya, and Heiji Enomoto. "Production of Lactic Acid from Glucose by Alkaline Hydrothermal Reaction." *Journal of Materials Science*, 2007. doi:10.1007/s10853-007-2012-0.
73. Kong, Lingzhao, Guangming Li, Hua Wang, Wenzhi He, and Fang Ling. "Hydrothermal Catalytic Conversion of Biomass for Lactic Acid Production." *Journal of Chemical Technology and Biotechnology*, 2008. doi:10.1002/jctb.1797.
74. Jin, Fangming, Atsushi Kishita, Takehiko Moriya, Heiji Enomoto, and Naohiro Sato. "The Production of Acetate for Use as a Roadway Deicer by Wet Oxidation of Organic Waste." In *ACS Division of Environmental Chemistry*, Preprints, 2000.
75. Jin, Fangming, Zhouyu Zhou, Takehiko Moriya, Hisanori Kishida, Hisao Higashijima, and Heiji Enomoto. "Controlling Hydrothermal Reaction Pathways to Improve Acetic Acid Production from Carbohydrate Biomass." *Environmental Science and Technology*, 2005. doi:10.1021/es048867a.
76. Jin, Fangming, Atsushi Kishita, Takehiko Moriya, Heiji Enomoto, and Naohiro Sato. "A New Process for Producing Ca/Mg Acetate Deicer with Ca/Mg Waste and Acetic Acid Produced by Wet Oxidation of Organic Waste." *Chemistry Letters*, 2002. doi:10.1246/cl.2002.88.
77. Jin, Fangming, Junchao Zheng, Heiji Enomoto, Takehiko Moriya, Naohiro Sato, and Hisao Higashijima. "Hydrothermal Process for Increasing Acetic Acid Yield from Lignocellulosic Wastes." *Chemistry Letters*, 2002. doi:10.1246/cl.2002.504.
78. Jin, F., Z. Zhou, A. Kishita, H. Enomoto, H. Kishida, and T. Moriya. "A New Hydrothermal Process for Producing Acetic Acid from Biomass Waste." *Chemical Engineering Research and Design*, 2007. doi:10.1205/cherd06020.
79. Putten, Robert Jan Van, Jan C. Van Der Waal, Ed De Jong, Carolus B. Rasrendra, Hero J. Heeres, and Johannes G. De Vries. "Hydroxymethylfurfural, a Versatile Platform Chemical Made from Renewable Resources." *Chemical Reviews*, 2013. doi:10.1021/cr300182k.
80. Kuster, B. F.M. "5-Hydroxymethylfurfural (HMF). A Review Focussing on Its Manufacture." *Starch - Stärke*, 1990. doi:10.1002/star.19900420808.
81. Asghari, Feridoun Salak, and Hiroyuki Yoshida. "Acid-Catalyzed Production of 5-Hydroxymethyl Furfural from D-Fructose in Subcritical Water." *Industrial and Engineering Chemistry Research*, 2006. doi:10.1021/ie051088y.
82. Asghari, Feridoun Salak, and Hiroyuki Yoshida. "Dehydration of Fructose to 5-Hydroxymethylfurfural in Sub-Critical Water over Heterogeneous Zirconium Phosphate Catalysts." *Carbohydrate Research*, 2006. doi:10.1016/j.carres.2006.06.025.
83. Saha, Basudeb, and Mahdi M. Abu-Omar. "Advances in 5-Hydroxymethylfurfural Production from Biomass in Biphasic Solvents." *Green Chemistry*, 2014. doi:10.1039/c3gc41324a.

84. Caruso, Tonino, and Ermanno Vasca. "Electrogenerated Acid as an Efficient Catalyst for the Preparation of 5-Hydroxymethylfurfural." *Electrochemistry Communications*, 2010. doi:10.1016/j.elecom.2010.05.040.

85. Signoretto, Michela, Somayeh Taghavi, Elena Ghedini, and Federica Menegazzo. "Catalytic Production of Levulinic Acid (LA) from Actual Biomass." *Molecules (Basel, Switzerland)*, 2019. doi:10.3390/molecules24152760.

86. Pileidis, Filoklis D., and Maria Magdalena Titirici. "Levulinic Acid Biorefineries: New Challenges for Efficient Utilization of Biomass." *ChemSusChem*, 2016. doi:10.1002/cssc.201501405.

87. Rackemann, D. W., and W. O. S. Doherty. "A Review on the Production of Levulinic Acid and Furanics from Sugars." In *34th Annual Conference of the Australian Society of Sugar Cane Technologists (ASSCT) 2012*, 2012.

88. Fang, Qi, and Milford A. Hanna. "Experimental Studies for Levulinic Acid Production from Whole Kernel Grain Sorghum." *Bioresource Technology*, 2002. doi:10.1016/S0960-8524(01)00144-4.

89. Chen, Hongzhang, Bin Yu, and Shengying Jin. "Production of Levulinic Acid from Steam Exploded Rice Straw via Solid Superacid, S_2O_{82}-/ZrO_2-SiO_2-Sm_2O_3." *Bioresource Technology*, 2011. doi:10.1016/j.biortech.2010.10.018.

90. Kumar, Vijay Bhooshan, Indra Neel Pulidindi, Rahul Kumar Mishra, and Aharon Gedanken. "Development of Ga Salt of Molybdophosphoric Acid for Biomass Conversion to Levulinic Acid." *Energy and Fuels*, 2016. doi:10.1021/acs.energyfuels.6b02403.

91. Khan, Amir Sada, Zakaria Man, Mohamad Azmi Bustam, Asma Nasrullah, Zahoor Ullah, Ariyanti Sarwono, Faiz Ullah Shah, and Nawshad Muhammad. "Efficient Conversion of Lignocellulosic Biomass to Levulinic Acid Using Acidic Ionic Liquids." *Carbohydrate Polymers*, 2018. doi:10.1016/j.carbpol.2017.10.064.

92. Alipour, Siamak, and Hamid Omidvarborna. "Enzymatic and Catalytic Hybrid Method for Levulinic Acid Synthesis from Biomass Sugars." *Journal of Cleaner Production*, 2017. doi:10.1016/j.jclepro.2016.12.086.

93. Luijkx, Gerard C.A., Fred van Rantwijk, and Herman van Bekkum. "Hydrothermal Formation of 1,2,4-Benzenetriol from 5-Hydroxymethyl-2-Furaldehyde and d-Fructose." *Carbohydrate Research*, 1993. doi:10.1016/0008-6215(93)80027-C.

94. "Methods in Lignin Chemistry." *Comparative Biochemistry and Physiology Part B: Comparative Biochemistry*, 1993. doi:10.1016/0305–0491(93)90261-3.

95. Wahyudiono, Takayuki Kanetake, Mitsuru Sasaki, and Motonobu Goto. "Decomposition of a Lignin Model Compound under Hydrothermal Conditions." *Chemical Engineering and Technology*, 2007. doi:10.1002/ceat.200700066.

96. Zhang, Bo, Hua Jiang Huang, and Shri Ramaswamy. "Reaction Kinetics of the Hydrothermal Treatment of Lignin." *Applied Biochemistry and Biotechnology*, 2008. doi:10.1007/s12010-007-8070-6.

97. Liu, Aiguo, Yoon Kook Park, Zhiliang Huang, Baowu Wang, Ramble O. Ankumah, and Prosanto K. Biswas. "Product Identification and Distribution from Hydrothermal Conversion of Walnut Shells." *Energy and Fuels*, 2006. doi:10.1021/ef050192p.

98. Seehar, Tahir H, Saqib Sohail Toor, Kamaldeep Sharma, Asbjørn H Nielsen, Thomas Helmer Pedersen, and Lasse Aistrup Rosendahl. "Influence of Process Conditions on Hydrothermal Liquefaction of Eucalyptus Biomass for Biocrude Production and Investigation of the Inorganics Distribution." *Sustainable Energy & Fuels*, 2021. doi:10.1039/d0se01634a.

99. Mathanker, Ankit, Deepak Pudasainee, Amit Kumar, and Rajender Gupta. "Hydrothermal Liquefaction of Lignocellulosic Biomass Feedstock to Produce Biofuels: Parametric Study and Products Characterization." *Fuel*, 2020. doi:10.1016/j.fuel.2020.117534.

100. Möller, Maria, Peter Nilges, Falk Harnisch, and Uwe Schröder. "Subcritical Water as Reaction Environment: Fundamentals of Hydrothermal Biomass Transformation." *ChemSusChem*, 2011. doi:10.1002/cssc.201000341.

101. Xue, Yuan, Hongyan Chen, Weina Zhao, Chao Yang, Peng Ma, and Sheng Han. "A Review on the Operating Conditions of Producing Bio-Oil from Hydrothermal Liquefaction of Biomass." *International Journal of Energy Research*, 2016. doi:10.1002/er.3473.

102. Chan, Yi Herng, Suzana Yusup, Armando T. Quitain, Yoshimitsu Uemura, and Mitsuru Sasaki. "Bio-Oil Production from Oil Palm Biomass via Subcritical and Supercritical Hydrothermal Liquefaction." *Journal of Supercritical Fluids*, 2014. doi:10.1016/j. supflu.2014.10.014.

103. Feng, Shanghuan, Zhongshun Yuan, Matthew Leitch, and Chunbao Xu. "Hydrothermal Liquefaction of Barks into Bio-Crude Effects of Species and Ash Content/Composition." *Fuel*, 2014. doi:10.1016/j.fuel.2013.07.096.

104. Zhong, Chongli, and Xiaomin Wei. "A Comparative Experimental Study on the Liquefaction of Wood." *Energy*, 2004. doi:10.1016/j.energy.2004.03.096.

105. Demirbaş, A. "Effect of Lignin Content on Aqueous Liquefaction Products of Biomass." *Energy Conversion and Management*, 2000. doi:10.1016/S0196-8904(00)00013-3.

106. Zhang, Bo, Marc von Keitz, and Kenneth Valentas. "Thermochemical Liquefaction of High-Diversity Grassland Perennials." *Journal of Analytical and Applied Pyrolysis*, 2009. doi:10.1016/j.jaap.2008.09.005.

107. Watanabe, Masaru, Takafumi Sato, Hiroshi Inomata, Richard Lee Smith, Kunio Arai, Andrea Kruse, and Eckhard Dinjus. "Chemical Reactions of C1 Compounds in Near-Critical and Supercritical Water." *Chemical Reviews*, 2004. doi:10.1021/cr020415y.

108. Zhu, Zhe, Lasse Rosendahl, Saqib Sohail Toor, Donghong Yu, and Guanyi Chen. "Hydrothermal Liquefaction of Barley Straw to Bio-Crude Oil: Effects of Reaction Temperature and Aqueous Phase Recirculation." *Applied Energy*, 2015. doi:10.1016/j. apenergy.2014.10.005.

109. Xu, Chunbao, and Timothy Etcheverry. "Hydro-Liquefaction of Woody Biomass in Sub- and Super-Critical Ethanol with Iron-Based Catalysts." *Fuel*, 2008. doi:10.1016/j. fuel.2007.05.013.

110. Sugano, Motoyuki, Hirokazu Takagi, Katsumi Hirano, and Kiyoshi Mashimo. "Hydrothermal Liquefaction of Plantation Biomass with Two Kinds of Wastewater from Paper Industry." In *Journal of Materials Science*, 2008. doi:10.1007/s10853-007-2106-8.

111. Knežević, D., W. P.M. Van Swaaij, and S. R.A. Kersten. "Hydrothermal Conversion of Biomass: I, Glucose Conversion in Hot Compressed Water." *Industrial and Engineering Chemistry Research*, 2009. doi:10.1021/ie801387v.

112. Biller, P., and A. B. Ross. "Potential Yields and Properties of Oil from the Hydrothermal Liquefaction of Microalgae with Different Biochemical Content." *Bioresource Technology*, 2011. doi:10.1016/j.biortech.2010.06.028.

113. Biller, P., R. Riley, and A. B. Ross. "Catalytic Hydrothermal Processing of Microalgae: Decomposition and Upgrading of Lipids." *Bioresource Technology*, 2011. doi:10.1016/j. biortech.2010.12.113.

114. Garcia Alba, Laura, Cristian Torri, Chiara Samorì, Jaapjan Van Der Spek, Daniele Fabbri, Sascha R. A. Kersten, and Derk W. F. Brilman. "Hydrothermal Treatment (HTT) of Microalgae: Evaluation of the Process as Conversion Method in an Algae Biorefinery Concept." In *Energy and Fuels*, 2012. doi:10.1021/ef201415s.

115. Karagöz, Selhan, Thallada Bhaskar, Akinori Muto, Yusaku Sakata, and Md Azhar Uddin. "Low-Temperature Hydrothermal Treatment of Biomass: Effect of Reaction Parameters on Products and Boiling Point Distributions." *Energy and Fuels*, 2004. doi:10.1021/ef030133g.

116. Boocock, D. G. B., and K. M. Sherman. "Further Aspects of Powdered Poplar Wood Liquefaction by Aqueous Pyrolysis." *The Canadian Journal of Chemical Engineering*, 1985. doi:10.1002/cjce.5450630415.

117. Anastasakis, Konstantinos, and A. B. Ross. "Hydrothermal Liquefaction of the Brown Macro-Alga Laminaria Saccharina: Effect of Reaction Conditions on Product Distribution and Composition." *Bioresource Technology*, 2011. doi:10.1016/j.biortech.2011.01.031.

118. Singh, Rawel, Bhavya Balagurumurthy, Aditya Prakash, and Thallada Bhaskar. "Catalytic Hydrothermal Liquefaction of Water Hyacinth." *Bioresource Technology*, 2015. doi:10.1016/j.biortech.2014.08.119.

119. Wang, Chao, Jingxue Pan, Jinhua Li, and Zhengyu Yang. "Comparative Studies of Products Produced from Four Different Biomass Samples via Deoxy-Liquefaction." *Bioresource Technology*, 2008. doi:10.1016/j.biortech.2007.06.023.

120. Xiu, Shuangning, Abolghasem Shahbazi, Vestel Shirley, and Dan Cheng. "Hydrothermal Pyrolysis of Swine Manure to Bio-Oil: Effects of Operating Parameters on Products Yield and Characterization of Bio-Oil." *Journal of Analytical and Applied Pyrolysis*, 2010. doi:10.1016/j.jaap.2010.02.011.

121. Liu, Hua Min, Ming Fei Li, and Run Cang Sun. "Hydrothermal Liquefaction of Cornstalk: 7-Lump Distribution and Characterization of Products." *Bioresource Technology*, 2013. doi:10.1016/j.biortech.2012.09.125.

122. Goudnaan, F., B. van de Beld, F. R. Boerefijn, G. M. Bos, J. E. Naber, S. van der Wal, and J. A. Zeevalkink. "Thermal Efficiency of the HTU® Process for Biomass Liquefaction." In *Progress in Thermochemical Biomass Conversion*, 2008. doi:10.1002/9780470694954.ch108.

123. Kersten, Sascha R. A., Biljana Potic, Wolter Prins, and Wim P. M. Van Swaaij. "Gasification of Model Compounds and Wood in Hot Compressed Water." *Industrial and Engineering Chemistry Research*, 2006. doi:10.1021/ie0509490.

124. Kabyemela, Bernard M., Tadafumi Adschiri, Roberto M. Malaluan, and Kunio Arai. "Kinetics of Glucose Epimerization and Decomposition in Subcritical and Supercritical Water." *Industrial and Engineering Chemistry Research*, 1997. doi:10.1021/ie960250h.

125. Nagappan, Senthil, Rahul R. Bhosale, Dinh Duc Nguyen, Nguyen Thuy Lan Chi, Vinoth Kumar Ponnusamy, Chang Soon Woong, and Gopalakrishnan Kumar. "Catalytic Hydrothermal Liquefaction of Biomass into Bio-Oils and Other Value-Added Products – A Review." *Fuel*, 2021. doi:10.1016/j.fuel.2020.119053.

126. Minowa, Tomoaki, Fang Zhen, and Tomoko Ogi. "Cellulose Decomposition in Hot-Compressed Water with Alkali or Nickel Catalyst." *Journal of Supercritical Fluids*, 1998. doi:10.1016/S0896-8446(98)00059-X.

127. Duan, Peigao, and Phillip E. Savage. "Hydrothermal Liquefaction of a Microalga with Heterogeneous Catalysts." *Industrial and Engineering Chemistry Research*, 2011. doi:10.1021/ie100758s.

128. Behrendt, Frank, York Neubauer, Michael Oevermann, Birgit Wilmes, and Nico Zobel. "Direct Liquefaction of Biomass." *Chemical Engineering and Technology*, 2008. doi:10.1002/ceat.200800077.

129. Lindemuth, T. E. "Investigations of the PERC Process for Biomass Liquefaction at the Department of Energy, Albany, Oregon Experimental Facility." In *ACS Symposium Series*, 1978. doi:10.1021/bk-1978-0076.ch019.

130. Schaleger, Larry L., Carlos Figueroa, and Hubert G. Davis. "Direct Liquefaction of Biomass: Results from Operation of Continuous Bench-Scale Unit in Liquefaction of Water Slurries of Douglas Fir Wood." In *Biotechnology Bioengineering Symposium*, LBL-14019; CONF-820580-1, ON: DE82015703, 1982.

131. He, Wenzhi, Guangming Li, Lingzhao Kong, Hua Wang, Juwen Huang, and Jingcheng Xu. "Application of Hydrothermal Reaction in Resource Recovery of Organic Wastes." *Resources, Conservation and Recycling*, 2008. doi:10.1016/j.resconrec.2007.11.003.

132. Behrendt, F., Y. Neubauer, K. Schulz-Tonnies, B. Wilmes, and N. Zobel. "Direktverflussigung von Biomasse- Reaktionsmechanismen und Produktverteilungen. Studie und Bewertung." *Technische Universitat Berlin*, Berlin. 8th June 2006. Available at-http://www.fnr-server.de/ftp/pdf/literatur/pdf_253studie_zur_direktverfluessigung_final_komprimiert.pdf

133. Stadlbauer, E. A. Thermokatalytische Niedertemperaturkonvertierung (NTK) von tierische und mikrobieller Biomasse unter Gewinnung von Wertstoffen und Energieträgern im Pilotmaßstab, Final Report of Project 18153, *Deutsche Bundesstiftung Umwelt*, 2005.

134. Bhatia, S. C. "Introduction to Advanced Renewable Energy Systems." Woodhead Publishing Ltd-libgen.lc, 2014.

5 Role of Systems Biology in Enhancing Efficiency of Biocatalysts

Trunil Desai, Ahmad Ahmad,
and Shireesh Srivastava
International Centre for Genetic Engineering
and Biotechnology (ICGEB), New Delhi

CONTENTS

5.1 Introduction 128
5.2 Need for Systems Biology 130
5.3 Reconstruction of Genome-Scale Metabolic Model 130
    5.3.1 KEGG 132
    5.3.2 BioCyc 132
    5.3.3 MetaCyc 132
    5.3.4 ExPASy 133
    5.3.5 UniProt 133
    5.3.6 BRENDA 133
    5.3.7 ModelSEED 133
    5.3.8 BiGG 133
    5.3.9 BioModels 133
5.4 Refinement of the Draft Metabolic Model 133
5.5 Different Optimization Criteria 135
5.6 Experimental Data Used for Model Reconstruction 137
5.7 Toolboxes Available for Metabolic Modelling and FBA 138
    5.7.1 COBRA Toolbox 138
    5.7.2 COBRApy 139
    5.7.3 ScrumPy – Metabolic Modelling in Python 139
    5.7.4 Sybil 139
    5.7.5 Pathway Tools 139
    5.7.6 ModelSEED 139
    5.7.7 RAVEN 140
    5.7.8 MERLIN 140
    5.7.9 CoReCo 140
    5.7.10 CarveMe 140

DOI: 10.1201/9781003158486-5

5.8   Methods of Analyses of GSMM for Metabolic Engineering 140
5.9   Pathway Prediction for Synthetic Biology Applications............................. 141
5.10   Genetic Engineering Tools 142
5.11   Improving Tolerance to Inhibitors through Systems Biology Analyses....... 144
5.12   Enzyme Discovery.. 145
5.13   Future Directions for Systems Biology – Integrative Analyses................... 145
References.. 145

5.1   INTRODUCTION

About 80% of fuel consumed in the world is fossil fuel (World Bank data 2015).
Liquid petroleum fuels are used for most of the transportation. It is one of the major
causes for greenhouse gas emissions. Since 1980, the atmospheric CO_2 levels have
increased from 337 parts per million (ppm) to 410 ppm (www.climate.gov). Ethanol
and biodiesel blending with petroleum fuels has been implemented to mitigate it
(Pandiyan et al. 2019). The increased sensitization about the harmful effects of global
warming caused primarily due to our reliance on petroleum-derived fuels has led to
the search for alternatives to petroleum-derived fuels and chemicals. Human beings
have forever used biomass for energy and materials purposes, albeit with poor efficiency. One of the major sources of biomass is agricultural residues. While a large
fraction of the total agricultural residues can be reused as fodder and to be burnt in
furnaces, a large portion remains unused or burnt in fields and offers an alternative
feedstock for production of biofuels and biochemicals. However, in order to make
cost-effective fuels and commodity chemicals, the process must be extremely efficient. Therein lies the challenge with second-generation biofuels and bioproducts.
Although several biofuel compounds such as ethanol, butanol, isobutanol and isoprene have been industrially produced, they are the product of fermentation of sugars
derived from edible biomass such as corn and sugarcane (Lee et al. 2019). There have
been efforts to produce biofuel molecules from non-edible carbon sources such as
lignocellulosic agricultural waste. Photosynthetic microorganisms such as cyanobacteria are also being developed for the production sugars for fermentation processes
(Gupta et al. 2020).

Agricultural residues are part of lignocellulosic (LC) biomass due to their content
of lignin and cellulose. Lignin is a polymer composed of aromatic residues, while
cellulose contains repeating glucose subunits joined by $\beta(1\rightarrow4)$ glycosidic bonds.
The LC biomass also contains a third polymer of C5 sugars called hemicellulose.
The intertwining of these polymers gives the LC biomass its strength and stability.
However, the same properties make it a very difficult substrate for microorganisms
to degrade. To harness fermentable sugars, the LC biomass has to be deconstructed
first. To achieve this in reasonable time typically involves multiple steps. The first
is the pre-treatment to remove lignin and make the sugar polymers more accessible.
Then comes an enzymatic hydrolysis to release the sugars from the polymers. The
sugars released upon hydrolysis still pose a significant challenge for efficient fermentation for the production of biofuels and biochemicals. This is because (a) the
hydrolysate is a mixture of C5 and C6 sugars and (b) it contains several inhibitors

that affect cellular growth and metabolism. Generally, while the C6 sugar(s) is(are) easier to ferment, the C5 sugars (xylose, ribose, arabinose, etc.) are poor substrates for organism growth and fermentation.

In nature, the organisms are exposed to a variety of substrates and conditions. All organisms have developed robust metabolic networks in order to be able to utilize the various carbon and nitrogen sources available to them, while perhaps minimizing the proteomic costs to form the "smallest" metabolic network for this purpose. However, the metabolic robustness also requires sophisticated control of the various pathways in order to effectively allocate the (protein) resources. Additionally, under a given condition, most organisms produce a variety of products. This variety is determined by the need to maximize the growth or ATP production for a given metabolic network. For example, under the anaerobic conditions, *E. coli* produces a mixture of acids (acetic, formic, succinic and lactic acids) and ethanol. However, for any microbial process to be commercially viable, it should be operating to maximize potential rates, yields and titres (RYT). A common approach for improving the yields is to knock out reactions producing competing products. Often, this results in a reduced ATP yield or altered redox balance and affects cellular growth rate. This, in turn, may affect the rate of the bioprocess. Thus, the aim of metabolic engineering is to improve yield while minimizing the effect on rates of the process. This is typically done by analysing the intracellular metabolic network and analysing the flow (of C) through various pathways, also known as flux distribution. Traditionally, the reliance was on flux analysis of central metabolic pathways as most of the flux goes through them and they are generally better characterized than the peripheral metabolic pathways. However, as the field of metabolic engineering progressed and as the synthetic biology developed where complete pathways were transferred from one organism to another, there was an increased appreciation to include a detailed metabolic map of most of the metabolic reactions occurring in the organism of interest. This led to the development of the field of genome-scale metabolic modelling. A typical genome-scale metabolic model (GSMM, also abbreviated as GEM in literature) would account for hundreds or thousands of metabolic genes and may have thousands of metabolic reactions. Importantly for metabolic engineering applications, the model will have gene-protein-reaction (GPR) association information, which lists which gene encodes for which (subunit of a) protein and which reaction the protein (or a complex) catalyses. What distinguishes a GSMM from a genome-scale metabolic network is the ability to simulate cellular responses and growth.

While GSMMs are useful tools for metabolic modelling, it requires considerable effort to build a functional, curated model that faithfully reproduces cellular metabolic responses. Additionally, special tools for the analysis of these models are required, both for basic analysis to simulate flux distribution and growth under different conditions and for identifying metabolic engineering targets for overproduction of native metabolite of interest or a heterologous product upon introduction of synthetic pathways. In this chapter, we will cover (a) the process of reconstruction of GSMM from its annotated genome sequence, (b) basic and advanced methods of analysis of GSMMs and (c) applications of the models and their analyses to improve the yields of bioproducts.

5.2   NEED FOR SYSTEMS BIOLOGY

Traditional metabolic engineering approach to making cell factories industrially feasible for the production of metabolites is slow and costly. Especially for the production of high-volume and low-cost products such as biofuels (d'Espaux et al. 2017; Liao et al. 2016) and organic acids (Choi et al. 2016), it is important to achieve high titres (the higher the better, typically over 100 g/L) with high yield and productivity (Choi et al. 2019). In order to make fermentation more efficient, the traditional approach has been pathway-centric engineering in order to (a) increase the precursor supply or (b) delete competing by-products. While the pathway-centric approach has served well in many cases, a thorough understanding of the metabolism of organism at a systems level is likely to provide a more robust set of targets. Systems metabolic engineering, which integrates systems biology, synthetic biology and evolutionary engineering with classical metabolic engineering, is used to analyse and develop high-performance cell factories. A first step in systems metabolic engineering is to generate a GSMM of the organism in order to understand its metabolic capabilities and limitations.

5.3   RECONSTRUCTION OF GENOME-SCALE METABOLIC MODEL

A GSMM is a collection of all relevant metabolic reactions operating inside an organism. The reconstruction of GSMM involves collection of all relevant genomic and metabolic data to explore the metabolic behaviour of the organism (Thiele and Palsson 2010). It is an iterative process which includes several steps as shown in Figure 5.1. The first step in model reconstruction is obtaining an annotated genome of the organism. In case of an un-annotated genome, the first step is to annotate the genome. Genome annotation refers to assigning functions to the sequenced genes.

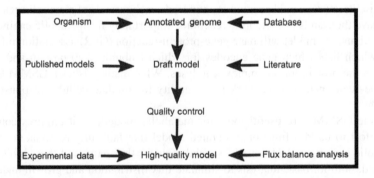

FIGURE 5.1   The steps for reconstruction of genome-scale metabolic models. An organism's genome is sequenced and annotated using various tools. A draft metabolic model of the organism is then reconstructed using the annotated sequence, previously published models and the literature. Various qualitative control analyses are performed on the draft model. After the qualitative control, integration of experimental data and flux balance analysis gives a high-quality model.

TABLE 5.1
List of Some Popular Tools for Genome Annotation

Tool Name	Description	Link
GeneMark	Free, used for both prokaryotes and eukaryotes	http://exon.gatech.edu/GeneMark/
AUGUSTUS	Free, primarily used for eukaryotes	http://bioinf.uni-greifswald.de/webaugustus/index
EuGene	Free, used for both prokaryotes and eukaryotes	http://eugene.toulouse.inra.fr/
GeneID	Free, primarily used for eukaryotes	https://genome.crg.cat/geneid.html
MAKER	Free for academic, primarily used for eukaryotes	http://www.yandell-lab.org/software/maker.html
RAST	Free for academic, used for prokaryotes	https://rast.nmpdr.org/

Accurate annotation of genes coding for metabolic enzymes is very crucial for metabolic model reconstruction. Table 5.1 shows a list of popular tools for genome annotation.

Using the annotated genome, a draft model is obtained as explained below. A list of all enzyme commission (EC) numbers associated with the metabolic genes of the genome is prepared. The EC numbers relate enzymes to the enzymatic reactions they catalyse. Subsequently, the draft model is prepared by obtaining the reactions catalysed by all the enzymes operating in the organism. Once the draft model is obtained, the subsequent steps involve manually refining the draft model. The draft model generally includes some reactions that involve various non-specific metabolites, e.g. DNA, RNA, amino acids, proteins, etc. The reactions involving these non-specific metabolites may create uncertainties in the model and hence are removed. A mass balance analysis is performed for all reactions of the draft model and corrected wherever required. The "AtomImbal" module of the ScrumPy (Poolman 2006) metabolic modelling tool or the "CheckBals" function of "COBRA Toolbox" (Heirendt et al. 2019) can be used to check carbon (C), nitrogen (N), phosphorus (P) and sulphur (S) balances in all the reactions of the model. Reaction directionalities are assigned based on biological database (such as MetaCyc and BioCyc) and information from the literature. Reaction directionalities can also be determined by calculating the ΔG of the reaction. The algorithm "von Bertalanffy 1.0" (Fleming and Thiele 2011) utilizes the Gibbs free energy of the compounds and calculates the ΔG of the reactions, finally revealing the directionality of the model reactions. Reactions such as ATP synthase and ATPase are assigned as irreversible.

Finally, the model is tested for the presence of any thermodynamically infeasible cycles (also known as "futile cycles") and the reaction directionalities are adjusted to resolve them. Information from various biological databases is incorporated during the course of model reconstructions. Table 5.2 shows a list of various databases utilized during the model reconstruction. The databases are further explained in the paragraphs below.

TABLE 5.2

A List of Some Databases Used for Genome-Scale Metabolic Model Reconstruction

Name	Link
KEGG	https://www.genome.jp/kegg/
NCBI	https://www.ncbi.nlm.nih.gov/
BioCyc	https://biocyc.org/
MetaCyc	https://metacyc.org/
ExPASy	https://www.expasy.org/
ModelSEED	https://modelseed.org/
BiGG	http://bigg.ucsd.edu/
BRENDA	https://www.brenda-enzymes.org/
UniProt	https://www.uniprot.org/
CheBI	https://www.ebi.ac.uk/chebi/

5.3.1 KEGG

KEGG (Kyoto Encyclopedia of Genes and Genomes) database is a repository of various types of biological information, e.g. information related to genes, reactions and pathways. The KEGG database is one of the most important databases required for the reconstruction of metabolic models. The information of EC numbers can be obtained from KEGG database. KEGG reaction database has a collection of various enzymatic reactions. KEGG pathways database gives a visual representation of the pathways operating inside a microorganism.

5.3.2 BIOCYC

BioCyc database is a collection of organism-specific databases called pathway genome database (PGDB). It also has Pathway Tools, a software tool to visualize, navigate and analyse the organism-specific PGDB. Omics analysis can also be performed using BioCyc tools. The BioCyc database also provides an option to simulate metabolic models and perform various types of analyses, e.g. comparative analysis and flux balance analysis (FBA). Metabolic route search tool can be utilized to search for the organism-specific PGDB.

5.3.3 METACYC

MetaCyc database is a collection of metabolic pathways from all domains of life. It is a curated database having over 2800 pathways from more than 3100 organisms. It has a repository of pathways related to primary and secondary metabolisms along with genes, enzymes and reactions. The PGDB of MetaCyc database can be downloaded from the server and used for subsequent analyses.

5.3.4 ExPASy

The ExPASy (the Expert Protein Analysis System) is a bioinformatics resource Web portal developed by Swiss Institute of Bioinformatics. It provides an integrative access to various bioinformatics tools and resources for different domains of life science, such as genomics, proteomics, phylogenetics and systems biology. It primarily focuses on recourses related to proteins and proteomics.

5.3.5 UniProt

The Universal Protein (UniProt) database is a widely used database that provides free access to information related to protein sequence and functional information. It contains information on over 120 million proteins from a variety of sources.

5.3.6 BRENDA

BRENDA (The Comprehensive Enzyme Information System) is one of the most comprehensive repositories related to enzymes. It contains measured values of various enzyme parameters (K_m, V_{max}) for a large collection of enzymes from different organisms.

5.3.7 ModelSEED

ModelSEED is a biological resource portal for reconstruction and analysis of GSMMs. It is generally used for automated reconstruction of GSMMs for microbes and plants.

5.3.8 BiGG

The BiGG Models is a repository of GSMMs. Apart from a database of models, the database integrates about 70 high-quality published models to create a single database using standardized identifiers for metabolites and reactions. It is created and maintained by University of California, San Diego. The BiGG database can be used to explore and compare various published GSMMs.

5.3.9 BioModels

The BioModels is also a database of metabolic models. It has a collection of manually curated as well as non-curated models and automatically generated model. The database provides an option to browse organism-wise models.

5.4 REFINEMENT OF THE DRAFT METABOLIC MODEL

The draft model obtained after the initial refining is usually unable to synthesize all essential biomass precursors, indicating the absence of one or more key reactions or pathways that participate in the synthesis of the corresponding precursor. The

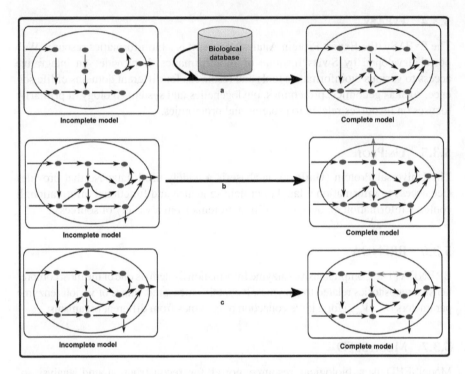

FIGURE 5.2 **Figure showing different possible methods for gap-filling process**. Gaps in metabolic networks are typically filled by (a) addition of reactions from database, (b) addition of exchange reactions of gapped metabolites and (c) changing reversibility of model reaction(s).

absence of such key reaction(s) results in gaps in the network. These gaps in the draft model are manually identified by performing sequential optimizations for the production of individual biomass precursors. Those precursors which are not produced in the draft model are termed as gapped metabolites. The appearance of gapped metabolites in a metabolic network indicates the presence of some missing links (reactions/pathways), also known as gapped reactions/pathways, in the network.

There are three possible approaches generally applied for gap-filling process as shown in Figure 5.2 below (Thiele and Palsson 2010). The first approach involves searching various resources and databases for missing pathway(s) or reaction(s) that will enable the production or consumption of the gapped metabolites. This will ensure the biosynthesis of biomass precursor(s) (Figure 5.2a). Occasionally, the biosynthetic routes for production or consumption of a gapped metabolite are unknown. Under these circumstances, addition of exchange reaction(s) for the gapped metabolite(s) ensures biosynthesis of biomass precursor(s) (Figure 5.2b). Finally, assigning proper reversibility of a few reactions may complete the gapped pathway (Figure 5.2c), resulting in the formation of desired biomass precursor(s). The COBRA Toolbox (Heirendt et al. 2019) has an inbuilt tool for finding the gapped metabolites by the "gapFind" tool (Satish Kumar, Dasika, and Maranas 2007). Also, the inbuilt function "fastGapFill" (Thiele, Vlassis, and Fleming 2014) can be used to fill the gaps in the model.

BLAST searches are then performed for the corresponding genes of the gap-filling reactions with the genome of the concerned organism. A gene locus is assigned if the BLAST result shows similarity. Else, the gap-filling reactions are left as orphan (i.e. without any associated GPR).

Finally, the model is screened for the presence of thermodynamically infeasible cycles, also known as futile cycles. In a metabolic model, there could be groups of reactions that run together so as to make a cycle that can form or dissipate ATP without any carbon intake (Figure 5.3). There may be more than one futile cycle present in a metabolic model. Futile cycle, operating in a metabolic model, can give rise to inconsistent results. It is, therefore, essential to ensure elimination of all such futile cycles present in the model. Futile cycles are detected by assigning some positive value (e.g. 1) to the ATP synthase reaction while fixing all the exchange reactions to zero during the optimization. If the optimization results in a feasible solution, then the reactions with non-zero fluxes are the infeasible loop reactions. The draft model is enquired and subsequently resolved for the presence of futile cycles by carefully assigning the directionalities of various reactions. The directionalities of reactions in the model can be assigned as given in the BioCyc and MetaCyc databases, and previously published models as well as in the literature.

5.5 DIFFERENT OPTIMIZATION CRITERIA

FBA has become one of the most popular methods for predicting internal metabolic fluxes under various conditions. But the selection of the objective function for optimizations is a subjective decision. There are various opinions regarding the

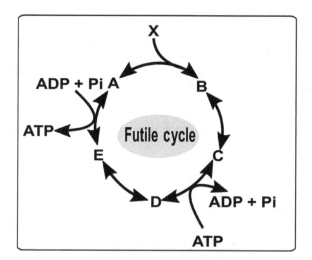

FIGURE 5.3 A schematic diagram showing a futile cycle. A futile cyclic pathway may be formed when a metabolite A is reformed via (say) metabolites E, D, C and B. An ATP molecule is produced in reaction converting A → E and consumed in the reaction D → C.

TABLE 5.3
Examples of GSMMs Reconstructed for Biotechnologically Important Organisms

S. No.	Organism Name	Model Name	References
1	*Geobacillus thermoglucosidasius*	*i*GT736	Ahmad et al. (2017)
2	*Escherichia coli*	*i*JO1366	Orth et al. (2011)
3	*Clostridium butyricum*	*i*Cbu641	Serrano-Bermúdez et al. (2017)
4	*Zymomonas mobilis*	ZmoMBEL601	Lee et al. (2010)
5	*Corynebacterium glutamicum*	*i*CW773	Zhang et al. (2017)
6	*Synechococcus* sp. PCC 7002	*i*syp702	Hendry et al. (2016)
7	*Chlorella vulgaris*	*i*CZ843	Zuñiga et al. (2016)
8	*Thalassiosira pseudonana*	*i*Thaps987	Ahmad, Tiwari, and Srivastava (2020)
9	*Saccharomyces cerevisiae*	ecYeast8	Lu et al. (2019)
10	*Pichia pastoris*	*i*LC915	Tomàs-Gamisans, Ferrer, and Albiol (2016)
11	*Methylocystis hirsuta CSC1*	*M. hirsuta*	Bordel et al. (2019)
12	*Bifidobacterium adolescentis*	*i*Bif452	El-Semman et al. (2014)
13	*Mycobacterium tuberculosis*	*i*NJ661	Jamshidi and Palsson (2007)
14	*Salmonella*	*Multiple models*	Seif et al. (2018)
15	*Human*	Recon 2.2	Swainston et al. (2016)

TABLE 5.4
List of Different Objective Functions Used in the Literature

Objective Function	Model	References
Maximization of growth (biomass)	*i*JO1366, *i*AF1260	Orth et al. (2011) and Feist et al. (2007)
Minimization of total fluxes	*i*Thaps980, *i*GT736	Ahmad et al. (2017) and Ahmad, Tiwari, and Srivastava (2020)
Minimization of ATP production	*E. coli* core model	Savinell and Palsson (1992)
Maximization of a product	*i*Thaps980	Ahmad, Tiwari, and Srivastava (2020)

assumption of cellular objective during the growth of the organisms/cells. Table 5.4 shows a list of various objective functions and studies that used these objective functions.

The most widely used objective function for FBA is maximization of growth rate (or biomass production). This objective function assumes that the primary objective of the organisms/cells is growth or production of biomass. In this case, the biomass reaction, which incorporates all the biomass precursors as reactants, is chosen as objective function and is maximized during optimizations (Orth et al. 2011; Feist et al. 2007). The carbon intake rate is fixed (to the experimentally determined value) during the optimizations involving maximization of the biomass. This optimization results in a solution(s) with highest possible growth using the fixed carbon intake rate. However, the growth rate increases if the carbon intake rate increases.

The other relevant objective function is minimization of total fluxes (Ahmad et al. 2017; Ahmad, Tiwari, and Srivastava 2020), which is based on the assumption that the primary cellular objective is to minimize the production and utilization cost of enzymes. During the optimizations where minimization of total fluxes is chosen as objective function, the growth rate must be fixed to the experimentally determined value. The minimization of ATP production is another objective function based on the assumption that the organisms/cells strive to conserve energy (Savinell and Palsson 1992). In this case, the ATP synthase reaction is chosen as objective function and the flux through this reaction is minimized. In this case, the growth rate is fixed to the experimentally determined value. Another related objective function is minimization of NADH production.

The maximization of flux through a product (such as ethanol, butanol, etc.) is another objective function used (Ahmad, Tiwari, and Srivastava 2020). Simulations using this criterion identify the (hypothetical) flux distribution that maximizes the production of a particular metabolite. In this case, the exchange reaction of the product is assigned as objective function while the growth rate and carbon uptake rate are fixed during maximization optimization.

5.6 EXPERIMENTAL DATA USED FOR MODEL RECONSTRUCTION

FBA of metabolic models requires various experimentally measured data to constrain the model during the optimizations. Mathematically, application of constraints in a metabolic model reduces the solution space of optimizations (Orth, Thiele, and Palsson 2010). The reduced solution space predicts more accurate flux distributions based on experimental data. These constraints can be applied very easily on the reactions of metabolic models during FBA. The experimental data required for modelling are given in Table 5.5.

Each organism/cell requires some input of macronutrients/micronutrients for their growth and metabolism and excretes/releases various metabolites as by-products. The growth rate is fixed when "minimization of total fluxes" is used as the objective function during the simulations. It should be noted that the growth rate cannot' be fixed (or constrained) when biomass reaction is chosen as objective function, e.g. in maximization

---

TABLE 5.5
List of Experimental Data Typically Used in Reconstruction Process and Flux Balance Analysis

S. No.	Description
1	Growth rate
2	Carbon source uptake rate
3	Product formation rate
4	Light absorption rate (for photosynthetic organisms)
5	Oxygen evolution rate (for photosynthetic organisms)
6	Nitrogen uptake rate
7	Growth information on a variety of substrates for heterotrophs, e.g. in Biolog plates, if available
8	Growth and product formation profiles of knockouts, if available

of biomass. The uptake rates of carbon sources such as glucose, xylose or CO_2 (in case of photosynthetic organism) is constrained by fixing the lower and upper bounds of the respective exchange reactions to the observed uptake rates. Photosynthetic organisms require photons for photolysis of water leading to evolution of oxygen. Therefore, the rate of photon absorption can also be fixed or constrained based on experimentally measured data. Similarly, the rate of oxygen evolution for photosynthetic organisms can be constrained by fixing the lower and upper bounds of the reactions representing photosystem II (PSII) to the observed oxygen evolution rates. The rates of other micronutrients such as nitrogen sources (NO_2^-, NO_3^-, NH_4^+) or phosphorus sources can also be constrained, but that is generally left unconstrained in majority of the studies. If the organism releases some by-products, e.g. *E. coli* releases formate or acetate as by-products during growth, the fluxes of these products can also be constrained by fixing the lower and upper bounds of the respective exchange reactions.

5.7   TOOLBOXES AVAILABLE FOR METABOLIC MODELLING AND FBA

Several tools have been developed to assist the process of metabolic modelling and FBA. These tools facilitate and speed up the process of reconstruction of metabolic models by automating the steps of model reconstructions using inbuilt scripts. Following is the list of some popular tools, along with their Web URLs, used for the reconstruction of metabolic models and FBA (Table 5.6).

5.7.1   COBRA TOOLBOX

The **CO**nstraint-**B**ased **R**econstruction and **A**nalysis Toolbox (COBRA Toolbox) (Heirendt et al. 2019) is one of the most widely used tools for the reconstruction of metabolic models and FBA. It is a toolbox based on MATLAB suite. The COBRA Toolbox is a collection MATLAB scripts for model generation and FBA. It imports

TABLE 5.6
A List of Popular Toolboxes for Analyses of GSMMs and Metabolic Modelling and Their URLs

S. No.	Tool/Platform/Software	Web URL
1	COBRA Toolbox	https://opencobra.github.io/cobratoolbox/stable/
2	COBRApy	https://opencobra.github.io/cobrapy/
3	ScrumPy	https://mudshark.brookes.ac.uk/ScrumPy
4	Sybil	https://cran.r-project.org/web/packages/sybil/index.html
5	Pathway Tools	http://bioinformatics.ai.sri.com/ptools/
6	ModelSEED	https://modelseed.org/
7	RAVEN	https://github.com/SysBioChalmers/RAVEN
8	MERLIN	https://merlin-sysbio.org/index.php/Home
9	CoReCo	https://www.simulationstore.com/node/1177
10	CarveMe	https://github.com/cdanielmachado/carveme

metabolic models primarily in Systems Biology Markup Language (SBML) format, but can also import models written properly in MS Excel format. The COBRA Toolbox is compatible with various linear programming solvers including, but not restricted to, GLPK, Gurobi and CPLEX. Although the COBRA Toolbox is freely available, the MATLAB suite is a proprietary platform.

5.7.2   COBRAPY

The **COBRApy** (**CO**nstraint-**B**ased **R**econstruction and Analysis-**Py**thon) is a Python package for model reconstruction and FBA (Ebrahim et al. 2013). It is the Python-based COBRA Toolbox readily available for free use. It can import models in SBML as well as JSON formats. It has almost all functions of MATALB-based COBRA Toolbox, and new functionalities are being added over time.

5.7.3   SCRUMPY – METABOLIC MODELLING IN PYTHON

The ScrumPy is one of the oldest metabolic modelling tools (Poolman 2006). It is a Python-based metabolic modelling tool. It has inbuilt scripts which can be used to generate a draft model from PGDB very easily. This draft model is a further refined reconstruction of robust metabolic model. ScrumPy can also be used for kinetic modelling and analysis of kinetic models. Flux control analysis can be performed very efficiently using the ScrumPy toolbox.

5.7.4   SYBIL

The sybil package is an R-based metabolic modelling tool (Gelius-Dietrich et al. 2013). It reads metabolic models in the form of CSV as well as in SBML formats. The sybil package can be used for performing various simulations such as maximization of biomass, flux variability analysis, gene essentiality analysis, etc. It is primarily compatible for Clp, GLPK and CPLEX linear programming solvers.

5.7.5   PATHWAY TOOLS

The Pathway Tools is versatile bioinformatics software for metabolic modelling and omics analysis (Karp et al. 2010). It is free for academics. It allows creating organism-specific PGDB that can be used by other tools for modelling purposes. It can be used to perform various simulations based on FBA. Pathway Tools can also be used for generating regulatory, network and signalling pathways. These networks can be visualized using the software. Furthermore, it can be used to predict pathway hole fillers and operons. The Pathway Tools can also be utilized to search for routes between metabolites and identify possible drug targets as choke points.

5.7.6   MODELSEED

The ModelSEED is a user-friendly, Web-based tool for reconstruction and analysis of metabolic models (Devoid et al. 2013). It uses genome sequence in FASTA file format, automatically annotates and generates SBML format of the reconstructed metabolic model. It has a database of reconstructed models which can be utilized for further analysis. The ModelSEED provides option to create a customized media composition according to the need of the organism.

5.7.7 RAVEN

The RAVEN (Reconstruction, Analysis and Visualization of Metabolic Networks) is also a toolbox of MATLAB software suite (Wang et al. 2018). It has inbuilt scripts that connect directly to various databases to reconstruct a draft metabolic model. It also utilizes previously published metabolic models besides various biochemical and biological databases. It provides tools for system-wide analysis of metabolic models. It also has various functions based on FBA. The RAVEN toolbox is freely available for users having MATLAB license.

5.7.8 MERLIN

The MERLIN is a user-friendly, graphical user interface (GUI) base software for the reconstruction of GSMMs and FBA (Dias et al. 2015). It is a Java-based software platform. It automatically performs BLAST searches, annotations and information retrieval from GenBank, Entrez and KEGG databases and provides several tools for improving the reconstruction and generating the SBML model.

5.7.9 CoReCo

The CoReCo (**Co**mparative **ReCo**nstruction) is another metabolic model reconstruction tool (Pitkänen et al. 2014). It has a special feature of reconstruction of multiple related species simultaneously. It takes full protein sequences as input and performs various homology and similarity searches followed by incorporation of relevant reactions in the draft model. The draft model is subsequently refined to generate complete gapless model.

5.7.10 CarveMe

CarveMe is a Python-based tool for the reconstruction of GSMMs (Machado et al. 2018). It is compatible with Gurobi as well as CPLEX solvers and imports/exports models in SBML format. It can be used to generate an ensemble of models and community-level models.

5.8   METHODS OF ANALYSES OF GSMM FOR METABOLIC ENGINEERING

Systems biology helps in the understanding of complex biological systems by utilizing computational mathematical methods. FBA is a computational technique that is used to analyse GSMMs and may be used to predict the metabolic behaviour of cells (Orth, Thiele, and Palsson 2010). FBA can also be used to devise metabolic engineering strategies for the production of metabolites. For example, OptKnock is an FBA-based algorithm for identifying gene knockout strategies that couple cellular growth with metabolite production (Burgard, Pharkya, and Maranas 2003). While

OptKnock gives the most optimistic prediction of a product formation, RobustKnock was designed to provide the worst-case prediction of product formation (Tepper and Shlomi 2010). FBA, combined with genetic algorithm, has been used to develop OptGene to rapidly find gene knockout targets for a desired metabolic objective. Potential engineering targets for improved production of succinic acid, vanillin and glycerol were identified using OptGene (Patil et al. 2005). A neighbourhood search algorithm was employed to find the gene knockout targets (Lun et al. 2009). Transcriptional and metabolic networks could be integrated using OptORF to search regulatory and metabolic perturbations which couple growth and metabolite production (Kim and Reed 2010). OptReg was developed to identify both upregulation and downregulation targets (Pharkya and Maranas 2006). In light of several tools available for knockout identification, OptPipe was developed to combine solutions by common knockout identification tools and to rank the predicted mutants according to production, growth and new adaptability measure (Hartmann et al. 2017).

Along with the development of tools for the identification of genetic intervention targets, tools to predict mutant flux distributions have also been developed. The first such tool was minimization of metabolic adjustments (MOMA), which minimized the sum of squared differences between wild type and mutant flux distributions (Segrè, Vitkup, and Church 2002). Regulatory on/off minimization (ROOM) is an alternative method that minimizes the number of changes in fluxes in mutant strains (Shlomi, Berkman, and Ruppin 2005). Both MOMA and ROOM use a reference flux distribution to calculate the perturbed state flux distribution. The reference flux distribution is generally calculated using FBA which could be greatly improved by incorporating experimental flux data. RELATive CHange (RELATCH) was developed so that the reference flux distribution can be calculated using experimental data such as metabolic flux analysis and gene expression data (Kim and Reed 2010). RELATCH also avoids large fold changes in the mutant network. It was successfully shown to more accurately predict flux distributions in genetically and environmentally perturbed *E. coli*, *Saccharomyces cerevisiae* and *Bacillus subtilis* strains. Flux scanning based on enforced objective flux (FSEOF) method identifies the reaction fluxes that increase or decrease after flux through a product is enforced during FBA. Heterologous expression of lycopene synthesis pathway in *E. coli* along with overexpression of *idi* and *mdh* with the *dxs* gene identified using FSEOF method increased the lycopene titres by 13.2 mg/L fold. Further, MOMA (Segrè, Vitkup, and Church 2002) was used to identify the gdhA and gpmB double knockout strain. The strain incorporating both the suggested overexpression and knockout strategies produced 283 mg/L lycopene in fed-batch fermentation (Choi et al. 2010).

5.9 PATHWAY PREDICTION FOR SYNTHETIC BIOLOGY APPLICATIONS

Biosynthetic pathways for many metabolites have not been identified yet. Several pathway prediction tools such as BNICE (Hatzimanikatis et al. 2005), DESHARKY (Rodrigo et al. 2008), RetroPath (Carbonell et al. 2011), SimPheny (www.genomatica.com) and GEM-Path (Campodonico et al. 2014) have been developed to design synthetic routes to metabolites.

BNICE identifies every possible biochemical route to a metabolite given a set of enzyme rules. For example, BNICE found 75,000 novel biochemical routes from chorismate to phenylalanine (Hatzimanikatis et al. 2005). DESHARKY employs a Monte Carlo method to find routes from a target metabolite using enzymatic reaction database (Rodrigo et al. 2008). Rahnuma computes all pathways between two or more metabolites by representing metabolic networks as hypergraphs (Mithani, Preston, and Hein 2009). It focuses on evolutionary differences between organisms allowing comparison of organisms in terms of both metabolic network and phylogeny. RetroPath applies reverse chemical transformations from product to the source metabolites and also determines which route might be the best to engineer in particular host (Carbonell et al. 2011). MAPPS (Metabolic network Analysis and Pathway Prediction Server) is a Web-based tool that allows pathway prediction and network comparison, and in silico metabolic engineering simulations. MAPPS allows the user to use custom data for analysis on custom genomes (Riaz, Preston, and Mithani 2020). Recently, a deep learning architecture has been developed for predicting classes of pathways in which a given metabolite participates. It correctly predicted the metabolic pathway class of 95.16% of the tested metabolites (Baranwal et al. 2020).

Naringenin biosynthesis pathway was expressed in *Saccharomyces cerevisiae*. By deregulating the feedback mechanism and inhibition of by-product formation, naringenin titre of 400 μM was achieved (Koopman et al. 2012). Complete biosynthetic pathway of artemisinic acid was constructed in *S. cerevisiae* after the genome analysis of *Artemisia annua*. An accumulation of 25 g/L artemisinic acid was achieved after multiple optimizations (Paddon et al. 2013). A constraint-based optimization approach (Poolman 2006) was used to identify seven reaction knockout targets (*ppsA*, *ldhA*, *poxB*, *pta*, *pflB*, *edd* and *aceA*) and one overexpression target (zwf) in *E. coli* GSMM with two heterologously expressed genes from cyanobacteria for alkane production (Fatma et al. 2018). A titre of 425 mg/L of alkanes, with pentadecane (249 mg/L) and heptadecane (160 mg/L) as the major components, was achieved by engineering this strain. Genes from *Clostridium* were expressed in *E. coli* for the production of 1-butanol (Atsumi et al. 2008). Random mutagenesis and selection for leucine production yielded an *E. coli* strain that produced 3-methyl-1-butanol with titres reaching 9.5 g/L (Connor, Cann, and Liao 2010). Engineering photosynthetic microorganisms would directly convert CO_2 into the desired products. An amount of 2.38 g/L 2,3-butanediol was produced by engineering an oxygen-insensitive and cofactor-matched pathway in cyanobacteria, *Synechococcus elongatus* PCC7942 (Oliver et al. 2013). Figure 5.4 shows different modes and pathways for the production of biofuel molecules. Selenzyme is a free online tool for pathway designing by enzyme selection (Carbonell et al. 2018). The graphical interface provided information about the enzyme using existing databases. It also provides evolutionary distance between the source and the host organisms. Along with highlighting conserved regions and predicted active sites, it also provides predicted solubilities and transmembrane regions.

5.10 GENETIC ENGINEERING TOOLS

DNA assembly tools such as Gibson Assembly (Casini et al. 2014), Golden Gate assembly (Potapov et al. 2018), transformation-associated recombination (TAR)

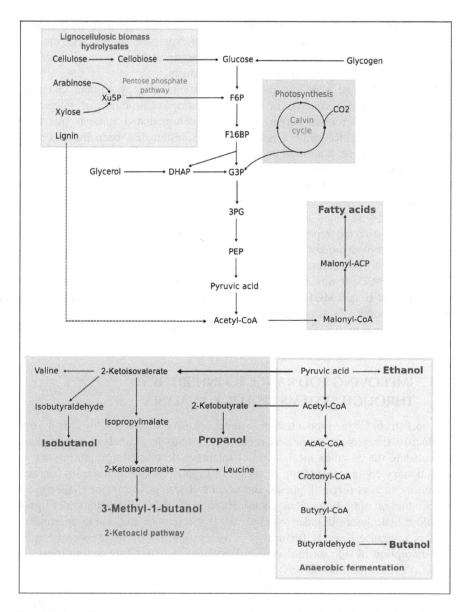

FIGURE 5.4   Different precursors for production of various biofuel molecules. Glucose is formed from hydrolysis of lignocellulosic materials. Also, fructose 6-phosphate is synthesized via pentose phosphate pathway from xylose. Calvin cycle fixes atmospheric carbon dioxide into glyceraldehyde 3-phosphate. These three molecules are metabolized via glycolysis to form pyruvate. Pyruvate is the precursor of ethanol. Fatty acids can be synthesized from pyruvate via acetyl coenzyme A. Acetyl coenzyme A can also be metabolized into butyraldehyde and, eventually, butanol. Biosynthetic pathway intermediates of valine and leucine, namely isobutyraldehyde and isopropylmalate, are precursors of isobutanol and 3-methy-1-butanol, respectively. 2-Acetolactate synthesized from pyruvate can be metabolized into 2,3-butanediol.

cloning (Ross et al. 2015) and Uracil-Specific Excision Reagent (USER) cloning (Lund et al. 2014) have enabled the construction of expression systems with multiple genes. Advances in gene synthesis have allowed the expression of synthetic pathways with codon-optimized genes for specific host organism.

For plasmid-based gene clusters, the fluctuations in copy number and plasmid instability are a concern in industrial settings. Chromosomal integration using homologous recombination and transposon-mediated random insertions is time-consuming. Recently, a recombineering system has been introduced in *Pseudomonas putida*. It allowed knockout lengths spanning 0.6–101.7 kb. Targeted chromosomal insertion was also demonstrated by inserting biosynthetic gene clusters for the production of proteins, polyketides, isoprenoids and amino acid derivatives (Choi et al. 2018). A clustered regularly interspaced short palindromic repeats (CRISPR/Cas9)-mediated genome editing via non-homologous end joining (NHEJ) or homology-directed repair (HDR) mechanism has recently been described in yeast. The complete process from target sequence selection to verification of mutant was reported to be achieved within 3 weeks in yeast (Kumari, Yusuf, and Gaur 2019). CRISPR/Cas9 was also used to engineer SSK42, an engineered ethanologenic strain of E. coli MG1655, for the production of butanol from xylose. Butanol biosynthetic pathway was inserted after the deletion of ethanol-producing pathway. The resultant strain, ASAO2, produced 4.32 g/L of butanol from xylose (Abdelaal, Jawed, and Yazdani 2019).

5.11 IMPROVING TOLERANCE TO INHIBITORS THROUGH SYSTEMS BIOLOGY ANALYSES

The host strain for the production of metabolites often is sensitive towards the end product or the growth conditions required. Systems-level analysis is necessary to elucidate the complex mechanism of tolerance developed through rational or evolutionary engineering (Ling et al. 2014). Lignocellulose hydrolysate contains inhibitors such as furfural, phenol and acetate. These pose a major challenge for the production of biofuel and biochemicals from microbial fermentation of lignocellulose hydrolysate (Sandström et al. 2014). Exogenous addition of spermidine confers tolerance towards furfural in *S. cerevisiae*. By overexpression of *spe3*, and disruption of *odc* (ornithine decarboxylase) and *tpo1* (polyamine transport protein) genes, spermidine was produced in *S. cerevisiae*. The engineered strain had 60% and 33% shorter lag phase when grown in medium containing furan derivatives and acetate, respectively (Kim et al. 2015). Transcriptomics analysis of strains developed by evolutionary engineering revealed mechanism for tolerance to hydrolysates and potential adaptation to oxidative stress (Almario, Reyes, and Kao 2013).

Genome sequencing, transcriptome and metabolic flux analysis techniques were employed to reveal mechanisms underlying thermotolerance in *S. cerevisiae*. It was found that non-sense mutations in C-5 sterol desaturase gene ERG3 changed the composition of sterol from ergosterol to fucosterol. Strains engineered with the ERG mutation grew faster at 40°C as compared to the wild type strain (Caspeta et al. 2014).

5.12 ENZYME DISCOVERY

Gut microbiota of obligate herbivores and termites can be potential sources of enzymes that degrade lignocellulosic biomass. Metagenomic analysis from adult elephant faecal samples revealed the presence of bacteria belonging to phylum Proteobacteria. A total of 55,000 ORFs had either catalytic domains or carbohydrate-binding modules in the metagenomic dataset (Jakeer et al. 2020). Secretome produced by *Aspergillus flavus* (isolated from elephant faeces), when used in 1:1 ratio with the commercial enzyme CTec2, increased saccharification of acid-pre-treated paddy straw by ~68% at 50°C (Kumar et al. 2020). Proteomics analysis of *Penicillium funiculosum* revealed an abundance of carbon active enzymes which act synergistically in saccharification of biomass (Ogunmolu et al. 2015).

5.13 FUTURE DIRECTIONS FOR SYSTEMS BIOLOGY – INTEGRATIVE ANALYSES

A systems-level understanding of cellular metabolism and its regulation is necessary to create efficient cell factories for biofuel production. Integration of regulatory mechanisms in new-generation GSMMs can further advance systems metabolic engineering. Techniques such as CRISPR/Cas9 system have enabled the implementation of various engineering strategies. Identification of metabolic engineering strategies has now become the bottleneck in strain engineering. ^{13}C-metabolic flux analysis (^{13}C-MFA) is a powerful method to understand intracellular metabolism (Zamboni et al. 2009; Long and Antoniewicz 2019). Several ^{13}C-MFA tools are available for accurate estimation of intracellular fluxes in cells (Desai and Srivastava 2018; Shupletsov et al. 2014; Young 2014). Integration of ^{13}C-MFA and omics data in constraint-based optimization methods for finding new engineering strategies could give more efficient and robust production of strains.

REFERENCES

Abdelaal, Ali Samy, Kamran Jawed, and Syed Shams Yazdani. 2019. "CRISPR/Cas9-Mediated Engineering of Escherichia Coli for n-Butanol Production from Xylose in Defined Medium." *Journal of Industrial Microbiology and Biotechnology* 46 (7): 965–75. doi:10.1007/s10295-019-02180-8.

Ahmad, Ahmad, Archana Tiwari, and Shireesh Srivastava. 2020. "A Genome-Scale Metabolic Model of Thalassiosira Pseudonana CCMP 1335 for a Systems-Level Understanding of Its Metabolism and Biotechnological Potential." *Microorganisms* 8 (9). Multidisciplinary Digital Publishing Institute: 1396. doi:10.3390/microorganisms8091396.

Ahmad, Ahmad, Hassan B. Hartman, S. Krishnakumar, David A. Fell, Mark G. Poolman, and Shireesh Srivastava. 2017. "A Genome Scale Model of Geobacillus Thermoglucosidasius (C56-YS93) Reveals Its Biotechnological Potential on Rice Straw Hydrolysate." *Journal of Biotechnology* 251 (June): 30–37. doi:10.1016/j.jbiotec.2017.03.031.

Almario, María P., Luis H. Reyes, and Katy C. Kao. 2013. "Evolutionary Engineering of Saccharomyces Cerevisiae for Enhanced Tolerance to Hydrolysates of Lignocellulosic Biomass." *Biotechnology and Bioengineering* 110 (10): 2616–23. doi:10.1002/bit.24938.

Atsumi, Shota, Anthony F. Cann, Michael R. Connor, Claire R. Shen, Kevin M. Smith, Mark P. Brynildsen, Katherine J. Y. Chou, Taizo Hanai, and James C. Liao. 2008. "Metabolic

Engineering of Escherichia Coli for 1-Butanol Production." *Metabolic Engineering*, Engineering Metabolic Pathways for Biofuels Production, 10 (6): 305–11. doi:10.1016/j.ymben.2007.08.003.

Baranwal, Mayank, Abram Magner, Paolo Elvati, Jacob Saldinger, Angela Violi, and Alfred O. Hero. 2020. "A Deep Learning Architecture for Metabolic Pathway Prediction." *Bioinformatics* 36 (8): 2547–53. doi:10.1093/bioinformatics/btz954.

Bordel, Sergio, Yadira Rodríguez, Anna Hakobyan, Elisa Rodríguez, Raquel Lebrero, and Raúl Muñoz. 2019. "Genome Scale Metabolic Modeling Reveals the Metabolic Potential of Three Type II Methanotrophs of the Genus Methylocystis." *Metabolic Engineering* 54 (July): 191–99. doi:10.1016/j.ymben.2019.04.001.

Burgard, Anthony P., Priti Pharkya, and Costas D. Maranas. 2003. "Optknock: A Bilevel Programming Framework for Identifying Gene Knockout Strategies for Microbial Strain Optimization." *Biotechnology and Bioengineering* 84 (6): 647–57. doi:10.1002/bit.10803.

Campodonico, Miguel A., Barbara A. Andrews, Juan A. Asenjo, Bernhard O. Palsson, and Adam M. Feist. 2014. "Generation of an Atlas for Commodity Chemical Production in Escherichia Coli and a Novel Pathway Prediction Algorithm, GEM-Path." *Metabolic Engineering* 25 (September): 140–58. doi:10.1016/j.ymben.2014.07.009.

Carbonell, Pablo, Anne-Gaëlle Planson, Davide Fichera, and Jean-Loup Faulon. 2011. "A Retrosynthetic Biology Approach to Metabolic Pathway Design for Therapeutic Production." *BMC Systems Biology* 5 (August): 122. doi:10.1186/1752-0509-5-122.

Carbonell, Pablo, Jerry Wong, Neil Swainston, Eriko Takano, Nicholas J. Turner, Nigel S. Scrutton, Douglas B. Kell, Rainer Breitling, and Jean-Loup Faulon. 2018. "Selenzyme: Enzyme Selection Tool for Pathway Design." *Bioinformatics* 34 (12): 2153–54. doi:10.1093/bioinformatics/bty065.

Casini, Arturo, James T. MacDonald, Joachim De Jonghe, Georgia Christodoulou, Paul S. Freemont, Geoff S. Baldwin, and Tom Ellis. 2014. "One-Pot DNA Construction for Synthetic Biology: The Modular Overlap-Directed Assembly with Linkers (MODAL) Strategy." *Nucleic Acids Research* 42 (1): e7. doi:10.1093/nar/gkt915.

Caspeta, Luis, Yun Chen, Payam Ghiaci, Amir Feizi, Steen Buskov, Björn M. Hallström, Dina Petranovic, and Jens Nielsen. 2014. "Altered Sterol Composition Renders Yeast Thermotolerant." *Science* 346 (6205). American Association for the Advancement of Science: 75–78. doi:10.1126/science.1258137.

Choi, Hyung Seok, Sang Yup Lee, Tae Yong Kim, and Han Min Woo. 2010. "In Silico Identification of Gene Amplification Targets for Improvement of Lycopene Production." *Applied and Environmental Microbiology* 76 (10): 3097–105. doi:10.1128/AEM.00115-10.

Choi, Kyeong Rok, Jae Sung Cho, In Jin Cho, Dahyeon Park, and Sang Yup Lee. 2018. "Markerless Gene Knockout and Integration to Express Heterologous Biosynthetic Gene Clusters in Pseudomonas Putida." *Metabolic Engineering* 47 (May): 463–74. doi:10.1016/j.ymben.2018.05.003.

Choi, Kyeong Rok, Woo Dae Jang, Dongsoo Yang, Jae Sung Cho, Dahyeon Park, and Sang Yup Lee. 2019. "Systems Metabolic Engineering Strategies: Integrating Systems and Synthetic Biology with Metabolic Engineering." *Trends in Biotechnology* 37 (8): 817–37. doi:10.1016/j.tibtech.2019.01.003.

Choi, Sol, Hyohak Song, Sung Won Lim, Tae Yong Kim, Jung Ho Ahn, Jeong Wook Lee, Moon-Hee Lee, and Sang Yup Lee. 2016. "Highly Selective Production of Succinic Acid by Metabolically Engineered Mannheimia Succiniciproducens and Its Efficient Purification." *Biotechnology and Bioengineering* 113 (10): 2168–77. doi:10.1002/bit.25988.

Connor, Michael R., Anthony F. Cann, and James C. Liao. 2010. "3-Methyl-1-Butanol Production in Escherichia Coli: Random Mutagenesis and Two-Phase Fermentation." *Applied Microbiology and Biotechnology* 86 (4): 1155–64. doi:10.1007/s00253-009-2401-1.

Desai, Trunil S., and Shireesh Srivastava. 2018. "FluxPyt: A Python-Based Free and Open-Source Software for 13C-Metabolic Flux Analyses." *PeerJ* 6: e4716. doi:10.7717/peerj.4716.

Devoid, Scott, Ross Overbeek, Matthew DeJongh, Veronika Vonstein, Aaron A. Best, and Christopher Henry. 2013. "Automated Genome Annotation and Metabolic Model Reconstruction in the SEED and Model SEED." In *Systems Metabolic Engineering: Methods and Protocols*, edited by Hal S. Alper, 17–45. Methods in Molecular Biology. Totowa, NJ: Humana Press. doi:10.1007/978-1-62703-299-5_2.

Dias, Oscar, Miguel Rocha, Eugénio C. Ferreira, and Isabel Rocha. 2015. "Reconstructing Genome-Scale Metabolic Models with Merlin." *Nucleic Acids Research* 43 (8): 3899–910. doi:10.1093/nar/gkv294.

Ebrahim, Ali, Joshua A. Lerman, Bernhard O. Palsson, and Daniel R. Hyduke. 2013. "COBRApy: COnstraints-Based Reconstruction and Analysis for Python." *BMC Systems Biology* 7 (1): 74. doi:10.1186/1752-0509-7-74.

El-Semman, Ibrahim E., Fredrik H. Karlsson, Saeed Shoaie, Intawat Nookaew, Taysir H. Soliman, and Jens Nielsen. 2014. "Genome-Scale Metabolic Reconstructions of Bifidobacterium Adolescentis L2–32 and Faecalibacterium Prausnitzii A2–165 and Their Interaction." *BMC Systems Biology* 8 (1): 41. doi:10.1186/1752-0509-8-41.

Espaux, Leo d', Amit Ghosh, Weerawat Runguphan, Maren Wehrs, Feng Xu, Oliver Konzock, Ishaan Dev, et al. 2017. "Engineering High-Level Production of Fatty Alcohols by Saccharomyces Cerevisiae from Lignocellulosic Feedstocks." *Metabolic Engineering* 42 (July): 115–25. doi:10.1016/j.ymben.2017.06.004.

Fatma, Zia, Hassan Hartman, Mark G. Poolman, David A. Fell, Shireesh Srivastava, Tabinda Shakeel, and Syed Shams Yazdani. 2018. "Model-Assisted Metabolic Engineering of Escherichia Coli for Long Chain Alkane and Alcohol Production." *Metabolic Engineering* 46 (March): 1–12. doi:10.1016/j.ymben.2018.01.002.

Feist, Adam M, Christopher S. Henry, Jennifer L. Reed, Markus Krummenacker, Andrew R. Joyce, Peter D. Karp, Linda J. Broadbelt, Vassily Hatzimanikatis, and Bernhard Ø. Palsson. 2007. "A Genome-Scale Metabolic Reconstruction for Escherichia Coli K-12 MG1655 That Accounts for 1260 ORFs and Thermodynamic Information." *Molecular Systems Biology* 3 (1). John Wiley & Sons, Ltd: 121. doi:10.1038/msb4100155.

Fleming, Ronan M. T., and Ines Thiele. 2011. "Von Bertalanffy 1.0: A COBRA Toolbox Extension to Thermodynamically Constrain Metabolic Models." *Bioinformatics* 27 (1): 142–43. doi:10.1093/bioinformatics/btq607.

Gelius-Dietrich, Gabriel, Abdelmoneim Amer Desouki, Claus Jonathan Fritzemeier, and Martin J Lercher. 2013. "Sybil – Efficient Constraint-Based Modelling in R." *BMC Systems Biology* 7 (November): 125. doi:10.1186/1752-0509-7-125.

Gupta, Jai Kumar, Preeti Rai, Kavish Kumar Jain, and Shireesh Srivastava. 2020. "Overexpression of Bicarbonate Transporters in the Marine Cyanobacterium Synechococcus Sp. PCC 7002 Increases Growth Rate and Glycogen Accumulation." *Biotechnology for Biofuels* 13 (1): 17. doi:10.1186/s13068-020-1656-8.

Hartmann, András, Ana Vila-Santa, Nicolai Kallscheuer, Michael Vogt, Alice Julien-Laferrière, Marie-France Sagot, Jan Marienhagen, and Susana Vinga. 2017. "OptPipe - a Pipeline for Optimizing Metabolic Engineering Targets." *BMC Systems Biology* 11 (1): 143. doi:10.1186/s12918-017-0515-0.

Hatzimanikatis, Vassily, Chunhui Li, Justin A. Ionita, Christopher S. Henry, Matthew D. Jankowski, and Linda J. Broadbelt. 2005. "Exploring the Diversity of Complex Metabolic Networks." *Bioinformatics (Oxford, England)* 21 (8): 1603–09. doi:10.1093/bioinformatics/bti213.

Heirendt, Laurent, Sylvain Arreckx, Thomas Pfau, Sebastián N. Mendoza, Anne Richelle, Almut Heinken, Hulda S. Haraldsdóttir, et al. 2019. "Creation and Analysis of

Biochemical Constraint-Based Models Using the COBRA Toolbox v.3.0." *Nature Protocols* 14 (3). Nature Publishing Group: 639–702. doi:10.1038/s41596-018-0098-2.

Hendry, John I., Charulata B. Prasannan, Aditi Joshi, Santanu Dasgupta, and Pramod P. Wangikar. 2016. "Metabolic Model of Synechococcus Sp. PCC 7002: Prediction of Flux Distribution and Network Modification for Enhanced Biofuel Production." *Bioresource Technology*, International Conference on New Horizons in Biotechnology (NHBT-2015), 213 (August): 190–97. doi:10.1016/j.biortech.2016.02.128.

Jakeer, Shaik, Mahendra Varma, Juhi Sharma, Farnaz Mattoo, Dinesh Gupta, Joginder Singh, Manoj Kumar, and Naseem A. Gaur. 2020. "Metagenomic Analysis of the Fecal Microbiome of an Adult Elephant Reveals the Diversity of CAZymes Related to Lignocellulosic Biomass Degradation." *Symbiosis* 81 (3): 209–22. doi:10.1007/s13199-020-00695-8.

Jamshidi, Neema, and Bernhard Ø. Palsson. 2007. "Investigating the Metabolic Capabilities of Mycobacterium Tuberculosis H37Rv Using the in Silico Strain INJ 661 and Proposing Alternative Drug Targets." *BMC Systems Biology* 1 (1): 26. doi:10.1186/1752-0509-1-26.

Karp, Peter D., Suzanne M. Paley, Markus Krummenacker, Mario Latendresse, Joseph M. Dale, Thomas J. Lee, Pallavi Kaipa, et al. 2010. "Pathway Tools Version 13.0: Integrated Software for Pathway/Genome Informatics and Systems Biology." *Briefings in Bioinformatics* 11 (1): 40–79. doi:10.1093/bib/bbp043.

Kim, Joonhoon, and Jennifer L. Reed. 2010. "OptORF: Optimal Metabolic and Regulatory Perturbations for Metabolic Engineering of Microbial Strains." *BMC Systems Biology* 4 (April): 53. doi:10.1186/1752-0509-4-53.

Kim, Sun-Ki, Yong-Su Jin, In-Geol Choi, Yong-Cheol Park, and Jin-Ho Seo. 2015. "Enhanced Tolerance of Saccharomyces Cerevisiae to Multiple Lignocellulose-Derived Inhibitors through Modulation of Spermidine Contents." *Metabolic Engineering* 29 (May): 46–55. doi:10.1016/j.ymben.2015.02.004.

Koopman, Frank, Jules Beekwilder, Barbara Crimi, Adele van Houwelingen, Robert D. Hall, Dirk Bosch, Antonius J. A. van Maris, Jack T. Pronk, and Jean-Marc Daran. 2012. "De Novo Production of the Flavonoid Naringenin in Engineered Saccharomyces Cerevisiae." *Microbial Cell Factories* 11 (1): 155. doi:10.1186/1475-2859-11-155.

Kumar, Mohit, Ajay Kumar Pandey, Sonam Kumari, Shahid Ali Wani, Shaik Jakeer, Rameshwar Tiwari, Rajendra Prasad, and Naseem A. Gaur. 2020. "Secretome Produced by a Newly Isolated Aspergillus Flavus Strain in Engineered Medium Shows Synergy for Biomass Saccharification with a Commercial Cellulase." *Biomass Conversion and Biorefinery*, August. doi:10.1007/s13399-020-00935-3.

Kumari, Priya, Farnaz Yusuf, and Naseem A. Gaur. 2019. "Novel Microbial Modification Tools to Convert Lipids into Other Value-Added Products." In *Microbial Lipid Production: Methods and Protocols*, edited by Venkatesh Balan, 161–71. Methods in Molecular Biology. New York, NY: Springer. doi:10.1007/978-1-4939-9484-7_10.

Lee, Kyung Yun, Jong Myoung Park, Tae Yong Kim, Hongseok Yun, and Sang Yup Lee. 2010. "The Genome-Scale Metabolic Network Analysis of Zymomonas Mobilis ZM4 Explains Physiological Features and Suggests Ethanol and Succinic Acid Production Strategies." *Microbial Cell Factories* 9 (1): 94. doi:10.1186/1475-2859-9-94.

Lee, Sang Yup, Hyun Uk Kim, Tong Un Chae, Jae Sung Cho, Je Woong Kim, Jae Ho Shin, Dong In Kim, Yoo-Sung Ko, Woo Dae Jang, and Yu-Sin Jang. 2019. "A Comprehensive Metabolic Map for Production of Bio-Based Chemicals." *Nature Catalysis* 2 (1). Nature Publishing Group: 18–33. doi:10.1038/s41929-018-0212-4.

Liao, James C., Luo Mi, Sammy Pontrelli, and Shanshan Luo. 2016. "Fuelling the Future: Microbial Engineering for the Production of Sustainable Biofuels." *Nature Reviews. Microbiology* 14 (5): 288–304. doi:10.1038/nrmicro.2016.32.

Ling, Hua, Weisuong Teo, Binbin Chen, Susanna Su Jan Leong, and Matthew Wook Chang. 2014. "Microbial Tolerance Engineering toward Biochemical Production: From

Lignocellulose to Products." *Current Opinion in Biotechnology*, Cell and Pathway Engineering, 29 (October): 99–106. doi:10.1016/j.copbio.2014.03.005.

Long, Christopher P., and Maciek R. Antoniewicz. 2019. "High-Resolution 13C Metabolic Flux Analysis." *Nature Protocols* 14 (10): 2856–77. doi:10.1038/s41596-019-0204-0.

Lu, Hongzhong, Feiran Li, Benjamín J. Sánchez, Zhengming Zhu, Gang Li, Iván Domenzain, Simonas Marcišauskas, et al. 2019. "A Consensus S. Cerevisiae Metabolic Model Yeast8 and Its Ecosystem for Comprehensively Probing Cellular Metabolism." *Nature Communications* 10 (1). Nature Publishing Group: 3586. doi:10.1038/s41467-019-11581-3.

Lun, Desmond S., Graham Rockwell, Nicholas J. Guido, Michael Baym, Jonathan A. Kelner, Bonnie Berger, James E. Galagan, and George M. Church. 2009. "Large-Scale Identification of Genetic Design Strategies Using Local Search." *Molecular Systems Biology* 5: 296. doi:10.1038/msb.2009.57.

Lund, Anne Mathilde, Helene Faustrup Kildegaard, Maja Borup Kjær Petersen, Julie Rank, Bjarne Gram Hansen, Mikael Rørdam Andersen, and Uffe Hasbro Mortensen. 2014. "A Versatile System for USER Cloning-Based Assembly of Expression Vectors for Mammalian Cell Engineering." *PLOS ONE* 9 (5). Public Library of Science: e96693. doi:10.1371/journal.pone.0096693.

Machado, Daniel, Sergej Andrejev, Melanie Tramontano, and Kiran Raosaheb Patil. 2018. "Fast Automated Reconstruction of Genome-Scale Metabolic Models for Microbial Species and Communities." *Nucleic Acids Research* 46 (15): 7542–53. doi:10.1093/nar/gky537.

Mithani, Aziz, Gail M. Preston, and Jotun Hein. 2009. "Rahnuma: Hypergraph-Based Tool for Metabolic Pathway Prediction and Network Comparison." *Bioinformatics* 25 (14): 1831–32. doi:10.1093/bioinformatics/btp269.

Ogunmolu, Funso Emmanuel, Inderjeet Kaur, Mayank Gupta, Zeenat Bashir, Nandita Pasari, and Syed Shams Yazdani. 2015. "Proteomics Insights into the Biomass Hydrolysis Potentials of a Hypercellulolytic Fungus Penicillium Funiculosum." *Journal of Proteome Research* 14 (10). American Chemical Society: 4342–58. doi:10.1021/acs.jproteome.5b00542.

Oliver, John W. K., Iara M. P. Machado, Hisanari Yoneda, and Shota Atsumi. 2013. "Cyanobacterial Conversion of Carbon Dioxide to 2,3-Butanediol." *Proceedings of the National Academy of Sciences* 110 (4). National Academy of Sciences: 1249–54. doi:10.1073/pnas.1213024110.

Orth, Jeffrey D., Ines Thiele, and Bernhard Ø. Palsson. 2010. "What Is Flux Balance Analysis?" *Nature Biotechnology* 28 (3): 245–48. doi:10.1038/nbt.1614.

Orth, Jeffrey D., Tom M. Conrad, Jessica Na, Joshua A. Lerman, Hojung Nam, Adam M. Feist, and Bernhard Ø. Palsson. 2011. "A Comprehensive Genome-Scale Reconstruction of Escherichia Coli Metabolism--2011." *Molecular Systems Biology* 7 (October): 535. doi:10.1038/msb.2011.65.

Paddon, C. J., P. J. Westfall, D. J. Pitera, K. Benjamin, K. Fisher, D. McPhee, M. D. Leavell, et al. 2013. "High-Level Semi-Synthetic Production of the Potent Antimalarial Artemisinin." *Nature* 496 (7446). Nature Publishing Group: 528–32. doi:10.1038/nature12051.

Pandiyan, K., Arjun Singh, Surender Singh, Anil Kumar Saxena, and Lata Nain. 2019. "Technological Interventions for Utilization of Crop Residues and Weedy Biomass for Second Generation Bio-Ethanol Production." *Renewable Energy* 132 (March): 723–41. doi:10.1016/j.renene.2018.08.049.

Patil, Kiran Raosaheb, Isabel Rocha, Jochen Förster, and Jens Nielsen. 2005. "Evolutionary Programming as a Platform for in Silico Metabolic Engineering." *BMC Bioinformatics* 6 (1): 308. doi:10.1186/1471-2105-6-308.

Pharkya, Priti, and Costas D. Maranas. 2006. "An Optimization Framework for Identifying Reaction Activation/Inhibition or Elimination Candidates for Overproduction in Microbial Systems." *Metabolic Engineering* 8 (1): 1–13. doi:10.1016/j.ymben.2005.08.003.

Pitkänen, Esa, Paula Jouhten, Jian Hou, Muhammad Fahad Syed, Peter Blomberg, Jana Kludas, Merja Oja, et al. 2014. "Comparative Genome-Scale Reconstruction of Gapless Metabolic Networks for Present and Ancestral Species." *PLoS Computational Biology* 10 (2): e1003465. doi:10.1371/journal.pcbi.1003465.

Poolman, M. G. 2006. "ScrumPy: Metabolic Modelling with Python." *IEE Proceedings - Systems Biology* 153 (5). IET Digital Library: 375–78. doi:10.1049/ip-syb:20060010.

Potapov, Vladimir, Jennifer L. Ong, Rebecca B. Kucera, Bradley W. Langhorst, Katharina Bilotti, John M. Pryor, Eric J. Cantor, et al. 2018. "Comprehensive Profiling of Four Base Overhang Ligation Fidelity by T4 DNA Ligase and Application to DNA Assembly." *ACS Synthetic Biology* 7 (11). American Chemical Society: 2665–74. doi:10.1021/acssynbio.8b00333.

Riaz, Muhammad Rizwan, Gail M. Preston, and Aziz Mithani. 2020. "MAPPS: A Web-Based Tool for Metabolic Pathway Prediction and Network Analysis in the Postgenomic Era." *ACS Synthetic Biology* 9 (5). American Chemical Society: 1069–82. doi:10.1021/acssynbio.9b00397.

Rodrigo, Guillermo, Javier Carrera, Kristala Jones Prather, and Alfonso Jaramillo. 2008. "DESHARKY: Automatic Design of Metabolic Pathways for Optimal Cell Growth." *Bioinformatics (Oxford, England)* 24 (21): 2554–56. doi:10.1093/bioinformatics/btn471.

Ross, Avena C., Lauren E. S. Gulland, Pieter C. Dorrestein, and Bradley S. Moore. 2015. "Targeted Capture and Heterologous Expression of the Pseudoalteromonas Alterochromide Gene Cluster in Escherichia Coli Represents a Promising Natural Product Exploratory Platform." *ACS Synthetic Biology* 4 (4): 414–20. doi:10.1021/sb500280q.

Sandström, Anders G., Henrik Almqvist, Diogo Portugal-Nunes, Dário Neves, Gunnar Lidén, and Marie F. Gorwa-Grauslund. 2014. "Saccharomyces Cerevisiae: A Potential Host for Carboxylic Acid Production from Lignocellulosic Feedstock?" *Applied Microbiology and Biotechnology* 98 (17): 7299–318. doi:10.1007/s00253-014-5866-5.

Satish Kumar, Vinay, Madhukar S. Dasika, and Costas D. Maranas. 2007. "Optimization Based Automated Curation of Metabolic Reconstructions." *BMC Bioinformatics* 8 (1): 212. doi:10.1186/1471-2105-8-212.

Savinell, Joanne M., and Bernhard O. Palsson. 1992. "Network Analysis of Intermediary Metabolism Using Linear Optimization. I. Development of Mathematical Formalism." *Journal of Theoretical Biology* 154 (4): 421–54. doi:10.1016/S0022-5193(05)80161-4.

Segrè, Daniel, Dennis Vitkup, and George M. Church. 2002. "Analysis of Optimality in Natural and Perturbed Metabolic Networks." *Proceedings of the National Academy of Sciences of the United States of America* 99 (23): 15112–17. doi:10.1073/pnas.232349399.

Seif, Yara, Erol Kavvas, Jean-Christophe Lachance, James T. Yurkovich, Sean-Paul Nuccio, Xin Fang, Edward Catoiu, Manuela Raffatellu, Bernhard O. Palsson, and Jonathan M. Monk. 2018. "Genome-Scale Metabolic Reconstructions of Multiple Salmonella Strains Reveal Serovar-Specific Metabolic Traits." *Nature Communications* 9 (1). Nature Publishing Group: 3771. doi:10.1038/s41467-018-06112-5.

Serrano-Bermúdez, Luis Miguel, Andrés Fernando González Barrios, Costas D. Maranas, and Dolly Montoya. 2017. "Clostridium Butyricum Maximizes Growth While Minimizing Enzyme Usage and ATP Production: Metabolic Flux Distribution of a Strain Cultured in Glycerol." *BMC Systems Biology* 11 (1): 58. doi:10.1186/s12918-017-0434-0.

Shlomi, Tomer, Omer Berkman, and Eytan Ruppin. 2005. "Regulatory on/off Minimization of Metabolic Flux Changes after Genetic Perturbations." *Proceedings of the National Academy of Sciences of the United States of America* 102 (21): 7695–700. doi:10.1073/pnas.0406346102.

Shupletsov, Mikhail S., Lyubov I. Golubeva, Svetlana S. Rubina, Dmitry A. Podvyaznikov, Shintaro Iwatani, and Sergey V. Mashko. 2014. "OpenFLUX2: (13)C-MFA Modeling Software Package Adjusted for the Comprehensive Analysis of Single and Parallel Labeling Experiments." *Microbial Cell Factories* 13 (November): 152. doi:10.1186/s12934-014-0152-x.

Swainston, Neil, Kieran Smallbone, Hooman Hefzi, Paul D. Dobson, Judy Brewer, Michael Hanscho, Daniel C. Zielinski, et al. 2016. "Recon 2.2: From Reconstruction to Model of Human Metabolism." *Metabolomics* 12 (7): 109. doi:10.1007/s11306-016-1051-4.

Tepper, Naama, and Tomer Shlomi. 2010. "Predicting Metabolic Engineering Knockout Strategies for Chemical Production: Accounting for Competing Pathways." *Bioinformatics (Oxford, England)* 26 (4): 536–43. doi:10.1093/bioinformatics/btp704.

Thiele, Ines, and Bernhard Ø. Palsson. 2010. "A Protocol for Generating a High-Quality Genome-Scale Metabolic Reconstruction." *Nature Protocols* 5 (1). Nature Publishing Group: 93–121. doi:10.1038/nprot.2009.203.

Thiele, Ines, Nikos Vlassis, and Ronan M. T. Fleming. 2014. "FastGapFill: Efficient Gap Filling in Metabolic Networks." *Bioinformatics* 30 (17): 2529–31. doi:10.1093/bioinformatics/btu321.

Tomàs-Gamisans, Màrius, Pau Ferrer, and Joan Albiol. 2016. "Integration and Validation of the Genome-Scale Metabolic Models of Pichia Pastoris: A Comprehensive Update of Protein Glycosylation Pathways, Lipid and Energy Metabolism." *PLOS ONE* 11 (1). Public Library of Science: e0148031. doi:10.1371/journal.pone.0148031.

Wang, Hao, Simonas Marcišauskas, Benjamín J. Sánchez, Iván Domenzain, Daniel Hermansson, Rasmus Agren, Jens Nielsen, and Eduard J. Kerkhoven. 2018. "RAVEN 2.0: A Versatile Toolbox for Metabolic Network Reconstruction and a Case Study on Streptomyces Coelicolor." *PLoS Computational Biology* 14 (10): e1006541. doi:10.1371/journal.pcbi.1006541.

Young, Jamey D. 2014. "INCA: A Computational Platform for Isotopically Non-Stationary Metabolic Flux Analysis." *Bioinformatics (Oxford, England)* 30 (9): 1333–35. doi:10.1093/bioinformatics/btu015.

Zamboni, Nicola, Sarah-Maria Fendt, Martin Rühl, and Uwe Sauer. 2009. "(13)C-Based Metabolic Flux Analysis." *Nature Protocols* 4 (6): 878–92. doi:10.1038/nprot.2009.58.

Zhang, Yu, Jingyi Cai, Xiuling Shang, Bo Wang, Shuwen Liu, Xin Chai, Tianwei Tan, Yun Zhang, and Tingyi Wen. 2017. "A New Genome-Scale Metabolic Model of Corynebacterium Glutamicum and Its Application." *Biotechnology for Biofuels* 10 (1): 169. doi:10.1186/s13068-017-0856-3.

Zuñiga, Cristal, Chien-Ting Li, Tyler Huelsman, Jennifer Levering, Daniel C. Zielinski, Brian O. McConnell, Christopher P. Long, et al. 2016. "Genome-Scale Metabolic Model for the Green Alga Chlorella Vulgaris UTEX 395 Accurately Predicts Phenotypes under Autotrophic, Heterotrophic, and Mixotrophic Growth Conditions." *Plant Physiology* 172 (1). American Society of Plant Biologists: 589–602. doi:10.1104/pp.16.00593.

6 Enzyme-Based Saccharification

D. Sathish, Shivam Aggarwal,
Manasa Nagesh Hegde, and Nidhi Adlakha
Regional Centre for Biotechnology

CONTENTS

6.1 Introduction 153
6.2 Advent of Cellulase Producers 154
    6.2.1 Fungus – Workhorse for Industrial Production 156
    6.2.2 Bacterial Cellulase Systems... 157
        6.2.2.1 Cell-Bound System – Cellulosome 157
        6.2.2.2 Cell-Free Cellulase Enzymes... 159
6.3 Molecular and Biochemical Approaches for Enhanced Cellulase and
    Hemicellulase Production... 160
    6.3.1 Modulation of Expression of Transcriptional Activators and
        Repressors.. 160
    6.3.2 Combating Carbon Catabolite Repression 160
6.4 Other Factors Influencing Cellulase Production and Downstream
    Saccharification 162
    6.4.1 Biochemical Factors – Effect of Carbon Sources............................ 162
    6.4.2 Impact of Different Nitrogen Sources 163
    6.4.3 Influence of pH and Temperature 164
6.5 Enzyme-Based Saccharification at Laboratory and Industrial Scales 164
6.6 Enzyme Cocktails for Enhanced Saccharification 165
6.7 Applications of Cellulases 167
6.8 Case Studies – Industrial Success Stories in India...................................... 168
    6.8.1 India Glycols Ltd. 168
    6.8.2 PRAJ Industries, Pune.. 169
    6.8.3 DBT-IOCL, Faridabad 169
References.. 169

6.1 INTRODUCTION

Plant biomass is an excellent substrate for biofuel production, but its recalcitrant nature with extensive cross-links prevents its direct use. As detailed in previous chapters, cellulose, hemicellulose, and lignin are extensively cross-linked and its saccharification for fermentable sugars is conceivable only after pre-treatment. Pre-treatment offers two main advantages: (a) extraction of pure lignin for bioconversion reactions

DOI: 10.1201/9781003158486-6

and (b) exposure of hemicellulose and cellulose for further disintegration to monomeric residues (Jayasekara and Ratnayake 2019).

Various methods are exploited for the pre-treatment of complex plant biomass including physical – hydrothermolysis, comminution; chemical – alkali, solvents, acid; ozone; physico-chemical – ammonia fibre explosion, steam explosion; or enzymatic methods (Wahlström and Suurnäkki 2015, Blanch, Simmons, and Klein-Marcuschamer 2011). Chemical-based disintegration involves the use of acid and alkali at very high temperature (above 80°C) and pressure (15 psi) (Silverstein et al. 2007).

Acid treatment uses H_2SO_4, HNO_3 or HCl to remove/solubilize hemicellulose, but has little effect on lignin. Most commonly, concentrated acid (70%–80%) is added to destroy intra- and inter-chain hydrogen bonds present in cellulosic biomass, leading to generation of monomeric sugars with 90% efficiency (Zhou et al. 2011). Similar to acid treatment, stronger alkali treatments have been performed using ammonia in conjunction with high pressure and temperature for woody plant structure (Silverstein et al. 2007). Nonetheless, chemical treatment is still highly discouraged owing to its various drawbacks, such as difficult handling of concentrated acids and bases at high temperature and pressure, uncontrolled degradation and/or charring of sugars, and most importantly, it is non-environment-friendly (Pancha et al. 2016). Therefore, as an efficient alternative, researchers have now diverted their focus towards enzyme-based disintegration. Along with its eco-friendly benefits, it additionally offers other advantages as follows: (a) it uses milder conditions (neutral pH, lower temperature (40°C–50°C) and normal pressure, which prevent sugar charring); (b) it is efficient and cheaper; (c) it gives high sugar yield; and (d) it does not produce any non-sugar or inhibitory compounds during hydrolysis, such as furfural, hydroxymethylfurfural, formic acid, oxalic acid, acetic acid, levulinic acid or phenol, which might interfere in downstream processing (Jayasekara and Ratnayake 2019).

Cellulolytic enzymes include cellulase, hemicellulase and laccases. The word "cellulase" encompasses a group of enzymes referred to as carbohydrate-active enzymes (CAZymes), which differ in their mode of action. To detail a few, endoglucanases are a set of cellulolytic enzymes which usually attack at internal bonds of cellulose creating new chain ends, whereas exoglucanases trim from either reducing or non-reducing ends of polysaccharide chains. Lastly, β-glucosidase cleaves dimers to fermentable glucose residues (Singh, Verma, and Kumar 2015) (Figures 6.1 and 6.2).

Here, in this chapter, we will elaborate cellulase producers and their mode of action. This chapter will also highlight the techniques employed to augment enzyme-based saccharification, using both media optimization and strain engineering approaches. We also represent case studies highlighting the importance of cellulolytic enzymes at industrial scale.

6.2   ADVENT OF CELLULASE PRODUCERS

Various microbes ranging from protists, bacteria and fungi are known to produce cellulolytic enzymes; however, the composition varies. Depending on end-use, researchers have mined nature's reservoir to find novel strains with better and

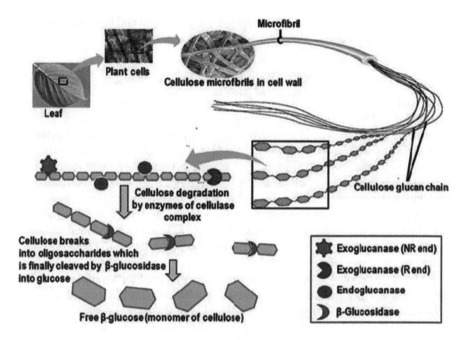

FIGURE 6.1 Schematic representation of sequential stages in cellulolysis (Singh, Verma, and Kumar 2015).

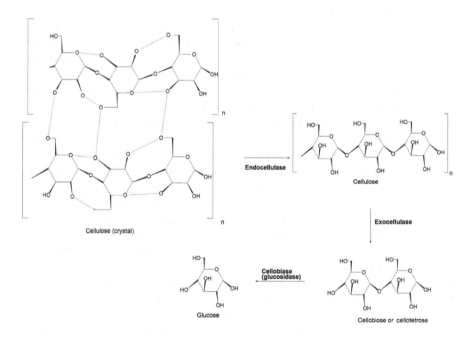

FIGURE 6.2 Cellulases and their action.

improved efficiency. Here, we have discussed majorly fungal and bacterial cellulases and elaborated their applications in diverse industrial sectors.

6.2.1 Fungus – Workhorse for Industrial Production

It was Sachs in 1862 who first described the presence of cellulose hydrolysing process in the seed of *Phoenix dactylifera* during germination. Others followed the suit and described the enzyme action in different parts of plants. The research in cellulose degradation took a major turn in 1888, when Marshall Ward first time reported secretion of cellulolytic enzymes from fungal sp., *Botrytis sp.* Other wood rotting fungi were also reported to have cellulases (Ward 1888). In early 1930s, Grassmann et al. (1933) found that supernatant from *Aspergillus oryzae* could hydrolyse multiple polymeric substrates including filter papers, hydrocellulose, mannan, and xylan (Boswell 1941, Grassmann, Stadler, and Bender 1933). In 1948, Woodward and Freeman independently found *Aspergillus niger* and *Aspergillus flavus* that could liquefy soluble cellulose derivatives (Freeman 1948).

Research in cellulases was fuelled by the increasing need for alternatives to fossil fuel. A destructive organism was causing rotting of the cotton canvas of a US army tent at Bougainville Island during the Second World War. But soon enough, the potential of exploiting its cellulolytic properties was identified (Reese, Siu, and Levinson 1950). In the early 1950s, the group at Biology Branch, Pioneering Research Division of US Army Quartermaster General Laboratories, Philadelphia, Pennsylvania, consisting of Siu, Levinson, Mandels and others, was focused on cellulases. With Elwyn T. Reese spearheading them, they found other cellulolytic organisms such as *Myrothecium verrucaria* and *Trichoderma viride* (later named *Trichoderma reesei* after Reese) that could grow on cellulose derivatives such as carboxymethyl cellulose (CMC). On comparing their enzyme activities as a measure of reducing sugars released from cotton duck strips, it was found that *T. reesei* QM6a, the strain isolated from the rotting army equipment, had the greatest activity when taking into account the amount of enzyme produced. They were also able to shed light on how the enzymes work as a group, and not individually to degrade cellulose (Reese and Levinson 1952).

T. reesei QM6a is the parent strain that was used for further mutations and genetic engineering because of its ability to produce copious amounts of "true cellulase" which could degrade cellulose as well as its derivatives completely (Bischof, Ramoni, and Seiboth 2016). For its industrial applicability, strain was mutagenized and rationally screened to identify the hyperproducer. An important step towards applying *T. reesei* cellulases industrially was the development of efficient strain mutagenesis and screening procedures. The most successful mutant, QM9414, had an extracellular protein and cellulase production level approximately four times to that of native QM6a; however, it remained catabolite-repressed. At Rutgers University, a combination of UV irradiation and chemical mutagens such as Nitrosoguanidine was used in three steps. First, mutagenesis by UV light and screening for catabolite derepression led to the isolation of strain M7. Another few rounds of mutagenesis by Nitrosoguanidine and screening led to the isolation of RUT-C30 strain. This strain produced 19 mg extracellular protein per mL and displayed a cellulase activity of 14 filter paper units/mL under controlled fermenter conditions. This was approximately 2.7 times the extracellular protein, 2.8 times the

TABLE 6.1

Cellulase Production by Mutant Strains of *T. reesei* following Growth on 6% (*w/v*) Roll-Milled Cotton for 14 Days

Strain	Soluble Protein (mg/mL)	FPU (U/mL)	Productivity (FPU/L h)	CMC (U/mL)	βGl (U/mL)
QM6a (parent)	7	5	15	88	0.3
QM9414 (Natick)	14	10	30	109	0.6
MCG77 (Natick)	16	11	33	104	0.9
NG-14 (Rutgers)	21	15	45	133	0.6
RUT-C30 (Rutgers)	19	14	42	150	0.3

Source: Ryu and Mandels (1980).

FPU, filter paper units; CMC, carboxymethyl cellulase (endoglucanase); βGl, β – glucosidase; U, enzyme units (mmol glucose produced/min in standard assay).

FPase, and twice the β-glucosidase activity of QM6a (Table 6.1). In addition, RUT-C30 was catabolite-derepressed and displayed a cellulase activity in the presence of 5% (*v/v*) glycerol that was almost equivalent to that produced on cellulose alone and was also resistant to catabolite repression by 5% (*w/v*) glucose. This meant that the fungus could be cultured easily on glucose and it would still produce cellulase (Peterson and Nevalainen 2012).

Overall, fungi are the most efficient cellulose decomposers accounting for approximately 80% cellulose breakdown on Earth and are preferred workhorses for the production of cellulases due to their copious extracellular production, but many bacterial species were discovered with exceptional tendency to secrete cellulases. They are exceptional in encoding cellulases that work optimally at alkaline pH range.

6.2.2 BACTERIAL CELLULASE SYSTEMS

The sum of cellulolytic enzymes produced from any bacterial strain is drastically less than that by fungal hyperproducers, yet they are much explored owing to few of their unique characteristics: (a) being stable over wide pH range; (b) being highly thermostable; and (c) being easy to engineer and express in heterologous system.

The bacterial cellulases are broadly classified into two categories based on their location and mode of action as cell-bound and cell-free systems.

6.2.2.1 Cell-Bound System – Cellulosome

Cellulosomes are multi-enzyme structures that act synergistically to catalyse cellulose degradation (Artzi, Bayer, and Moraïs 2017). The cellulosome structure is made up of a range of cellulases and related enzymes, which are grouped together into a complex through unique type of scaffolding. The cellulosomes are the systematic cellular machinery that plays a key role in breaking down the complex polymers in cell wall. Each cellulosome possesses structural subunits such as dockerin, scaffoldin and cohesin and catalytic subunits including surface layer homology (SLH) domain, carbohydrate-binding module (CBM) and enzymes that are responsible for catalysis. The structure of various components of a cellulosome on the bacterial cell surface is represented in Figure 6.3. Special type I and type II interactions hold the cohesins and dockerins together. Type I interaction connects the dockerins and group of scaffoldins. In addition, scaffoldin and dockerins also interact with cohesins on

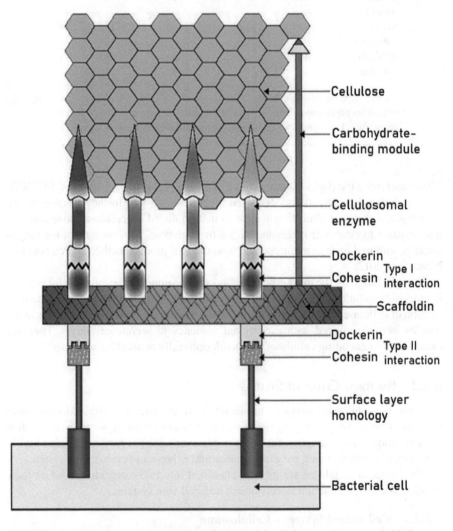

FIGURE 6.3 Structure and composition of a bacterial cellulosome (Arora et al. 2015).

the cell surface. This other mode of interaction is called type II interaction (Smith and Bayer 2013).

The cellulosomes offer an additional proximity advantage wherein the synergistically acting cellulolytic enzymes (Blanchette et al. 2012) are brought together. Importantly, the consortia of enzymes in any cellulosomal complex are directed by the composition of available substrate (Cho et al. 2010). In this direction, Kahn et al. (2019) proposed designer cellulosomes as a powerful tool in the production of value-added products and waste management.

Kahn et al. (2019) tried to create hyperthermostable designer cellulosomes with higher activity and efficiency. Enzymes from a hyperthermophilic, cellulolytic and anaerobic bacterium *Caldicellulosiruptor bescii* were incorporated to cellulosomal modules in *Clostridium thermocellum*. The designed cellulosome worked efficiently over 75°C for 72 h. This could be used for industrial purposes at high temperatures (Kahn et al. 2019).

In a similar study done by Galanopoulou et al. (2016), a tetravalent scaffoldin having cohesins of mesophilic origin and a CBM derived from *Clostridium thermocellum* were combined. The scaffoldins fused with thermophilic enzymes derived from *Geobacillus* and *Caldicellulosiruptor* species were fused to dockerins. The improved designer cellulosome demonstrated higher thermostability, i.e. 60°C for 6 h. Additionally, the hydrolytic efficiency of the complex enhanced by 50% (Galanopoulou et al. 2016).

Integration of lytic polysaccharide monooxygenases (LPMOs) into cellulosomes in bacterial system also demonstrated enhanced degradation efficiency. Cloned LPMOs were self-assembled with exo- and endocellulases to drive the synthesis of designer cellulosomal complex, which showed a 2.6-fold increase in conversion efficiency (Arfi et al. 2014). From the above studies, it has been found that the designer cellulosome enhances the efficiency of cellulose hydrolysis and, further, the development of designer cellulosomes may improve the enzymatic saccharification of various complex lignocellulosic substrates.

6.2.2.2   Cell-Free Cellulase Enzymes

Like fungi, bacteria also harbour cell-free enzyme system wherein the cellulolytic enzymes are secreted into the medium. These secretory cocktails of enzymes act synergistically to degrade complex recalcitrant biomass. Total cellulolytic enzyme comprises of exoglucanase, endoglucanase and β-D-glucosidases. The enzymes could act either individually or in combination. *Bacillus* sp., *Paenibacillus* sp. and *Pseudomonas* sp. are typical examples of bacterial cellulase producers (Prasanna, Ramanjaneyulu, and Rajasekhar Reddy 2016).

However, bacterial cellulases are far behind fungal cocktail of enzymes. Fungi produce copious amount of cellulases into the medium accounting for almost 80% of total secretory proteins. Therefore, for sustainable degradation of recalcitrant polysaccharide to monomeric residues, fungal enzyme cocktail is of choice.

In view of this fact, researchers diverted their focus towards engineering fungus for improving economics underlying saccharification of biomass. Various genetic tools are employed to engineer fungal strains for increased production.

6.3 MOLECULAR AND BIOCHEMICAL APPROACHES FOR ENHANCED CELLULASE AND HEMICELLULASE PRODUCTION

Several transcription factors have been characterized in various filamentous fungi regulating the expression of hydrolytic enzyme-encoding genes. The Cre1 and Xyr1 are widely studied transcription factors regulating cellulase and hemicellulases in filamentous fungi.

6.3.1 Modulation of Expression of Transcriptional Activators and Repressors

The XlnR was the first transcription factor found in the filamentous fungi *A. niger* (van Peij et al. 1998) that contains zinc cluster DNA-binding domain. It recognizes the binding motif with a core of 5'-GGCTAR-3' in the promoter regions of various cellulase and hemicellulase genes in *A. niger* (de Vries et al. 2002, Gielkens et al. 1999, van Peij et al. 1998, van Peij, Visser, and de Graaff 1998). XlnR is one of the well-studied transcription factors involved in the regulation of cellulase and hemicellulase genes in *A. niger* (Gielkens et al. 1999, van Peij et al. 1998), and it is referred to as Xyr1 in *T. reesei* (Mach-Aigner et al. 2008, Seiboth et al. 2012). In *T. reesei*, Xyr1 is known to differentially regulate the expression of hydrolytic genes in the presence of various small molecule inducers such as xylose, xylobiose and sophorose (Stricker et al. 2006). *XlnR* mutant showed reduced growth on xylan and downregulation of hydrolytic genes in *A. niger* (Khosravi et al. 2019, van Peij et al. 1998) and *T. reesei* (Derntl et al. 2013, Stricker et al. 2006). Overexpression of Xyr1 in *T. reesei* has revealed an inducer-independent expression of cellulase-encoding genes (Derntl, Mach, and Mach-Aigner 2019, Lv et al. 2015). Similarly, overexpression of Xyr1 has shown elevated production of cellulase and hemicellulase enzymes in the presence of inducers in various filamentous fungi including *A. niger* (van Peij et al. 1998), *T. reesei* (Derntl, Mach, and Mach-Aigner 2019, Zhang et al. 2018) and *T. harzianum* (da Silva Delabona et al. 2017).

Hakkinen et al. (2014) identified another transcriptional activator, ACE3, in *T. reesei*. The study highlighted an increment in the cellulase and hemicellulase production upon *ACE3* gene overexpression, whereas the deletion of *ACE3* gene completely aborted the cellulase production while decreased hemicellulase expression was observed. It was reported that ACE3 and Xyr1 work in a synergistic manner in such a way that the expression of ACE3 influences the expression of Xyr1 and vice versa (Hakkinen et al. 2014, Zhang et al. 2019). ACE2 and HAP2/3/5 complex are other transcriptional activators which influence the expression of hydrolytic enzyme-encoding genes (Aro et al. 2001, Zeilinger et al. 2001). Aro et al. (2001) reported about 30%–70% reduced cellulase activity in *ACE2*-deleted *T. reesei* grown in Solka-Floc cellulose.

6.3.2 Combating Carbon Catabolite Repression

The fungal cells sideline the expression of hydrolytic enzymes to utilize easily metabolizing monomers such as glucose by employing a range of transcription factors. Supplementation of glucose to the culture medium precludes the expression of the cellulase and hemicellulase genes in fungi due to carbon catabolite repression (CCR). The CCR mechanism is well conserved in various organisms from bacteria to human. This phenomenon has been explained by two mechanisms: inhibition of inducer consumption by glucose and glucose repression (Ilmen et al. 1997, Ilmen, Thrane, and Penttila 1996, Kubicek 1993, Zeilinger et al. 2003).

A concerted effort has been made to unravel the CCR mechanism and to find the effect of CCR disruption on hydrolytic enzymes production. Earlier attempts were made during 1970s to get the carbon catabolite-derepressed strains of *T. reesei* by treatment with UV irradiation and N-nitroguanidine and selection with 2-deoxyglucose to eliminate CCR (Montenecourt and Eveleigh 1977b, a, Montenecourt and Eveleigh 1979). The resulted *T. reesei* strain RUT-C30 produced about 20 g/L extracellular proteins and 150 U/L of FPase which is twice the amount of extracellular proteins produced by NG14 (Eveleigh and Montenecourt 1979) and 15–20 times higher with respect to its parent strain QM6a in shake flask conditions (Bisaria and Ghose 1981, Eveleigh 1982). Besides, *T. reesei* RUT-C30 was carbon catabolite-derepressed unlike its parental strains such as QM6a, NG14 or QM9414 and showed cellulase production in the presence of glucose or glycerol (Bisaria and Ghose 1981, Montenecourt and Eveleigh 1979, Nakari-Setala et al. 2009). *T. reesei* RUT-C30 produced about 2.7-fold higher proteins and FPase activity compared to QM6a when fermented in the presence of roll-milled cotton as carbon source (Ryu and Mandels 1980). The *Cre1* gene in *T. reesei* RUT-C30 has been truncated by random mutagenesis encoding only one of the two commonly occurring zinc finger regions of Cre1 protein (Ilmen, Thrane, and Penttila 1996). Ilmen, Thrane and Penttila (1996) observed steady state production of cellulase transcripts in the presence of glucose by *T. reesei* RUT-C30. In addition, they also observed that the replacement of truncated *Cre1* gene in RUT-C30 with full-length *Cre1* gene resulted in restored glucose repression which confirms the involvement of Cre1 in carbon catabolite repression. Similarly, various mutants of *T. reesei* such as RL-P37 (Sheir-Neiss and Montenecourt 1984) and CL847 (Durand, Clanet, and Tiraby 1988) have been developed for the enhanced cellulase production with carbon catabolite derepression. van der Veen et al. (1994) observed elevated arabinofuranosidase B activity in *CreA*-mutated *A. nidulans* when grown in a medium supplemented with L-arabitol and L-arabinose. Similarly, an increment in the cellulolytic and xylolytic enzyme production has been reported in various *Cre1*-deleted strains such as *Neurospora crassa* (Sun and Glass 2011), *Talaromyces cellulolyticus* (Fujii, Inoue, and Ishikawa 2013) and *A. oryzae* (Ichinose et al. 2018).

Cre1/CreA regulates the expression of target genes including cellulase-encoding (Nakari-Setala et al. 2009) and xylanase-encoding (Mach et al. 1996) genes by binding to the consensus motif (5′-SYGGRG-3′) of promoter region of target genes. The role of the Cre1/CreA binding motif has been studied in various filamentous fungi including *T. reesei* (Mach et al. 1996, Portnoy, Margeot, Linke, et al. 2011, Takashima et al. 1996) and *A. nidulans* (Cubero and Scazzocchio 1994, Panozzo, Cornillot, and Felenbok 1998). de Graaff et al. (1994) observed that deletion analysis of 158 bp region of promoter of *xlnA* gene containing four CreA binding motifs (GTGGGG, CCCCAG, CCCCAC and CCCCGC) resulted in increased *xlnA* transcripts in *Aspergillus tubingensis*. In addition, they also developed a reporter system with glucose oxidase gene driven by promoter region with CreA binding motifs which showed no activity of glucose oxidase in the presence of glucose and a higher activity in the presence of xylan which confirms the involvement of CreA and its binding motifs in the regulation of cellulolytic genes. In *T. reesei,* Zou et al. (2012) demonstrated that the replacement of the Cre1 binding sites with the binding sites of transcription activators ACE2 and HAP2/3/5 complex in promoter region of *cel7a* (*cbhI*) gene showed about 5.5-fold and 7.4-fold increased expression of *egfp* reporter gene in inducing media containing wheat bran and cellulose and in repressing media containing 2% glucose, respectively.

The function of Cre1 is beyond the CCR alone, and it is the main regulator of carbon metabolism in fungi. Transcription profiling by genome-wide microarray analysis of Cre1 mutant *T. reesei* grown on glucose showed that 47.3% of the genes were repressed and 29% of the genes were induced by Cre1. In addition to the carbon catabolite repression, Cre1 also plays an essential role in the regulation of developmental processes, transport proteins, nitrogenous substances uptake, components of chromatin remodelling and nucleosome positioning (Portnoy, Margeot, Linke, et al. 2011, Ries et al. 2014). Another mode of increasing enzyme production for enzyme-based saccharification is optimization of process at the level of media and culture conditions.

6.4 OTHER FACTORS INFLUENCING CELLULASE PRODUCTION AND DOWNSTREAM SACCHARIFICATION

Over the decades, rational enzyme engineering and factorial media design have drastically improved the production of cellulolytic enzymes, making enzymes amenable to industrial utilization. Media composition for the cellulolytic enzymes production is specific and has to be optimized for each organism. Along with the media components, various other factors influence the production of cellulases, among which carbon sources, nitrogen sources, pH and temperature alone or in combination play an essential role in cellulase productivity.

6.4.1 BIOCHEMICAL FACTORS – EFFECT OF CARBON SOURCES

Carbon source is one of the most important factors which affect the cellulolytic enzymes production, attributing to the fact that various carbon sources such as lactose, galactose, and cellobiose act as main inducers underlying the production of cellulose-degrading enzymes. Oligosaccharides such as sophorose (Mandels, Parrish, and Reese 1962), lactose (Morikawa et al. 1995), D-galactose (Karaffa et al. 2006) and xylobiose (Stricker et al. 2006) and polysaccharides such as cellulose (Kubicek 1993) and xylan (Zeilinger and Mach 1998) have been reported to induce cellulolytic and xylanolytic enzyme production in *T. reesei*. Lactose present in whey and cheese has been identified as an inducer for the production of cellulase in *T. reesei* MCG 80 (Allen and Andreotti 1982, Sternberg and Mandels 1979).

Role of other inducing sugar molecules has been deciphered in *Neurospora crassa*. Disaccharides such as cellobiose have been shown to induce β-glucosidases such as cellobiase and aryl-β-glucosidase (Eberhart 1961, Eberhart and Beck 1973). Such disaccharides have been shown to potently increase cellulose-degrading potential of secretory proteins. Ilmen et al. (1997) reported that the medium supplemented with sophorose and cellobiose produced the highest cellulase activity followed by cellobiose or lactose alone. In addition, they observed the inhibition effect of glucose on cellulase induction even at low concentrations and also inhibited the inducing effect of sophorose at high concentrations. Mrudula and Murugammal (2011) observed that cellulase production was highest in *Aspergillus niger* when lactose was supplemented as a carbon source. In addition, lactose also was shown to be the best inducer of FPase and CMCase activities in both solid state and liquid cultures. Among the various carbon sources tested, lactose was reported as the best carbon source for the production of cellulase in *Aspergillus* sp. (Devanathan et al. 2007, Kathiresan and Manivannan 2006), *T. reesei* C5 (Muthuvelayudham and Viruthagiri 2006) and *Penicillium* sp. (Prasanna, Ramanjaneyulu, and Rajasekhar Reddy 2016).

On the contrary, easily metabolizing monomers such as glucose repress the hydrolytic enzyme production by carbon catabolite repression through transcription factor *creA* (Dowzer and Kelly 1989, Fujii, Inoue, and Ishikawa 2013, Ilmen, Thrane, and Penttila 1996, Wang et al. 2015). The addition of glucose alone or in combination with other carbon sources influenced the cellulase production in *T. reesei* (Ilmen et al. 1997, Nogawa et al. 2001), *N. crassa* (Eberhart and Beck 1973) and *A. niger* (Hanif, Yasmeen, and Rajoka 2004, Nazir et al. 2010).

Gautam et al. (2011) observed that 1% cellulose and 1% sucrose were best carbon sources among the various carbon sources tested for the cellulase production in *A. niger* and *Trichoderma* sp., respectively. Sucrose was reported as the best carbon source for the cellulase production in *Paecilomyces variotii* (Asma et al. 2012) and *Trichoderma* sp., (Kilikian et al. 2014). Similarly, Abd-Elrsoul and Bakhiet (2018) obtained maximum cellulase by *Fusarium solani* with the supplementation of sucrose in the growth media.

Agro-industrial residues such as sugarcane bagasse, wheat straw, wheat bran and soybean bran alone or in combination with cellulose were used as cost-effective carbon sources for the production of cellulose-degrading enzymes in various filamentous fungi including *Acremonium strictum* (Goldbeck et al. 2013), *Trichoderma* sp. and *Myceliophthora thermophila* M77 (Kilikian et al. 2014) and *A. niger* (Salihu et al. 2015).

Overall, the identity of a broad range of inducers is still not known and a rational screening of optimal carbon source is necessary to harvest the real potential of fungal species.

6.4.2 IMPACT OF DIFFERENT NITROGEN SOURCES

Various inorganic nitrogen sources such as sodium nitrate, ammonium hydrogen phosphate, ammonium nitrate, potassium nitrate, ammonium chloride and ammonium hydrogen sulphate have been used for the cellulase production (Chopra and Mehta 1985, Gokhale, Patil, and Bastawde 1991, Jourdier et al. 2013). Liquid ammonia has widely been used to maintain pH, simultaneously providing a nitrogen source in fed-batch fermentations (Bailey et al. 2007, Bailey and Tähtiharju 2003, Jourdier et al. 2013, Portnoy, Margeot, Seidl-Seiboth, et al. 2011). Organic sources such as yeast extract (Ahamed and Vermette 2009, Ellilä et al. 2017, Nazir et al. 2010) and urea (Deswal, Khasa, and Kuhad 2011) have been used as nitrogen sources for the successful production of cellulase in various fungal systems.

6.4.3 INFLUENCE OF pH AND TEMPERATURE

The fungal growth and production of cellulolytic enzymes are also influenced by pH and temperature of the culture medium. In *T. reesei*, Sternberg and Mandels (1979) observed slower uptake of carbon source and cellulase production by ambient pH. The regulation of pH signalling was discovered and well studied in *A. nidulans* involving Pal proteins and PacC transcription factor which regulate various genes in both acidic and alkaline conditions (Caddick, Brownlee, and Arst 1986, Kunitake et al. 2016, Tilburn et al. 1995). Generally, pH and temperature for the cellulase production vary from 5 to 7 and 30°C to 50°C, respectively, for most fungal strains. Gautam et al. (2011) observed the best cellulase production at pH 6–7 and 40°C from *A. niger* and pH 6.5 and 45°C from *Trichoderma* sp. El-Hadi et al. (2014) observed a maximum CMCase activity at pH 7.0 and 37°C for *A. hortai*. Prasanna, Ramanjaneyulu, and Rajasekhar Reddy (2016) tested various pH and temperature values for *Penicillium* sp. and found that pH 5.0 and 30°C was the best for the production of cellulases.

6.5 ENZYME-BASED SACCHARIFICATION AT LABORATORY AND INDUSTRIAL SCALES

After the production of cellulases by various techniques in laboratory scale, commercial-scale production of cellulases was attempted at pilot and industrial scales. Commercial production of cellulases has been standardized in various filamentous fungi using (a) range of substrates and (b) fermentation processes. Some of them are solid state fermentation (SSF), submerged solid state fermentation, batch culture, submerged fermentation and flask culture. The method employed depends on the purity of the target enzyme, cost-effectiveness and availability of resources (Chaudhary and Padhiar 2020).

Deshpande et al. (2008) used water hyacinth as an alternate substrate and Toyoma Ogowa (TO) medium in a liquid–solid ratio of 2.5 for maximum cellulase production through solid state fermentation method using *T. reesei* mutant strain QM9414. The authors also showed that the experiment carried out in solid state cabinet fermenter (SSCF) in TO medium with whey (40%) and peptone (0.15%) produced hydrolytic enzymes about two- to three-fold with respect to flask culture. In addition, a significant increase in the cellulase and xylanase production was observed when *A. niger* was cocultured along with *T. reesei*.

Similarly, the effect of coculturing three fungi such as *T. reesei*, *A. niger* and *Phanerochaete chrysosporium* for the optimum cellulase production has been studied by Lio and Wang (2012) through SSF using soybean cotyledon fibre and dried distillers grains with solubles (DDGS) as substrates. The authors achieved maximum cellulase and xylanase activities (3.2 IU/g and 757.4 IU/g, respectively) by coculturing *T. reesei* along with *Phanerochaete chrysosporium* in the SSF of soybean cotyledon fibre. In addition, they also observed highest xylanase activity (399.2 IU/g) with the same inoculation method with respect to other combinations of fungi in the SSF of DDGS.

T. reesei RUT-C30 is known to produce endo- β-1,4-glucanase, exo-β-1,4-glucanase and β-glucosidase in higher levels. Pilot-scale production of cellulases from *T. reesei* RUT-C30 was tried under fed-batch conditions to improve enzyme activities in a 50-L fermenter. The cellulase product showed 273 U/mL of CMCase activity, 35 U/mL of FPase activity and 162 FPU/L of filter paper activity. The pH of the medium was balanced using ammonium hydroxide. Ammonium sulphate precipitation of the product was done after several rounds of filtration and drying to get the powder of cellulase (Kang et al. 2004).

Production of cellulases by *T. reesei* QM9414 in fed-batch and continuous-flow cultures was tried in a total of 3 L capacity. In fed-batch cultures, the addition of higher concentrations of cellulose improves the enzyme productivity compared to batch process. Rest of the parameters were kept the same for both the processes. The medium was supplemented with 0.5% cellulose. Optimal dilution ratio of 1:2 and dilution rate of 0.025/h were found. Growth time of 100 h was found to be optimum (Ghose and Sahai 1979).

Cellulase production was optimized in laboratory and pilot scales. *T. reesei* mutants MCG 80, SVG, MCG 77, CL-847, VTT-D, RUT-C30 and QM9414 are known to completely hydrolyse crystalline cellulose. These types produce cellulolytic proteins such as cellobiase, exocellulases and endocellulases. Type and initial optimization were done in immersed shake culture fermentation. Stirred tank reactor was used for fermentation up to 3000 L in pilot scale. On average, 250 mg of cellulases was formed per gram of carbon source. At the industrial level, *Trichoderma* cellulases are produced by solid state and immersed fermentation. The selection of *Trichoderma* strain depends on the nature and the purpose of cellulases required (Esterbauer et al. 1991).

Other than these pilot-scale plants, many companies such as Sigma-Aldrich, Novozymes and Merck have been manufacturing cocktails of cellulases. It could further be used in industrial and laboratory scales for cellulose degradation.

6.6   ENZYME COCKTAILS FOR ENHANCED SACCHARIFICATION

The recalcitrant nature of cellulosic microfibrils is a major hurdle in enzyme-based saccharification. Multiple CAZymes including exo- and endoglucanases work in a synergistic manner to hydrolyse the complex cellulosic substrates into cellobiose which is then hydrolysed into monomeric glucose by the action of β-glucosidases, and thus the composition of cellulase cocktail is a key behind efficient disintegration. It has been reported that the stoichiometry of hydrolytic enzymes in the cocktail is critical not only for minimizing product-driven feedback inhibition but also for enhancing the degradation efficiency (Wang and Lu 2016). The optimization of the existing cellulase cocktail for efficient enzyme-based saccharification is majorly done by:

1. Supplementing the available cocktails with additional hydrolytic weapons such as LPMO and swollenin (Gonçalves et al. 2015, Selig et al. 2008, Xiao et al. 2004)

2. Optimizing the ratio of endoglucanase to glucosidase in the cocktail to minimize the cellobiose-driven feedback inhibition (Sanchez-Cantu et al. 2018,
Sanhueza et al. 2018).
3. Adding accessory enzymes such as hemicellulases, acetylxylan esterase,
α-L-arabinofuranosidase, feruloyl esterase and p-coumaroyl esterase. This
plays an essential role in hydrolysing the complex polymers that is shielded
by hemicellulose (Qing, Yang, and Wyman 2010).

TABLE 6.2
List of Various Commercial Enzyme Preparations

Enzyme Preparations and Their Supplier	Fungal Strains Produced From	Protein Content and Enzyme Activity Characteristics	References
Accellerase-1500 (Genencor)	T. reesei	Protein – 114.0 mg/mL; FPA – 0.50 FPU/mg; BGL – 1.31 CBU/mg; xylanase – 0.66 IU/mg; BXL – 0.06 IU/mg	Alvira, Negro, and Ballesteros (2011)
GC-220 (Genencor)	T. longibrachiatum/T. reesei	Protein – 64 mg/mL; FPA – 116 FPU/mL; cellobiase – 215 U/ml; xylanase – 677 U/mL	Kabel et al. (2006)
Celluclast (Novozymes)	T. longibrachiatum and A. niger	Protein – 52.0 mg/mL; FPA – 99.6 FPU/mL; CMCase – 42.7 U/mL; BGL – 11.52; xylanase – 238.0 U/mL	Agrawal et al. (2018)
Cellic CTec1 (Novozymes)	Myceliophthora thermophila, T. reesei	Protein – 160.4 mg; FPA – 23.1 FPU/g; xylanase – 0.77 U/mg protein; CMCase – 0.19 U/mg protein; BGL – 1.32 U/mg protein; CBHI – 0.36 U/mg protein; BXL – 0.15U/mg protein	Sun et al. (2015)
Cellic CTec2 (Novozymes)	M. thermophila, T. reesei	Protein – 193.5; FPA – 137.0 FPU/g; xylanase – 21.17 U/mg protein; CMCase – 2.26 U/mg protein; BGL – 6.04 U/mg protein; CBHI – 0.98 U/mg protein; BXL – 0.28 U/mg protein	Sun et al. (2015)
Cellic CTec3 (Novozymes)	T. reesei, A. oryzae and Thielavia terrestris	Protein – 233.7; FPA – 165.2 FPU/g; xylanase – 14.93 U/mg protein; CMCase – 1.24 U/mg protein; BGL – 10.16 U/mg protein; CBHI – 3.08 U/mg protein; BXL – 1.15 U/mg protein	Sun et al. (2015)

BGL – β-glucosidase; BXL – β-xylosidase.

Various commercial enzyme preparations with their protein content and enzyme activity are listed in Table 6.2. These formulations differ in the protein content and enzyme activity. Sun et al. (2015) observed that the protein content and FPA of CTec3 was higher with respect to the previous enzyme preparations, i.e. CTec2 and CTec1. The authors also reported that CTec2 showed highest specific activity for xylanase and CMCase compared to CTec3, whereas CTec3 showed highest activity for β-glucosidase, CBHI and β-xylosidase. To compare the hydrolytic efficiency, the same authors also tested these three enzyme preparations for the digestion of steam-pre-treated poplar and sweet sorghum bagasse. They observed a maximum of 94% glucan hydrolysis of pre-treated poplar with CTec3 followed by CTec2 (86%) and CTec1 (80%), whereas for the pre-treated sweet sorghum bagasse, CTec3 achieved 66% of glucan hydrolysis followed by CTec2 (55%) and CTec1 (47%) with the same enzyme loading at 24 h. Similarly, Kabel et al. (2006) compared the hydrolytic activities of 14 commercial enzymes' mixtures for the digestion of pre-treated wheat bran and grass as substrates and found a vast difference between the actual and claimed cellulase and xylanase activities. Alvira, Negro, and Ballesteros (2011) observed that the addition of accessory enzymes such as xylanase and α-l-arabinofuranosidase to the Novozymes' two commercial enzyme preparations, i.e. NS 50013 and NS 50010, showed 22.1%, 19.5% and 18.6% higher production of arabinose, glucose and xylose, respectively, on hydrolysis of pre-treated wheat straw.

Another major challenge in the development of efficient enzyme cocktail is to make a robust consortium which can handle different varieties of plant biomass substrate as each varies in their composition. For example, agro-industrial waste, forestry residue, municipal waste and paper waste have different ratios of cellulose and hemicellulose polymers. The hydrolysis efficiency of commercial enzymes cannot be assessed by just FPA, when it comes for the degradation of different complex lignocellulosic materials. Hence, the tailoring of enzyme preparations has given boost to industries involved in bioconversion of cellulose to monomeric glucose residue and sincere efforts are required towards the development of tailor-made enzyme cocktail containing various concentrations of cellulases and other accessory enzymes.

6.7 APPLICATIONS OF CELLULASES

Cellulases are one of most vital industrial enzymes on the worldwide market. The cellulose hydrolysate is a rich source of sugars, which may function as the raw material for the assembly of various bio-products of economic importance such as antibiotics, animal feeds and bioethanol.

The enzymatic hydrolysis is preferred than chemical methods of hydrolysis due to the recovery rate, low energy requirements and eco-friendliness. Cellulase has a wide spectrum of applications in industries such as textile, detergent, paper and pulp, and bioethanol industries (Figure 6.4). The applications of cellulases have been discussed much in detail in Chapter 9.

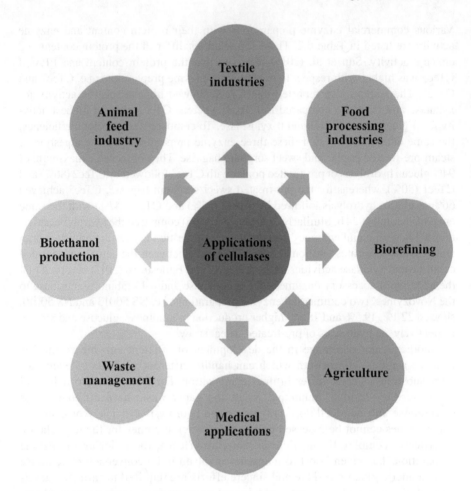

FIGURE 6.4   Applications of cellulases.

6.8   CASE STUDIES – INDUSTRIAL SUCCESS STORIES IN INDIA

Here we discuss some of the successful industrial cellulase production stories which have demonstrated their ability to produce a cocktail of cellulases cost effectively due to their innovative approaches and National Biofuel Policy.

6.8.1   India Glycols Ltd.

The Department of BioTechnology in association with Institute for Chemical Technology has developed 2G-Ethanol Technology. It has been validated at a scale of 10 ton biomass/day at India Glycols Ltd. site at Kashipur, Uttarakhand, India. The second-generation technology is a multi-step multi-enzyme depolymerization or hydrolysis process system that uses enzyme combinations or cocktails separately and sequentially. The ratio of different cellulases in the cocktail along with the fungal system used, the substrate and conditions employed for fermentation determines the activity of the cocktail. The ratio of activities in the cellulase preparation determines the extent of saccharification.

This functional plant can use all types of lignocellulose feedstocks, including but not limited to agricultural residue, herbaceous material, municipal solid waste, forestry residue, paper waste, pulp and paper mill residue or any other source with varying content of cellulose, hemicellulose and lignin. The conversion happens in two or more steps with the first step using at least one enzyme that can break down different polysaccharides to produce short-chain oligosaccharides and subsequent steps using enzymes that can break down the oligosaccharides that are released by endo-acting and exo-acting hydrolytic enzymes, into monomeric sugars (Lali et al. 2010).

A membrane separation unit ensures that the costly enzyme cocktail is recovered completely, making the plant economically feasible. The sugars so obtained can be used for fermentation to ethanol or any other purposes.

6.8.2 PRAJ INDUSTRIES, PUNE

PRAJ's end-to-end 2G Smart Bio-Refinery cellulosic ethanol technology is named Enfinity. This technology enables the plant to process various types of biomass such as rice and wheat straw, cane trash, corn cobs and stover, empty fruit bunches, cane bagasse and cotton stalk very efficiently.

The plant can produce bioethanol and biogas with provision for set-up of bio-CNG unit and has a capacity of 1 million litre/annum. Complete end-to-end offering from feedstock processing till end product and integrated processes have resulted in energy and water consumption optimization with zero process liquid discharge.

6.8.3 DBT-IOCL, FARIDABAD

A joint venture by Department of Biotechnology, Ministry of Science and Technology and Research and Development Centre, Indian Oil Corporation Limited has resulted in setting up of a pilot plant – Bioenergy Research Centre (DBT-IOC Centre) – for the development of 2G ethanol and other chemicals. The project was drafted and scaled up indigenously. After a rigorous study on databases generated by using different substrates and catalysts, plans for commercialization are underway.

The centre has also developed an indigenous cellulase enzyme technology after screening more than 85,000 fungal strains for different enzymatic activities such as β-glucosidase, endo-/exoglucanase and filter paper activity. This is the first attempt in India to develop a large-scale enzyme production process at 5000-L level. The indigenous system has been tested at a 1 MT/day pilot plant and has shown efficiency comparable to commercial enzyme cocktails at 30% cost-effectiveness (ARTFuels Forum 2020).

REFERENCES

Abd-Elrsoul, R. M. M. A., and S. E. A Bakhiet. 2018. "Optimization of factors influencing cellulase production by some indigenous isolated fungal species." *Jordan Journal of Biological Sciences* 11: 31–36.

Agrawal, Ruchi, Surbhi Semwal, Ravindra Kumar, Anshu Mathur, Ravi Prakash Gupta, Deepak K. Tuli, and Alok Satlewal. 2018. "Synergistic Enzyme Cocktail to Enhance Hydrolysis of Steam Exploded Wheat Straw at Pilot Scale." *Frontiers in Energy Research* 6 (122). doi 10.3389/fenrg.2018.00122.

Ahamed, A., and P. Vermette. 2009. "Effect of culture medium composition on Trichoderma reesei's morphology and cellulase production." *Bioresource Technology* 100 (23):5979–87. doi 10.1016/j.biortech.2009.02.070.

Allen, A. L., and R. F. Andreotti. 1982. "Cellulase production in continuous and fed cultures by Trichoderma reesei MCG 80." *Symposium on Biotechnology in Energy Production and Conservation*, Gatlinburg, TN, USA.

Alvira, P., M. J. Negro, and M. Ballesteros. 2011. "Effect of endoxylanase and α-l-arabinofuranosidase supplementation on the enzymatic hydrolysis of steam exploded wheat straw." *Bioresource Technology* 102 (6):4552–58. doi 10.1016/j.biortech.2010.12.112.

Arfi, Y., M. Shamshoum, I. Rogachev, Y. Peleg, and E. A. Bayer. 2014. "Integration of bacterial lytic polysaccharide monooxygenases into designer cellulosomes promotes enhanced cellulose degradation." *Proceedings of the National Academy of Sciences of the United States of America* 111 (25):9109–14. doi 10.1073/pnas.1404148111.

Aro, N., A. Saloheimo, M. Ilmen, and M. Penttila. 2001. "ACEII, a novel transcriptional activator involved in regulation of cellulase and xylanase genes of Trichoderma reesei." *Journal of Biological Chemistry* 276 (26):24309–14. doi 10.1074/jbc.M003624200.

Arora, Richa, Shuvashish Behera, Nilesh Kumar Sharma, and Sachin Kumar. 2015. "Bioprospecting thermostable cellulosomes for efficient biofuel production from lignocellulosic biomass." *Bioresources and Bioprocessing* 2 (1):38. doi 10.1186/s40643-015-0066-4.

ARTFuels Forum, European Commission. 2020. "Success stories of advanced biofuels in transport." *IEA Bioenergy* 1–95. https://artfuelsforum.eu/.

Artzi, L., E. A. Bayer, and S. Moraïs. 2017. "Cellulosomes bacterial nanomachines for dismantling plant polysaccharides." *Nature Review Microbiology* 15 (2):83–95. doi 10.1038/nrmicro.2016.164.

Asma, H., S. Archana, K. Sudhir, R. K. Jainb, S. Baghelc, R. Mukesh, and K. Agrawale. 2012. "Cellulolytic enzymatic activity of soft rot filamentous fungi Paecilomyces variotii." *Advances in Bioresearch* 3 (3):10–17.

Bailey, Michael J., Bernhard Adamitsch, Jari Rautio, Niklas von Weymarn, and Markku Saloheimo. 2007. "Use of a growth-associated control algorithm for efficient production of a heterologous laccase in Trichoderma reesei in fed-batch and continuous cultivation." *Enzyme and Microbial Technology* 41 (4):484–91. doi 10.1016/j.enzmictec.2007.04.002.

Bailey, M. J., and J. Tähtiharju. 2003. "Efficient cellulase production by Trichoderma reesei in continuous cultivation on lactose medium with a computer-controlled feeding strategy." *Applied Microbiology and Biotechnology* 62 (2–3):156–62. doi 10.1007/s00253-003-1276-9.

Bisaria, Virendra S., and Tarun K. Ghose. 1981. "Biodegradation of cellulosic materials Substrates, microorganisms, enzymes and products." *Enzyme and Microbial Technology* 3 (2):90–104. doi 10.1016/0141-0229(81)90066-1.

Bischof, Robert H., Jonas Ramoni, and Bernhard Seiboth. 2016. "Cellulases and beyond the first 70 years of the enzyme producer Trichoderma reesei." *Microbial Cell Factories* 15 (1):106. doi 10.1186/s12934-016-0507-6.

Blanch, Harvey W., Blake A. Simmons, and Daniel Klein-Marcuschamer. 2011. "Biomass deconstruction to sugars." *Biotechnology Journal* 6 (9):1086–1102. doi 10.1002/biot.201000180.

Blanchette, Craig, Catherine I. Lacayo, Nicholas O. Fischer, Mona Hwang, and Michael P. Thelen. 2012. "Enhanced cellulose degradation using cellulase-nanosphere complexes." *PLOS ONE* 7 (8):e42116. doi 10.1371/journal.pone.0042116.

Boswell, J. G. 1941. "The biological decomposition of cellulose." *New Phytologist* 40 (1):20–33. doi 10.1111/j.1469-8137.1941.tb07026.x.

Caddick, M. X., A. G. Brownlee, and H. N. Arst, Jr. 1986. "Regulation of gene expression by pH of the growth medium in Aspergillus nidulans." *Molecular and General Genetics* 203 (2):346–53. doi 10.1007/BF00333978.

Chaudhary, K., and A. Padhiar. 2020. "Cellulase production by fungi from agro wastes under solid state fermentation." *Bioscience Biotechnology Research Communications* 13 (3): 1495–1501. doi: 10.21786/bbrc/13.3/76.

Cho, W., S. D. Jeon, H. J. Shim, R. H. Doi, and S. O. Han. 2010. "Cellulosomic profiling produced by Clostridium cellulovorans during growth on different carbon sources explored by the cohesin marker." *Journal of Biotechnology* 145 (3):233–39. doi 10.1016/j.jbiotec.2009.11.020.

Chopra, S., and P. Mehta. 1985. "Influence of various nitrogen and carbon sources on the production of pectolytic, cellulolytic and proteolytic enzymes by Aspergillus niger." *Folia Microbiologica* 30 (2):117–25. doi 10.1007/BF02922204.

Cubero, B., and C. Scazzocchio. 1994. "Two different, adjacent and divergent zinc finger binding sites are necessary for CREA-mediated carbon catabolite repression in the proline gene cluster of Aspergillus nidulans." *EMBO Journal* 13 (2):407–15.

da Silva Delabona, P., G. N. Rodrigues, M. P. Zubieta, J. Ramoni, C. A. Codima, D. J. Lima, C. S. Farinas, J. G. da Cruz Pradella, and B. Seiboth. 2017. "The relation between xyr1 overexpression in Trichoderma harzianum and sugarcane bagasse saccharification performance." *Journal of Biotechnology* 246:24–32 doi 10.1016/j.jbiotec.2017.02.002.

de Graaff, L. H., H. C. van den Broeck, A. J. van Ooijen, and J. Visser. 1994. "Regulation of the xylanase-encoding xlnA gene of Aspergillus tubigensis." *Molecular Microbiology* 12 (3):479–90. doi 10.1111/j.1365-2958.1994.tb01036.x.

de Vries, R., P. van de Vondervoort, L. Hendriks, M. van de Belt, and J. Visser. 2002. "Regulation of the α-glucuronidase-encoding gene (aguA) from Aspergillus niger." *Molecular Genetics and Genomics* 268 (1):96–102. doi 10.1007/s00438-002-0729-7.

Derntl, C., L. Gudynaite-Savitch, S. Calixte, T. White, R. L. Mach, and A. R. Mach-Aigner. 2013. "Mutation of the xylanase regulator 1 causes a glucose blind hydrolase expressing phenotype in industrially used Trichoderma strains." *Biotechnology for Biofuels* 6 (1):62. doi 10.1186/1754-6834-6-62.

Derntl, Christian, Robert L. Mach, and Astrid R. Mach-Aigner. 2019. "Fusion transcription factors for strong, constitutive expression of cellulases and xylanases in Trichoderma reesei." *Biotechnology for Biofuels* 12 (1):231. doi 10.1186/s13068-019-1575-8.

Deshpande, S. K., M. G. Bhotmange, T. Chakrabarti, and P. N. Shastri. 2008. "Production of cellulase and xylanase by Trichoderma reesei (QM 9414 mutant), Aspergillus niger and mixed culture by solid state fermentation (SSF) of water hyacinth (Eichhornia crassipes)." *Indian Journal of Chemical Technology* 15 (5):449–456.

Deswal, D., Y. P. Khasa, and R. C. Kuhad. 2011. "Optimization of cellulase production by a brown rot fungus Fomitopsis sp. RCK2010 under solid state fermentation." *Bioresource Technology* 102 (10):6065–72. doi 10.1016/j.biortech.2011.03.032.

Devanathan, G., A. Shanmugam, T. Balasubramanian, and S. Manivannan. 2007. "Cellulase production by Aspergillus niger isolated from coastal mangrove debris." *Trends in Applied Sciences Research* 2 (1):23–27. doi 10.3923/tasr.2007.23.27.

Dowzer, C. E., and J. M. Kelly. 1989. "Cloning of the creA gene from Aspergillus nidulans a gene involved in carbon catabolite repression." *Current Genetics* 15 (6):457–59. doi 10.1007/bf00376804.

Durand, Henri, Marc Clanet, and Gérard Tiraby. 1988. "Genetic improvement of Trichoderma reesei for large scale cellulase production." *Enzyme and Microbial Technology* 10 (6):341–46. doi 10.1016/0141-0229(88)90012-9.

Eberhart, B. M. 1961. "Exogenous enzymes of Neurospora conidia and mycelia." *Journal of Cellular and Comparative Physiology* 58:11–16 doi 10.1002/jcp.1030580103.

Eberhart, B. M., and R. S. Beck. 1973. "Induction of beta-glucosidases in Neurospora crassa." *Journal of Bacteriology* 116 (1):295–303. doi 10.1128/JB.116.1.295-303.1973.

El-Hadi, Abeer A., Salwa Abu El-Nour, Ali Hammad, Zeinat Kamel, and Mai Anwar. 2014. "Optimization of cultural and nutritional conditions for carboxymethylcellulase production by Aspergillus hortai." *Journal of Radiation Research and Applied Sciences* 7 (1):23–28. doi 10.1016/j.jrras.2013.11.003.

Ellilä, Simo, Lucas Fonseca, Cristiane Uchima, Junio Cota, Gustavo Henrique Goldman, Markku Saloheimo, Vera Sacon, and Matti Siika-aho. 2017. "Development of a low-cost cellulase production process using Trichoderma reesei for Brazilian biorefineries." *Biotechnology for Biofuels* 10 (1):30. doi 10.1186/s13068-017-0717-0.

Esterbauer, H., W. Steiner, I. Labudova, A. Hermann, and M. Hayn. 1991. "Production of Trichoderma cellulase in laboratory and pilot scale." *Bioresource Technology* 36 (1):51–65. doi 10.1016/0960-8524(91)90099-6.

Eveleigh, D. E.. 1982. "Reducing the cost of cellulase production - selection of the hyper-cellulolytic Trichoderma reesei RUT-C30 mutant." PhD diss., New Brunswick Rutgers University.

Eveleigh, D. E., and B. S. Montenecourt. 1979. "Increasing yields of extracellular enzymes." *Advances in Applied Microbiology* 25:57–74 doi 10.1016/s0065-2164(08)70146-1.

Freeman, G. G., A. J. Baillie, and C. A. Macinnes. 1948. "Bacterial degradation of CMC and methyl ethyl cellulose." *Chemistry & Industry* 279–282.

Fujii, T., H. Inoue, and K. Ishikawa. 2013. "Enhancing cellulase and hemicellulase production by genetic modification of the carbon catabolite repressor gene, creA, in Acremonium cellulolyticus." *AMB Express* 3 (1):73. doi 10.1186/2191-0855-3-73.

Galanopoulou, A. P., S. Moraïs, A. Georgoulis, E. Morag, E. A. Bayer, and D. G. Hatzinikolaou. 2016. "Insights into the functionality and stability of designer cellulosomes at elevated temperatures." *Applied Microbiology and Biotechnology* 100 (20):8731–43. doi 10.1007/s00253-016-7594-5.

Gautam, S. P., P. S. Bundela, A. K. Pandey, J. Khan, M. K. Awasthi, and S. Sarsaiya. 2011. "Optimization for the production of cellulase enzyme from municipal solid waste residue by two novel cellulolytic fungi." *Biotechnology Research International* 2011:810425. doi 10.4061/2011/810425.

Ghose, T. K., and V. Sahai. 1979. "Production of cellulases by Trichoderma reesei QM 9414 in fed-batch and continuous-flow culture with cell recycle." *Biotechnology and Bioengineering* 21 (2):283–96. doi 10.1002/bit.260210213.

Gielkens, M. M., E. Dekkers, J. Visser, and L. H. de Graaff. 1999. "Two cellobiohydrolase-encoding genes from Aspergillus niger require D-xylose and the xylanolytic transcriptional activator XlnR for their expression." *Applied and Environmental Microbiology* 65 (10):4340–45. doi 10.1128/AEM.65.10.4340-4345.1999.

Gokhale, D. V., S. G. Patil, and K. B. Bastawde. 1991. "Optimization of cellulase production by Aspergillus niger NCIM 1207." *Applied Biochemistry and Biotechnology* 30 (1):99–109. doi 10.1007/BF02922026.

Goldbeck, Rosana, Mayla M. Ramos, Gonçalo A. G. Pereira, and Francisco Maugeri-Filho. 2013. "Cellulase production from a new strain Acremonium strictum isolated from the Brazilian Biome using different substrates." *Bioresource Technology* 128:797–803 doi 10.1016/j.biortech.2012.10.034.

Gonçalves, G. A., Y. Takasugi, L. Jia, Y. Mori, S. Noda, T. Tanaka, H. Ichinose, and N. Kamiya. 2015. "Synergistic effect and application of xylanases as accessory enzymes to enhance the hydrolysis of pretreated bagasse." *Enzyme and Microbial Technology* 72:16–24 doi 10.1016/j.enzmictec.2015.01.007.

Grassmann, W., R. Stadler, and R. Bender. 1933. "Zur Spezifitat cellulose- und hemicellulose-spaltender enzyme." *Justus Liebigs Annalen der Chemie* 502 (1):20–40.

Grassmann, W., L. Zechmeister, G. Toth, and R. Stadler. 1933. "On the enzymic breakdown of cellulose and its cleav age products. II. Enzymic cleavage of polysaccharides." *Justus Liebigs Annalen der Chemie* 503:167–179.

Hakkinen, M., M. J. Valkonen, A. Westerholm-Parvinen, N. Aro, M. Arvas, M. Vitikainen, M. Penttila, M. Saloheimo, and T. M. Pakula. 2014. "Screening of candidate regulators for cellulase and hemicellulase production in Trichoderma reesei and identification of a factor essential for cellulase production." *Biotechnology Biofuels* 7 (1):14. doi 10.1186/1754-6834-7-14.

Hanif, A., A. Yasmeen, and M. I. Rajoka. 2004. "Induction, production, repression, and de-repression of exoglucanase synthesis in Aspergillus niger." *Bioresource Technology* 94 (3):311–19. doi 10.1016/j.biortech.2003.12.013.

Ichinose, S., M. Tanaka, T. Shintani, and K. Gomi. 2018. "Increased production of biomass-degrading enzymes by double deletion of creA and creB genes involved in carbon catabolite repression in Aspergillus oryzae." *Journal of Bioscience Bioengineering* 125 (2):141–47. doi 10.1016/j.jbiosc.2017.08.019.

Ilmen, M., A. Saloheimo, M. L. Onnela, and M. E. Penttila. 1997. "Regulation of cellulase gene expression in the filamentous fungus Trichoderma reesei." *Applied and Environmental Microbiology* 63 (4):1298–306. doi 10.1128/AEM.63.4.1298-1306.1997.

Ilmen, M., C. Thrane, and M. Penttila. 1996. "The glucose repressor gene cre1 of Trichoderma isolation and expression of a full-length and a truncated mutant form." *Molecular and General Genetics* 251 (4):451–60. doi 10.1007/BF02172374.

Jayasekara, S., and R. Ratnayake. 2019. "Microbial cellulases an overview and applications." In *Cellulose*, ed. A. R. Pascual and M. E. E. Martin, 1–21. London: IntechOpen. doi 10.5772/intechopen.84531.

Jourdier, E., C. Cohen, L. Poughon, C. Larroche, F. Monot, and F. B. Chaabane. 2013. "Cellulase activity mapping of Trichoderma reesei cultivated in sugar mixtures under fed-batch conditions." *Biotechnology Biofuels* 6 (1):79. doi 10.1186/1754-6834-6-79.

Kabel, M. A., M. J. van der Maarel, G. Klip, A. G. Voragen, and H. A. Schols. 2006. "Standard assays do not predict the efficiency of commercial cellulase preparations towards plant materials." *Biotechnology Bioengineering* 93 (1):56–63. doi 10.1002/bit.20685.

Kahn, A., S. Morais, A. P. Galanopoulou, D. Chung, N. S. Sarai, N. Hengge, D. G. Hatzinikolaou, M. E. Himmel, Y. J. Bomble, and E. A. Bayer. 2019. "Creation of a functional hyperthermostable designer cellulosome." *Biotechnology for Biofuels* 12:44. doi 10.1186/s13068-019-1386-y.

Kang, S. W., Y. S. Park, J. S. Lee, S. I. Hong, and S. W. Kim. 2004. "Production of cellulases and hemicellulases by Aspergillus niger KK2 from lignocellulosic biomass." *Bioresource Technology* 91 (2):153–56. doi 10.1016/s0960-8524(03)00172-x.

Karaffa, L., E. Fekete, C. Gamauf, A. Szentirmai, C. P. Kubicek, and B. Seiboth. 2006. "D-Galactose induces cellulase gene expression in Hypocrea jecorina at low growth rates." *Microbiology (Reading)* 152 (Pt 5):1507–14. doi 10.1099/mic.0.28719-0.

Kathiresan, K., and S. Manivannan. 2006. "Cellulase production by Penicillium fellutanum isolated from coastal mangrove rhizosphere soil." *Research Journal of Microbiology* 1 (5):438–42. doi 10.3923/jm.2006.438.442.

Khosravi, Claire, Joanna E. Kowalczyk, Tania Chroumpi, Evy Battaglia, Maria-Victoria Aguilar Pontes, Mao Peng, Ad Wiebenga, Vivian Ng, Anna Lipzen, Guifen He, Diane Bauer, Igor V. Grigoriev, and Ronald P. de Vries. 2019. "Transcriptome analysis of Aspergillus niger xlnR and xkiA mutants grown on corn Stover and soybean hulls reveals a highly complex regulatory network." *BMC Genomics* 20 (1):853. doi 10.1186/s12864-019-6235-7.

Kilikian, B. V., L. C. Afonso, T. F. Souza, R. G. Ferreira, and I. R. Pinheiro. 2014. "Filamentous fungi and media for cellulase production in solid state cultures." *Brazilian Journal of Microbiology* 45 (1):279–86. doi 10.1590/S1517-83822014005000028.

Kubicek, C. P. 1993. "From cellulose to cellulase inducers facts and fiction." *Proceedings of the 2nd Tricel Symposium on Trichoderma reesei Cellulases and Other Hydrolases*, Espoo, Finland.

Kunitake, E., D. Hagiwara, K. Miyamoto, K. Kanamaru, M. Kimura, and T. Kobayashi. 2016. "Regulation of genes encoding cellulolytic enzymes by Pal-PacC signaling in Aspergillus nidulans." *Applied Microbiology and Biotechnology* 100 (8):3621–35. doi 10.1007/s00253-016-7409-8.

Lali, A. M., P. D. Nagwekar, J. S. Varavadekar, P. C. Wadekar, S. S. Gujrathi, R. D. Valthe, S. H. Birhade, and A. Odaneth. 2010. "Method for production of fermentable sugars from biomass." U.S. Patent No. 8338139B2. U.S Patent and Trademark. Lio, J., and T. Wang. 2012. "Solid-state fermentation of soybean and corn processing coproducts for potential feed improvement." *Journal of Agricultural and Food Chemistry* 60 (31):7702–09. doi 10.1021/jf301674u.

Lv, X., F. Zheng, C. Li, W. Zhang, G. Chen, and W. Liu. 2015. "Characterization of a copper responsive promoter and its mediated overexpression of the xylanase regulator 1 results in an induction-independent production of cellulases in Trichoderma reesei." *Biotechnology Biofuels* 8:67. doi 10.1186/s13068-015-0249-4.

Mach, R. L., J. Strauss, S. Zeilinger, M. Schindler, and C. P. Kubicek. 1996. "Carbon catabolite repression of xylanase I (xyn1) gene expression in Trichoderma reesei." *Molecular Microbiology* 21 (6):1273–81. doi 10.1046/j.1365-2958.1996.00094.x.

Mach-Aigner, A. R., M. E. Pucher, M. G. Steiger, G. E. Bauer, S. J. Preis, and R. L. Mach. 2008. "Transcriptional regulation of xyr1, encoding the main regulator of the xylanolytic and cellulolytic enzyme system in Hypocrea jecorina." *Applied and Environmental Microbiology* 74 (21):6554–62. doi 10.1128/AEM.01143-08.

Mandels, M., F. W. Parrish, and E. T. Reese. 1962. "Sophorose as an inducer of cellulase in Trichoderma viride." *Journal of Bacteriology* 83 (2):400–08. doi 10.1128/jb.83.2.400-408.1962.

Montenecourt, B. S., and D. E. Eveleigh. 1977a. "Preparation of mutants of Trichoderma reesei with enhanced cellulase production." *Applied and Environmental Microbiology* 34 (6):777–82. doi 10.1128/AEM.34.6.777-782.1977.

Montenecourt, B. S., and D. E. Eveleigh. 1977b. "Semiquantitative plate assay for determination of cellulase production by Trichoderma viride." *Applied and Environmental Microbiology* 33 (1):178–83. doi 10.1128/AEM.33.1.178-183.1977.

Montenecourt, Bland S., and Douglas E. Eveleigh. 1979. "Selective screening methods for the isolation of high yielding cellulase mutants of *Trichoderma reesei.*" In *Hydrolysis of Cellulose Mechanisms of Enzymatic and Acid Catalysis*, ed., R. D. Brown, and L. Jurasek, 289–301. Washington: American Chemical Society.

Morikawa, Y., T. Ohashi, O. Mantani, and H. Okada. 1995. "Cellulase induction by lactose in Trichoderma reesei PC-3-7." *Applied Microbiology and Biotechnology* 44 (1):106–11. doi 10.1007/BF00164488.

Mrudula, Soma, and Rangasamy Murugammal. 2011. "Production of cellulase by Aspergillus niger under submerged and solid state fermentation using coir waste as a substrate." *Brazilian Journal of Microbiology* 42:1119–27.

Muthuvelayudham, R., and T. Viruthagiri. 2006. "Fermentative production and kinetics of cellulase protein on Trichoderma reesei using sugarcane bagasse and rice straw." *African Journal of Biotechnology* 5 (20):1873–1881.

Nakari-Setala, T., M. Paloheimo, J. Kallio, J. Vehmaanpera, M. Penttila, and M. Saloheimo. 2009. "Genetic modification of carbon catabolite repression in Trichoderma reesei for improved protein production." *Applied and Environmental Microbiology* 75 (14):4853–60. doi 10.1128/AEM.00282-09.

Nazir, A., R. Soni, H. S. Saini, A. Kaur, and B. S. Chadha. 2010. "Profiling differential expression of cellulases and metabolite footprints in Aspergillus terreus." *Applied Biochemistry and Biotechnology* 162 (2):538–47. doi 10.1007/s12010-009-8775-9.

Nogawa, Masahiro, Masahiro Goto, Hirofumi Okada, and Yasushi Morikawa. 2001. "l-Sorbose induces cellulase gene transcription in the cellulolytic fungus Trichoderma reesei." *Current Genetics* 38 (6):329–34. doi 10.1007/s002940000165.

Pancha, Imran, Kaumeel Chokshi, Rahulkumar Maurya, Sourish Bhattacharya, Pooja Bachani, and Sandhya Mishra. 2016. "Comparative evaluation of chemical and enzymatic saccharification of mixotrophically grown de-oiled microalgal biomass for reducing sugar production." *Bioresource Technology* 204:9–16. doi 10.1016/j.biortech.2015.12.078.

Panozzo, C., E. Cornillot, and B. Felenbok. 1998. "The CreA repressor is the sole DNA-binding protein responsible for carbon catabolite repression of the alcA gene in Aspergillus nidulans via its binding to a couple of specific sites." *Journal of Biological Chemistry* 273 (11):6367–72. doi 10.1074/jbc.273.11.6367.

Peterson, Robyn, and Helena Nevalainen. 2012. "Trichoderma reesei RUT-C30 – thirty years of strain improvement." *Microbiology* 158 (1):58–68. doi 10.1099/mic.0.054031-0.

Portnoy, T., A. Margeot, R. Linke, L. Atanasova, E. Fekete, E. Sandor, L. Hartl, L. Karaffa, I. S. Druzhinina, B. Seiboth, S. Le Crom, and C. P. Kubicek. 2011. "The CRE1 carbon catabolite repressor of the fungus Trichoderma reesei a master regulator of carbon assimilation." *BMC Genomics* 12:269. doi 10.1186/1471-2164-12-269.

Portnoy, T., A. Margeot, V. Seidl-Seiboth, S. Le Crom, F. Ben Chaabane, R. Linke, B. Seiboth, and C. P. Kubicek. 2011. "Differential regulation of the cellulase transcription factors XYR1, ACE2, and ACE1 in Trichoderma reesei strains producing high and low levels of cellulase." *Eukaryot Cell* 10 (2):262–71. doi 10.1128/ec.00208-10.

Prasanna, H. N., G. Ramanjaneyulu, and B. Rajasekhar Reddy. 2016. "Optimization of cellulase production by Penicillium sp." *3 Biotech* 6 (2):162. doi 10.1007/s13205-016-0483-x.

Qing, Q., B. Yang, and C. E. Wyman. 2010. "Xylooligomers are strong inhibitors of cellulose hydrolysis by enzymes." *Bioresource Technology* 101 (24):9624–30. doi 10.1016/j.biortech.2010.06.137.

Reese, Elwyn T., and H. S. Levinson. 1952. "A comparative study of the breakdown of cellulose by microorganisms." *Physiologia Plantarum* 5 (3):345–66. doi 10.1111/j.1399-3054.1952.tb07530.x.

Reese, E. T., R. G. H. Siu, and H. S. Levinson. 1950. "The biological degradation of soluble cellulose derivatives and its relationship to the mechanism of cellulose hydrolysis." *Journal of Bacteriology* 59 (4):485–97. doi 10.1128/JB.59.4.485-497.1950.

Ries, L., N. J. Belshaw, M. Ilmén, M. E. Penttilä, M. Alapuranen, and D. B. Archer. 2014. "The role of CRE1 in nucleosome positioning within the cbh1 promoter and coding regions of Trichoderma reesei." *Applied Microbiology and Biotechnology* 98 (2):749–62. doi 10.1007/s00253-013-5354-3.

Ryu, Dewey D. Y., and Mary Mandels. 1980. "Cellulases Biosynthesis and applications." *Enzyme and Microbial Technology* 2 (2):91–102. doi 10.1016/0141-0229(80)90063-0.

Salihu, A., O. Abbas, A. B. Sallau, and M. Z. Alam. 2015. "Agricultural residues for cellulolytic enzyme production by Aspergillus niger effects of pretreatment." *3 Biotech* 5 (6):1101–06. doi 10.1007/s13205-015-0294-5.

Sanchez-Cantu, M., L. Ortiz-Moreno, M. E. Ramos-Cassellis, M. Marin-Castro, and C. De la Cerna-Hernandez. 2018. "Solid-state treatment of castor cake employing the enzymatic cocktail produced from Pleurotus djamor fungi." *Applied Biochemistry and Biotechnology* 185 (2):434–49. doi 10.1007/s12010-017-2656-4.

Sanhueza, C., G. Carvajal, J. Soto-Aguilar, M. E. Lienqueo, and O. Salazar. 2018. "The effect of a lytic polysaccharide monooxygenase and a xylanase from Gloeophyllum trabeum on the enzymatic hydrolysis of lignocellulosic residues using a commercial cellulase." *Enzyme and Microbial Technology* 113:75–82. doi 10.1016/j.enzmictec.2017.11.007.

Seiboth, B., R. A. Karimi, P. A. Phatale, R. Linke, L. Hartl, D. G. Sauer, K. M. Smith, S. E. Baker, M. Freitag, and C. P. Kubicek. 2012. "The putative protein methyltransferase LAE1 controls cellulase gene expression in Trichoderma reesei." *Molecular Microbiology* 84 (6):1150–64. doi 10.1111/j.1365-2958.2012.08083.x.

Selig, M. J., E. P. Knoshaug, W. S. Adney, M. E. Himmel, and S. R. Decker. 2008. "Synergistic enhancement of cellobiohydrolase performance on pretreated corn stover by addition of xylanase and esterase activities." *Bioresource Technology* 99 (11):4997–5005. doi 10.1016/j.biortech.2007.09.064.

Sheir-Neiss, G., and B. S. Montenecourt. 1984. "Characterization of the secreted cellulases of Trichoderma reesei wild type and mutants during controlled fermentations." *Applied Microbiology and Biotechnology* 20 (1):46–53. doi 10.1007/BF00254645.

Silverstein, Rebecca A., Ye Chen, Ratna R. Sharma-Shivappa, Michael D. Boyette, and Jason Osborne. 2007. "A comparison of chemical pretreatment methods for improving saccharification of cotton stalks." *Bioresource Technology* 98 (16):3000–11. doi 10.1016/j.biortech.2006.10.022.

Singh, Gopal, A. K. Verma, and Vinod Kumar. 2015. "Catalytic properties, functional attributes and industrial applications of β-glucosidases." *3 Biotech* 6 (1):3. doi 10.1007/s13205-015-0328-z.

Smith, S. P., and E. A. Bayer. 2013. "Insights into cellulosome assembly and dynamics from dissection to reconstruction of the supramolecular enzyme complex." *Current Opinion in Structural Biology* 23 (5):686–94. doi 10.1016/j.sbi.2013.09.002.

Sternberg, D., and G. R. Mandels. 1979. "Induction of cellulolytic enzymes in Trichoderma reesei by sophorose." *Journal of Bacteriology* 139 (3):761–69. doi 10.1128/jb.139.3.761-769.1979.

Stricker, Astrid R., Karin Grosstessner-Hain, Elisabeth Würleitner, and Robert L. Mach. 2006. "Xyr1 (xylanase regulator 1) regulates both the hydrolytic enzyme system and D-xylose metabolism in Hypocrea jecorina." *Eukaryotic Cell* 5 (12):2128–37. doi 10.1128/EC.00211-06.

Sun, F. F., J. Hong, J. Hu, J. N. Saddler, X. Fang, Z. Zhang, and S. Shen. 2015. "Accessory enzymes influence cellulase hydrolysis of the model substrate and the realistic lignocellulosic biomass." *Enzyme and Microbial Technology* 79–80:42–48. doi 10.1016/j.enzmictec.2015.06.020.

Sun, J., and N. L. Glass. 2011. "Identification of the CRE-1 cellulolytic regulon in Neurospora crassa." *PLoS One* 6 (9):e25654. doi 10.1371/journal.pone.0025654.

Takashima, S., A. Nakamura, H. Iikura, H. Masaki, and T. Uozumi. 1996. "Cloning of a gene encoding a putative carbon catabolite repressor from Trichoderma reesei." *Bioscience, Biotechnology, and Biochemistry* 60 (1):173–76. doi 10.1271/bbb.60.173.

Tilburn, J., S. Sarkar, D. A. Widdick, E. A. Espeso, M. Orejas, J. Mungroo, M. A. Peñalva, and H. N. Arst, Jr. 1995. "The Aspergillus PacC zinc finger transcription factor mediates regulation of both acid- and alkaline-expressed genes by ambient pH." *The EMBO Journal* 14 (4):779–90.

van der Veen, P., H. N. Arst, Jr., M. J. Flipphi, and J. Visser. 1994. "Extracellular arabinases in Aspergillus nidulans the effect of different cre mutations on enzyme levels." *Archives of Microbiology* 162 (6):433–40. doi 10.1007/BF00282109.

van Peij, N. N., M. M. Gielkens, R. P. de Vries, J. Visser, and L. H. de Graaff. 1998. "The transcriptional activator XlnR regulates both xylanolytic and endoglucanase gene expression in Aspergillus niger." *Applied and Environmental Microbiology* 64 (10):3615–19. doi 10.1128/aem.64.10.3615-3619.1998.

van Peij, N. N., J. Visser, and L. H. de Graaff. 1998. "Isolation and analysis of xlnR, encoding a transcriptional activator co-ordinating xylanolytic expression in Aspergillus niger." *Molecular Microbiology* 27 (1):131–42. doi 10.1046/j.1365-2958.1998.00666.x.

Wahlström, R. M., and A. Suurnäkki. 2015. "Enzymatic hydrolysis of lignocellulosic polysaccharides in the presence of ionic liquids." *Green Chemistry* 17 (2):694–714. doi 10.1039/C4GC01649A.

Wang, M., and X. Lu. 2016. "Exploring the synergy between cellobiose dehydrogenase from Phanerochaete chrysosporium and cellulase from Trichoderma reesei." *Front Microbiology* 7:620. doi 10.3389/fmicb.2016.00620.

Wang, Q., H. Lin, Q. Shen, X. Fan, N. Bai, and Y. Zhao. 2015. "Characterization of cellulase secretion and Cre1-mediated carbon source repression in the potential lignocellulose-degrading strain Trichoderma asperellum T-1." *PLoS One* 10 (3):e0119237. doi 10.1371/journal.pone.0119237.

Ward, H. Marshall. 1888. "A lily-disease." *Annals of Botany* os-2 (3):319–82. doi 10.1093/aob/os-2.3.319.

Xiao, Z., X. Zhang, D. J. Gregg, and J. N. Saddler. 2004. "Effects of sugar inhibition on cellulases and beta-glucosidase during enzymatic hydrolysis of softwood substrates." *Applied Biochemistry and Biotechnology* 113–116:1115–26. doi 10.1385/abab:115: 1-3:1115.

Zeilinger, S., A. Ebner, T. Marosits, R. Mach, and C. Kubicek. 2001. "The Hypocrea jecorina HAP 2/3/5 protein complex binds to the inverted CCAAT-box (ATTGG) within the cbh2 (cellobiohydrolase II-gene) activating element." *Molecular Genetics and Genomics* 266 (1):56–63. doi 10.1007/s004380100518.

Zeilinger, S., and R. L. Mach. 1998. "Xylanolytic enzymes of Trichoderma reesei properties and regulation of expression." *Topics in Current Chemistry* 1:27–35.

Zeilinger, S., M. Schmoll, M. Pail, R. L. Mach, and C. P. Kubicek. 2003. "Nucleosome transactions on the Hypocrea jecorina (Trichoderma reesei) cellulase promoter cbh2 associated with cellulase induction." *Molecular Genetics and Genomics* 270 (1):46–55. doi 10.1007/s00438-003-0895-2.

Zhang, J., Y. Chen, C. Wu, P. Liu, W. Wang, and D. Wei. 2019. "The transcription factor ACE3 controls cellulase activities and lactose metabolism via two additional regulators in the fungus Trichoderma reesei." *Journal of Biological Chemistry* 294 (48):18435–50. doi 10.1074/jbc.RA119.008497.

Zhang, Jiajia, Chuan Wu, Wei Wang, Wei Wang, and Dongzhi Wei. 2018. "Construction of enhanced transcriptional activators for improving cellulase production in Trichoderma reesei RUT C30." *Bioresources and Bioprocessing* 5 (1):40. doi 10.1186/s40643-018-0226-4.

Zhou, Na, Yimin Zhang, Xiaobin Wu, Xiaowu Gong, and Qinhong Wang. 2011. "Hydrolysis of Chlorella biomass for fermentable sugars in the presence of HCl and $MgCl_2$." *Bioresource Technology* 102 (21):10158–61. doi 10.1016/j.biortech.2011.08.051.

Zou, Gen, Shaohua Shi, Yanping Jiang, Joost van den Brink, Ronald P. de Vries, Ling Chen, Jun Zhang, Liang Ma, Chengshu Wang, and Zhihua Zhou. 2012. "Construction of a cellulase hyper-expression system in Trichoderma reesei by promoter and enzyme engineering." *Microbial Cell Factories* 11 (1):21. doi 10.1186/1475-2859-11-21.

7 Enhancement of Biomass for Deconstruction

Lavi Rastogi, Deepika Singh,
Rajan Kumar Sah, Aniket Anant Chaudhari,
and Prashant Anupama-Mohan Pawar
NCR Biotech Science Cluster

CONTENTS

7.1   Introduction 180
7.2   *In planta* Modification of Cellulose Structure.. 180
    7.2.1   Biosynthesis of Cellulose.. 181
        7.2.1.1   Enhancement of the SuSy Activity 181
        7.2.1.2   Disrupting Native Cellulose Biosynthesis Pathway 182
        7.2.1.3   Altering Crystallinity of Cell Wall 183
        7.2.1.4   *In planta* Expression of Cellulose-Degrading Enzymes.... 184
7.3   *In planta* Modification of Xylan Structure 184
    7.3.1   Structural Feature and Biosynthesis of Xylan 185
        7.3.1.1   Modulating Xylan Synthetic Complexes.......................... 187
        7.3.1.2   Altering Xylan Reducing End Sequence and Methyl
                Glucuronic Acid Level 187
        7.3.1.3   Modifying Polysaccharide Acetyltransferase Expression... 188
        7.3.1.4   *In planta* Expression of Xylan Hydrolytic Enzymes 189
7.4   *In planta* Modification of Pectin Structure.. 190
    7.4.1   Types of Pectin-Degrading Enzymes 191
    7.4.2   Bioengineering of Pectin 191
7.5   *In planta* Modification of Mannan 193
    7.5.1   Structural Features of Different Types of Mannan 193
    7.5.2   Introduction to Mannan Biosynthesis... 193
        7.5.2.1   Mannan Acetylation in Plants... 194
        7.5.2.2   Mannan-Degrading Enzymes.. 194
7.6   *In planta* Modification of Lignin.. 195
    7.6.1   Introduction to Lignin Biosynthesis 195
        7.6.1.1   Reducing the Lignin Content.. 197
        7.6.1.2   Fine-tuning the Lignin Monomer Composition................. 198

DOI: 10.1201/9781003158486-7

       7.6.1.3   Disrupting Cross-Linkages between Lignin and Other
               Cell Wall Components 199
       7.6.1.4   Addition of Labile Monolignols 200
       7.6.1.5   Modifying Lignin Polymerisation 200
7.7   Conclusions 201
References 201

7.1   INTRODUCTION

The plant cell wall is a major determining factor while processing lignocellulosic biomass to value-added products. Cell wall mainly consists of cellulose
and matrix components (mannan, xyloglucan, xylan, and pectin) which are
sugar-based polysaccharides, and monosugars derived from these polysaccharides are widely used as a substrate for biofuel production. However, biomass
properties such as high molecular weight lignin, crystalline cellulose, and
branched hemicelluloses make it recalcitrant and limit the extraction of sugar.
Furthermore, these cell wall components are interconnected by hydrophilic,
hydrophobic, and covalent interactions, making the cell wall tough to break during physical, chemical, or enzymatic treatment. Plant cell wall with altered cellulose, lignin, and non-cellulosic polysaccharide structure can overcome hurdles
faced during biofuel production. According to model-based prediction, reducing lignin content and polysaccharide acetylation increases biomass digestibility
and ethanol production (Klein-Marcuschamer et al. 2010). Also, *in planta* lignin
engineering via reducing lignin content, altering lignin composition and its polymerisation, decreases biomass recalcitrance in different plant species (Chanoca,
de Vries, and Boerjan 2019, Ralph, Lapierre, and Boerjan 2019). A recent review
by Donev and colleagues has summarised the engineering of non-cellulosic polysaccharides for bioenergy applications (Donev et al. 2018). This chapter will further provide an overview of wall components synthesis and how their *in vivo*
modification affects cell wall polymer properties.

7.2   *IN PLANTA* MODIFICATION OF CELLULOSE STRUCTURE

Cellulose, a homopolymer of β-1,4 glucose subunits, forms glucan chains (GCs), and
18–50 such chains combine to form cellulose microfibrils (CMs) (Fernandes et al.
2011, Jarvis 2018). The inter- and intramolecular hydrogen bonds and Van der Waals
interaction in CM make cellulose a primary load-bearing structure of the plant cell
wall. Cross-linking with other cell wall components and the crystalline nature of
the GC make it a rigid structure resistant to pretreatment and enzymatic hydrolysis
during lignocellulosic biomass processing. Altering cellulose structure affects plant
growth and development but improves cellulose accessibility. Defect in cellulose biosynthesis often led to the collapse of xylem vessel cells, which are water-transporting
cells inside the plant (Taylor 2008). Spatio-temporal modification of cellulose biosynthetic or hydrolysing enzymes might be a better approach to tailoring the cellulose production in plants.

7.2.1 BIOSYNTHESIS OF CELLULOSE

Glucose is the primary cellulose unit, obtained via cleavage of sucrose, catalysed by the sucrose synthase (SuSy) enzyme. SuSy belongs to a glycosyltransferase (GT) family which catalyses sucrose's breakdown into monomeric units, UDP-fructose and UDP-glucose. Cellulose chain initiation and elongation happens at plasma membrane by cellulose synthase complexes (CSCs) (McFarlane, Döring, and Persson 2014, Jarvis 2018). CSCs are organised protein structures of cellulose synthases (CESAs), embedded in the plasma membrane, and each complex is of 25–30 nm diameter (Figure 7.1). Out of 10 Arabidopsis CESA isoforms, CESA1, CESA3, and CESA6 are required for primary wall cellulose biosynthesis, whereas CESA4, CESA7, and CESA8 are involved in secondary wall cellulose synthesis (F. Li et al. 2018). CSCs movement and direction of cellulose synthesis are further regulated by microtubules mediated by POM2 or CSI1 (Bringmann et al. 2012). Primary cell wall cellulose is non-crystalline (amorphous) and less rigid than crystalline secondary cell wall cellulose. Researchers have studied cellulose biosynthesis and altered its structure to simplify cell wall structure for improving biomass properties.

7.2.1.1 Enhancement of the SuSy Activity

Sucrose synthase (SuSy) catalyses the breakdown of sucrose into fructose and UDP-glucose and makes it readily available to the CESAs. SuSy function is well investigated in different plant species, and its upregulation leads to an overall increase in biomass yield (Table 7.1). Increasing SuSy expression leads to enhanced growth, xylem cell area, wall thickness, and cellulose content. Overexpression of *Panicum virgatum SuSy1* leads to a 37% increase in total plant biomass and an unexpected increase in lignin level (Poovaiah et al. 2015). Another *SuSy3* from rice was expressed under the influence of Arabidopsis CESA promoter, resulting in a decrease in the crystallinity of cellulose and an increase in plant biomass of transgenic lines

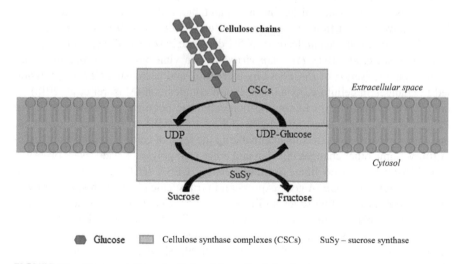

FIGURE 7.1 Representation of cellulose biosynthesis in plants.

TABLE 7.1

Effect of SuSy Manipulation on Plant Biomass Properties

Gene	Species	Gene Expression	Effect on Plant Biomass and Cell Wall Properties	References
PvSuSy1	*Panicum virgatum*	Overexpression	14% increase in biomass	Poovaiah et al. (2015)
OsSuSy3	*Oryza sativa*	Overexpression	Increase in biomass	Fan et al. (2017)
GhSuSy	*Gossypium hirsutum*	Heterologous expression in poplar	2%–6% increase biomass	Coleman, Yan, and Mansfield (2009)
PsnSuSy2	*Populus sp.*	Heterologous expression in Nicotiana	Increase in cellulose content and 25% thicker cell wall	Wei et al. (2015)

(Fan et al. 2017). Overexpression *GhSuSy* in poplar leads to a 2%–6% increase in total biomass. Also, isolation and introduction of poplar SuSy in *Nicotiana benthamiana* resulted in increased cellulose biosynthesis and cell wall thickness (Wei et al. 2015). Manipulating SuSy level affects glucose and cell wall metabolism. These studies suggest that altering SuSy expression in planta increases the plant biomass without affecting the plant growth.

7.2.1.2 Disrupting Native Cellulose Biosynthesis Pathway

Manipulating cellulose biosynthetic machinery via genetic engineering can tailor the cellulose content and other cell wall components. CESA is involved in cellulose chain elongation and a potential target to alter cellulose structure (Table 7.2). Enhanced cellulose levels are observed in transgenic lines expressing *Pinus massoniana CESA* in hybrid poplar. In addition to this, anatomical studies showed an increase in cell wall thickening and plant biomass production (Maleki et al. 2020). However, constitutive expression aspen *CESA8* in hybrid *Populus deltoides × Populus euramericana* leads to a 75% decrease in secondary cell wall cellulose content (Joshi et al. 2011). *Panicum virgatum* is a vital bioenergy crop, and functional characterisation of *PvCESA4* by increasing and decreasing *CESA* expression led to changes in cellulose, xylan, and lignin composition (Mazarei et al. 2018). A compensatory increase in lignin and non-cellulosic polysaccharides was found in *PvCESA4* transgenic lines. Switchgrass CESA4 and CESA6 have been identified, and their overexpression results in enhancing the level of cellulose content and crystallinity. Some other changes in xylan and other wall polysaccharides were observed in CESA-overexpressed lines. Similarly, barley CESA (*HvCESA*) overexpression led to suppression of *CESA* gene expression in transgenic plants with severe growth defect and reduced cellulose level (Tan et al. 2015). Crystallinity was also reduced by 34% in Arabidopsis mutant having a defect in *CESA3* gene. Still, it increases the accessibility of polysaccharide-degrading enzymes by 151% compared to wild type Arabidopsis plants (Harris, Stork, and Debolt 2009). Similarly, expressing the modified version of Arabidopsis *CESA3* under the control of fused promoter increases

TABLE 7.2
Effect of CESA Overexpression on Cellulose Content

Gene	Species	Expression in Other Plant Species	Effect on Cellulose Content	References
PtdCesA8	Aspen	Populus tremuloides	75% decrease	Joshi et al. (2011)
PmCESA2	Pinus massoniana	Populus deltoides Populus euramericana	Increased	Maleki et al. (2020)
PvCESA4 and PvCESA6	Panicum virgatum L.	Switchgrass	Decreased	Mazarei et al. (2018)
HvCESA	Hordeum vulgare	Barley	Decrease in cellulose content	Tan et al. (2015)

cellulose susceptibility for hydrolytic enzymes by 40%–66% in stem and leaf tissue (Sahoo et al. 2013). The unintended plant cell wall composition changes after CESA disruption suggest a specialised homeostatic mechanism to counteract alteration in cellulose biosynthesis. Spatio-temporal expression of CESA could be a useful approach to tailoring cellulose structure without compromising plant growth.

7.2.1.3 Altering Crystallinity of Cell Wall

Crystalline cellulose is a highly ordered and rigid structure mainly found in the secondary cell wall. This structure is resistant to enzymatic and chemical hydrolysis. Reducing crystallinity in planta positively affects cell wall assembly and can reduce the cost of bioprocessing. KORRIGAN1 (KOR1) is a membrane-associated cellulase involved in cellulose biosynthesis. KOR1 is identified, characterised in different plant species, and plays a vital role in cellulose biosynthesis. Functional characterisation of poplar KOR1 was done by expressing it in Arabidopsis. The Arabidopsis transgenic lines expressing KOR1 have decreased crystalline cellulose and accumulated more non-crystalline cellulose. However, heterologous expression of *Eucalyptus tereticornis* in poplar increases cellulose content (Aggarwal, Kumar, and Reddy 2015). Functional characterisation of poplar *KOR1* revealed a role in secondary cell wall biosynthesis (Yu et al. 2014). Downregulation of *Populus KOR1* and *KOR2* by RNAi approach resulted in decreased cellulosic and hemicellulosic sugars. Additional anatomical studies revealed a reduction in xylem fibre thickness in RNAi lines as compared to wild type poplar trees. In summary, the role of KOR in cellulose biosynthesis is not still clear, but altering its expression affects cellulose content and cell wall formation.

Another strategy to engineer cellulose is introducing labile primary cell wall cellulose in plants because it is non-crystalline and easily breakable (Sakamoto et al. 2018). Arabidopsis double mutant *nst1nst3* lacks a secondary cell wall in fibre cells and leads to a growth defect. But the expression of a chimeric activator of VP16-ERF035 in *nst1nst3* mutants rescues growth because of enhanced synthesis of the primary cell wall. Because of upregulation in primary cell wall, specific *CESA1*, *CESA3,* and *CESA6* genes led to an increase in cellulose level. This strategy needs to be tested and implemented in bioenergy trees or crops.

7.2.1.4 *In planta* Expression of Cellulose-Degrading Enzymes

High production cost of cellulolytic enzyme is one of the major limitations in biofuel production. Cellulolytic enzymes can be expressed in planta to convert cellulose into monomeric glucose units by expressing endo-1,4-β-glucanases, exo-1,4-β-glucanases, and β-glucosidases. Initial prediction was that expressing cellulolytic enzymes might inhibit the plant growth in ambient temperature. A better approach would be to express thermophilic cellulolytic enzymes that are not active during plant development but during lignocellulosic biomass processing. Using this approach, plant growth will be not impacted and the polysaccharides will be easily degradable during bioprocessing. The expression of thermophilic cellulases into the apoplast, vacuoles, and endoplasmic reticulum improves saccharification efficiency without affecting plant growth. In planta expression of endoglucanases and exoglucanases (cellobiohydrolases) has been reported in many dicot and monocot species. Increased cellulose accessibility was found by expressing the endocellulases from *Acidothermus cellulolyticus* heterologously in mitochondria or endoplasmic reticulum of maize (Mei et al. 2009). Similar results were obtained when one of the widely studied endoglucanases, E1 (Cel5A), was expressed heterologously under the control of a constitutive promoter in the cell wall of *Nicotiana tabacum* and *Zea mays*. And these transgenic plants were less recalcitrant than wild type plants when wall saccharide was subjected to pre- and post-enzymatic hydrolysis (Brunecky et al. 2011). Additionally, enhanced saccharification efficiency was observed in transgenic lines expressing archean endoglucanase driven by *35S* promoter as compared to wild type Arabidopsis plants. A phenotypic study of the transgenic plants showed the highest enzymatic activities in the dried stems of the Arabidopsis. This study proved that endoglucanases were active during the later stage of plant development (Mir et al. 2017). Also, expressing cellulase can affect stem length and germination. This was observed when *Trichoderma reesei* endoglucanases targeted the apoplastic region. But when this enzyme was expressed using ER-localised KDEL peptide, the growth was normal (Klose et al. 2015). Hybrid poplar lines expressing thermophilic *Thermotoga neapolitana* endoglucanases have a defect in plant development. Cell wall analysis of transgenic poplar lines depicted a reduction in cellulose and lignin content of leaf and stem tissue (Xiao et al. 2018). Post-harvest bioprocessing caused activation of endoglucanases at 100°C, and transgenic polysaccharides were more digestible than those of wild type plants.

Cellulose modification using different approaches has manifested many off-target or pleiotropic effects on plant cell wall and morphology. Also, a similar improvement in different species leads to various changes in the cellulose content. The detailed understanding of cellulosic biosynthetic machinery in model and bioenergy crops is necessary to fine-tune cellulose structure.

7.3 *IN PLANTA* MODIFICATION OF XYLAN STRUCTURE

Xylan, a polymer of β-1,4-linked D-xylose, is the dominant matrix polysaccharide of the primary and secondary cell walls of dicot and monocot plants. Xylan is a major factor for the lignocellulosic biomass recalcitrance during the industrial processing of biomass to value-added chemicals. Thus, understanding the xylan biosynthesis and structure in different plant species will aid in fine-tuning the plant cell wall structure and improve its properties for bioenergy applications.

Xylan is abundantly present in both cell wall layers and consists of 30% of the total dry weight (Pauly and Keegstra 2010). It is estimated that annually 10 billion tons of carbon is incorporated into xylan polysaccharides by terrestrial plants (Smith et al. 2017). Therefore, with such a high amount of terrestrial biomass, xylan of plant cell wall becomes an important biomass resource with significant economic roles. But xylan is resistant to enzymatic, chemical, and thermal degradation due to its branched structure and interaction with other cell wall components. Due to this, it would be challenging to produce useful products, majorly in the bioethanol industry. Fermentation of xylose and glucose requires activation of different sets of metabolic pathways due to differences in the carbon numbers. Many established useful industrial microbes can efficiently utilise the hexose sugars from cellulose, but lack such a system for pentose sugars from xylan (York and O'Neill 2008). Moreover, complex xylan side chains and linkages demand advanced and specialised hydrolytic enzymes for the complete xylan hydrolysis. Hence, to make lignocellulosic biomass fermentation processes more efficient and considerably lower their expenses, researchers are interested in in planta xylan modification.

7.3.1 STRUCTURAL FEATURE AND BIOSYNTHESIS OF XYLAN

The xylan backbone is the repeating unit of β-1,4 xylose (Xyl$p$) monomers, highly decorated with acetyl, arabinose, glucuronic acid, and (methyl)glucuronic acid substitutions. Short oligosaccharide sequences, i.e. reducing end sequence (RES) consisting of xylose, rhamnose, and galacturonic acid, are also attached to the xylan backbone (Deniaud et al. 2003).

Xylan substituents vary significantly among species, between cell types, and at different developmental stages and divide into three classes (Figure 7.2) (Jacobs, Larsson, and Dahlman 2001). O-Acetyl glucuronoxylans (AcGXs) structure is predominantly observed in the secondary walls of hardwood and herbaceous dicot plants such as Arabidopsis and poplar. Xylan backbone contains around 100 β-1, 4-linked xylosyl residues, and one of every ten residues is substituted with (4-O-methyl)-α-D-glucuronic acid ((Me)GlcA) at O-2 position. The xylan backbone and side chain are frequently substituted with O-acetyl groups; approximately 50% of the xyloses in the backbone are O-acetylated in poplar and Arabidopsis; and their percentage varies among species and tissue types. AcGX also contains a unique tetrasaccharide RES, but its function is not explored (Peña et al. 2007, York and O'Neill 2008). Gymnosperm xylan is substituted with MeGlcA at O-2 and one α-arabinofuranose (Araf) unit at O-3 position. O-Acetyl substitution is absent in gymnosperm xylan (Sjöholm et al. 2000). The main substitution on grass primary cell wall xylan is a substitution of disaccharide Araf residue, MeGlcA, or GlcA at O-2 position. As compared to the dicot xylan, the monocots arabinoglucuronoxylan (AGX) and O-acetylglucuronoarabinoxylan (AcGAX) are less acetylated.

Araf of AGX and AcGX are frequently esterified with p-coumaric or ferulic acids at O-5, which is cross-linked with lignin via these esters (Grabber, Ralph, and Hatfield 2000, Bunzel et al. 2003). The RES is not identified in grasses. But other

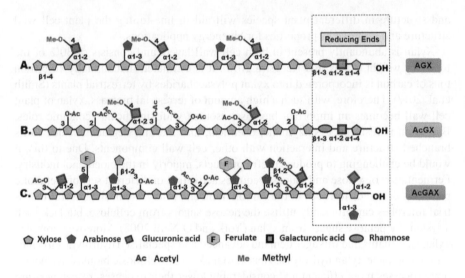

FIGURE 7.2 Representation of different types of xylan.

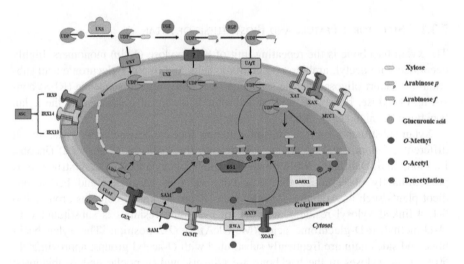

FIGURE 7.3 Xylan is synthesised in the Golgi apparatus. Substrates are synthesised both in the cytosol and in the lumen (please see the text for details). (Modified from Rennie and Scheller 2014.)

different monomers are found at the RES of grass AcGAX and AcAX, with precisely decorated xylose, suggesting greater structural diversity in monocot xylan (Ratnayake et al. 2014, Scheller and Ulvskov 2010).

Synthesis of xylan polysaccharides takes place in the lumen of Golgi bodies (Figure 7.3). Activated nucleotide sugars, methyl groups, and acetyl groups are the substrates for their synthesis. These units are transported from the cytoplasm to Golgi via transporters. The xylan backbone is catalysed by glycosyltransferase (GT) families, i.e. GT43 and GT47. Irregular xylem (IRX) 9, IRX14, and IRX10 are xylosyltransferases, and mutation in these genes leads to reduced xylose content, short xylan chain, and abnormal plant growth. These GTs can work individually or in complexes, known as xylan synthesis complex (XSC), but the exact mechanism of this complex formation in xylan chain elongation is unknown. Starting or ending of xylan synthesis probably depends on the formation of the RES. Mutation in genes from the GT47 and GT8 families leads to a reduction in the xylan chain elongation and xylose content, but the biochemical characterisation of these proteins is lacking. UDP-GlcA is decorated on xylan chains by the glucuronic acid substitution of xylan (GUX) enzymes (Saez-Aguayo et al. 2017), which can be 4-O-methyl esterified by glucuronoxylan methyltransferase (GXMT). Araf and xylosyl residues are transferred to xylan by xylosyl arabinosyltransferase (XAT) and xylosyl arabinosyl substitution of xylan 1 (XAX1), respectively. The xylan is also substituted with β-xylosyl residues at O-2 position in Arabidopsis seed mucilage, and the xylosyltransferase MUC1 catalyses the reaction. Acetyl groups are translocated by Reduced Wall Acetylation (RWA) proteins, and acetyl donors could be cytoplasmic acetyl-CoA or an unidentified donor. O-acetylation of the xylan polysaccharide occurs via xylan acetyltransferases (XOAT) from the trichome birefringence-like (TBL) family with different regiospecificities. Golgi-localised acetyl xylan esterase (AXE) is involved in the de-esterification of acetylated xylan to maintain the balance of the O-acetylation pool on the xylan chain. Brittle Sheath 1 (BS1) and DEACETYLASE ON ARABINOSYL SIDECHAIN OF XYLAN1 (DARX1) have recently characterised xylan esterase in rice and have yet been identified in other species (Zhang et al. 2017, Zhang 2019).

7.3.1.1 Modulating Xylan Synthetic Complexes

Modification of XSC genes can also be a suitable biotechnological route to altering the cell wall for industrial processes. RNAi lines of IRX9 and IRX14 homologs in poplar result in the upregulation of genes related to the cell cycle that caused the taller plants to have increased stem volume and diameter (Ratke et al. 2018). And downregulation of the same poplar genes in another study leads to an increase in cellulose digestion (Lee et al. 2011). Like CSCs, the stoichiometry of XSC varies among plants and at the tissue level, and its composition decides the activity of xylosyltransferase (XylT). Thus, it is appealing to suggest the differences in XylT activity on tissue and species levels (Song et al. 2015, Jiang et al. 2016) based on XSC composition differences. Overexpression of cotton XylT (IRX9 and IRX14) resulted in an increase in xylose yield due to the upregulation of other xylan biosynthetic genes (L. Li et al. 2014). An increase in xylose yield might lead to less deposition of cellulose and lignin, reducing cell wall complexity.

7.3.1.2 Altering Xylan Reducing End Sequence and Methyl Glucuronic Acid Level

The RES acts as a primer or terminator as its role is unclear. Mutations in genes involved in xylan biosynthesis led to differences in the chain length. Although it was observed that plants get severely dwarfed when these RES formation genes are mutated (Rennie and Scheller 2014), a more comprehensive understanding of RES genes and advanced biotechnological tools are needed for the alterations of xylan content more efficiently. In poplar, overexpressed lines of *IRX8* showed higher recalcitrance with defect in growth, whereas the downregulation resulted in the opposite effect (Biswal et al. 2015). But, till now, whether the downregulation or overexpression of other RES-related genes leads to this type of opposite outputs is unknown.

Methylation at the *O*-4 position of α-1,2-linked D-GlcA substitutions (MeGlcA) is the only xylan modification used to discriminate between the primary and secondary cell walls. Depending on the spacing of substitution, xylan is divided into minor and major domains of xylan. This modification is associated with secondary cell wall characteristics and biomass recalcitrance. GUX1 and GUX2 are involved in the transfer of GlcA or MeGlcA acid to xylan backbone in evenly and oddly manner, respectively, while xylan of primary wall is modified by GUX3 (Bromley et al. 2013). A reduction in the glucuronic acid level in *gux1 gux2* mutant increases xylan hydrolysis and efficiency of polysaccharide extraction. Nevertheless, the decrease in MeGlcA substitutions result in increased xylose release (Urbanowicz et al. 2012). These modifications can be taken as the targets for altering the xylan–cellulose interactions. And that may be attained by using secondary wall tissue-specific promoters, which would have a more effective result than the constitutive *35S* promoter.

7.3.1.3 Modifying Polysaccharide Acetyltransferase Expression

The substitutions such as acetyl esters and methyl esters in the polysaccharides lead to an increase in the biomass resistance because these modifications hamper the accessibility of the cell wall-degrading enzymes. Moreover, the fermentation process is subsequently inhibited by the release of acetate in the medium. The acetic acid release from the lignocellulosic mass inhibits yeast fermentation (Pawar et al. 2013). Therefore, reducing the acetyl level could be an optimum strategy to re-engineer plant cell walls.

As explained earlier, RWA is involved in transportation of unknown acetyl donor (Gille and Pauly 2012), and mutation in all Arabidopsis RWA genes leads to lesser acetylation in the wall polysaccharides, including xylan (Lee et al. 2012, Manabe et al. 2011). This led to an increase in xylan accessibility, but Arabidopsis *rwa* mutants have a defect in growth. However, decreasing expression of all RWA genes under the influence of xylem-specific promoter reduces the acetylation level by 25% and increases glucose and xylose conversion by 14% and 40%, respectively. TBL29 is an acetyltransferase that particularly *O*-acetylates the xylan, and inactivation of TBL29 decreases the xylan acetylation level drastically and then results in plant growth retardation. Remarkably, when the xylan glucuronosyltransferase expression is increased in Arabidopsis *tbl29* mutants, the plants show normal growth and low acetylation level in the original mutant plant (Xiong, Dama, and Pauly 2015). And these Arabidopsis lines showed a significant increase in saccharification efficiency. Xylan of grasses is less acetylated as compared to the xylan of hardwood, as alternative grass xylans are substituted with ferulates. Dimerised esters of ferulates in the walls are responsible for producing intermolecular and intramolecular cross-links in xylans that add to the biomass recalcitrance of grasses. The mechanism and pathway for feruloyl esterification are unclear, except for some studies in rice. Grasses have many BAHD enzymes present in the cytoplasm and have a role in xylan feruloyl esterification. However, rice BAHD RNAi lines have 20% less ferulic acid in the cell walls (Park et al. 2010), but whether it is a direct or indirect effect of BAHD modification is not known. Thus, another way to reduce ferulic acids in the cell wall is to enhance ferulic acid esterases exogenously in plants. Such strategies have been utilised for limited species with limited success.

7.3.1.4 *In planta* Expression of Xylan Hydrolytic Enzymes

The plant cell wall polymers' recalcitrant nature needs efficient enzymatic machinery for the proper conversion into fermentative sugars. And to develop such a strategy remains to be a challenge for biofuel industries as the production of these enzymes is costly. Generation of xylan-digesting enzymes directly within the plants may offer more cost-effective and less capital-intensive alternatives than independent microbial fermentation. As discussed before, side chain complexity is the determining factor of conformation, solubility, and interactions of xylans with other cell wall polymers. Thus, it significantly affects the extent and mode of enzymatic actions. Therefore, diverse hydrolytic enzymes are required with variable specificities and modes of cleavage. The xylan hydrolysis is carried out by multiple hydrolytic enzymes containing endoxylanase, β-xylosidase, α-glucuronidase, α-arabinofuranosidase, and AXE (Mazumder et al. 2012). Expression of these cell wall-degrading enzymes does not affect plant growth, but changes cell wall structure and improves polysaccharide digestibility (Table 7.3). Stable expression of the endo-1,4-xylanases (*xynA1*) in rice seed grains and straw displays no phenotypic effect, but its enzyme activity was

TABLE 7.3

Effect of Expression of Xylan-Degrading Enzymes on Plant Morphology and Saccharification

Enzyme	Gene and Species of Origin	Species	Effect on Growth and Saccharification	References
Xylanase	*xynA1*	Rice	NE and NA	Kimura et al. (2003)
Xylanase	*xynA*	Barely	NE and NA	Patel et al. (2000)
Xylanase	*xynZ* from *Clostridium thermocellum*	Tobacco	NE and NA	Herbers et al. (1995)
Acetylxylan esterase	*AnAXE* from *Aspergillus niger*	Arabidopsis	NE and +	Pawar et al. (2016) and Pogorelko et al. (2013)
α-Glucuronidase	*ScAGU115*	Arabidopsis	NE	Chong et al. (2015)
Ferulic acid esterase	*AnFAE*	Wheat	NE and NA	Harholt et al. (2010)
Ferulic acid esterase	*AnFAE or faeA*	Arabidopsis	NE and +	Buanafina et al. (2010)
Ferulic acid esterase	*AnFAE or faeA*	*Festuca arundinacea*	NE and +	Morris et al. (2017)

NE – no effect, NA – not available, + – increase in saccharification.

intact at 60°C for 24 h (Zhang et al. 2017). Also, endoxylanase can be expressed in the barley endosperm without any detrimental effect on growth and development (Patel et al. 2000). But changes in seed morphology and weight were reported. These negative phenotypic changes in plants can be resolved by expressing thermophilic enzymes as they are not active during plant growth. These thermophilic bacterial xylanases can be expressed in potato and tobacco successfully without any negative effects on growth (Herbers, Wilke, and Sonnewald 1995). Decreased xylan molecular weight was observed in Arabidopsis line expressing xylanases in apoplastic space from *Dictyoglomus thermophilum* (Borkhardt et al. 2010). Also, synergistic action xylanases and AXE facilitate efficient degradation of xylan. The expression of *An*AXE in Arabidopsis, poplar, and Brachypodium improved the saccharification process of stem lignocellulose (Pawar et al. 2016, 2017, Pogorelko et al. 2013). Importantly, these plants do not have any effect on plant growth, suggesting a potential approach to improving xylan and cellulose degradation.

Ferulic acid esterase (FAE) was expressed successfully under the control of maize ubiquitin promoter in *Festuca arundinacea,* with no major morphological changes (Buanafina et al. 2010). This led to a decreased ferulate level in cell wall, and in vitro dry matter digestibility was also increased compared to wild type plants (Buanafina et al. 2010). Similarly, cell wall digestion was increased in FAE overexpressed in Arabidopsis (Morris et al. 2017). This is because decreasing xylan ferulate level decreases xylan–lignin cross-linking, and polysaccharides become more accessible for enzymatic hydrolysis.

In planta expression of these degrading enzymes could loosen polysaccharide with more accessibility and can reduce the load of hydrolytic enzymes. This can help in reducing the cost of bioethanol production during pretreatment, hydrolysis, and fermentation process.

7.4   *IN PLANTA* MODIFICATION OF PECTIN STRUCTURE

Pectin is a natural heteropolysaccharide mainly present between the middle lamella regions of the primary cell wall. It accounts for about 35% (dicotyledons), 10% (grasses), and about 5% (woody tissue) of the dry weight of cell wall biomass (Edwards and Doran-Peterson 2012). The heterogeneous polymer of pectin is divided mainly into homogalacturonan (HG), rhamnogalacturonan I (RGI), and rhamnogalacturonan II (RGII) depending on backbone and side chain composition. The quantity of each of these polysaccharides is spatio-temporally variable in different species. The galacturonan backbone of these polysaccharides is esterified by a methyl group at C-6 carbonyl position and O-acetyl group at C-2 or C-3 position (Harholt, Suttangkakul, and Scheller 2010). Biosynthesis of pectin is carried out in the Golgi complex and then transported to the cell wall via membrane vesicle. Activated nucleotide monosugars are combined by membrane-associated Golgi-localised glycosyltransferases (GTs). GTs transfer glycosyl residues to monosaccharide, oligosaccharide, or polysaccharide acceptors from nucleotide sugars. A total of 67 GTs are required for the formation of all types of pectin. Further modifications such as O-methylation and O-acetylation on pectic polysaccharides are catalysed by methyltransferase and acetyltransferase. Some pectins are decorated with feruloylation, catalysed by feruloyl transferases. The substrates used for these modifications are S-adenosylmethionine (SAM), acetyl-CoA, and feruloyl-CoA, respectively (Mohnen 2008).

Pectin resides in both primary and secondary cell walls. The primary cell wall is a pectinaceous matrix containing cellulose, hemicelluloses, and a small amount of lignin. The secondary cell wall also contains a small proportion of pectin, making accessibility of polysaccharide difficult for cellulolytic enzymes during biomass to biofuel conversion. Thus, in planta pectin degradation is an effective strategy to alter cell structure (Xiao and Anderson 2013).

7.4.1 TYPES OF PECTIN-DEGRADING ENZYMES

Different bacteria and fungi secrete various pectinolytic enzymes. Those are mainly divided into three classes – protopectinases, esterases, and depolymerases. Plants may secrete them for a specialised purpose during different stages of plant development. Protopectinases (cleave protopectin) are classified as 'A-Type' protopectinases that react in the inner region of polygalacturonic acid. These enzymes are mainly found in fungi and yeasts. 'B-Type' protopectinases from bacterial sources such as *Bacillus subtilis* react on the outer side of the polygalacturonic acid chain. Similarly, the enzyme polygalacturonase (PG) cleaves polygalacturonic acid by hydrolytic cleavage at oxygen bridge. The subclass of this enzyme, i.e. endo-polygalacturonases from various bacteria, fungi, yeast, parasitic nematodes, and some higher plants, cuts the galacturonic acid polymer from inside, and the exo-polygalacturonases secreted by bacteria such as *A. tumefaciens* and *F. oxysporum* cleaves from outside. Pectin lyases perform the non-hydrolytic cleavage of pectic acid. They produce double-bond unsaturated products by cleaving glycosidic linkage at the C-4 position. De-esterification of methyl and acetyl esters linked to galacturonic acid chain can be carried out by pectin esterases. Pectin esterases from fungi can act randomly and cleave methyl and acetyl groups. However, esterases originated from plants act on non-reducing ends next to the carboxyl group. After pectin's de-esterification, they can be further cleaved by other pectinolytic enzymes (Kameshwar and Qin 2018). These varieties of pectin-degrading enzymes can be potentially used to degrade pectin and cell wall structure for various industrial applications.

7.4.2 BIOENGINEERING OF PECTIN

Cell wall remodelling can be done either by expressing fungal and bacterial pectinolytic enzymes in plants or by interfering with the pectin biosynthesis pathway (Table 7.4). It was observed that pectin acetylesterase (PAE1) from *P. trichocarpa* deacetylates pectin when expressed in *B. vulgaris* and *S. tuberosum*. The overexpression of PAE from *P. trichocarpa*, i.e. *PtPAE1*, in *N. tabacum* deacetylates pectin, but not the xylan (Gou et al. 2012). It was also shown that deacetylation hinders the cell elongation of floral parts such as style and filament, pollen germination, and pollen tube elongation. Also, the overexpression of *PAE1* results in severe male sterility. It was also observed that the overexpression of *PtPAE1* surprisingly reduces the digestibility of pectin by pectinase of the microbial source. The overexpression of fungal gene encoding polygalacturonase (PG) or pectin methylesterase

TABLE 7.4

Effect of *In planta* Expression of Pectinases on Plant Phenotype

Enzymes	Mode of Action	Plant Phenotype	References
Polygalacturonase (pga2) from *A. niger*	Breakdown of polygalacturonan polymer and increase in saccharification efficiency	Stunted growth and reduced biomass	Tomassetti et al. (2015)
Pectate lyase (PL1) from *P. carotovorum*	Cleavage of pectate polymer. Increase in saccharification efficiency	Reduced growth	Tomassetti et al. (2015)
Protopectinases	Cleavage of protopectin	NA	Sista Kameshwar and Qin (2018)
Pectin methylesterase (PME)	De-esterification of methyl group from pectin polymer	NA	Jenkins et al. (2001)
Pectin acetylesterase (PAE)	De-esterification of acetyl group from pectin polymer	NA	Sista Kameshwar and Qin (2018)

NA – not available.

inhibitor (PMEI) results in the reduction of de-esterified HG. Also, transgenic plants which are designed to overexpress either *PMEI* or *PG* show altered pectin content, resulting in a threefold increase in saccharification efficiency (Q. Li et al. 2014). In poplar, the overexpression of *pectate lyase* degrades homogalacturonan and shows improved saccharification of wood (Biswal et al. 2014). But the *polygalacturonase (PG)* expression in transgenic plants also results in less biomass formation (Xiao and Anderson 2013). The expression of fungal *polygalacturonase 2 (PGA2)* from *A. niger* leads to de-esterification of homogalacturonan and increases saccharification efficiency in Arabidopsis and tobacco. But it also results in loss of pectin integrity and stunted plant growth. However, when the *PGA2* is expressed under the control of the senescence-specific SAG12 promoter, the transgene will only express at the plant's late developmental phases without affecting the biomass yield (Tomassetti et al. 2015). Although a smaller amount of pectin is present in the cell wall, its modification can lead to cell wall changes and a positive effect on saccharification.

The expression of homogalacturonan-specific pectate lyase facilitates the solubility of wood matrix polysaccharides. The overexpression of this pectate lyase in poplar increases the solubility of pectin and xylans, and other hemicellulosic sugars (Biswal et al. 2014). This indicates that the homogalacturonan acts as a limiting factor in the solubility of major cell wall polysaccharides and degradation, which improves the saccharification efficiency of biomass. Similar to this, the knockdown of a gene from the pectin biosynthesis pathway *galacturonosyltransferase 4,* i.e. *GAUT4,* in poplar and switchgrass increases the biomass yield and saccharification efficiency. *GAUT4* is responsible for HG synthesis. The downregulation of *GAUT4* lowers the HG and RGII content in cell wall and reduces the calcium and boron concentration. All these increase the extraction efficiency of cell wall polysaccharides because of the decreased recalcitrance (Biswal, Atmodjo, Li, et al. 2018). The downregulation of *GAUT12.1* from *P. deltoides* results in increased biomass saccharification and plant growth. Still, opposite to that, the overexpression of *PtGAUT12.1* in *P. deltoides* showed an adverse effect on overall biomass production and ultimately on saccharification efficiency, and cell wall recalcitrance increased by 9%–15% (Biswal, Atmodjo, Li, et al. 2018). Hence, engineering the pectin content allows the maximum conversion of recalcitrant cell wall biomass into biofuel with minimal pre-processing.

7.5 *IN PLANTA* MODIFICATION OF MANNAN

7.5.1 STRUCTURAL FEATURES OF DIFFERENT TYPES OF MANNAN

Mannans are hemicellulosic cell wall polysaccharides and divided into homomannan, glucomannan, galactomannan, and galactoglucomannan (GGM) depending on their chemical structure (Schröder, Atkinson, and Redgwell 2009, Zhong, Cui, and Ye 2019). They are considered as storage and structural polysaccharides. HM and galactomannan are found in seeds, whereas glucomannan is found in bulbs and tubers. They are utilised by growing embryo and shoot. The hydrophobic nature of galactomannan shields the developing shoot from variation in water balance. Similar to xyloglucan, glucomannan interacts with cellulose. There is evidence that glucomannan–lignin–xylan complexes are present in spruce wood. GGMs participate in cross-linking with CMs and hemicellulose–lignin complex (Schröder, Atkinson, and Redgwell 2009,). GGMs also show an affinity for cellulose and can be found deposited in cell wall of seed endosperm (Schröder, Atkinson, and Redgwell 2009). Mannans are major hemicellulosic polysaccharides in softwood plants and some lower-order plants such as bryophytes. Softwood mannans contain mainly galactoglucomannan. Mannans in hardwood are about 1%–4% of the cell wall. They are primarily glucomannan with substitution of few galactose residues. Homomannan is made up of a linear chain of β-1,4-mannosyl residues, while glucomannan is similar to homomannan but intermixed with glucosyl residues at β-1,4 position. Similarly, galactomannan contains homomannan with α-galactose side chain on *O*-6 position of few mannosyl residues. Galactoglucomannan is β-1,4-glucomannan with α-galactose on *O*-6 position of some mannosyl residues. Mannans from all plant sources are acetylated at *O*-2 and *O*-3 positions of mannosyl residues. The range of mannan acetylation varies in different plant species. Glucomannan shows high affinity for cellulose. Acetylation of mannans hinders their affinity for cellulose (Zhong, Cui, and Ye 2019).

7.5.2 INTRODUCTION TO MANNAN BIOSYNTHESIS

Mannan synthesis is carried out by cellulose synthase-like A (CSLA) family which is a subgroup of glycosyltransferase (GT) superfamily 2. GDP-mannose and GDP-glucose serve as substrate for mannan backbone formation. Galactomannan α-galactosyltransferase (GMGT) of family GT34 mediates α-galactosyl side chain transfer on mannose backbone. In developing endosperms, galactomannans are synthesised by GDP-Man-dependent mannosyltransferase and UDP-Gal-dependent galactosyltransferase (GalT). Mannan synthesis-related (MSR) proteins may be required for glucomannan biosynthesis. It is reported that mutation in the two MSR genes in Arabidopsis results in reduced levels of both mannan synthase and glucomannan level. Few GTs involved in mannan biosynthesis are characterised, and further investigation is necessary to understand mannan biosynthesis in more detail.

Some members of Arabidopsis *CSLA* gene family show essential function in plant development. The *csla9* mutant shows reduced glucomannan content; however, triple mutant *csla2csla3csla9* is observed with negligible glucomannan content in Arabidopsis stem. This also shows no phenotypic deformities with respect to wild type. Opposite to this, the overexpression of *CSLA2, CSLA7* and *CSLA9* increases the glucomannan content affecting the embryogenesis process (Goubet et al. 2009). More studies are necessary to understand mannan biosynthesis and its role in modifying plant lignocellulosic biomass for bioenergy purpose.

7.5.2.1 Mannan Acetylation in Plants

Mannan O-acetyltransferases (MOATs) of DUF231 family from Arabidopsis and konjac shows O-acetyltransferase activity and shifts acetyl group on O-2 and O-3 positions on mannosyl residues in oligomers of mannan. The downregulation of these MOATs in Arabidopsis results in sudden reduction in acetylation of mannan (Zhong, Cui, and Ye 2019). Arabidopsis MOAT1, MOAT2, MOAT3, and MOAT4 and their homolog from *A. konjac AkMOAT1* show O-acetyltransferase activity. They execute transferase activity by transferring acetyl group from acetyl-CoA to mannohexaose acceptor. It was observed that mutation in conserved residues of GDS and DXXH motif of MOAT3 terminates its acetyltransferase activity. Inhibition of *MOAT 1, 2, 3,* and *4* genes by RNAi shows a drastic reduction in acetylation of glucomannan (Zhong, Cui, and Ye 2018). Reducing mannan acetylation might change interaction with cellulose and other cell wall properties, but this has not been investigated further.

7.5.2.2 Mannan-Degrading Enzymes

Mannan-degrading enzymes are divided into two glycoside hydrolase families, namely GH5 and GH26. They mainly include β-mannanase (cleaves internal β-1,4-linkages), β-glucosidase (exo-hydrolytic enzyme that breaks 1,4-β-glucopyranose at non-reducing end), and β-mannosidase (cuts β-1,4-linked mannan from non-reducing end). Furthermore, side chains on mannan can be cleaved by using acetyl mannan esterase and α-galactosidase. These enzymes are extracted from various bacterial, fungal, and yeast sources such as *A. niger, S. cerevisiae, B. subtilis,* etc. It was observed that the combined treatment of β-mannosidase and β-mannanase from *Sclerotium rolfsii* resulted in an overall increase in the production of monomers from galactomannan. Similarly, thermostable α-galactosidase, β-mannosidase, and β-mannanase from *Thermotoga neapolitana* hydrolyse the galactomannan polysaccharides into monosaccharides (Moreira and Filho 2008). Because of their diverse activity on mannan substrate, endo-β-mannanase hydrolyses mannans by cleaving β-1,4-mannose backbone to facilitate cell expansion, softening fruits during ripening . *Lycopersicon esculentum* beta-mannanase 4a (*Le*MAN4a) enzyme from tomato exhibits endo-β-mannanase and mannan transglucosylase activities (Schröder, Atkinson, and Redgwell 2009).

To the best of our knowledge, mannan-degrading enzymes are not expressed in plants to study the effect on saccharification properties of plant biomass; further attention in this field is necessary to exploit mannan structure for bioenergy applications.

7.6   *IN PLANTA* MODIFICATION OF LIGNIN

7.6.1   INTRODUCTION TO LIGNIN BIOSYNTHESIS

Lignin is a major structural component of cell wall, mainly of terrestrial plants, and acts as a recalcitrant polymer during plant biomass processing for biofuel production. Lignin is a complex heterogeneous polymer composed of phenolic alcohol derivatives known as monolignols, mainly of three types, i.e. *p*-coumaryl alcohol, coniferyl alcohol, and sinapyl alcohol. Oxidative polymerisation of these monolignols produces *p*-hydroxyphenyl (H) units, guaiacyl (G) units, and syringyl (S) units, respectively. These subunits are synthesised in the cytoplasm via phenylpropanoid pathway and get polymerised in the apoplastic region by laccases and peroxidases. In recent years, a new type of lignin, known as catechyl (C) lignin, has been produced from caffeoyl alcohol, reported in seed coats of vanilla orchid and Brazilian cactus (Chen et al. 2012). The proportion of lignin units differ in plant system: In angiosperm, lignin exists as a heteropolymer with G and S units found prevalently in almost equal proportions with very less amounts of H units, whereas in gymnosperm, lignin is devoid of S units. Lignin is mainly deposited in secondary cell walls after plant cell growth cessation. The amount, composition, and structure of lignin is highly diverse across plant taxa, even across cell types and cell wall layers. The most important characteristic of lignin is that it interacts with other cell wall components such as CMs and matrix polysaccharides in cell wall region via various covalent and non-covalent linkages. At the time of deposition, lignin displaces water from the cell wall matrix, forms a hydrophobic network with CMs, and covalently links to side groups of matrix polysaccharide. This interaction adds significant mechanical strength to cell walls and reduces plant material digestibility by making cell wall recalcitrant. Besides being the main factors determining the recalcitrance to enzymatic digestion during biomass processing, lignin also imparts some positive roles to plants such as acting as a physical barrier to pathogen attack, strengthening the vasculature for efficient transport of water and minerals, and conferring rigidity to cell walls which allows plants to stand by providing mechanical and structural support. This sturdiness of cell wall necessitates the need of loosening the complex cell wall structure for lignin degradation and proper accessibility to hydrolytic enzymes during industrial processing of lignocellulosic biomass. Genetic and metabolic strategies have been used to deconstruct the cell wall by focusing mainly on the different critical enzymes involved in lignin biosynthesis. Therefore, it is important for lignin engineering to understand the overall pathway through which lignin monomers are synthesised and polymerised to produce recalcitrant lignin polymer.

In plants, lignin monomers are synthesised in the cytosol as a part of the well-known phenylpropanoid pathway (Figure 7.4). Phenylalanine (Phe) is a main precursor of this pathway in dicot. Phe is synthesised via shikimate pathway in the

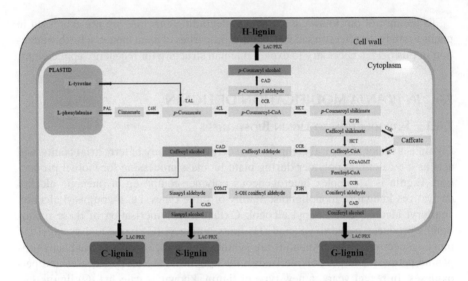

FIGURE 7.4 Representation of phenylpropanoid pathway in plants. (Please refer text for more details.)

plastid, is transported to cytoplasm, and undergoes deamination catalysed by phenyl-alanine ammonia lyase (PAL) cinnamate.

Cinnamate undergoes hydroxylation and CoA ligation by cinnamate 4-hydroxylate (C4H) and 4-coumarate-CoA ligase (4CL) to form p-coumarate and p-coumaroyl-CoA, respectively. p-Coumaroyl-CoA serves as a common intermediate for the syn-thesis of different types of lignin monomers. In grasses, aromatic amino acid tyrosine acts as a main precursor, which undergoes deamination by tyrosine ammonia lyase (TAL) enzyme and directly forms p-coumarate, ligated with CoA via 4CL to pro-duce p-coumaroyl-CoA. p-Coumaroyl-CoA is directly reduced by cinnamoyl-CoA reductase (CCR) to produce p-coumaryl aldehyde, converted to p-coumaryl alco-hol by cinnamyl-alcohol dehydrogenase (CAD). p-Coumaryl alcohol is polymerised in cell wall, named as H-lignin. p-Coumaroyl group from p-coumaroyl-CoA is transferred to shikimate by hydroxycinnamoyl CoA:shikimate hydroxycinnamoyl transferase (HCT) to form p-coumaroyl shikimate. It then undergoes hydroxylation reaction by p-coumaroyl shikimate 3'-hydroxylase (C3'H) to produce caffeoyl shi-kimate. Caffeoyl moiety from caffeoyl shikimate is either directly transferred to CoA via HCT to form caffeoyl-CoA, or caffeoyl shikimate esterase (CSE) can cleave shikimate and ligate with CoA by 4CL to produce caffeoyl-CoA. Caffeoyl-CoA is then methoxylated by caffeoyl-CoA O-methyltransferase (CCoAOMT) to pro-duce feruloyl-CoA, which is reduced to coniferyl aldehyde by CCR. CAD converts coniferyl aldehyde to coniferyl alcohol, a monolignol which gets polymerised by lac-cases or peroxidases to produce G-lignins. For the synthesis of S-lignins, coniferyl aldehyde is further hydroxylated by ferulate 5-hydroxylase (F5H) and methoxylated by caffeic acid O-methyltransferase (COMT) to produce sinapyl aldehyde. CAD con-verts sinapyl aldehyde to sinapyl alcohol, which gets polymerised to form S-lignins in apoplastic region. Moreover, an additional branch is added to this pathway that led to the synthesis of a new type of lignin called catechyl lignin (C-lignins) formed via polymerisation of caffeoyl alcohol. Caffeoyl alcohol is produced by a reduction in caffeoyl-CoA by CCR to produce caffeoyl aldehyde, which is converted to caffeoyl alcohol. Certain enzymes of the lignin biosynthetic pathway such as C4H, C3'H, and F5H which are mainly involved in hydroxylation reactions are monooxygenases which belong to cytochrome P450 superfamily. These enzymes are localised in ER membrane with their active side protruding in the cytosol where monolignols are synthesised (Ro et al. 2001). The overall mechanism of monolignols transport to the apoplastic region is unclear, but three different types of mechanism are proposed for monolignol transport, i.e. through passive diffusion, by vesicle-associated exocytosis, using ABC transporters, and/or through proton-coupled antiporter (Barros et al. 2015, Alejandro et al. 2012). For example, AtABCG29 is found to be a p-coumaroyl alcohol transporter (Alejandro et al. 2012). Once excreted in the apoplastic space, lignin monomers undergo oxidative polymerisation by laccases and peroxidases. Laccases are copper-containing oxidoreductases, belong to a multigenic family, and use O_2 to oxidise all types of monolignols. Till now, 17 and 39 laccase genes have been identified in Arabidopsis and *Populus trichocarpa*, respectively (Barros et al. 2015). However, some peroxidases use H_2O_2 and monolignols as substrates for the synthesis of lignins (Barros et al. 2015). In plants, peroxidases belong to a large multigenic family, represented by class III type peroxidases, having a cell wall signal peptide that targets the enzyme to apoplastic space. In Arabidopsis, 73 peroxidase genes have been identified (Barros et al. 2015). Although both laccases and peroxidases are involved in lignin polymerisation, these enzymes appear to show cell type-specific activity. *Arabidopsis thaliana* laccases (At-LAC4, At-LAC11, and At-LAC17) catalyse lignin polymerisation in xylem treachery elements and xylem fibres (Zhao et al. 2013), whereas peroxidase 64 (PER64) catalyses lignin polymerisation in Casparian strip (Lee et al. 2013).

7.6.1.1 Reducing the Lignin Content

Efforts to alter the total cell wall lignin content through genetic engineering have improved the biomass digestibility, but such modification imposes plant growth penalty and hence compromise the lignocellulose biomass yield. Reducing the gene expression of key lignin biosynthetic genes such as *C4H* (Bjurhager et al. 2010), *4CL* (Voelker et al. 2011), *C3'H* (Coleman et al. 2008), and *CCR* (Acker et al. 2014) affects total lignin content. But the growth and biomass yield of some lignin-modified plants were hampered. Thus, instead of reducing overall plant lignin content, cell type-specific or tissue-specific lignin content and composition have become a promising approach without compromising plant growth.

It is observed that lignin-modified plants are severely affected with multiple growth abnormalities, and one of the prominent phenotypes is dwarf plant due to collapse in water transporting xylem vessel. Weaker xylem vessel gets collapsed because of negative pressure generated during transpiration pull, leading to growth abnormalities. Controlling xylem vessel lignin deposition using specific promoters overcomes the adverse effects of lignin reduction on plant growth. For instance, the dwarf phenotype in *c4h* mutant is partially restored by the reintroduction of *C4H* gene under the control of VASCULAR-RELATED NAC-DOMAIN 6 (VND6) promoter, resulting in lignin restoration of lignin deposition in xylem vessels and normal transport of water and minerals (Yang et al. 2013). However, the saccharification was not improved in *c4h*ProVND6:C4H plants and showed re-deposition of lignin in the interfascicular fibre region, indicating that this promoter was not specific to vessel cells. A similar strategy has been used with *cse* lignin mutant in Arabidopsis where the *CSE* gene was reintroduced under the control of *VND6* and *VND7* promoters, which partially restored the growth pattern vessel cell wall integrity (Vargas et al. 2016). However, *cse*ProVND6:CSE and *cse*ProVND7:CSE plants did not show any improvement in saccharification efficiency. These studies indicate that VND6 and VND7 promoters are not strong and/or specific to fully restoring the dwarf phenotype of lignin mutant plants. Secondary wall NAC binding element of the xylem cysteine protease 1 (ProSNBE) is an artificial promoter used for complementation in *ccr1* mutant background (De Meester et al. 2018). ProSNBE:CCR1 line showed full recovery in growth pattern and vessel cell wall integrity and also improved saccharification efficiency in comparison with the wild type. In poplar, a new strategy has been implemented by targeting a lignin biosynthesis-related transcription factor 1 (LTF1) to modify the lignin deposition (Gui et al. 2020). LTF1 binds to the promoter of lignin biosynthetic genes and suppresses its synthesis. Under the control of vessel-specific and fibre-specific promoters, the phosphorylation-null LTF1 was introduced in *Populus* to observe lignin deposition and biomass yield changes. The engineered LTF1 under fibre-specific promoter suppressed lignin biosynthesis in fibre cells and is found to have increased biomass without growth defects and improvement in the cell wall digestibility and sugar release.

Apart from reducing overall lignin content in plants or relocating it to specific cell types, lignin composition and polymerisation can be altered for generating easily digestible lignocellulosic biomass yield without affecting plant stature (Mahon and Mansfield 2019).

It can be achieved by:

1. altering the ratio of different lignin monomers,
2. modifying the cross-linkages between lignin and other cell wall components,
3. adding labile monolignols, and
4. reducing the degree of polymerisation.

7.6.1.2 Fine-tuning the Lignin Monomer Composition

Multiple strategies have been implemented to alter lignin composition in plants mainly by producing one type of lignin, either S-, G-, or H-lignin using genetic manipulation. The proportion of these lignins vary in plants. G-lignin and S-lignin are found in almost equal ratios, whereas H-lignin is found in minor amounts. There are certain lignin mutants in Arabidopsis, where mutating and/or altering expression in one or many genes of lignin biosynthetic pathway has shown to deviate the overall flux towards synthesis of one type of lignin (Table 7.5). These mutants such as *fah1*, *C4H:F5H fah1*, *med-ref8*, *cse* are very well known. F5H-deficient *fah1* (*ferulic acid hydroxylase1*) lignin mutant blocks S-lignin biosynthesis pathway and dominates in

TABLE 7.5
Arabidopsis Lignin Mutants Showing Lignin Composition and Effect on Saccharification

Lignin Mutants	S	G	H	Saccharification	References
fah1	3%	95%	Very low	Decreased sugar release	Meyer et al. (1998)
C4H:F5H fah1	92%	Very low	Very low	Enhanced sugar release	Ruegger et al. (1999)
med5a/5bref8	3%	2%	95%	Enhanced sugar release	Bonawitz et al. (2014)
cse	19%	37%	44%	Enhanced sugar release	Vanholme et al. (2013)

G-lignin (Meyer et al. 1998), resulting in less digestible cell walls than wild types. These mutant plants when complemented with C4H:F5H deposits mainly S-lignin (Ruegger et al. 1999) that results in better cell wall digestibility. Other lignin mutants such as *ref8* (*reduced epidermal fluorescence8*), having mutation in *C3'H*, an important key enzyme involved in the biosynthesis of G- and S-lignins, showed dwarf phenotype. Interestingly, mutating MED5a and MED5b subunits of mediator (a transcriptional co-regulatory complex) in *ref8-1* mutant rescues growth and this *med5a/5b ref-8* looks like wild type plant and results in approx. 95% H-lignin accumulation, resulting in increased glucose release after digestion (Bonawitz et al. 2014). Likewise, *Arabidopsis cse* lignin mutant also accumulated H-lignin but in higher level, around 30-fold more, in comparison with wild type plant (Vanholme et al. 2013). However, some of these mutants produce lesser biomass, but increase glucose release after enzymatic digestion.

The known fact is that out of three different lignin types, S-lignin is found to have high levels of β-ether bonds, most labile bonds in native lignin, producing more linear lignin that is easily extractable. Increasing S-to-G ratio in plants has been more effective in terms of biomass digestibility and sugar extractability, ultimately for better and enhanced biofuel production (Studer et al. 2011, Mansfield, Kang, and Chapple 2012).

7.6.1.3 Disrupting Cross-Linkages between Lignin and Other Cell Wall Components

CMs are embedded in dehydrated matrix of hemicellulose and lignin where these polymers interact with each other via several covalent and non-covalent interactions: Hemicelluloses interact with CMs with hydrogen bonds, whereas lignin was found to be covalently linked to hemicelluloses via mainly three possible linkages: through 4-O methyl-glucuronic acid on xylan backbone, benzyl ether linkages between C2 or C5 hydroxyl groups on xylan and lignin, and phenyl glycosides between xylose C1 and lignin. Removal of 4-O methyl-glucuronic acid from xylan backbone disrupts linkage between lignin and xylan and has been shown to improve the extractability of cellulose and xylan (Lyczakowski et al. 2017). Benzyl ether and phenyl glycoside linkages are present on lesser or highly acetylated xylan. Increasing xylan acetylation may reduce the frequency of covalent linkages between lignin and xylan (Giummarella and Lawoko 2016). Some evidence showed that modification in the xylan and pectin biosynthesis genes also results in improved cell wall digestibility, possibly through disrupting the linkages with lignin (Biswal et al. 2015, Biswal, Atmodjo, Pattathil, et al. 2018). There are evidences where increasing peptide cross-linkages in lignin increased the number of sites for protease action, thereby enhancing the loosening of cell wall structure that significantly improves saccharification efficiency of biomass (Liang et al. 2008).

7.6.1.4 Addition of Labile Monolignols

Although the majority of lignin polymer is made from incorporation of majorly three monolignols, i.e. *p*-coumaryl alcohol, coniferyl alcohol, and sinapyl alcohol, the lignin polymerisation process has shown a high degree of plasticity as many alternative monomers have been discovered, derived from lignin monolignol biosynthetic pathway to be incorporated in this process. Out of those, labile monolignols are easily breakable monolignol conjugates when introduced into lignin polymer and are capable of introducing chemically labile β-ether bonds into lignin chains. The most studied one is monolignol ferulate conjugate, produced by adding ferulate moieties on different monolignols. Ferulate moiety has an additional methoxyl group making it compatible for incorporation into lignin polymer and introduces easily cleavable bonds. These conjugates are not found in plants naturally, but monocots have been found with a similar monolignol conjugate, i.e. monolignol *p*-coumarate conjugate, that is produced by the addition of *p*-coumarate onto monolignols. Incorporation of such conjugates in lignin has been shown to enhance enzymatic biomass digestibility after pretreatment and significantly improve fermentable sugars release. An enzyme ferulate monolignol transferase (FMT) from *Angelica sinensis* has been introduced in hybrid poplar that incorporated monolignol ferulate conjugates in lignin backbone to increase the frequency of easily breakable bonds in lignin resulting in improved lignin extractability (Wilkerson et al. 2014). Likewise, Smith et al. (2015) expressed a rice *p*-coumaroyl-CoA monolignol transferase (PMT) in hybrid poplar that incorporated *p*-coumarate conjugates (*p*CA) in lignin backbone and resulted in improved biomass digestibility (Smith et al. 2015).

7.6.1.5 Modifying Lignin Polymerisation

Lignin monomers are polymerised in cell wall region via enzymes such as laccases and peroxidases which are well-known enzymes functioning in oxidative reactions. Several laccases are found in different isoforms, are involved in lignin polymerisation, have also become a suitable target for altering lignin composition. For instance, in *Arabidopsis*, *lac4-2/17* double mutant showed a 40% decrease in total lignin content in stem, resulting in higher saccharification efficiency (Berthet et al. 2011). In poplar, *lac14* mutant resulted in an increased S-to-G ratio, which led to enhancement in biomass enzymatic digestibility and improved sugar release for biofuel production (Qin et al. 2020). Additionally, the introduction of some potential initiators or terminators of lignin polymerisation can be used as an effective strategy to produce new lignin with lesser recalcitrance. In grasses, compounds such as tricin, a flavonoid component derived from a combination of biosynthetic pathways such as shikimate and acetate-/malonate-derived polyketide pathways, attach at the ends of lignin polymer and shorten lignin chains, possibly by acting as a chain terminator (Lan et al. 2015). In Arabidopsis, the expression of a bacterial hydroxycinnamoyl-CoA hydratase-lyase (HCHL) produces hydroxybenzaldehyde and hydroxybenzoate derivatives by cleaving propanoid side chain of hydroxycinnamoyl-CoA lignin precursors. These derivatives act as lignin initiators and lead to more polymer initiation events that results in the formation of shorter chains, thereby helping in reducing recalcitrance created by lignin and enhancing sugar release for biofuel production (Eudes et al. 2012).

7.7 CONCLUSIONS

Cell wall formation, composition, and polymer mutual interactions, as well as the efficiency of the deconstruction of lignocellulosic biomass for getting the bio-industry products, are driven by the plant cell wall biosynthesis. But how cell wall synthesis, deposition, and modifications affect saccharification and fermentation still needs to be explored further. An in-depth understanding of cell wall biosynthesis and connected metabolic pathways in diverse bioenergy plants can give new ideas for the alternations of wall composition in lignocellulosic biomass. These customised compositions of lignocellulosic biomass can be useful for the generation of value-added products in the biorefinery. In this chapter, we reviewed some of the wall modifications and their effect on biomass recalcitrance and plant growth. But these studies are limited to model plant species, and conclusions are based on plants grown in controlled growth conditions. Thus, translating these strategies into multiple bioenergy crops in field and stress conditions is necessary to understand whether such modifications benefit plant growth and positively affect lignocellulosic biomass processing in a cost-effective manner.

REFERENCES

Acker, Rebecca Van, Jean Charles Leple, Dirk Aerts, Veronique Storme, Geert Goeminne, Bart Ivens, Frédéric Légée, et al. 2014. "Improved Saccharification and Ethanol Yield from Field-Grown Transgenic Poplar Deficient in Cinnamoyl-CoA Reductase." *Proceedings of the National Academy of Sciences of the United States of America* 111 (2): 845–50. https://doi.org/10.1073/pnas.1321673111.

Aggarwal, Diwakar, Anil Kumar, and M. Sudhakara Reddy. 2015. "Genetic Transformation of Endo-1,4-β-Glucanase (Korrigan) for Cellulose Enhancement in Eucalyptus Tereticornis." *Plant Cell, Tissue and Organ Culture* 122 (2): 363–71. https://doi.org/10.1007/s11240-015-0774-7.

Alejandro, Santiago, Yuree Lee, Takayuki Tohge, Damien Sudre, Sonia Osorio, Jiyoung Park, Lucien Bovet, et al. 2012. "AtABCG29 Is a Monolignol Transporter Involved in Lignin Biosynthesis." *Current Biology* 22 (13): 1207–12. https://doi.org/10.1016/j.cub.2012.04.064.

Barros, Jaime, Henrik Serk, Irene Granlund, and Edouard Pesquet. 2015. "The Cell Biology of Lignification in Higher Plants." *Annals of Botany* 115 (7): 1053–74. https://doi.org/10.1093/aob/mcv046.

Berthet, Serge, Nathalie Demont-Caulet, Brigitte Pollet, Przemyslaw Bidzinski, Laurent Cézard, Phillipe le Bris, Jonathan Herve, et al. 2011. "Disruption of LACCASE4 and 17 Results in Tissue-Specific Alterations to Lignification of *Arabidopsis thaliana* Stems." *Plant Cell* 23 (3): 1124–37. https://doi.org/10.1105/tpc.110.082792.

Biswal, Ajaya K., Melani A. Atmodjo, Mi Li, Holly L. Baxter, Chang Geun Yoo, Yunqiao Pu, Yi Ching Lee, et al. 2018. "Sugar Release and Growth of Biofuel Crops Are Improved by Downregulation of Pectin Biosynthesis." *Nature Biotechnology* 36 (3): 249–57. https://doi.org/10.1038/nbt.4067.

Biswal, Ajaya K., Melani A. Atmodjo, Sivakumar Pattathil, Robert A. Amos, Xiaohan Yang, Kim Winkeler, Cassandra Collins, et al. 2018. "Biotechnology for Biofuels Working towards Recalcitrance Mechanisms : Increased Xylan and Homogalacturonan Production by Overexpression of GAlactUronosylTransferase12 (GAUT12) Causes Increased Recalcitrance and Decreased Growth in Populus." *Biotechnology for Biofuels*, 1–26. https://doi.org/10.1186/s13068-017-1002-y.

Biswal, Ajaya K., Zhangying Hao, Sivakumar Pattathil, Xiaohan Yang, Kim Winkeler, Cassandra Collins, Sushree S. Mohanty, et al. 2015. "Downregulation of GAUT12 in *Populus deltoides* by RNA Silencing Results in Reduced Recalcitrance, Increased Growth and Reduced Xylan and Pectin in a Woody Biofuel Feedstock." *Biotechnology for Biofuels* 8 (1). https://doi.org/10.1186/s13068-015-0218-y.

Biswal, Ajaya K., Kazuo Soeno, Madhavi Latha Gandla, Peter Immerzeel, Sivakumar Pattathil, Jessica Lucenius, Ritva Serimaa, et al. 2014. "Aspen Pectate Lyase PtxtPL1-27 Mobilises Matrix Polysaccharides from Woody Tissues and Improves Saccharification Yield." *Biotechnology for Biofuels* 7 (1): 1–13. https://doi.org/10.1186/1754-6834-7-11.

Bjurhager, Ingela, Anne-mari Olsson, Bo Zhang, Lorenz Gerber, Manoj Kumar, Lars A. Berglund, Ingo Burgert, and Lennart Salme. 2010. "Ultrastructure and Mechanical Properties of Populus Wood with Reduced Lignin Content Caused by Transgenic Down-Regulation of Cinnamate 4-Hydroxylase." *Biomacromolecules* 2359–65.

Bonawitz, Nicholas D., Jeong Im Kim, Yuki Tobimatsu, Peter N. Ciesielski, Nickolas A. Anderson, Eduardo Ximenes, Junko Maeda, et al. 2014. "Disruption of Mediator Rescues the Stunted Growth of a Lignin-Deficient Arabidopsis Mutant." *Nature* 509 (7500): 376–80. https://doi.org/10.1038/nature13084.

Borkhardt, Bernhard, Jesper Harholt, Peter Ulvskov, Birgitte K. Ahring, Bodil Jørgensen, and Henrik Brinch-Pedersen. 2010. "Autohydrolysis of Plant Xylans by Apoplastic Expression of Thermophilic Bacterial Endo-Xylanases." *Plant Biotechnology Journal*. https://doi.org/10.1111/j.1467-7652.2010.00506.x.

Bringmann, Martin, Eryang Li, Arun Sampathkumar, Tomas Kocabek, Marie Theres Hauser, and Staffan Perssona. 2012. "POM-POM$_2$/CELLULOSE SYNTHASE INTERACTING1 Is Essential for the Functional Association of Cellulose Synthase and Microtubules in Arabidopsis." *Plant Cell*. https://doi.org/10.1105/tpc.111.093575.

Bromley, Jennifer R., Marta Busse-Wicher, Theodora Tryfona, Jennifer C. Mortimer, Zhinong Zhang, David M. Brown, and Paul Dupree. 2013. "GUX1 and GUX2 Glucuronyltransferases Decorate Distinct Domains of Glucuronoxylan with Different Substitution Patterns." *Plant Journal* 74 (3): 423–34. https://doi.org/10.1111/tpj.12135.

Brunecky, Roman, Michael J. Selig, Todd B. Vinzant, Michael E. Himmel, David Lee, Michael J. Blaylock, and Stephen R. Decker. 2011. "In Planta Expression of A. Cellulolyticus Cel5A Endocellulase Reduces Cell Wall Recalcitrance in Tobacco and Maize." *Biotechnology for Biofuels*. https://doi.org/10.1186/1754-6834-4-1.

Buanafina, Marcia M. de O., Tim Langdon, Barbara Hauck, Sue Dalton, Emma Timms-Taravella, and Phillip Morris. 2010. "Targeting Expression of a Fungal Ferulic Acid Esterase to the Apoplast, Endoplasmic Reticulum or Golgi Can Disrupt Feruloylation of the Growing Cell Wall and Increase the Biodegradability of Tall Fescue (Festuca Arundinacea)." *Plant Biotechnology Journal*. https://doi.org/10.1111/j.1467-7652.2009.00485.x.

Bunzel, Mirko, John Ralph, Carola Funk, and Hans Steinhart. 2003. "Isolation and Identification of a Ferulic Acid Dehydrotrimer from Saponified Maize Bran Insoluble Fiber." *European Food Research and Technology* 217 (2): 128–33. https://doi.org/10.1007/s00217-003-0709-0.

Chanoca, Alexandra, Lisanne de Vries, and Wout Boerjan. 2019. "Lignin Engineering in Forest Trees." *Frontiers in Plant Science*. https://doi.org/10.3389/fpls.2019.00912.

Chen, Fang, Yuki Tobimatsu, Daphna Havkin-Frenkel, Richard A. Dixon, and John Ralph. 2012. "A Polymer of Caffeyl Alcohol in Plant Seeds." *Proceedings of the National Academy of Sciences of the United States of America* 109 (5): 1772–77. https://doi.org/10.1073/pnas.1120992109.

Chong, Sun Li, Marta Derba-Maceluch, Sanna Koutaniemi, Leonardo D. Gómez, Simon J. McQueen-Mason, Maija Tenkanen, and Ewa J. Mellerowicz. 2015. "Active Fungal GH115 Aα-Glucuronidase Produced in Arabidopsis Thaliana Affects Only the UX1-Reactive Glucuronate Decorations on Native Glucuronoxylans." *BMC Biotechnology*. https://doi.org/10.1186/s12896-015-0154-8.

Coleman, Heather D., Ji Young Park, Ramesh Nair, Clint Chapple, and Shawn D. Mansfield. 2008. "RNAi-Mediated Suppression of p-Coumaroyl-CoA 3'-Hydroxylase in Hybrid Poplar Impacts Lignin Deposition and Soluble Secondary Metabolism." *Proceedings of the National Academy of Sciences of the United States of America* 105 (11): 4501–06. https://doi.org/10.1073/pnas.0706537105.

Coleman, Heather D., Jimmy Yan, and Shawn D. Mansfield. 2009. "Sucrose Synthase Affects Carbon Partitioning to Increase Cellulose Production and Altered Cell Wall Ultrastructure." *Proceedings of the National Academy of Sciences of the United States of America* 106 (31): 13118–23. https://doi.org/10.1073/pnas.0900188106.

Deniaud, Estelle, Bernard Quemener, Joël Fleurence, and Marc Lahaye. 2003. "Structural Studies of the Mix-Linked β-(1 → 3)/β-(1 → 4)-D-Xylans from the Cell Wall of Palmaria Palmata (Rhodophyta)." *International Journal of Biological Macromolecules* 33 (1–3): 9–18. https://doi.org/10.1016/S0141–8130(03)00058-8.

Donev, Evgeniy, Madhavi Latha Gandla, Leif J. Jönsson, and Ewa J. Mellerowicz. 2018. "Engineering Non-Cellulosic Polysaccharides of Wood for the Biorefinery." *Frontiers in Plant Science* 9 (October). https://doi.org/10.3389/fpls.2018.01537.

Edwards, Meredith C., and Joy Doran-Peterson. 2012. "Pectin-Rich Biomass as Feedstock for Fuel Ethanol Production." *Applied Microbiology and Biotechnology* 95 (3): 565–75. https://doi.org/10.1007/s00253-012-4173-2.

Eudes, Aymerick, Anthe George, Purba Mukerjee, Jin S. Kim, Brigitte Pollet, Peter I. Benke, Fan Yang, et al. 2012. "Biosynthesis and Incorporation of Side-Chain-Truncated Lignin Monomers to Reduce Lignin Polymerization and Enhance Saccharification." *Plant Biotechnology Journal* 10 (5): 609–20. https://doi.org/10.1111/j.1467-7652.2012.00692.x.

Fan, Chunfen, Shengqiu Feng, Jiangfeng Huang, Yanting Wang, Leiming Wu, Xukai Li, Lingqiang Wang, et al. 2017. "AtCesA8-Driven OsSUS3 Expression Leads to Largely Enhanced Biomass Saccharification and Lodging Resistance by Distinctively Altering Lignocellulose Features in Rice." *Biotechnology for Biofuels* 10 (1): 1–12. https://doi.org/10.1186/s13068-017-0911-0.

Fernandes, Anwesha N., Lynne H. Thomas, Clemens M. Altaner, Philip Callow, V. Trevor Forsyth, David C. Apperley, Craig J. Kennedy, and Michael C. Jarvis. 2011. "Nanostructure of Cellulose Microfibrils in Spruce Wood." *Proceedings of the National Academy of Sciences of the United States of America*. https://doi.org/10.1073/pnas.1108942108.

Gille, Sascha, and Markus Pauly. 2012. "O-Acetylation of Plant Cell Wall Polysaccharides." *Frontiers in Plant Science* 3 (JAN): 1–7. https://doi.org/10.3389/fpls.2012.00012.

Giummarella, Nicola, and Martin Lawoko. 2016. "Structural Basis for the Formation and Regulation of Lignin-Xylan Bonds in Birch." *ACS Sustainable Chemistry and Engineering* 4 (10): 5319–26. https://doi.org/10.1021/acssuschemeng.6b00911.

Gou, Jin Ying, Lisa M. Miller, Guichuan Hou, Xiao Hong Yu, Xiao Ya Chen, and Chang Jun Liu. 2012. "Acetylesterase-Mediated Deacetylation of Pectin Impairs Cell Elongation, Pollen Germination, and Plant Reproduction." *Plant Cell* 24 (1): 50–65. https://doi.org/10.1105/tpc.111.092411.

Goubet, Florence, Christopher J. Barton, Jennifer C. Mortimer, Xiaolan Yu, Zhinong Zhang, Godfrey P. Miles, Jenny Richens, Aaron H. Liepman, Keith Seffen, and Paul Dupree. 2009. "Cell Wall Glucomannan in Arabidopsis Is Synthesised by CSLA Glycosyltransferases, and Influences the Progression of Embryogenesis." *Plant Journal* 60 (3): 527–38. https://doi.org/10.1111/j.1365-313X.2009.03977.x.

Grabber, John H., John Ralph, and Ronald D. Hatfield. 2000. "Cross-Linking of Maize Walls by Ferulate Dimerization and Incorporation into Lignin." *Journal of Agricultural and Food Chemistry* 48 (12): 6106–13. https://doi.org/10.1021/jf0006978.

Gui, Jinshan, Pui Ying Lam, Yuki Tobimatsu, Jiayan Sun, Cheng Huang, Shumin Cao, Yu Zhong, Toshiaki Umezawa, and Laigeng Li. 2020. "Fibre-Specific Regulation of Lignin Biosynthesis Improves Biomass Quality in Populus." *New Phytologist* 226 (4): 1074–87. https://doi.org/10.1111/nph.16411.

Harholt, Jesper, Inga C. Bach, Solveig Lind-Bouquin, Kylie J. Nunan, Susan M. Madrid, Henrik Brinch-Pedersen, Preben B. Holm, and Henrik V. Scheller. 2010. "Generation of Transgenic Wheat (Triticum Aestivum L.) Accumulating Heterologous Endo-Xylanase or Ferulic Acid Esterase in the Endosperm." *Plant Biotechnology Journal.* https://doi.org/10.1111/j.1467-7652.2009.00490.x.

Harris, Darby, Jozsef Stork, and Seth Debolt. 2009. "Genetic Modification in Cellulose-Synthase Reduces Crystallinity and Improves Biochemical Conversion to Fermentable Sugar." *GCB Bioenergy.* https://doi.org/10.1111/j.1757-1707.2009.01000.x.

Herbers, K., I. Wilke, and U. Sonnewald. 1995. "A Thermostable Xylanase from Clostridium Thermocellum Expressed at High Levels in the Apoplast of Transgenic Tobacco Has No Detrimental Effects and Is Easily Purified." *Bio/Technology.* https://doi.org/10.1038/nbt0195-63.

Jacobs, A., P. T. Larsson, and O. Dahlman. 2001. "Distribution of Uronic Acids in Xylans from Various Species of Soft- and Hardwood as Determined by MALDI Mass Spectrometry." *Biomacromolecules* 2 (3): 979–90. https://doi.org/10.1021/bm010062x.

Jarvis, Michael C. 2018. "Structure of Native Cellulose Microfibrils, the Starting Point for Nanocellulose Manufacture." *Philosophical Transactions of the Royal Society A: Mathematical, Physical and Engineering Sciences.* https://doi.org/10.1098/rsta.2017.0045.

Jenkins, J., O. Mayans, D. Smith, K. Worboys, and R. W. Pickersgill. 2001. "Three-Dimensional Structure of Erwinia Chrysanthemi Pectin Methylesterase Reveals a Novel Esterase Active Site." *Journal of Molecular Biology* 305 (4): 951–60. https://doi.org/10.1006/jmbi.2000.4324.

Jiang, Nan, Richard E. Wiemels, Aaron Soya, Rebekah Whitley, Michael Held, and Ahmed Faik. 2016. "Composition, Assembly, and Trafficking of a Wheat Xylan Synthase Complex." *Plant Physiology* 170 (4): 1999–2023. https://doi.org/10.1104/pp.15.01777.

Joshi, Chandrashekhar P., Shivegowda Thammannagowda, Takeshi Fujino, Ji Qing Gou, Utku Avci, Candace H. Haigler, Lisa M. McDonnell, et al. 2011. "Perturbation of Wood Cellulose Synthesis Causes Pleiotropic Effects in Transgenic Aspen." *Molecular Plant* 4 (2): 331–45. https://doi.org/10.1093/mp/ssq081.

Kameshwar, Ayyappa Kumar Sista, and Wensheng Qin. 2018. "Structural and Functional Properties of Pectin and Lignin–Carbohydrate Complexes de-Esterases: A Review." *Bioresources and Bioprocessing* 5 (1): 1–16. https://doi.org/10.1186/s40643-018-0230-8.

Kimura, T., T. Mizutani, T. Tanaka, T. Koyama, K. Sakka, and K. Ohmiya. 2003. "Molecular Breeding of Transgenic Rice Expressing a Xylanase Domain of the XynA Gene from Clostridium Thermocellum." *Applied Microbiology and Biotechnology.* https://doi.org/10.1007/s00253-003-1301-z.

Klein-Marcuschamer, Daniel, Piotr Oleskowicz-Popiel, Blake A. Simmons, and Harvey W. Blanch. 2010. "Technoeconomic Analysis of Biofuels: A Wiki-Based Platform for Lignocellulosic Biorefineries." *Biomass and Bioenergy* 34 (12): 1914–21. https://doi.org/10.1016/j.biombioe.2010.07.033.

Klose, Holger, Markus Günl, Bjorn Usadel, Rainer Fischer, and Ulrich Commandeur. 2015. "Cell Wall Modification in Tobacco by Differential Targeting of Recombinant Endoglucanase from Trichoderma Reesei." *BMC Plant Biology.* https://doi. org/10.1186/s12870-015-0443-3.

Lan, Wu, Fachuang Lu, Matthew Regner, Yimin Zhu, Jorge Rencoret, Sally A. Ralph, Uzma I. Zakai, Kris Morreel, Wout Boerjan, and John Ralph. 2015. "Tricin, a Flavonoid Monomer in Monocot Lignification." *Plant Physiology* 167 (4): 1284–95. https://doi. org/10.1104/pp.114.253757.

Lee, Chanhui, Quincy Teng, Ruiqin Zhong, and Zheng Hua Ye. 2012. "Arabidopsis GUX Proteins Are Glucuronyltransferases Responsible for the Addition of Glucuronic Acid Side Chains onto Xylan." *Plant and Cell Physiology* 53 (7): 1204–16. https://doi. org/10.1093/pcp/pcs064.

Lee, Chanhui, Quincy Teng, Ruiqin Zhong, and Zheng Hua Ye. 2011. "Molecular Dissection of Xylan Biosynthesis during Wood Formation in Poplar." *Molecular Plant.* https://doi. org/10.1093/mp/ssr035.

Lee, Yuree, Maria C. Rubio, Julien Alassimone, and Niko Geldner. 2013. "A Mechanism for Localised Lignin Deposition in the Endodermis." *Cell* 153 (2): 402–12. https://doi.org/ 10.1016/j.cell.2013.02.045.

Li, Fengcheng, Sitong Liu, Hai Xu, and Quan Xu. 2018. "A Novel FC17/CESA4 Mutation Causes Increased Biomass Saccharification and Lodging Resistance by Remodeling Cell Wall in Rice." *Biotechnology for Biofuels* 11 (1): 1–13. https://doi. org/10.1186/s13068-018-1298-2.

Li, Long, Junfeng Huang, Lixia Qin, Yuying Huang, Wei Zeng, Yue Rao, Juan Li, Xuebao Li, and Wenliang Xu. 2014. "Two Cotton Fiber-Associated Glycosyltransferases, GhGT43A1 and GhGT43C1, Function in Hemicellulose Glucuronoxylan Biosynthesis during Plant Development." *Physiologia Plantarum.* https://doi.org/10.1111/ppl.12190.

Li, Quanzi, Jian Song, Shaobing Peng, Jack P. Wang, Guan Zheng Qu, Ronald R. Sederoff, and Vincent L. Chiang. 2014. "Plant Biotechnology for Lignocellulosic Biofuel Production." *Plant Biotechnology Journal* 12 (9): 1174–92. https://doi.org/10.1111/pbi.12273.

Li, Shundai, Logan Bashline, Yunzhen Zheng, Xiaoran Xin, Shixin Huang, Zhaosheng Kong, Seong H. Kim, Daniel J. Cosgrove, and Ying Gu. 2016. "Cellulose Synthase Complexes Act in a Concerted Fashion to Synthesise Highly Aggregated Cellulose in Secondary Cell Walls of Plants." *Proceedings of the National Academy of Sciences of the United States of America* 113 (40): 11348–53. https://doi.org/10.1073/pnas.1613273113.

Liang, Haiying, Christopher J. Frost, Xiaoping Wei, Nicole R. Brown, John E. Carlson, and Ming Tien. 2008. "Improved Sugar Release from Lignocellulosic Material by Introducing a Tyrosine-Rich Cell Wall Peptide Gene in Poplar." *Clean - Soil, Air, Water* 36 (8): 662–68. https://doi.org/10.1002/clen.200800079.

Lyczakowski, Jan J., Krzysztof B. Wicher, Oliver M. Terrett, Nuno Faria-Blanc, Xiaolan Yu, David Brown, Kristian B. R. M. Krogh, Paul Dupree, and Marta Busse-Wicher. 2017. "Removal of Glucuronic Acid from Xylan Is a Strategy to Improve the Conversion of Plant Biomass to Sugars for Bioenergy." *Biotechnology for Biofuels* 10 (1): 1–11. https:// doi.org/10.1186/s13068-017-0902-1.

Mahon, Elizabeth L., and Shawn D. Mansfield. 2019. "Tailor-Made Trees: Engineering Lignin for Ease of Processing and Tomorrow's Bioeconomy." *Current Opinion in Biotechnology* 56: 147–55. https://doi.org/10.1016/j.copbio.2018.10.014.

Maleki, Samaneh Sadat, Kourosh Mohammadi, Ali Movahedi, Fan Wu, and Kong Shu Ji. 2020. "Increase in Cell Wall Thickening and Biomass Production by Overexpression of PmCesA2 in Poplar." *Frontiers in Plant Science* 11 (February): 1–11. https://doi. org/10.3389/fpls.2020.00110.

Manabe, Yuzuki, Majse Nafisi, Yves Verhertbruggen, Caroline Orfila, Sascha Gille, Carsten Rautengarten, Candice Cherk, et al. 2011. "Loss-of-Function Mutation of REDUCED

WALL ACETYLATION2 in Arabidopsis Leads to Reduced Cell Wall Acetylation and Increased Resistance to Botrytis Cinerea." *Plant Physiology* 155 (3): 1068–78. https://doi.org/10.1104/pp.110.168989.

Mansfield, Shawn D., Kyu-young Kang, and Clint Chapple. 2012. "Designed for Deconstruction – Poplar Trees Altered in Cell Wall Lignification Improve the Efficacy of Bioethanol Production." *New Phytologist* 91–101.

Mazarei, Mitra, Holly L. Baxter, Mi Li, Ajaya K. Biswal, Keonhee Kim, Xianzhi Meng, Yunqiao Pu, et al. 2018. "Functional Analysis of Cellulose Synthase CesA4 and CesA6 Genes in Switchgrass (Panicum Virgatum) by Overexpression and RNAi-Mediated Gene Silencing." *Frontiers in Plant Science* 9. https://doi.org/10.3389/fpls.2018.01114.

Mazumder, Koushik, Maria J. Peña, Malcolm A. O'Neill, and William S. York. 2012. "Structural Characterisation of the Heteroxylans from Poplar and Switchgrass." *Methods in Molecular Biology* 908: 215–28. https://doi.org/10.1007/978-1-61779-956-3_19.

McFarlane, Heather E., Anett Döring, and Staffan Persson. 2014. "The Cell Biology of Cellulose Synthesis." *Annual Review of Plant Biology*. https://doi.org/10.1146/annurev-arplant-050213-040240.

Meester, Barbara De, Lisanne De Vries, Merve Özparpucu, Notburga Gierlinger, Sander Corneillie, Andreas Pallidis, Geert Goeminne, et al. 2018. "Vessel-Specific Reintroduction of CINNAMOYL-COA REDUCTASE1 (CCR1) in Dwarfed ccr1 Mutants Restores Vessel and Xylary Fiber Integrity and Increases Biomass." *Plant Physiology* 176 (1): 611–33. https://doi.org/10.1104/pp.17.01462.

Mei, Chuansheng, Sang Hyuck Park, Robab Sabzikar, Chunfang Qi, Callista Ransom, and Mariam Sticklen. 2009. "Green Tissue-Specific Production of a Microbial Endo-Cellulase in Maize (Zea Mays L.) Endoplasmic-Reticulum and Mitochondria Converts Cellulose into Fermentable Sugars." *Journal of Chemical Technology and Biotechnology*. https://doi.org/10.1002/jctb.2100.

Meyer, Knut, Amber M. Shirley, Joanne C. Cusumano, Dolly A. Bell-Lelong, and Clint Chapple. 1998. "Lignin Monomer Composition Is Determined by the Expression of a Cytochrome P450-Dependent Monooxygenase in Arabidopsis." *Proceedings of the National Academy of Sciences of the United States of America* 95 (12): 6619–23. https://doi.org/10.1073/pnas.95.12.6619.

Mir, Bilal Ahmad, Alexander A. Myburg, Eshchar Mizrachi, and Don A. Cowan. 2017. "In Planta Expression of Hyperthermophilic Enzymes as a Strategy for Accelerated Lignocellulosic Digestion." *Scientific Reports*. https://doi.org/10.1038/s41598-017-11026-1.

Mohnen, Debra. 2008. "Pectin Structure and Biosynthesis." *Current Opinion in Plant Biology* 11 (3): 266–77. https://doi.org/10.1016/j.pbi.2008.03.006.

Moreira, L. R. S., and E. X. F. Filho. 2008. "An Overview of Mannan Structure and Mannan-Degrading Enzyme Systems." *Applied Microbiology and Biotechnology* 79 (2): 165–78. https://doi.org/10.1007/s00253-008-1423-4.

Morris, Phillip, Sue Dalton, Tim Langdon, Barbara Hauck, and Marcia M. O. de Buanafina. 2017. "Expression of a Fungal Ferulic Acid Esterase in Suspension Cultures of Tall Fescue (Festuca Arundinacea) Decreases Cell Wall Feruloylation and Increases Rates of Cell Wall Digestion." *Plant Cell, Tissue and Organ Culture*. https://doi.org/10.1007/s11240-017-1168-9.

Park, Jong In, Takeshi Ishimizu, Keita Suwabe, Keisuke Sudo, Hiromi Masuko, Hirokazu Hakozaki, Ill Sup Nou, Go Suzuki, and Masao Watanabe. 2010. "UDP-Glucose Pyrophosphorylase Is Rate Limiting in Vegetative and Reproductive Phases in Arabidopsis Thaliana." *Plant and Cell Physiology*. https://doi.org/10.1093/pcp/pcq057.

Patel, Minesh, Jennifer S. Johnson, Richard I. S. Brettell, Jake Jacobsen, and Gang Ping Xue. 2000. "Transgenic Barley Expressing a Fungal Xylanase Gene in the Endosperm of the Developing Grains." *Molecular Breeding*. https://doi.org/10.1023/A:1009640427515.

Pauly, Markus, and Kenneth Keegstra. 2010. "Plant Cell Wall Polymers as Precursors for Biofuels." *Current Opinion in Plant Biology* 13 (3): 304–11. https://doi.org/10.1016/j.pbi.2009.12.009.

Pawar, Prashant Mohan Anupama, Marta Derba-Maceluch, Sun Li Chong, Madhavi Latha Gandla, Shamrat Shafiul Bashar, Tobias Sparrman, Patrik Ahvenainen, et al. 2017. "In Muro Deacetylation of Xylan Affects Lignin Properties and Improves Saccharification of Aspen Wood." *Biotechnology for Biofuels*. https://doi.org/10.1186/s13068-017-0782-4.

Pawar, Prashant Mohan Anupama, Marta Derba-Maceluch, Sun Li Chong, Leonardo D. Gómez, Eva Miedes, Alicja Banasiak, Christine Ratke, et al. 2016. "Expression of Fungal Acetyl Xylan Esterase in Arabidopsis thaliana Improves Saccharification of Stem Lignocellulose." *Plant Biotechnology Journal* 14 (1): 387–97. https://doi.org/10.1111/pbi.12393.

Pawar, Prashant Mohan Anupama, Sanna Koutaniemi, Maija Tenkanen, and Ewa J. Mellerowicz. 2013. "Acetylation of Woody Lignocellulose: Significance and Regulation." *Frontiers in Plant Science* 4 (MAY): 1–8. https://doi.org/10.3389/fpls.2013.00118.

Peña, Maria J., Ruiqin Zhong, Gong Ke Zhou, Elizabeth A. Richardson, Malcolm A. O'Neill, Alan G. Darvill, William S. York, and Zheng Hua Yeb. 2007. "Arabidopsis Irregular Xylem8 and Irregular Xylem9: Implications for the Complexity of Glucuronoxylan Biosynthesis." *Plant Cell* 19 (2): 549–63. https://doi.org/10.1105/tpc.106.049320.

Pogorelko, Gennady, Vincenzo Lionetti, Oksana Fursova, Raman M. Sundaram, Mingsheng Qi, Steven A. Whitham, Adam J. Bogdanove, Daniela Bellincampi, and Olga A. Zabotina. 2013. "Arabidopsis and Brachypodium Distachyon Transgenic Plants Expressing Aspergillus Nidulans Acetylesterases Have Decreased Degree of Polysaccharide Acetylation and Increased Resistance to Pathogens." *Plant Physiology*. https://doi.org/10.1104/pp.113.214460.

Poovaiah, Charleson R., Mitra Mazarei, Stephen R. Decker, Geoffrey B. Turner, Robert W. Sykes, Mark F. Davis, and C. Neal Stewart. 2015. "Transgenic Switchgrass (Panicum Virgatum L.) Biomass Is Increased by Overexpression of Switchgrass Sucrose Synthase (PvSUS1)." *Biotechnology Journal* 10 (4): 552–63. https://doi.org/10.1002/biot.201400499.

Qin, Shifei, Chunfen Fan, Xiaohong Li, Yi Li, Jian Hu, Chaofeng Li, and Keming Luo. 2020. "LACCASE14 Is Required for the Deposition of Guaiacyl Lignin and Affects Cell Wall Digestibility in Poplar." *Biotechnology for Biofuels* 13 (1): 1–14. https://doi.org/10.1186/s13068-020-01843-4.

Ralph, John, Catherine Lapierre, and Wout Boerjan. 2019. "Lignin Structure and Its Engineering." *Current Opinion in Biotechnology*. https://doi.org/10.1016/j.copbio.2019.02.019.

Ratke, Christine, Barbara K. Terebieniec, Sandra Winestrand, Marta Derba-Maceluch, Thomas Grahn, Bastian Schiffthaler, Thomas Ulvcrona, et al. 2018. "Downregulating Aspen Xylan Biosynthetic GT43 Genes in Developing Wood Stimulates Growth via Reprograming of the Transcriptome." *New Phytologist* 219 (1): 230–45. https://doi.org/10.1111/nph.15160.

Ratnayake, Sunil, Cherie T. Beahan, Damien L. Callahan, and Antony Bacic. 2014. "The Reducing End Sequence of Wheat Endosperm Cell Wall Arabinoxylans." *Carbohydrate Research* 386 (1): 23–32. https://doi.org/10.1016/j.carres.2013.12.013.

Rennie, Emilie A., and Henrik Vibe Scheller. 2014. "Xylan Biosynthesis." *Current Opinion in Biotechnology* 26: 100–07. https://doi.org/10.1016/j.copbio.2013.11.013.

Ro, Dae Kyun, Nancy Mah, Brian E Ellis, and Carl J. Douglas. 2001. "Functional Characterisation and Subcellular Localisation of Poplar (Populus Trichocarpa ≠ Populus Deltoides)." *Plant Physiology* 126 (May): 317–29.

Ruan, Yong Ling. 2014. "Sucrose Metabolism: Gateway to Diverse Carbon Use and Sugar Signaling." *Annual Review of Plant Biology*. https://doi.org/10.1146/annurev-arplant-050213-040251.

Ruegger, Max, Knut Meyer, Joanne C. Cusumano, and Clint Chapple. 1999. "Regulation of Ferulate-5-Hydroxylase Expression in Arabidopsis in the Context of Sinapate Ester Biosynthesis." *Plant Physiology* 119 (1): 101–10. https://doi.org/10.1104/pp.119.1.101.

Saez-Aguayo, Susana, Carsten Rautengarten, Henry Temple, Dayan Sanhueza, Troy Ejsmentewicz, Omar Sandoval-Ibañez, Daniela Doñas, et al. 2017. "UUAT1 Is a Golgi-Localized UDP-Uronic Acid Transporter That Modulates the Polysaccharide Composition of Arabidopsis Seed Mucilage." *Plant Cell* 29 (1): 129–43. https://doi.org/10.1105/tpc.16.00465.

Sahoo, Dipak K., Jozsef Stork, Seth Debolt, and Indu B. Maiti. 2013. "Manipulating Cellulose Biosynthesis by Expression of Mutant ArabidopsisproM24:: CESA3ixr1–2 Gene in Transgenic Tobacco." *Plant Biotechnology Journal.* https://doi.org/10.1111/pbi.12024.

Sakamoto, Shingo, Marc Somssich, Miyuki T. Nakata, Faride Unda, Kimie Atsuzawa, Yasuko Kaneko, Ting Wang, et al. 2018. "Complete Substitution of a Secondary Cell Wall with a Primary Cell Wall in Arabidopsis." *Nature Plants* 4 (10): 777–83. https://doi.org/10.1038/s41477-018-0260-4.

Scheller, Henrik Vibe, and Peter Ulvskov. 2010. "Hemicelluloses." *Annual Review of Plant Biology* 61: 263–89. https://doi.org/10.1146/annurev-arplant-042809-112315.

Schröder, Roswitha, Ross G. Atkinson, and Robert J. Redgwell. 2009. "Re-Interpreting the Role of Endo-β-Mannanases as Mannan Endotransglycosylase/Hydrolases in the Plant Cell Wall." *Annals of Botany.* https://doi.org/10.1093/aob/mcp120.

Sjöholm, E., K. Gustafsson, F. Berthold, and A. Colmsjö. 2000. "Influence of the Carbohydrate Composition on the Molecular Weight Distribution of Kraft Pulps." *Carbohydrate Polymers* 41 (1): 1–7. https://doi.org/10.1016/S0144-8617(99)00066-1.

Smith, Peter J., Hsin Tzu Wang, William S. York, Maria J. Peña, and Breeanna R. Urbanowicz. 2017. "Designer Biomass for Next-Generation Biorefineries: Leveraging Recent Insights into Xylan Structure and Biosynthesis." *Biotechnology for Biofuels* 10 (1): 1–14. https://doi.org/10.1186/s13068-017-0973-z.

Smith, Rebecca A., Eliana Gonzales-Vigil, Steven D. Karlen, Ji Young Park, Fachuang Lu, Curtis G. Wilkerson, Lacey Samuels, John Ralph, and Shawn D. Mansfield. 2015. "Engineering Monolignol P-Coumarate Conjugates into Poplar and Arabidopsis Lignins." *Plant Physiology* 169 (4): 2992–3001. https://doi.org/10.1104/pp.15.00815.

Song, Lili, Wei Zeng, Aimin Wu, Kelsey Picard, Edwin R. Lampugnani, Roshan Cheetamun, Cherie Beahan, et al. 2015. "Asparagus Spears as a Model to Study Heteroxylan Biosynthesis during Secondary Wall Development." *PLoS ONE* 10 (4): 1–22. https://doi.org/10.1371/journal.pone.0123878.

Studer, Michael H., Jaclyn D. DeMartini, Mark F. Davis, Robert W. Sykes, Brian Davison, Martin Keller, Gerald A. Tuskan, and Charles E. Wyman. 2011. "Lignin Content in Natural Populus Variants Affects Sugar Release." *Proceedings of the National Academy of Sciences of the United States of America* 108 (15): 6300–05. https://doi.org/10.1073/pnas.1009252108.

Tan, Hwei Ting, Neil J. Shirley, Rohan R. Singh, Marilyn Henderson, Kanwarpal S. Dhugga, Gwenda M. Mayo, Geoffrey B. Fincher, and Rachel A. Burton. 2015. "Powerful Regulatory Systems and Post-Transcriptional Gene Silencing Resist Increases in Cellulose Content in Cell Walls of Barley." *BMC Plant Biology.* https://doi.org/10.1186/s12870-015-0448-y.

Taylor, Neil G. 2008. "Cellulose Biosynthesis and Deposition in Higher Plants." *New Phytologist.* https://doi.org/10.1111/j.1469-8137.2008.02385.x.

Tomassetti, Susanna, Daniela Pontiggia, Ilaria Verrascina, Ida Barbara Reca, Fedra Francocci, Gianni Salvi, Felice Cervone, and Simone Ferrari. 2015. "Controlled Expression of Pectic Enzymes in Arabidopsis Thaliana Enhances Biomass Conversion without Adverse Effects on Growth." *Phytochemistry* 112 (1): 221–30. https://doi.org/10.1016/j.phytochem.2014.08.026.

Urbanowicz, Breeanna R., Maria J. Peña, Supriya Ratnaparkhe, Utku Avci, Jason Backe, Heather F. Steet, Marcus Foston, et al. 2012. "4-O-Methylation of Glucuronic Acid in Arabidopsis Glucuronoxylan Is Catalysed by a Domain of Unknown Function Family 579 Protein." *Proceedings of the National Academy of Sciences of the United States of America* 109 (35): 14253–58. https://doi.org/10.1073/pnas.1208097109.

Vanholme, Ruben, Igor Cesarino, Katarzyna Rataj, Yuguo Xiao, Lisa Sundin, Geert Goeminne, Hoon Kim, et al. 2013. "Caffeoyl Shikimate Esterase (CSE) Is an Enzyme in the Lignin Biosynthetic Pathway in Arabidopsis." *Science* 341 (6150): 1103–06. https://doi.org/10.1126/science.1241602.

Vargas, Lívia, Igor Cesarino, Ruben Vanholme, Wannes Voorend, Marina De Lyra Soriano Saleme, Kris Morreel, and Wout Boerjan. 2016. "Improving Total Saccharification Yield of Arabidopsis Plants by Vessel-Specific Complementation of Caffeoyl Shikimate Esterase (Cse) Mutants." *Biotechnology for Biofuels* 9 (1): 1–16. https://doi.org/10.1186/s13068-016-0551-9.

Voelker, Steven L., Barbara Lachenbruch, Frederick C. Meinzer, and Steven H. Strauss. 2011. "Reduced Wood Stiffness and Strength, and Altered Stem Form, in Young Antisense 4CL Transgenic Poplars with Reduced Lignin Contents." *New Phytologist* 1096–1109.

Wei, Zhigang, Zanshuang Qu, Lijie Zhang, Shuanjing Zhao, Zhihong Bi, Xiaohui Ji, Xiaowen Wang, and Hairong Wei. 2015. "Overexpression of Poplar Xylem Sucrose Synthase in Tobacco Leads to a Thickened Cell Wall and Increased Height." *PLoS ONE* 10 (3): 1–20. https://doi.org/10.1371/journal.pone.0120669.

Wilkerson, C. G., S. D. Mansfield, F. Lu, S. Withers, J. Y. Park, S. D. Karlen, E. Gonzales-Vigil, et al. 2014. "Monolignol Ferulate Transferase Introduces Chemically Labile Linkages into the Lignin Backbone." *Science*. https://doi.org/10.1126/science.1250161.

Xiao, Chaowen, and Charles T. Anderson. 2013. "Roles of Pectin in Biomass Yield and Processing for Biofuels." *Frontiers in Plant Science* 4 (MAR): 1–7. https://doi.org/10.3389/fpls.2013.00067.

Xiao, Yao, Xuejun He, Yemaiza Ojeda-Lassalle, Charleson Poovaiah, and Heather D. Coleman. 2018. "Expression of a Hyperthermophilic Endoglucanase in Hybrid Poplar Modifies the Plant Cell Wall and Enhances Digestibility." *Biotechnology for Biofuels*. https://doi.org/10.1186/s13068-018-1224-7.

Xiong, Guangyan, Murali Dama, and Markus Pauly. 2015. "Glucuronic Acid Moieties on Xylan Are Functionally Equivalent to O-Acetyl-Substituents." *Molecular Plant*. https://doi.org/10.1016/j.molp.2015.02.013.

Yang, Fan, Prajakta Mitra, Ling Zhang, Lina Prak, Yves Verhertbruggen, Jin Sun Kim, Lan Sun, et al. 2013. "Engineering Secondary Cell Wall Deposition in Plants." *Plant Biotechnology Journal* 11 (3): 325–35. https://doi.org/10.1111/pbi.12016.

York, William S., and Malcolm A. O'Neill. 2008. "Biochemical Control of Xylan Biosynthesis - Which End Is Up?" *Current Opinion in Plant Biology* 11 (3): 258–65. https://doi.org/10.1016/j.pbi.2008.02.007.

Yu, Liangliang, Hongpeng Chen, Jiayan Sun, and Laigeng Li. 2014. "PtrKOR1 Is Required for Secondary Cell Wall Cellulose Biosynthesis in Populus." *Tree Physiology* 34 (11): 1289–1300. https://doi.org/10.1093/treephys/tpu020.

Zhang, Baocai, Lanjun Zhang, Feng Li, Dongmei Zhang, Xiangling Liu, Hang Wang, Zuopeng Xu, Chengcai Chu, and Yihua Zhou. 2017. "Control of Secondary Cell Wall Patterning Involves Xylan Deacetylation by a GDSL Esterase." *Nature Plants* 3 (March): 1–9. https://doi.org/10.1038/nplants.2017.17.

Zhang, Lanjun, Chengxu Gao, Frederic Mentink-Vigier, Lu Tang, Dongmei Zhang, Shaogan Wang, Shaoxue Cao, et al. 2019. "Arabinosyl Deacetylase Modulates the Arabinoxylan Acetylation Profile and Secondary Wall Formation." *Plant Cell*. https://doi.org/10.1105/tpc.18.00894.

Zhao, Qiao, Jin Nakashima, Fang Chen, Yanbin Yin, Chunxiang Fu, Jianfei Yun, Hui Shao, Xiaoqiang Wang, Zeng Yu Wang, and Richard A. Dixon. 2013. "LACCASE Is Necessary and Nonredundant with PEROXIDASE for Lignin Polymerization during Vascular Development in Arabidopsis." *Plant Cell* 25 (10): 3976–87. https://doi.org/10.1105/tpc.113.117770.

Zhong, Ruiqin, Dongtao Cui, and Zheng Hua Ye. 2019. "Secondary Cell Wall Biosynthesis." *New Phytologist* 221 (4): 1703–23. https://doi.org/10.1111/nph.15537.

Zhong, Ruiqin, Dongtao Cui, and Zheng Hua Ye. 2018. "Members of the DUF231 Family Are O-Acetyltransferases Catalyzing 2-O- and 3-O-Acetylation of Mannan." *Plant & Cell Physiology* 59 (11): 2339–49. https://doi.org/10.1093/pcp/pcy159.

8 Lignocellulosic Biorefineries – A Step towards a Carbon-Neutral Economy

Bhawna Madan, Prachi Varshney, Parmeshwar Patil, Vivek Rathore, Jaya Rawat, and Bharat Newalkar
Bharat Petroleum Corporation Ltd.

CONTENTS

8.1 Introduction 212
8.2 Lignocellulosic Biorefinery 214
    8.2.1 Fractionation Technologies and Their Importance............................ 214
    8.2.2 Evaluation of Sustainable Technology for LBM Feedstock
        Biorefinery 215
        8.2.2.1 Technology Efficiency 215
        8.2.2.2 Technology Flexibility.. 216
        8.2.2.3 Technology Maturity... 216
        8.2.2.4 Technology Profitability.. 217
        8.2.2.5 Robustness 217
    8.2.3 Expected Challenges in the Commercialization of LBM................... 217
8.3 Commodity Chemicals 218
    8.3.1 Opportunities, Uses, Market, Technology, and Challenges............. 218
    8.3.2 Lactic Acid.. 219
    8.3.3 Succinic Acid.. 220
    8.3.4 Diol Compounds 221
    8.3.5 Levulinic Acid 222
    8.3.6 2, 5-Furandicarboxylic Acid 223
    8.3.7 Isoprene... 224
    8.3.8 Lignin-Derived Chemicals 225
8.4 Valorization of LBM into Gaseous and Liquid Fuels..................................... 226
    8.4.1 Biogas 226
    8.4.2 Synthesis Gas.. 227
    8.4.3 Drop-in Fuels 229
8.5 Conclusions... 231
References... 231

DOI: 10.1201/9781003158486-8

8.1 INTRODUCTION

Globally, majority of energy carriers and petrochemicals are produced through crude oil refining; however, the uncontrollable use of fossil fuels is envisaged to enhance increased emission of greenhouse gases (GHGs) (Bessou et al., 2011; Raschka and Carus, 2012). To mitigate climate change, it is essential to attain net-zero emissions by 2050, which can be achieved by the transition towards the production of sustainable energy (Oshiro et al., 2018). Lignocellulosic biomass (LBM) is the most abundant renewable resource with a global annual production of 181.5 billion tonnes that can be utilized for the production of biofuels, platform chemicals, and value-added products (Paul and Dutta, 2018). LBM includes a range of different biomasses from forestry, agriculture, aquaculture (algae and seaweeds), and residues from industries and households including wood, organic residues (both plant- and animal-derived), etc., which can be utilized as feedstocks for sustainable biorefinery (Cheng and Wang, 2013). It is prudent to mention here that every year, a huge portion of agricultural residues is burned in developing countries, leading to excessive particulate matter emission and increasing air pollution (Bhuvaneshwari et al., 2019; Duque-Acevedo et al., 2020). More importantly, energy production through biomass resources has the advantage of a carbon-neutral cycle (Bessou et al., 2011; Kajaste, 2014).

LBM feedstock biorefinery offers a platform for sustainable processing of a variety of resources by various routes to generate a spectrum of biofuels and biochemical marketable products and energy as given in Figure 8.1 (Kamm et al., 2005; Takkellapati et al., 2018). It is analogous to today's petroleum refinery which produces multiple fuels/products by processing crude oil.

The biorefineries are classified based on several criteria: (a) type of feedstock: whole-crop biorefineries (WCBRs) using cereals as feedstock, oleochemical/-triglyceride biorefineries using oil from plants, animals, algae, etc., lignocellulosic

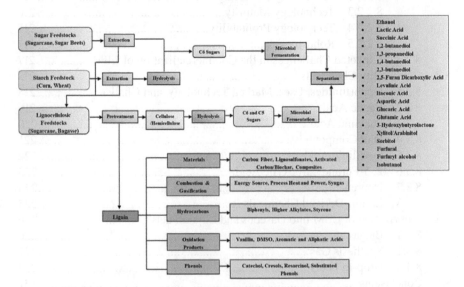

FIGURE 8.1   Schematic representation of LBM biorefinery.

feedstock biorefineries using lignocellulose feedstocks, green biorefineries using grasses and immature cereal, and marine biorefineries using marine biomass; (b) conversion processes applied: biochemical biorefineries using bioconversion/enzyme conversion route, which involves biomass fractionation into sugars and lignin components using a combination of chemical pretreatment and enzymatic hydrolysis, followed by microbial fermentation to convert sugars into biofuels and biochemicals, and thermochemical biorefineries using the thermochemical route, which involves biomass gasification followed by conversion of syngas to fuel and other products. These technologies are fairly developed and scaled up to commercial-scale production across the globe, especially in the USA, Europe, and Brazil (Choi et al., 2015). (c) Technology status: conventional and advanced biorefineries; first-, second-, and third-generation biorefineries; and (d) type of intermediates – syngas, sugar (Cherubini et al., 2009; Takkellapati et al., 2018).

Despite the huge potential of biorefinery, utilization of biomass for food or fuel has often become a point of debate. This perceived conflict between energy and food production is associated with the sugar-based conventional biorefineries that can be allayed by developing technologies based on lignocellulosic materials. However, due to the higher production cost of lignocellulosic ethanol, its prices cannot compete with petroleum products, which remains a crucial challenge. The major advantages and challenges of the lignocellulosic refinery are given in Table 8.1. Pretreatment of lignocellulosic biomass is an extremely expensive and energy-intensive step due to the recalcitrance of lignocellulosic biomass. This issue is being addressed by several companies globally by the development of cost-efficient technologies and

TABLE 8.1
Advantages and Challenges of LBM Biorefineries

Advantages	Challenges
• A sustainable supply of ethanol is ensured as the present source molasses cannot fulfil the legislative requirement of mandatory ethanol blending.	• Sustainable supply of lignocellulosic biomass feedstock.
• It leads to the reduction in crude oil imports.	• Volatile pricing of lignocellulosic biomass: As biomass accounts for ~30% of ethanol delivered cost, price variation will see a large impact.
• It provides additional income to farmers.	• High-CAPEX requirement.
• As agricultural waste will be directed towards lignocellulosic refinery instead of burning, it helps in combating pollution.	• The higher cost of cellulase enzymes will impact operating cost as these are tied to technology.
• It creates additional jobs in the rural sector and leads to the development of bio-based economy.	• Technologies are evolving at a faster pace; thus, there is a risk of technology obsolesce and investment.
• Bio-based chemicals have the potential to substitute petrochemicals.	
• It improves the octane number of gasoline and reduces GHG emissions.	

co-production of high-value–low-volume and low-value–high-volume products (Rosales-Calderon and Arantes, 2019; Zhang, 2008). The high-value products can enhance profitability, and the high-volume fuels will help meet the global energy/fuel demand. Section 8.3 of this chapter covers applications, technology, and recent advancements in the area of few commodity chemicals that have the potential to replace petroleum-based products.

As an alternative strategy, LBM is converted into gaseous vectors, i.e. biogas via anaerobic digestion and syngas via gasification, and liquid vectors, i.e. drop-in fuels by pyrolysis and thermochemical liquefaction. Biogas is commercially produced and utilized for the production of electricity, heat, and compressed natural gas (CNG) (Zheng et al., 2014). Syngas (consisting of H_2, CO, and CO_2) is converted into biofuels and biochemicals through gas fermentation by acetogenic bacteria and liquid fuels by Fischer–Tropsch synthesis (Ciliberti et al., 2020; Rauch et al., 2014). The drop-in fuels are gaining interest as their properties are functionally similar to fossil fuels and therefore are compatible with petroleum refinery infrastructure (Karatzos et al., 2014). The opportunities and challenges in the field of biogas, syngas, and drop-in fuels are covered in detail in Section 8.4.

The following section covers fractionation technologies of lignocellulosic biomass feedstock, evaluation of sustainable technology, and techno-economic challenges.

8.2 LIGNOCELLULOSIC BIOREFINERY

8.2.1 Fractionation Technologies and Their Importance

To achieve complete utilization of all three major components of lignocellulosic biomass, it is essential to design a pretreatment process that can fractionate and separate all three components with high purity. Each of these components is subsequently utilized in the value chain, viz. C6 and C5 streams are converted into bioenergy and fermentation products/chemicals and lignin is valorized into value-added products.

Pretreatment is the most complex and capital-intensive step that depolymerizes cellulose and hemicellulose and separates lignin. The pretreatment process for any biomass depends on its characteristics and targeted derived products. Various pretreatment technologies based on mechanical, physical, chemical, and biological methods have been developed. Figure 8.2 shows a comparison of different pretreatment methods employed for the fractionation of LBM. The removal of lignin during pretreatment gives a more accessible pore structure, exposes cellulose residues, and decreases the enzyme adsorption to lignin, thereby increasing the enzymatic hydrolysis significantly (Pihlajaniemi et al., 2016). Moreover, the structure and the reactivity of biorefinery lignin depend strongly on the pretreatment method. Alkaline pretreatment (including NaOH and NH_3) and Organosolv pretreatment have high delignification efficiency, whereas acid pretreatment results in better separation of C6 and C5 streams. The ideal pretreatment process would allow for a high conversion of the cellulose and hemicellulose to simple sugars, would minimize the degradation of these sugars to undesired forms that reduce fuel yields and inhibit fermentation, does not require, in particular, large or expensive reaction vessels, and is a

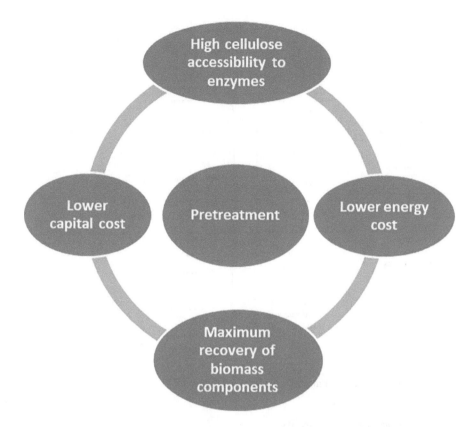

FIGURE 8.2 Parameters for optimization of pretreatment process.

simple, relatively robust, and cost-effective process (Figure 8.3) (Kumar et al., 2009; Santibañez-Aguilar et al., 2014).

8.2.2 Evaluation of Sustainable Technology for LBM Feedstock Biorefinery

LBM feedstock biorefineries are sustainable only when the production cost is minimized and all components of LBM are effectively utilized. The comprehensive assessment of technology is carried out based on the following five assessments.

8.2.2.1 Technology Efficiency

Efficiency is one of the most important indicators of perfection in technology development followed by commercialization. There are two key aspects of evaluating technology efficiency: (a) energy efficiency and carbon footprint and (b) material efficiency (Carriquiry et al., 2011). While the life cycle assessment (LCA) would play a key role in evaluating the former, fixing type(s) of feedstock would be crucial in evaluating the latter (Kim and Dale, 2005).

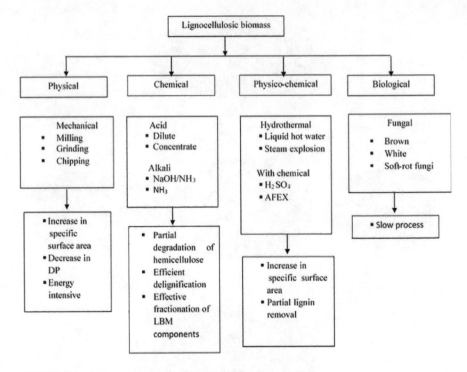

FIGURE 8.3   Pretreatment methods for fractionation of LBM.

8.2.2.2   Technology Flexibility

Three aspects of flexibility may be considered for assessment: (a) process inputs (feedstock options and enzyme options), (b) process output (products), and (c) plant configuration. A technology's ability to process multiple feedstocks would reduce risks arising out of dependency on a single biomass supply (Zhang et al., 2018). A process having a customizable product portfolio would also reduce the risks of product price variation. Similarly, a process with pure sugar streams has the potential for easy augmentation with future technologies, which are converting sugars to value-added biomaterials/biochemicals (Farzad et al., 2017). Furthermore, the capability of a process should have integration capability with an existing agri-process which would unleash operational synergies and significantly reduce cost and risk.

8.2.2.3   Technology Maturity

A highly efficient or fast process that is yet to gain reliable levels of technology readiness poses much higher risks associated with scale-up and plant operation than technologies that have travelled a significant learning curve at a commercial scale. Similarly, technologies that use a standard set of equipment tried, tested, and employed at commercial scale for significant time scale (e.g. equipment of paper and pulp industry) pose fewer risks than technologies using specially designed equipment as this equipment would need to gain, in due course of time, the necessary technological readiness and would have to handle many unanticipated challenges.

8.2.2.4 Technology Profitability

Freezing the battery limits, biomass types, and other relevant assumptions/-parameters would be crucial for appropriate profitability assessment and comparison. Multi-product schemes with better converging technologies and economic sustainability will be a game-changer in the upcoming time.

8.2.2.5 Robustness

Technological advancement with niche product development and their market development will bring more robustness in the bioenergy sector. There are two aspects of robustness: robustness of technology and robustness of profitability. The assessment of the technology robustness requires a detailed periodic technology audit and should only be taken up with selected technology(s) in the phase of technology assessment.

Also, the assessment of profitability and robustness requires evaluation of the sensitivity of profitability, i.e. internal rate of return (IRR), to factors such as enzyme and feedstock cost, plant's operating index, etc. Even, the technology which is based on the high cost of enzyme/per litre of ethanol production would be sensitive towards fluctuation in enzyme prices (Rajendran and Murthy, 2017).

However, this assessment has two major challenges: data availability and stakeholder participation. The data availability is limited due to the fewer maturity issues of the technology and confidentiality (Lindorfer et al., 2019). Also, it needs to be ensured that only economically viable "best-case" solutions from various technology providers are compared. For example, extreme enzyme dosage can result in a very short total hydrolysis time, but may not be a cost-effective strategy for producing ethanol. Therefore, while comparing enzyme dosage of various technologies, such extreme cases should not be considered.

From the perspective of sustainability and energy balance, better techno-economic viability is achieved by utilization of cellulose for fibre production, hemicellulose for co-production of fuels and chemicals, and valorization of lignin to value-added chemicals. The entire thermal and electric energy requirement of the process is met through bio-residues of the cellulosic ethanol plant (Melin and Hurme, 2011).

8.2.3 Expected Challenges in the Commercialization of LBM

The impediment in the commercialization of LBM is the high cost of production, which predominately depends on the feedstock supply chain, utilization of multiple feedstocks based on geographical location, and level of technology (Balan, 2014; Hassan and Kalam, 2013; Hoekman, 2009). The key challenges are as follows:

- *Supply chain optimization*: Availability of biomass at the required scale and cost in a sustainable manner is one of the major challenges while scaling up the operation. The steps in the biomass supply chain include collection, storage, pre-processing, transportation, and post-processing (Pantaleo et al., 2020; Sokhansanj and Hess, 2009). Texture variability, seasonal availability, distribution over a large area, and low bulk density leading to high transportation costs are major key factors required to be considered

(Caputo et al., 2005). This challenge is unique based on geographical conditions and varies with landholding. Therefore, the collection would need country-specific solutions that need to be adopted by taking all stakeholders in the value chain.

- *Multifeedstock biorefinery*: To ensure a consistent feedstock supply throughout the year, it is important to construct biorefinery that can process different feedstocks based on availability (Zhang et al., 2018). Typically, the US and European technologies are based on single feedstock such as corn stover/cobs as, due to its vast availability, production can be run throughout the year. However, in underdeveloped countries, plants need to be designed for processing multiple feedstocks.

- *Improvement of technology*: Lignocellulosic biorefinery requires very high capital and operating cost; therefore, technology should be improvised by the development of cost-effective pretreatment methods and the development of enzymes with improved hydrolytic efficiency, tolerance to inhibitors, ethanol accumulation, and thermotolerant microbes for efficient simultaneous saccharification and fermentation (Balan, 2014; Robak and Balcerek, 2018).

- *Zero waste technology*: Effective utilization of all streams of LBM biorefinery along with zero waste discharge can change the product portfolios of biorefineries which will be an opportunity for technological upgradations. This will also drive the path towards a cleaner environment.

It is always a challenge for biorefiners to make sustainable biorefinery, which is also evident from the previous sections. Therefore, integration of chemical complex with biorefinery is essential. Various biochemicals which can be produced through the biorefinery route have been discussed in the subsequent section, which will improve the economics of LBM biorefinery.

8.3 COMMODITY CHEMICALS

8.3.1 OPPORTUNITIES, USES, MARKET, TECHNOLOGY, AND CHALLENGES

Chemical production through the biological route has several advantages over the petrochemical route, which are environmental-friendliness, carbon-neutrality, energy efficiency, and sustainability (Winters, 2016). The market of bio-based chemicals is predicted to grow significantly in near future. Thus, bio-based chemicals have tremendous potential to replace petrochemicals (Takkellapati et al., 2018). The key market drivers for a higher growth rate of biochemicals are growing population, rising demand, and increasing customer awareness about bio-based products. Research on the sustainable production of biochemicals is also stimulated by policy legislation and COP21, Paris Agreement signed by 196 countries to reduce CO_2 emission. Among the bio-based chemicals, several chemicals such as lactic acid, ethylene, ethylene glycol, acetic acid, glycerol, propylene, 1,3-propanediol, isopropanol, isobutene, glutamic acid, xylitol, sorbitol, polyhydroxyalkanoates have already been commercialized. However, the commercialization of several other bio-based chemicals such as 2,5-furandicarboxylic acid (FDCA), glycolic acid, acrylic acid,

3-hydroxylpropionic acid, 1,4-butanediol, isobutene, isoprene, levulinic acid is currently in progress. However, the transition from fossil fuels to bio-based fuels has several challenges such as high production cost and difficulties in matching the product specifications as compared to petrochemicals (Ögmundarson et al., 2020). Some of the high-volume commodity chemicals have been discussed in detail below.

8.3.2 Lactic Acid

Lactic acid (LA, 2-hydroxypropanoic acid) is a commodity chemical valued at USD 2.64 billion in 2018 and is expected to grow annually at a compound annual growth rate (CAGR) of 18.7% from 2019 to 2025 (Alexandri et al., 2020). It is globally used for applications in food (e.g. yogurt, cheese), pharmaceuticals (e.g. supplements in the synthesis of dermatologic drugs and against osteoporosis), cosmetics, textile, and chemical industries. Currently, there is a huge demand for LA as a building block chemical for the production of polylactic acid (PLA), which is a biodegradable alternative to plastics derived from petrochemical and is also considered as a feedstock for the production of green solvents (e.g. ethyl lactate) (Bai et al., 2004; Sodergard and Stolt, 2010). LA can be produced either by chemical synthesis or by biotechnological processes using microbes. However, the major limitation of chemical synthesis results is the production of lactic acid in a racemic (50:50) mixture. Conversely, the biotechnological route leads to the production of a pure enantiomeric form of LA (Klotz et al., 2016; Datta and Henry, 2006) and also offers advantages in terms of lower energy consumption and utilization of renewable substrates as raw materials (Reddy et al., 2016). Presently, more than 95% of LA produced globally is manufactured by the fermentation process.

In industrial fermentations, LA is produced through first-generation feedstocks, i.e. corn starch and cane sugars, which represents one of the major challenges for the cost-effective production of LA. Therefore, several low-cost materials such as by-products or wastes of agricultural and food industries, and microalgal biomass have also been evaluated for the production of LA (Abdel-Banat B MA et al., 2010; Li et al., 2013; Mazzoli, 2020; Nguyen et al., 2012; Overbeck et al., 2016; Tang et al., 2016; Tashiro et al., 2011). However, LA yield using these substrates is extremely low because of the release of inhibitors during the pretreatment step. Further, during LA fermentation, the pH of fermentation broth continuously decreases, resulting in decreased productivity. Therefore, the pH of fermentation broth is controlled by the addition of lime, resulting in the formation of calcium lactate salt. Post-fermentation, LA is recovered from calcium lactate by the addition of strong acids such as sulphuric acid, which results in a significant increase in downstream processing cost along with the formation of by-product gypsum (calcium sulphate) (Joglekar et al., 2006). These challenges can be addressed by isolation of robust, stress-tolerant (including inhibitor, end product, i.e. LA, low pH, osmotic stress, and thermal tolerance) microbes producing enantiomeric lactic acid at high yields (Kuo et al., 2015; Ye et al., 2013, 2014).

LA is produced by a huge diversity of bacteria classified as lactic acid bacteria (LAB) belonging to genera *Enterococcus, Lactobacillus, Pediococcus, Lactococcus, Streptococcus, Weissella,* etc. LAB are classified as "generally recognized as safe (GRAS)" by the United States Food and Drug Administration (US-FDA) and the European Food Safety Agency (EFSA). These microbes are also genetically engineered for simultaneous utilization of glucose and xylose (Hu et al., 2016; Tarraran and Mazzoli, 2018; Yang et al., 2013). In another strategy, the lactic acid pathway is engineered in cellulase- and xylanase-producing strains, which would eliminate the requirement of the addition of cellulose and xylose degradation enzymes (Liaud et al., 2015). Further, metabolic engineering of microbes will improve both titre and yield (Liu et al., 2019; Papagianni, 2012).

Commercially, lactic acid is produced majorly by Corbion Purac Corporation (200 KTPY), Cargill, Incorporated (200 KTPY), Galactic (80 KTPY), Henan Jindan Lactic Acid Technology Co. Ltd (100 KTPY), Chongqing Bofei Biochemical Products (75 KTPA), and Archer Daniels Midland Company (Castillo Martinez et al., 2013; Jem and Tan, 2020; Smith et al., 2010) (https://www.lactic.com/en-us/news/corporate. aspx). Cargill, Incorporated, has developed a genetically engineered *S. cerevisiae* producing lactic acid at high titres (120 g/L) and yields at pH values <3.0, which is well below the pKa of lactic acid, i.e. 3.86. The low-pH lactic acid production process has several advantages such as reduced cost of production, improved product quality, easy downstream processing, reduced chemical usage, reduced nutrient costs, and reduced contamination (Miller et al., 2011).

8.3.3 Succinic Acid

Succinic acid (SA) (HOOC-CH$_2$-CH$_2$-COOH, also known as amber acid and butane-dioic acid) is an aliphatic, saturated four-carbon dicarboxylic acid. Its largest application is as a surfactant/detergent, additive, and foaming agent. It has also been recognized as a building block for many important chemicals that are used in the food, pharmaceutical, personal care, leather, and textile industries (Akhtar et al., 2014; Kumar et al., 2020; Lu et al., 2021). SA is a feedstock for several industrial products such as 1, 4-butanediol, tetrahydrofuran, adipic acid, maleic anhydride, polybutylene succinate (PBS), γ-butyrolactone, or various pyrrolidinone derivatives and also finds application as an ion chelator that prevents corrosion and pitting in the metal industry (Akhtar et al., 2014; Morales et al., 2016; Saxena et al., 2016). Due to increased industrial applications of SA, its worldwide demand has increased from 30,000–50,000 tons/year in 2014 to more than 700,000 tons/year in 2020 (Kumar et al., 2020).

To fulfil the demand, SA is produced via both chemical and biological pathways. Chemically, it is manufactured through hydrogenation of 1,4-dicarboxylic unsaturated C4 acids or anhydrides, oxidation of 1,4-butanediol, and oxidation of n-butane or benzene to maleic anhydride followed by hydration to maleic acid, which is later converted to succinic acid by hydrogenation (Jiang et al., 2013; Saxena et al., 2016; Xu et al., 2018). However, the chemical synthesis of SA leads to several challenges such as high production costs and environmental pollution. Depending on the purity, the market price of petroleum-based SA varies in the range of USD 5900–8800/ton (Kumar et al., 2020).

Alternatively, bio-based SA from renewable feedstocks has received increased attention and it is soon expected to replace the petroleum-based succinic acid. It has been recognized as one of the top-ten high-value-added biomass-derived chemicals by the United States Department of Energy (DOE) (Bozell and Petersen, 2010). Bio-SA is one of the intermediates of the TCA cycle (tricarboxylic acid cycle or citric acid cycle) and is obtained through anaerobic fermentation by microbes. Many of these microbes consume CO_2 during fermentation and thus also reduce GHG emissions (Bechthold et al., 2008; Saxena et al., 2016). Reports in the literature suggest that many different microorganisms such as anaerobic and facultative anaerobic bacteria (*Actinobacillus succinogenes, Mannheimia succiniciproducens, Ruminococcus flavefaciens, Anaerobiospirillum succiniciproducens, Corynebacterium crenatum,* and *E. coli*), fungi (*Aspergillus fumigatus, Aspergillus niger, Byssochlamys nivea, Penicillium viniferum, Lentinus degener, Paecilomyces variotii,* and *Trichoderma reesei*), and yeast (*Saccharomyces cerevisiae*) are used for the production of SA (Bechthold et al., 2008; Jiang et al., 2017). Among them, *Actinobacillus succinogenes* and *Anaerobiospirillum succiniciproducens* are recognized as the most efficient natural SA-producing strains. One advantage of using these strains over others is that they can utilize a wide spectrum of carbohydrates including xylose, lactose, arabinose, cellobiose, and other reducing sugars and form a fewer number of by-products (Jiang et al., 2013; Saxena et al., 2016). However, some limitations such as the utilization of cheap substrates, yield, productivity of SA, tolerance to stress conditions such as pH and other potential inhibitors, and sustainable separation and purification processes need to be addressed for the successful commercialization of bio-based SA processes (Salvachúa et al., 2016; Yang et al., 2020).

To this end, lignocellulosic wastes such as corn stover, corn stalk, sugarcane bagasse, and cotton stalk, which are available in abundance all year round, have been evaluated for the production of SA (Akhtar et al., 2014; Jiang et al., 2017). Moreover, genetically engineered microorganisms such as *E. coli* (Balzer et al., 2013; van Heerden and Nicol, 2013), *Aspergillus niger* (Yang et al., 2020), *Saccharomyces cerevisiae* (Raab et al., 2010) have been developed for the improved yield and productivity of SA (Jiang et al., 2014; Wang et al., 2014).

Various companies and the industrial consortium have already begun the commercial production of bio-SA. In 2012, Reverdia, a joint venture between Royal DSM and Roquette Frères started the production of bio-SA under the trademark of BIOSUCCINIUM® in Cassano, Italy. BIOSUCCINIUM® is produced using a low-pH, yeast-based technology that consumes less energy, locks CO_2 into the final SA molecule, and does not produce salts as a by-product when compared to other bacteria-based fermentations. Likewise, in 2014, Succinity GmbH, a joint venture between Corbion Purac and BASF started a bio-SA plant in Montmeló, Spain, with an annual capacity of 10,000 tons. Some of the other players in the bio-SA market include GC Innovation America (the USA), Nippon Shokubai (Japan), Kawasaki Kasei Chemicals (Japan), Mitsubishi Chemical Corporation (Japan), Anhui Sunsing Chemicals (China), and Gadiv Petrochemical Industries (Israel). These players are committed to establishing less time-consuming and cost-effective processes for sustainable production of bio-SA with the use of inexpensive next-generation feedstocks, and novel metabolic and process strategies.

8.3.4 Diol Compounds

Short-chain diols (propanediols, butanediols, and pentanediols) are bulk chemicals having a wide range of applications as fuels, solvents, polymer monomers, and pharmaceutical precursors. The chemical processes for the preparation of these diols are energy-intensive. Therefore, to meet the growing population demand, there is tremendous potential to synthesize these chemicals through biological route (Jiang et al., 2014; Sabra et al., 2016; Zeng and Sabra, 2011). 1,3-Propanediol (PDO) is a commodity chemical used in the synthesis of polymers, polyester polytrimethylene terephthalate (PTT), polyethers, and polyurethanes (Dan-Mallam et al., 2014; Hao et al., 2008; Kaur et al., 2012; Lin et al., 2005; Xu et al., 2009). 1,3-PDO is chemically synthesized from acrolein or ethylene oxide (Besson et al., 2003). Alternatively, it can also be synthesized by selective dihydroxylation of glycerol. However, biotechnological processes for the production of PDO are cost-effective, utilize renewable sources as feedstocks and are therefore preferred over chemical synthesis. Traditionally, PDO is produced by several microorganisms such as *Klebsiella, Citrobacter, Clostridium,* and *Lactobacillus* from glycerol under anaerobic conditions (Ji et al., 2009). DuPont and Genencor have developed engineered *E. coli,* which converts glycolysis-derived dihydroxyacetone phosphate (DHAP) to 1,3-propanediol aerobically, resulting in a yield of 135 g/L of PDO from glucose. The engineered *E. coli* strain has glycerol-3-phosphate dehydrogenase (DAR1) and glycerol-3-phosphate phosphatase (GPP2) genes, obtained from *Saccharomyces cerevisiae* and glycerol dehydratase (dhaB1, dhaB2, and dhaB3), and its reactivating factors (dhaBX and orfX), obtained from *Klebsiella pneumoniae* (Nakamura and Whited, 2003). Furthermore, the productivity of PDO is continuously improved by metabolic engineering of microorganisms and optimal fermentation conditions (Frazão et al., 2019; Huang et al., 2012; Jolly et al., 2014). DuPont Tate & Lyle Bio Products is currently the largest producer of bio-based PDO, and it requires 40% less energy compared to petrochemicals (Biddy et al., 2016).

2,3-Butanediol (2,3-BDO) has various industrial applications in the manufacture of printing inks, perfumes, fumigants, moistening and softening agents, explosives, and plasticizers, and as a carrier for pharmaceuticals (Ji et al., 2011; Song et al., 2019; Tsvetanova et al., 2014). Several bacterial strains *Klebsiella, Enterobacter, Serratia, Paenibacillus polymyxa, Bacillus licheniformis, Bacillus subtilis,* and *Bacillus amyloliquefaciens* can produce 2,3-BDO (Ji et al., 2011; Jurchescu et al., 2013). In bacterial metabolism, pyruvate is first converted to α-acetolactate by α-acetolactate synthase which is subsequently reduced to acetoin by α-acetolactate decarboxylase. The acetoin is further reduced to 2,3-BDO by butanediol dehydrogenase/diacetyl acetoin reductase in a reversible reaction (Ji et al., 2011). However, the production of 2,3-BDO is economically not viable due to several factors, i.e. lower yields, production of by-products (such as lactic acid, succinic acid, acetic acid, and ethanol), and high cost of raw materials. Because of the above, several strategies have been employed to increase the yield of 2,3-BDO such as the selection of high-yielding strains, mutagenesis (Ji et al., 2008), metabolic engineering of microbes (Tsvetanova et al., 2014), and optimization of conditions for production (Cheng et al., 2010). Recently, GS Caltex, a Korean petrochemical company, has successfully developed a technology for the purification of 2,3-BDO from the fermentation broth, resulting in 99.5% purity (Song et al., 2019).

8.3.5 LEVULINIC ACID

Levulinic acid (4-oxopentanoic acid/3-hydroxypropionic acid) has a broad range of applications and its market was $27.2 million in 2019, and it is expected to grow at a CAGR of 8.8% during 2020–2030 (PS market research; imarcgroup; verified market research). Levulinic acid is a feedstock for various industrial chemicals such as 1,4-pentanediol, methyl pent-4-enoate, γ-valerolactone, α-angelica lactone, acrylic acid, levulinates, diphenolic acid, 1-pentanol, succinic acid, 5-nonanone, 5-aminolevulinic acid, pentane, tetrahydrofuran, and methyltetrahydrofuran (Adeleye et al., 2019; van der Waal and de Jong, 2016). Sodium levulinate is used as a preservative and skin conditioning agent in cosmetics, and a preservative in food, especially fresh meat. Calcium levulinate is used for the preparation of pills and injections. δ-Aminolevulinic acid (DALA) is used as a photoactivation weedicide. Diphenolic acid is used in polymeride and other materials. It can also be utilized to form potential biofuel precursors such as methyltetrahydrofuran (used as an additive in gasoline), valerolactone, and ethyl levulinate (Adeleye et al., 2019). Levulinic acid can be produced from a diverse range of feedstocks including food grains, lignocellulosic biomass, and algal biomass (Rackemann and Doherty, 2019; Zheng et al., 2016). The process of levulinic acid production involves the isomerization of glucose to fructose using Lewis acid followed by dehydration of fructose to 5-HMF catalysed by bifunctional acid which, in turn, is followed by rehydration of 5-HMF to levulinic acid (Adeleye et al., 2019; Jeong et al., 2018). Furthermore, the metabolic pathway for levulinic acid production consisting of several enzymatic steps can be integrated into *Saccharomyces cerevisiae, Pichia stipites, Pseudomonas* sp., *Bacillus* sp., *Chrysosporium* sp., and *Escherichia coli.* (WO2012030860A1; USO9523105B2). GF Biochemicals acquired Segetis in 2016 and produces levulinic acid derivatives at a commercial scale from lignocellulosic biomass (de Jong et al., 2012).

8.3.6 2, 5-Furandicarboxylic Acid

2,5-Furandicarboxylic acid (FDCA) is another candidate from the list of top-ten biomass-derived chemicals, which can serve as a building block for future chemical production, as defined by the US-DOE (Bozell and Petersen, 2010). FDCA is reported to be able to substitute a variety of petrochemicals that are applicable in polyurethane, polyester, and polyamide industries. FDCA along with ethylene glycol can be utilized for the production of bio-based polyethylene furanoate (PEF), which has improved mechanical properties than polyethylene terephthalate (PET) that is produced from petroleum-derived terephthalic acid and adipic acid (Kim et al., 2020; Motagamwala et al., 2018). It is also an important component for the preparation of hexanoic acid, fungicides, macrocyclic ligands, corrosion inhibitors, and thiol-ene films. The diethyl esters of FDCA have strong anaesthetic properties analogous to cocaine and are mostly used in pharmacology. FDCA-derived anilides demonstrate strong antibacterial action, and a dilute solution of FDCA in tetrahydrofuran is used for making artificial veins for transplantation (Lewkowski, 2001; Rajesh et al., 2018). Besides, FDCA is also used in the synthesis of metal–organic frameworks (MOFs), which have several applications including drug delivery, gas/solvent storage, catalysis, and synthesis of new topological compounds (Rose et al., 2013). Although neither FDCA nor any of its derivatives have yet achieved commercial success, the improvement in their production process offers the potential for providing bio-based replacement of polymers.

The most common route for the production of FDCA consists of two steps: dehydration of carbohydrates to 5-hydroxymethylfurfural (5-HMF) and oxidation of 5-HMF to FDCA with air over different catalysts (Bozell and Petersen, 2010; Kim et al., 2020). However, technological challenges such as (a) instability of 5-HMF caused by undesirable condensation reactions at moderate temperatures, especially in acidic aqueous solutions, (b) poor solubility of FDCA in most commonly used solvents, (c) incomplete oxidation of 5-HMF to FDCA resulting in other intermediates such as 5-formylfuran-2-carboxylic acid (FFCA) and 5-(hydroxymethyl)furan-2-carboxylic acid (HMFCA), and (d) preferential production of 5-HMF by dehydration of fructose and/or glucose, thus competing with the food chain, need to be addressed for the sustainable production of FDCA (Jensen and Riisager, 2020; Motagamwala et al., 2018).

The company Avantium, along with global food and beverage giants Coca-Cola, Danone, and Carlsberg, has worked on developing PEF for soda bottles. Their proprietary YXY® plants-to-plastics technology is capable of catalytically converting fructose syrup from corn and wheat sugars into PEF. The company is setting up its first plant in the northern Netherlands and aiming to produce 5000 tons of PEF by 2023. Additionally, in 2016, Avantium and BASF formed a joint venture named Synvina to construct a commercial-scale plant of FDCA in Antwerp, Belgium; however, in 2019, BASF pulled out of Synvina and Avantium acquired full ownership of Synvina from BASF. Among Avantium's series of plans to commercialize its YXY® technology, it has also partnered with Japanese chemical company Toyobo Co., Ltd. to manufacture PEF polymers and films at Toyobo's Iwakuni plant (Jensen and Riisager, 2020). The companies DuPont Nutrition and Biosciences, and Archer Daniels Midland Company have jointly developed a process of converting fructose and methanol into furan dicarboxylic methyl ester (FDME) and plans to the open world's first FDME production plant in Decatur, IL, the USA. Further, Eastman Chemical Company has agreed to license its proprietary FDCA and derivatives production technology from renewable feedstocks to Origin Materials. Some of the other key players that are striving to move ahead in the FDCA and derivatives market are Corbion, Rennovia Inc., AVA Biochem, Petrobras, VTT Technical Research Center, Battelle Memorial Institute Inc., and Evonik Oxeno GmbH (Jensen and Riisager, 2020).

Reports in the literature suggest that biotransformation of HMF into FDCA is possible by using microbial as well as enzyme technologies; however, none of the major players have established a complete green synthesis process. Most of the companies are attempting to produce FDCA from lignocellulosic wastes by partial mild or toxic chemical treatment. This is because the microbial conversion of HMF to FDCA is a complex and time-consuming process (up to 2–3 days) and results in a low yield of the product. Strategies such as media optimization, genetic engineering, and metabolic engineering can help overcome these limitations (Rajesh et al., 2018).

8.3.7 ISOPRENE

Isoprene (2-methyl-1,3-butadiene) is a platform chemical commercially used for the production of polyisoprene rubber, styrenic thermoplastic elastomer block copolymers, and butyl rubber having a market value of $1–2 billion (Kim et al., 2016). Traditionally, the production of isoprene is carried out by the separation of the C5 stream from naphtha crackers (Senyek, 2008). Recently, naphtha cracking has been carried out in FCC to maximize light olefins, which has decreased isoprene yield, and therefore, it is a favourable opportunity to produce isoprene by biological route (Biddy et al., 2016). Moreover, biological production is driven by sustainability and reduction in the carbon footprint. The microbial production process of isoprene has several advantages such as tolerance of bacteria to isoprene and easy downstream process (isoprene is collected in upper gaseous phase in fermenter due to its lower boiling point of 34°C and lower solubility in water); due to its high reactivity, it can easily be converted into complicated products. Isoprene is synthesized by two naturally occurring pathways, i.e. mevalonate (MVA) pathway which is present in eukaryotes and mycobacteria and non-mevalonate, methylerythritol 4-phosphate (MEP/DXP) pathway present in eubacteria, green algae, and chloroplasts of higher plants (Eroglu and Melis, 2010; Seemann et al., 2006). *Escherichia coli* and *Bacillus subtilis* have been genetically engineered with MVA and DXP pathway genes, resulting in isoprene production (Kim et al., 2016; Li et al., 2018; Xue and Ahring, 2011; Yang et al., 2012). DuPont-Genencor along with Goodyear Tire & Rubber Company has developed genetically engineered microbes and processes for the production of isoprene at high yields. Similarly, Amyris is collaborating with Michelin, a petrochemical company, and Braskem for the development and commercialization of isoprene from plant sugars (Takkellapati et al., 2018).

8.3.8 LIGNIN-DERIVED CHEMICALS

Lignin is an extremely abundant raw material contributing 40% of the energy content of lignocellulosic biomass. In addition to LBM-to-biofuel technologies, pulp and paper industries also generate huge amounts of lignin. The price of lignin is based on its purity, ranging from ~280 USD/MT for low-purity lignin to 750 USD/MT for high-purity lignin. In commercial LBM biorefineries, lignin is mostly used in low-value commercial applications such as concrete additives or as low-grade fuel to provide heat and power to the process. However, lignin can be valorized into a huge array of value-added chemicals such as aromatics, hydrogels, carbon fibres, thermoplastic elastomers, and chemicals (Figure 8.1). Lignosulphonates produced during sulphite processing of LBM have applications as surfactants, animal feed, pesticides, dispersants, flocculants, concrete additives, and composites (Aro and Fatehi, 2017). Pyrolysis of lignin results in the production of pyro-oil, acetic acid, aldehydes, and aromatic compounds such as vanillin, vanillic acid, and syringic acid (Fan et al., 2017; Laurichesse and Avérous, 2014; Mu et al., 2013). Benzene, toluene, and xylene (BTX) produced from lignin have properties similar to the petrochemical production route and therefore offer huge potential to replace petrochemical-based BTX production (Hodásová et al., 2015). Esterified lignins are mainly used in the synthesis of polyesters, elastomeric materials, and epoxy resins (Laurichesse and Avérous, 2014). Lignin as a macromolecule can be blended with other polymers and used as an adsorbent for wastewater purification and preparation of hydrogels (Hodásová et al., 2015). Lignin possesses high mechanical strength and antibacterial and antioxidant properties and is therefore recognized as a promising material for the preparation of hydrogels. These hydrogels have extensive usage in drug delivery, tissue engineering, and antimicrobial materials (Asina et al., 2017; Yu and Kim, 2020; Zhang et al., 2019). Lignin has the potential to be utilized as activated carbon for water purification and production of carbon fibres (Duval and Lawoko, 2014). Carbon fibres have high specific stiffness and strength and have applications in lightweight composites. Lignin–PAN blend can also be utilized for the preparation of carbon fibre which will significantly reduce the cost of carbon fibre production (Bengtsson et al., 2020; Mainka et al., 2015; Souto et al., 2018). The technical challenges in the valorization of lignin are heterogeneity and complex and stable chemical bonds (Beckham et al., 2016).

8.4 VALORIZATION OF LBM INTO GASEOUS AND LIQUID FUELS

The waste stream of the LBM refinery has the potential for conversion into biogas by anaerobic digestion, syngas by gasification, and drop-in fuels by pyrolysis and thermochemical liquefaction. Even these gaseous fuels can be obtained from non-lignocellulosic biomass and industrial waste gas which is abundantly available. Biogas is used as a transportation fuel, whereas syngas can be processed for the production of chemicals and fuels through both chemical and biological routes (Griffin and Schultz, 2012; Mittal et al., 2018; Tirado-Acevedo et al., 2010). These energy forms have the potential to contribute to worldwide energy demand and have been discussed in the following section.

8.4.1 BIOGAS

Biogas is produced by the anaerobic digestion of organic components. Typically, biogas is composed of 60%–70% CH_4 and 30%–40% CO_2 along with trace amounts of other components such as hydrogen sulphide (H_2S), water vapour, and ammonia (NH_3) (Wang et al., 2013). Major biomass components such as carbohydrates, lignin, fat, and protein can be digested with the help of microbes to produce biogas. The process of anaerobic digestion takes place in four major steps (Figure 8.4). (a) Hydrolysis: Complex or polymeric molecules such as carbohydrates, lipids, and proteins are broken into simple or monomeric organic molecules. During this stage, hydrolytic microorganisms colonize on the surface of feedstock or agricultural waste, which results in the degradation of organic molecules (Ariunbaatar et al., 2014). (b) Acidogenesis: At this stage, products of hydrolysis are converted into volatile fatty acids and other products such as smaller amounts of ethanol and lactate. Acidogenesis proceeds at a faster rate compared to other stages of anaerobic digestion as acidogenic bacteria have a regeneration time lesser than 36h (Wu et al., 2019). (c) Acetogenesis: During acetogenesis, products of acidogenesis are converted to acetate by the reduction of CO_2. Several bacteria that contribute to acetogenesis are *Clostridium* spp., *Syntrophobacter wolinii*, *Peptococcus* anaerobes, etc. (Meegoda et al., 2018). (d) Methanogenesis: The final stage of anaerobic digestion is the methanogenesis phase, where methanogenic microorganisms convert the product of acetogenesis into methane. Methanogens have slower regeneration rates and require relatively higher pH compared to other stages of anaerobic digestion (Dahlgren, 2020).

Biomass anaerobic digestion is affected by different process parameters such as concentration of organic components, temperature, pH, and hydraulic retention time. Utilization of substrate and anaerobic digestion conditions are deciding factors for the

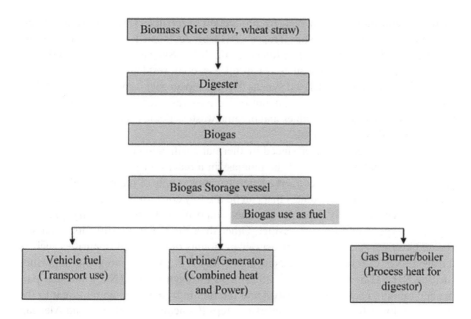

FIGURE 8.4 Schematic process of anaerobic digestion.

composition and heating value of produced biogas (Koniuszewska et al., 2020). The major bottleneck in anaerobic digestion is longer biomass digestion time due to inherent properties of lignocellulosic biomass such as chemical and structural orientation (Mao et al., 2015). Various enhancement techniques such as (a) pretreatment of biomass that improves the accessibility of organic components, (b) addition of enzymes to enhance hydrolysis process and the use of fungi to improve selective biodegradation of hemicellulose and lignin, and (c) partial composting and silage that improves the preservation of various biomass such as softwood biomass and hardwood biomass are used to improve anaerobic digestion process (Kainthola et al., 2019).

Biogas can be compressed after the removal of H_2S and CO_2 in the same way as natural gas is compressed to compressed natural gas (CNG). The compressed biogas (CBG) has calorific value and other properties similar to CNG and hence can be utilized as a green renewable automotive, transportation, or commercial fuel (Mittal et al., 2018). The key benefits of CBG are the availability of more affordable transport fuels, waste management, reduction in carbon emissions and thus pollution, additional revenue source for farmers, boost to entrepreneurship, rural economy and employment, support to national commitments in achieving climate change goals, energy security, and production of bio-manure (Scarlat et al., 2018; Sirothiya and Chavadi, 2020).

8.4.2 Synthesis Gas

Synthesis gas (Syngas) is a mixture predominantly of carbon monoxide (CO) and hydrogen gas (H_2) at different ratios and very often includes minor amounts of CO_2, CH_4, and water vapour. It is generally used at high pressures for the production of chemicals and fuels, such as NH_3 for production of fertilizers, H_2 for use in refineries, and methanol, and Fischer–Tropsch (FT) products such as synthetic gasoline and diesel (Adhikari et al., 2015; Rauch et al., 2014). Syngas can be produced from any carbonaceous feedstock such as natural gas, coal, residual oils, and waste biomass. However, biomass-derived syngas (termed biosyngas) is more sustainable than fossil-derived syngas as ample amount of biomass resources are available, ranging from agriculture crops to residues and organic wastes (Foulds et al., 1998; Goncalves dos Santos and Alencar, 2020).

Biosyngas are generally produced by thermal gasification of biomass, which is a complete thermal breakdown of the biomass in a reactor or gasifier. Thus, biomass gasification can effectively convert a heterogeneous supply of biomass feedstock into consistent gaseous intermediates that can be reliably converted into liquid fuels. The quality of syngas largely depends upon the composition of biomass, gasifying agents (e.g. air, oxygen), and type of gasifier. Other than the major constituents discussed above, biosyngas also presents measurable amounts of undesired impurities such as nitrogen (N_2), different hydrocarbons (e.g. C_2H_6, C_2H_4, C_2H_2), inorganic impurities such as NH_3, H_2S, sulphur oxides (SO_x), and nitrogen oxides (NO_x), tar, and particulate matter. These gas impurities and the tars represent a potential threat to the success of downstream syngas conversion steps (Goncalves dos Santos and Alencar, 2020; Munasinghe and Khanal, 2011). The tars have to be cracked or removed first, to enable the use of conventional low-temperature wet gas cleaning or advanced high-temperature dry gas cleaning of the remaining impurities. The syngas so obtained are then processed biologically or chemically to produce ethanol. However,

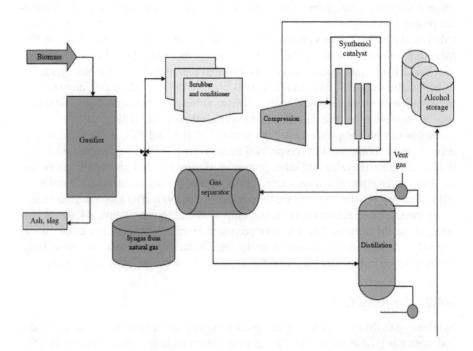

FIGURE 8.5   Process flow diagram for conversion of syngas to ethanol via chemical route.

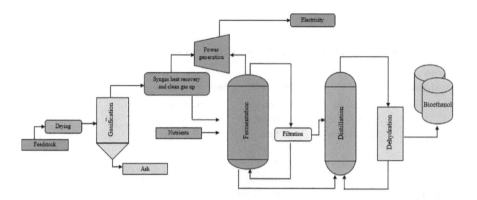

FIGURE 8.6 Process flow diagram for conversion of syngas to ethanol via fermentation (biological) route.

the biological route (using microbial catalysis) offers several advantages over the chemical route (using metal catalysis) such as independence from an expensive metal catalyst, elimination of the need of specific H_2/CO ratio for bioconversion, and issues related to noble metal poisoning (Munasinghe and Khanal, 2011). Figures 8.5–8.6 represent the schematic diagrams of biological and chemical routes. Currently, processes have been developed based on both these approaches and are being offered at the commercial level. A list of leading process developers is given in Table 8.2.

8.4.3 Drop-in Fuels

The key advantage of "drop-in" biofuel is that it can be readily integrated into existing petroleum refinery infrastructure including refinery, pipelines, blending terminals, and vehicle engines (Karatzos et al., 2014). The drop-in biofuels can be produced using LBM via thermochemical route, i.e. pyrolysis and gasification. Fast pyrolysis converts small lignocellulosic particles of approximately 3 mm size into bio-oil/pyrolysis oil under controlled reaction conditions at a temperature of 450°C–550°C in the absence of oxygen and short vapour residence time of less than 2 s, which results in 70%–75% of bio-oil production on a dry weight basis (Holladay et al., 2007; Zacher et al., 2014). Several reactor configurations such as fluid bed, rotating cone, and ablative reactor have been designed to maximize bio-oil production and quality (Jiang et al., 2015). The critical factors for obtaining high bio-oil yield are better temperature control, efficient heat transfer to LBM particles, and lower vapour residence time (Demirbas and Arin, 2002). Pyrolysis bio-oils contain a mixture of carboxylic acids, alcohols, esters, phenols, ketones, aldehydes, guaiacols, etc., and have up to 40% oxygen. Because of high oxygen content, bio-oils also have other adverse properties such as high viscosity, thermal instability, ignition and combustion difficulties, coking, and corrosiveness. Therefore, bio-oils need to be extensively upgraded to produce deoxygenated hydrocarbon drop-in biofuel blendstocks (Jones et al., 2009). The hydrogenation of pyro-oils can be achieved by hydrodeoxygenation (HDO) and catalytic cracking. HDO removes oxygen from

TABLE 8.2

List of Leading Process Developers of Syngas Technology

Process Developer	Route	Feedstock	Source	Remarks
M/s Coskata	Gasification followed by syngas fermentation	Natural gas Steel mill gas Biomass Waste Coal	http://www.coskata.com/ethanol/	–
M/s LanzaTech	Gas fermentation	Natural gas Steel mill gas Biomass Waste Coal	http://www.lanzatech.com/innovation/	Indian Oil has signed MoU with M/s LanzaTech
M/s INEOS Bio	Gasification followed by syngas fermentation	Biomass Municipal waste	http://www.ineosbio.com	–
M/s Celanese	Gasification followed by catalytic conversion	Petcoke	https://www.celanese.com/innovation/TCX.aspx	Indian Oil to set up JV to produce 1.1 million tons of ethanol at Paradip
M/s Synhenol Energy Corporation	Gasification followed by catalytic conversion	Biomass Municipal waste Natural gas	http://www.synthenol.com/html/technology.html	–

bio-oil at a temperature of 300°C–500°C in the presence of catalyst and hydrogen (Elliott, 2007). HDO process can be integrated into petroleum refinery by utilizing existing infrastructure; however, the major challenge is to process pyro-oil into feed quality that can be acceptable in biorefinery (Zacher et al., 2014). The upgradation of bio-oils can also be achieved through catalytic cracking using molecular sieves such as ZSM-5 and HZSM-5. However, this process results in lower yields as compared to HDO (Gayubo et al., 2004).

Liquefaction is another strategy where LBM is converted into liquid fuels in organic-/aqueous solutions (PEG, ethylene glycol, and glycerol) under pressure. The liquid fuel obtained through liquefaction has better fuel properties compared to pyro-oil (Elliott et al., 2015). Catalysts used during liquefaction are acidic and basic. The acidic catalysts result in a higher conversion yield up to 90%; however, these catalysts are corrosive. The basic catalysts such as NaOH and $Ca(OH)_2$ result in lower yields below 40%.

The thermochemical processes have been demonstrated worldwide; however, these processes are not commercially viable. The major challenge of thermochemical conversion of LBM to drop-in biofuels is the presence of different kinds of reactive groups in liquid fuels. This challenge can be addressed by the selection of highly specific and efficient catalysts or fractionation of cellulose and lignin before liquefaction (Jiang et al., 2015). Another challenge is LBM has a very low effective hydrogen-to-carbon ratio (H_{eff}/C); for example, sugar has a H_{eff}/C of 0 and drop-in biofuels require a H_{eff}/C of 2. Therefore, a large amount of hydrogen is required to produce energy-dense and reduced drop-in biofuels (Karatzos et al., 2014). This challenge can be addressed by an improvement in catalytic efficiency, stability, and regeneration yield of catalyst.

Furthermore, from a techno-economic perspective, efforts should be focused on the development of integrated process from LBM to fuels, improvement in the quality of pyro-oil feedstock, supply chain optimization, and improvement in the product quality that meet ASTM and functional specifications (Zacher et al., 2014).

8.5 CONCLUSIONS

The transition from fossil fuel economy to sustainable bioeconomy is the current demand for energy security and reduction in GHG emissions. Biorefinery aims towards the valorization of all components of lignocellulosic biomass into energy and commodity chemicals. This chapter highlights the profitable production of several commodity chemicals using renewable feedstocks. Although there has been considerable progress in the pretreatment technologies, hydrolytic efficiency of enzymes, and genetic and metabolic engineering of microbes for the production of value-added products, several challenges such as feedstock logistics, the possibility for processing multiple feedstocks, reducing the cost of production, and valorization of all components of biomass still need to be addressed concerning both technology and economic perspective.

REFERENCES

Abdel-Banat B M.A., Hoshida, H., Ano, A., Nonklang, S., Akada, R., 2010. High-temperature fermentation: How can processes for ethanol production at high temperatures become superior to the traditional process using mesophilic yeast? Appl. *Microbiol. Biotechnol.* 85, 861–867. https://doi.org/10.1007/s00253-009-2248-5

Adeleye, A.T., Louis, H., Akakuru, O.U., Joseph, I., Enudi, O.C., Michael, D.P., 2019. A review on the conversion of levulinic acid and its esters to various useful chemicals. *AIMS Energy* 7, 165–185. https://doi.org/10.3934/ENERGY.2019.2.165

Adhikari, U., Eikeland, M.S., Halvorsen, B.M., 2015. Gasification of biomass for production of syngas for biofuel, in: *Proceedings of the 56th Conference on Simulation and Modelling (SIMS 56).* Linköping, Sweden, pp. 255–260. https://doi.org/10.3384/ecp15119255

Akhtar, J., Idris, A., Aziz, R.A., 2014. Recent advances in production of succinic acid from lignocellulosic biomass. *Appl. Microbiol. Biotechnol.* 98, 987–1000. https://doi.org/10.1007/s00253-013-5319-6

Alexandri, M., Blanco-Catalá, J., Schneider, R., Turon, X., Venus, J., 2020. High L(+)-lactic acid productivity in continuous fermentations using bakery waste and lucerne green juice as renewable substrates. *Bioresour. Technol.* 316. https://doi.org/10.1016/j.biortech.2020.123949

Ariunbaatar, J., Panico, A., Esposito, G., Pirozzi, F., Lens, P.N.L., 2014. Pretreatment methods to enhance anaerobic digestion of organic solid waste. *Appl. Energy* 123, 143–156. https://doi.org/10.1016/j.apenergy.2014.02.035

Aro, T., Fatehi, P., 2017. Production and application of lignosulfonates and sulfonated lignin. *ChemSusChem* 10, 1861–1877. https://doi.org/10.1002/cssc.201700082

Asina, F.N.U., Brzonova, I., Kozliak, E., Kubátová, A., Ji, Y., 2017. Microbial treatment of industrial lignin: Successes, problems and challenges. *Renew. Sustain. Energy Rev.* 77, 1179–1205. https://doi.org/10.1016/j.rser.2017.03.098

Bai, D.M., Zhao, X.M., Li, X.G., Xu, S.M., 2004. Strain improvement of *Rhizopus oryzae* for over-production of L(+)-lactic acid and metabolic flux analysis of mutants. *Biochem. Eng. J.* 18, 41–48. https://doi.org/10.1016/S1369-703X(03)00126-8

Balan, V., 2014. Current challenges in commercially producing biofuels from lignocellulosic biomass. *ISRN Biotechnol.* 2014, 463074. https://doi.org/10.1155/2014/463074

Balzer, G.J., Thakker, C., Bennett, G.N., San, K.Y., 2013. Metabolic engineering of Escherichia coli to minimize byproduct formate and improving succinate productivity through increasing NADH availability by heterologous expression of NAD+-dependent formate dehydrogenase. *Metab. Eng.* 20, 1–8. https://doi.org/10.1016/j.ymben.2013.07.005

Bechthold, I., Bretz, K., Kabasci, S., Kopitzky, R., Springer, A., 2008. Succinic acid: A new platform chemical for biobased polymers from renewable resources. *Chem. Eng. Technol.* 31, 647–654. https://doi.org/10.1002/ceat.200800063

Beckham, G.T., Johnson, C.W., Karp, E.M., Salvachúa, D., Vardon, D.R., 2016. Opportunities and challenges in biological lignin valorization. *Curr. Opin. Biotechnol.* 42, 40–53. https://doi.org/10.1016/j.copbio.2016.02.030

Bengtsson, A., Hecht, P., Sommertune, J., Ek, M., Sedin, M., Sjöholm, E., 2020. Carbon fibers from lignin–Cellulose precursors: Effect of carbonization conditions. *ACS Sustain. Chem. Eng.* 8, 6826–6833. https://doi.org/10.1021/acssuschemeng.0c01734

Besson, M., Gallezot, P., Pigamo, A., Reifsnyder, S., 2003. Development of an improved continuous hydrogenation process for the production of 1,3-propanediol using titania supported ruthenium catalysts. *Appl. Catal. A Gen.* 250, 117–124. https://doi.org/10.1016/S0926-860X(03)00233-3

Bessou, C., Ferchaud, F., Gabrielle, B., Mary, B., 2011. Biofuels, greenhouse gases and climate change. *A review. Agron. Sustain. Dev.* 31, 1. https://doi.org/10.1051/agro/2009039

Bhuvaneshwari, S., Hettiarachchi, H., Meegoda, J.N., 2019. Crop residue burning in India: Policy challenges and potential solutions. *Int. J. Environ. Res. Public Health* 16, 832. https://doi.org/10.3390/ijerph16050832

Biddy, M.J., Scarlata, Christopher, A., Kinchin, C., 2016. Chemicals from biomass: A market assessment of bioproducts with near-term potential. United States. https://doi.org/10.2172/1244312

Bozell, J.J., Petersen, G.R., 2010. Technology development for the production of biobased products from biorefinery carbohydrates—The US Department of Energy's "top 10" revisited. *Green Chem.* 12, 539–555. https://doi.org/10.1039/b922014c

Caputo, A.C., Palumbo, M., Pelagagge, P.M., Scacchia, F., 2005. Economics of biomass energy utilization in combustion and gasification plants: Effects of logistic variables. *Biomass Bioenerg.* 28, 35–51. https://doi.org/10.1016/j.biombioe.2004.04.009

Carriquiry, M.A., Du, X., Timilsina, G.R., 2011. Second generation biofuels: Economics and policies. *Energy Policy* 39, 4222–4234. https://doi.org/10.1016/j.enpol.2011.04.036

Castillo Martinez, F.A., Balciunas, E.M., Salgado, J.M., Domínguez González, J.M., Converti, A., Oliveira, R.P. de S., 2013. Lactic acid properties, applications and production: A review. *Trends Food Sci. Technol.* 30, 70–83. https://doi.org/10.1016/j.tifs.2012.11.007

Cheng, H., Wang, L., 2013. Lignocelluloses feedstock biorefinery as petrorefinery substitutes. *Biomass Now - Sustain. Growth Use.* https://doi.org/10.5772/51491

Cheng, K.K., Liu, Q., Zhang, J.A., Li, J.P., Xu, J.M., Wang, G.H., 2010. Improved 2,3-butanediol production from corncob acid hydrolysate by fed-batch fermentation using Klebsiella oxytoca. *Process Biochem.* 45, 613–616. https://doi.org/10.1016/j.procbio.2009.12.009

Cherubini, F., Wellisch, M., Willke, T., 2009. Toward a common classification approach for biorefinery systems. https://doi.org/10.1002/bbb

Choi, S., Song, C.W., Shin, J.H., Lee, S.Y., 2015. Biorefineries for the production of top building block chemicals and their derivatives. *Metab. Eng.* 28, 223–239. https://doi.org/-10.1016/j.ymben.2014.12.007

Ciliberti, C., Biundo, A., Albergo, R., Agrimi, G., Braccio, G., de Bari, I., Pisano, I., 2020. Syngas derived from lignocellulosic biomass gasification as an alternative resource for innovative bioprocesses. *Processes* 8, 1–38. https://doi.org/10.3390/pr8121567

Dahlgren, S., 2020. Biogas-based fuels as renewable energy in the transport sector: An overview of the potential of using CBG, LBG and other vehicle fuels produced from biogas. *Biofuels* 0, 1–13. https://doi.org/10.1080/17597269.2020.1821571

Dan-Mallam, Y., Abdullah, M.Z., Yusoff, P.S.M.M., 2014. Mechanical properties of recycled kenaf/polyethylene terephthalate (PET) fiber reinforced polyoxymethylene (POM) hybrid composite. *J. Appl. Polym. Sci.* 131, 1–7. https://doi.org/10.1002/app.39831

Datta, R., Henry, M., 2006. Lactic acid: Recent advances in products, processes and technologies – A review. *J. Chem. Technol. Biotechnol.* 81, 1119–1129. https://doi.org/10.1002/jctb

de Jong, E., Higson, A., Walsh, P., Wellisch, M., 2012. Bio-based chemicals value added products from. IEA Bioenergy I Task 42 Biorefinery 1–33.

Demirbas, A., Arin, G., 2002. An overview of biomass pyrolysis. *Energ. Source* 24, 471–482. https://doi.org/10.1080/00908310252889979

Duque-Acevedo, M., Belmonte-Ureña, L.J., Cortés-García, F.J., Camacho-Ferre, F., 2020. Agricultural waste: Review of the evolution, approaches and perspectives on alternative uses. *Glob. Ecol. Conserv.* 22, e00902. https://doi.org/10.1016/j.gecco.2020.e00902

Duval, A., Lawoko, M., 2014. A review on lignin-based polymeric, micro- and nano-structured materials. *React. Funct. Polym.* 85, 78–96. https://doi.org/10.1016/j.reactfunctpolym.2014.09.017

Elliott, D.C., 2007. Historical developments in hydroprocessing bio-oils. *Energ. Fuel.* 21, 1792–1815. https://doi.org/10.1021/ef070044u

Elliott, D.C., Biller, P., Ross, A.B., Schmidt, A.J., Jones, S.B., 2015. Hydrothermal liquefaction of biomass: Developments from batch to continuous process. *Bioresour. Technol.* 178, 147–156. https://doi.org/10.1016/j.biortech.2014.09.132

Eroglu, E., Melis, A., 2010. Extracellular terpenoid hydrocarbon extraction and quantitation from the green microalgae *Botryococcus braunii* var. *Showa. Bioresour. Technol.* 101, 2359–2366. https://doi.org/10.1016/j.biortech.2009.11.043

Fan, L., Zhang, Y., Liu, S., Zhou, N., Chen, P., Cheng, Y., Addy, M., Lu, Q., Omar, M.M., Liu, Y., Wang, Y., Dai, L., Anderson, E., Peng, P., Lei, H., Ruan, R., 2017. Bio-oil from fast pyrolysis of lignin: Effects of process and upgrading parameters. *Bioresour. Technol.* 241, 1118–1126. https://doi.org/10.1016/j.biortech.2017.05.129

Farzad, S., Mandegari, M.A., Guo, M., Haigh, K.F., Shah, N., Görgens, J.F., 2017. Multiproduct biorefineries from lignocelluloses: A pathway to revitalisation of the sugar industry? *Biotechnol. Biofuels* 10, 87. https://doi.org/10.1186/s13068-017-0761-9

Foulds, G.A., Rigby, G.R., Leung, W., Falsetti, J., Jahnke, F., 1998. Synthesis gas production : Comparison of gasification with steam reforming for direct reduced iron production. *Stud. Surf. Sci. Catal.* 119, 889–894. https://doi.org/10.1016/S0167-2991(98)80544-2

Frazão, C.J.R., Trichez, D., Serrano-bataille, H., Dagkesamanskaia, A., Christopher, M.T., Walther, T., François, J.M., 2019. Construction of a synthetic pathway for the production of 1, 3-propanediol from glucose. *Sci. Rep. s,* 9(1), 11576. https://doi.org/10.1038/s41598-019-48091-7

Gayubo, A.G., Aguayo, A.T., Atutxa, A., Aguado, R., Olazar, M., Bilbao, J., 2004. Transformation of oxygenate components of biomass pyrolysis oil on a HZSM-5 Zeolite. II. Aldehydes, Ketones, and Acids. *Ind. Eng. Chem. Res.* 43, 2619–2626. https://doi.org/10.1021/ie030792g

Goncalves dos Santos, R., Alencar, A.C., 2020. Biomass-derived syngas production via gasification process and its catalytic conversion into fuels by Fischer Tropsch synthesis : A review. *Int. J. Hydrogen Energy* 45, 18114–18132. https://doi.org/10.1016/j.ijhydene.2019.07.133

Griffin, D.W., Schultz, M.A., 2012. Fuel and chemical products from biomass syngas: A comparison of gas fermentation to thermochemical conversion routes. *Environ. Prog. Sustain. Energy* 31, 219–224. https://doi.org/10.1002/ep.11613

Hao, J., Lin, R., Zheng, Z., Liu, H., Liu, D., 2008. Isolation and characterization of microorganisms able to produce 1,3-propanediol under aerobic conditions. *World J. Microbiol. Biotechnol.* 24, 1731–1740. https://doi.org/10.1007/s11274-008-9665-y

Hassan, M.H., Kalam, A., 2013. An overview of biofuel as a renewable energy source : Development and challenges. 56, 39–53. https://doi.org/10.1016/j.proeng.2013.03.087

Hodásová, Ľ., Jablonský, M., Škulcová, A.B., Ház, A., 2015. Lignin, potential products and their market value. *Wood Res.* 60, 973–986.

Hoekman, S.K., 2009. Biofuels in the U.S. - Challenges and opportunities. *Renew. Energy* 34, 14–22. https://doi.org/10.1016/j.renene.2008.04.030

Holladay, J.E., White, J.F., Bozell, J.J., Johnson, D., 2007. Top value-added chemicals from biomass, Results of screening for potential candidates from Biorefinery Lignin II, USDOE, PNNL-16983.

Hu, J., Lin, Y., Zhang, Z., Xiang, T., Mei, Y., Zhao, S., Liang, Y., & Peng, N. (2016). High-titer lactic acid production by Lactobacillus pentosus FL0421 from corn stover using fed-batch simultaneous saccharification and fermentation. *Bioresour. Technol. 214*, 74–80. https://doi.org/10.1016/j.biortech.2016.04.034

Huang, Y., Li, Z., Shimizu, K., Ye, Q., 2012. Bioresource technology simultaneous production of 3-hydroxypropionic acid and 1, 3-propanediol from glycerol by a recombinant strain of *Klebsiella pneumoniae. Bioresour. Technol.* 103, 351–359. https://doi.org/10.1016/j.biortech.2011.10.022

Jem, K.J., Tan, B., 2020. The development and challenges of poly (lactic acid) and poly (glycolic acid). *Adv. Ind. Eng. Polym. Res.* 3, 60–70. https://doi.org/10.1016/j.aiepr.2020.01.002

Jensen, M.H., Riisager, A., 2020. Chapter 5- Advances in the synthesis and application of 2,5-furandicarboxylic acid, in: Saravanamurugan, S., Pandey, A., Li, H., Riisager, A. (Eds.), *Biomass, Biofuels, Biochemicals, Biomass, Biofuels, Biochemicals.* Elsevier, pp. 135–170. https://doi.org/10.1016/B978-0-444-64307-0.00005-6

Jeong, H., Park, S.Y., Ryu, G.H., Choi, J.H., Kim, J.H., Choi, W.S., Lee, S.M., Choi, J.W., Choi, I.G., 2018. Catalytic conversion of hemicellulosic sugars derived from biomass to levulinic acid. *Catal. Commun.* 117, 19–25. https://doi.org/10.1016/j.catcom.2018.04.016

Ji, X.-J., Huang, H., Li, S., Du, J., Lian, M., 2008. Enhanced 2,3-butanediol production by altering the mixed acid fermentation pathway in *Klebsiella oxytoca. Biotechnol. Lett.* 30, 731–734. https://doi.org/10.1007/s10529-007-9599-8

Ji, X.J., Huang, H., Ouyang, P.K., 2011. Microbial 2,3-butanediol production: A state-of-the-art review. *Biotechnol. Adv.* 29, 351–364. https://doi.org/10.1016/j.biotechadv.2011.01.007

Ji, X.-J., Huang, H., Zhu, J.-G., Hu, N., Li, S., 2009. Efficient 1,3-propanediol Production by fed-batch culture of *Klebsiella pneumoniae*: The role of pH fluctuation. *Appl. Biochem. Biotechnol.* 159, 605. https://doi.org/10.1007/s12010-008-8492-9

Jiang, J., XU, J., Song, Z., 2015. Review of the direct thermochemical conversion of lignocellulosic biomass for liquid fuels. *Front. Agric. Sci. Eng.* 2, (1) 13–27.

Jiang, M., Ma, J., Wu, M., Liu, R., Liang, L., Xin, F., Zhang, W., Jia, H., Dong, W., 2017. Progress of succinic acid production from renewable resources: Metabolic and fermentative strategies. *Bioresour. Technol.* 245, 1710–1717. https://doi.org/10.1016/j.biortech.2017.05.209

Jiang, M., Xu, R., Xi, Y.L., Zhang, J.H., Dai, W.Y., Wan, Y.J., Chen, K.Q., Wei, P., 2013. Succinic acid production from cellobiose by *Actinobacillus succinogenes. Bioresour. Technol.* 135, 469–474. https://doi.org/10.1016/j.biortech.2012.10.019

Jiang, Y., Liu, W., Zou, H., Cheng, T., Tian, N., Xian, M., 2014. Microbial production of short chain diols. *Microb. Cell Fact.* 13, 1–17. https://doi.org/10.1186/s12934-014-0165-5

Joglekar, H.G., Rahman, I., Babu, S., Kulkarni, B.D., Joshi, A., 2006. Comparative assessment of downstream processing options for lactic acid. *Sep. Purif. Technol.* 52, 1–17. https://doi.org/10.1016/j.seppur.2006.03.015

Jolly, J., Hitzmann, B., Ramalingam, S., Ramachandran, K.B., 2014. Biosynthesis of 1, 3-propanediol from glycerol with *Lactobacillus reuteri*: Effect of operating variables. *J. Biosci. Bioeng.* xx, 1–7. https://doi.org/10.1016/j.jbiosc.2014.01.003

Jones, S., Valkenburg, C., Walton, C., 2009. Production of gasoline and diesel from biomass via fast pyrolysis, hydrotreating and hydrocracking: A design case. *Energy* 76. https://doi.org/PNNL-22684.pdf

Jurchescu, I.M., Hamann, J., Zhou, X., Ortmann, T., Kuenz, A., Prüße, U., Lang, S., 2013. Enhanced 2,3-butanediol production in fed-batch cultures of free and immobilized Bacillus licheniformis DSM 8785. *Appl. Microbiol. Biotechnol.* 97, 6715–6723. https://doi.org/10.1007/s00253-013-4981-z

Kainthola, J., Kalamdhad, A.S., Goud, V.V., 2019. Enhanced methane production from anaerobic co-digestion of rice straw and *Hydrilla verticillata* and its kinetic analysis. *Biomass Bioenerg.* 125, 8–16. https://doi.org/10.1016/j.biombioe.2019.04.011

Kajaste, R., 2014. Chemicals from biomass - Managing greenhouse gas emissions in biorefinery production chains - A review. *J. Clean. Prod.* 75, 1–10. https://doi.org/10.1016/j.jclepro.2014.03.070

Kamm, B., Kamm, M., Gruber, P. R., Kromus, S. (2005). Biorefinery Systems – An Overview. In *Biorefineries-Industrial Processes and Products* (pp. 1–40). John Wiley & Sons, Ltd. https://doi.org/https://doi.org/10.1002/9783527619849.ch1

Karatzos, S., Mcmillan, J.D., Saddler, J.N., 2014. " The potential and challenges of drop - in biofuels." IEA Bioenergy l Task 39 Rep. Biorefinery.

Kaur, G., Srivastava, A.K., Chand, S., 2012. Advances in biotechnological production of 1,3-propanediol. *Biochem. Eng. J.* 64, 106–118. https://doi.org/10.1016/j.bej.2012.03.002

Kim, H., Lee, S., Ahn, Y., Lee, J., Won, W., 2020. Sustainable production of bioplastics from lignocellulosic biomass: Technoeconomic analysis and life-cycle assessment. *ACS Sustain. Chem. Eng.* 8, 12419–12429. https://doi.org/10.1021/acssuschemeng.0c02872

Kim, J.H., Wang, C., Jang, H.J., Cha, M.S., Park, J.E., Jo, S.Y., Choi, E.S., Kim, S.W., 2016. Isoprene production by *Escherichia coli* through the exogenous mevalonate pathway with reduced formation of fermentation byproducts. *Microb. Cell Fact.* 15, 1–10. https://doi.org/10.1186/s12934-016-0612-6

Kim, S., Dale, B.E., 2005. Life cycle assessment of various cropping systems utilized for producing biofuels: Bioethanol and biodiesel. *Biomass and Bioenergy* 29, 426–439. https://doi.org/10.1016/j.biombioe.2005.06.004

Klotz, S., Kaufmann, N., Kuenz, A., Prüße, U., 2016. Biotechnological production of enantiomerically pure d-lactic acid. *Appl. Microbiol. Biotechnol.* 100, 9423–9437. https://doi.org/10.1007/s00253-016-7843-7

Koniuszewska, I., Korzeniewska, E., Harnisz, M., Czatzkowska, M., 2020. Intensification of biogas production using various technologies: A review. *Int. J. Energy Res.* 44, 6240–6258. https://doi.org/10.1002/er.5338

Kumar, P., Barrett, D.M., Delwiche, M.J., Stroeve, P., 2009. Methods for pretreatment of lignocellulosic biomass for efficient hydrolysis and biofuel production. *Ind. Eng. Chem. Res.* 48, 3713–3729. https://doi.org/10.1021/ie801542g

Kumar, R., Basak, B., Jeon, B.H., 2020. Sustainable production and purification of succinic acid: A review of membrane-integrated green approach. *J. Clean. Prod.* 277, 123954. https://doi.org/10.1016/j.jclepro.2020.123954

Kuo, Y.C., Yuan, S.F., Wang, C.A., Huang, Y.J., Guo, G.L., Hwang, W.S., 2015. Production of optically pure l-lactic acid from lignocellulosic hydrolysate by using a newly isolated and d-lactate dehydrogenase gene-deficient Lactobacillus paracasei strain. *Bioresour. Technol.* 198, 651–657. https://doi.org/10.1016/j.biortech.2015.09.071

Laurichesse, S., Avérous, L., 2014. Chemical modification of lignins: Towards bio-based polymers. *Prog. Polym. Sci.* 39, 1266–1290. https://doi.org/10.1016/j. progpolymsci.2013.11.004

Lewkowski, J., 2001. Synthesis, chemistry and applications of 5-hydroxymethyl-furfural and its derivatives. *Arkivoc* 2001, 17–54. https://doi.org/10.3998/ark.5550190.0002.102

Li, M., Nian, R., Xian, M., Zhang, H., 2018. Metabolic engineering for the production of isoprene and isopentenol by Escherichia coli. *Appl. Microbiol. Biotechnol.* 102, 7725–7738. https://doi.org/10.1007/s00253-018-9200-5

Li, Y., Wang, L., Ju, J., Yu, B., Ma, Y., 2013. Efficient production of polymer-grade d-lactate by *Sporolactobacillus laevolacticus* DSM442 with agricultural waste cottonseed as the sole nitrogen source. *Bioresour. Technol.* 142, 186–191. https://doi.org/10.1016/j. biortech.2013.04.124

Liaud, N., Rosso, M.N., Fabre, N., Crapart, S., Herpoël-Gimbert, I., Sigoillot, J.C., Raouche, S., Levasseur, A., 2015. L-lactic acid production by *Aspergillus brasiliensis* overexpressing the heterologous ldha gene from Rhizopus oryzae. *Microb. Cell Fact.* 14, 1–9. https://doi.org/10.1186/s12934-015-0249-x

Lin, R., Liu, H., Hao, J., Cheng, K., Liu, D., 2005. Enhancement of 1,3-propanediol production by *Klebsiella pneumoniae* with fumarate addition. *Biotechnol. Lett.* 27, 1755–1759. https://doi.org/10.1007/s10529-005-3549-0

Lindorfer, J., Lettner, M., Hesser, F., Fazeni, K., Rosenfield, D., Annevelink, B., Mandl, M., 2019. Technical, economic and environmental assessment of biorefinery concepts: Developing a practical approach for characterisation. IEA (International Energy Agency). Bioenergy: Task 2019, 42, 01.

Liu, J., Chan, S.H.J., Chen, J., Solem, C., Jensen, P.R., 2019. Systems biology – A guide for understanding and developing improved strains of lactic acid bacteria. *Front. Microbiol.* 10, 876. https://doi.org/10.3389/fmicb.2019.00876

Lu, J., Li, J., Gao, H., Zhou, D., Xu, H., Cong, Y., Zhang, W., Xin, F., Jiang, M., 2021. Recent progress on bio-succinic acid production from lignocellulosic biomass. *World J. Microbiol. Biotechnol.* 37, 1–8. https://doi.org/10.1007/s11274-020-02979-z

Mainka, H., Täger, O., Körner, E., Hilfert, L., Busse, S., Edelmann, F.T., Herrmann, A.S., 2015. Lignin – An alternative precursor for sustainable and cost-effective automotive carbon fiber. *J. Mater. Res. Technol.* 4, 283–296. https://doi.org/10.1016/j.jmrt.2015.03.004

Mao, C., Feng, Y., Wang, X., Ren, G., 2015. Review on research achievements of biogas from anaerobic digestion. *Renew. Sustain. Energy Rev.* 45, 540–555. https://doi.org/10.1016/j. rser.2015.02.032

Mazzoli, R., 2020. Metabolic engineering strategies for consolidated production of lactic acid from lignocellulosic biomass. *Biotechnol. Appl. Biochem.* 67, 61–72. https://doi. org/10.1002/bab.1869

Meegoda, J.N., Li, B., Patel, K., Wang, L.B., 2018. A review of the processes, parameters, and optimization of anaerobic digestion. *Int. J. Environ. Res. Public Health* 15. https://doi. org/10.3390/ijerph15102224

Melin, K., & Hurme, M. (2011). Lignocellulosic biorefinery economic evaluation. *Cellul. Chem. Technol.*, 45(7–8), 443–454.

Miller, C., Fosmer, A., Rush, B., McMullin, T., Beacom, D., Suominen, P., 2011. Industrial production of lactic acid, in: *Comprehensive Biotechnology.* pp. 179–188. https://doi. org/10.1016/B978-0-08-088504-9.00177-X

Mittal, S., Ahlgren, E.O., Shukla, P.R., 2018. Barriers to biogas dissemination in India: A review. *Energy Policy* 112, 361–370. https://doi.org/10.1016/j.enpol.2017.10.027

Morales, M., Ataman, M., Badr, S., Linster, S., Kourlimpinis, I., Papadokonstantakis, S., Hatzimanikatis, V., Hungerbühler, K., 2016. Sustainability assessment of succinic acid production technologies from biomass using metabolic engineering. *Energy Environ. Sci.* 9, 2794–2805. https://doi.org/10.1039/c6ee00634e

Motagamwala, A.H., Won, W., Sener, C., Alonso, D.M., Maravelias, C.T., Dumesic, J.A., 2018. Toward biomass-derived renewable plastics: Production of 2,5-furandicarboxylic acid from fructose. *Sci. Adv.* 4, 1–9. https://doi.org/10.1126/sciadv.aap9722

Mu, W., Ben, H., Ragauskas, A., Deng, Y., 2013. Lignin pyrolysis components and upgrading—Technology review. *BioEnergy Res.* 6, 1183–1204. https://doi.org/10.1007/s12155-013-9314-7

Munasinghe, P.C., Khanal, S.K., 2011. Biomass-derived syngas fermentation into biofuels, in: *Biofuels: Alternative Feedstocks and Conversion Processes*. Elsevier Inc., pp. 79–98. https://doi.org/10.1016/B978-0-12-385099-7.00004-8

Nakamura, C.E., Whited, G.M., 2003. Metabolic engineering for the microbial production of. *Curr. Opin. Biotechnol.* 14, 454–459. https://doi.org/10.1016/j.copbio.2003.08.005

Nguyen, C.M., Kim, J.S., Hwang, H.J., Park, M.S., Choi, G.J., Choi, Y.H., Jang, K.S., Kim, J.C., 2012. Production of l-lactic acid from a green microalga, *Hydrodictyon reticulum*, by Lactobacillus paracasei LA104 isolated from the traditional Korean food, makgeolli. *Bioresour. Technol.* 110, 552–559. https://doi.org/10.1016/j.biortech.2012.01.079

Ögmundarson, Ó., Sukumara, S., Laurent, A., Fantke, P., 2020. Environmental hotspots of lactic acid production systems. *GCB Bioenergy* 12, 19–38. https://doi.org/10.1111/gcbb.12652

Oshiro, K., Masui, T., Kainuma, M., 2018. Transformation of Japan's energy system to attain net-zero emission by 2050. *Carbon Manag.* 9, 493–501. https://doi.org/10.1080/17583004.2017.1396842

Overbeck, T., Steele, J.L., Broadbent, J.R., 2016. Fermentation of de-oiled algal biomass by Lactobacillus casei for production of lactic acid. *Bioprocess Biosyst. Eng.* 39, 1817–1823. https://doi.org/10.1007/s00449-016-1656-z

Pihlajaniemi, V., Sipponen, M. H., Liimatainen, H., Sirviö, J. A., Nyyssölä, A., & Laakso, S. (2016). Weighing the factors behind enzymatic hydrolyzability of pretreated lignocellulose. *Green Chem.*, 18(5), 1295–1305. https://doi.org/10.1039/C5GC01861G

Pantaleo, A., Villarini, M., Colantoni, A., Carlini, M., Santoro, F., Hamedani, S.R., 2020. Techno-economic modeling of biomass pellet routes: Feasibility in Italy. *Energies* 13, 1–15. https://doi.org/10.3390/en13071636

Papagianni, M., 2012. Metabolic engineering of lactic acid bacteria for the production of industrially important compounds. *Comput. Struct. Biotechnol. J.* 3, e201210003. https://doi.org/10.5936/csbj.201210003

Paul, S., Dutta, A., 2018. Challenges and opportunities of lignocellulosic biomass for anaerobic digestion. *Resour. Conserv. Recycl.* 130, 164–174. https://doi.org/10.1016/j.resconrec.2017.12.005

Raab, A.M., Gebhardt, G., Bolotina, N., Weuster-Botz, D., Lang, C., 2010. Metabolic engineering of *Saccharomyces cerevisiae* for the biotechnological production of succinic acid. *Metab. Eng.* 12, 518–525. https://doi.org/10.1016/j.ymben.2010.08.005

Rackemann, D., Doherty, W., 2019. A review on the production of levulinic acid and furanics from sugars. *J. Chem. Inf. Model.* 53, 1689–1699.

Rajendran, K., Murthy, G.S., 2017. How does technology pathway choice influence economic viability and environmental impacts of lignocellulosic biorefineries? *Biotechnol. Biofuels* 10, 1–19. https://doi.org/10.1186/s13068-017-0959-x

Rajesh, R.O., Pandey, A., Parameswaran B., 2018. Bioprocesses for the production of 2,5-Furandicarboxylic acid. https://doi.org/10.1007/978-981-10-7434-9_8

Raschka, A., Carus, M., 2012. *Industrial Material Use of Biomass Basic Data for Germany, Europe and the World*. Hürth: Nova-Institute.

Rauch, R., Hrbek, J., Hofbauer, H., 2014. Biomass gasification for synthesis gas production and applications of the syngas. *WIREs Energy Environ.* 3, 343–362. https://doi.org/10.1002/wene.97

Reddy, L.V., Kim, Y.M., Yun, J.S., Ryu, H.W., Wee, Y.J., 2016. L-Lactic acid production by combined utilization of agricultural bioresources as renewable and economical substrates through batch and repeated-batch fermentation of Enterococcus faecalis RKY1. *Bioresour. Technol.* 209, 187–194. https://doi.org/10.1016/j.biortech.2016.02.115

Robak, K., Balcerek, M., 2018. Review of second generation bioethanol production from residual biomass. *Food Technol. Biotechnol.* 56, 174–187. https://doi.org/10.17113/ftb.56.02.18.5428

Rosales-Calderon, O., Arantes, V., 2019. A review on commercial-scale high-value products that can be produced alongside cellulosic ethanol. *Biotechnology for Biofuels.* BioMed Central. https://doi.org/10.1186/s13068-019-1529-1

Rose, M., Weber, D., Lotsch, B. V., Kremer, R.K., Goddard, R., Palkovits, R., 2013. Biogenic metal-organic frameworks: 2,5-Furandicarboxylic acid as versatile building block. *Microporous Mesoporous Mater.* 181, 217–221. https://doi.org/10.1016/j.micromeso.2013.06.039

Sabra, W., Groeger, C., Zeng, A.P., 2016. Microbial cell factories for diol production. *Adv. Biochem. Eng. Biotechnol.* 155, 165–197. https://doi.org/10.1007/10_2015_330

Salvachúa, D., Mohagheghi, A., Smith, H., Bradfield, M.F.A., Nicol, W., Black, B.A., Biddy, M.J., Dowe, N., Beckham, G.T., 2016. Succinic acid production on xylose-enriched biorefinery streams by *Actinobacillus succinogenes* in batch fermentation. *Biotechnol. Biofuels* 9, 1–15. https://doi.org/10.1186/s13068-016-0425-1

Santibañez-Aguilar, J.E., González-Campos, J.B., Ponce-Ortega, J.M., Serna-González, M., El-Halwagi, M.M., 2014. Optimal planning and site selection for distributed multiproduct biorefineries involving economic, environmental and social objectives. *J. Clean. Prod.* 65, 270–294. https://doi.org/10.1016/j.jclepro.2013.08.004

Saxena, R.K., Saran, S., Isar, J., Kaushik, R., 2016. Production and applications of succinic acid. *Curr. Dev. Biotechnol. Bioeng. Prod. Isol. Purif. Ind. Prod.*, 601–630. https://doi.org/10.1016/B978-0-444-63662-1.00027-0

Scarlat, N., Dallemand, J.-F., Fahl, F., 2018. Biogas: Developments and perspectives in Europe. *Renew. Energy* 129, 457–472. https://doi.org/10.1016/j.renene.2018.03.006

Seemann, M., Tse Sum Bui, B., Wolff, M., Miginiac-Maslow, M., Rohmer, M., 2006. Isoprenoid biosynthesis in plant chloroplasts via the MEP pathway: Direct thylakoid/ferredoxin-dependent photoreduction of GcpE/IspG. *FEBS Lett.* 580, 1547–1552. https://doi.org/10.1016/j.febslet.2006.01.082

Senyek, M.L., 2008. Isoprene polymers, in: *Encyclopedia of Polymer Science and Technology.* American Cancer Society. https://doi.org/10.1002/0471440264.pst175

Sirothiya, M., Chavadi, C., 2020. Role of compressed biogas to assess the effects of perceived value on customer satisfaction and customer loyalty. *BIMTECH Bus. Perspect.* - I, 70–89.

Smith, A.D., Landoll, M., Falls, M., Holtzapple, M.T., 2010. Chemical production from lignocellulosic biomass: Thermochemical, sugar and carboxylate platforms. *Bioalcohol Prod. Biochem. Convers. Lignocellul. Biomass*, 391–414. https://doi.org/10.1533/9781845699611.5.391

Sodergard, A., Stolt, M., 2010. Industrial production of high molecular. *Poly(lactic acid) Synth. Sturucture, Prop. Process. Appl.*, 27–41.

Sokhansanj, S., Hess, J.R., 2009. Biomass supply logistics and infrastructure. *Methods Mol. Biol.* 581, 1–25. https://doi.org/10.1007/978-1-60761-214-8_1

Song, C.W., Park, J.M., Chung, S.C., Lee, S.Y., Song, H., 2019. Microbial production of 2,3-butanediol for industrial applications. *J. Ind. Microbiol. Biotechnol.* 46, 1583–1601. https://doi.org/10.1007/s10295-019-02231-0

Souto, F., Calado, V., Pereira, N., 2018. Lignin-based carbon fiber: A current overview. *Mater. Res. Express* 5, 72001. https://doi.org/10.1088/2053-1591/aaba00

Takkellapati, S., Li, T., Gonzalez, M.A., 2018. An overview of biorefinery-derived platform chemicals from a cellulose and hemicellulose biorefinery. *Clean Technol. Environ. Policy* 20, 1615–1630. https://doi.org/10.1007/s10098-018-1568-5

Tang, J., Wang, X., Hu, Y., Zhang, Y., Li, Y., 2016. Lactic acid fermentation from food waste with indigenous microbiota: Effects of pH, temperature and high OLR. *Waste Manag.* 52, 278–285. https://doi.org/10.1016/j.wasman.2016.03.034

Tarraran, L., Mazzoli, R., 2018. Alternative strategies for lignocellulose fermentation through lactic acid bacteria: The state of the art and perspectives. *FEMS Microbiol. Lett.* 365. https://doi.org/10.1093/femsle/fny126

Tashiro, Y., Kaneko, W., Sun, Y., Shibata, K., Inokuma, K., Zendo, T., Sonomoto, K., 2011. Continuous D-lactic acid production by a novelthermotolerant Lactobacillus delbrueckii subsp. lactis QU 41. Appl. Microbiol. Biotechnol. 89, 1741–1750. https://doi.org/10.1007/s00253-010-3011-7

Tirado-Acevedo, O., Chinn, M.S., Grunden, A.M., 2010. Production of biofuels from synthesis gas using microbial catalysts. *Adv. Appl. Microbiol.* 70, 57–92. https://doi.org/10.1016/S0065-2164(10)70002-2

Tsvetanova, F., Petrova, P., Petrov, K., 2014. 2,3-Butanediol production from starch by engineered *Klebsiella pneumoniae* G31-A. *Appl. Microbiol. Biotechnol.* 98, 2441–2451. https://doi.org/10.1007/s00253-013-5418-4

van der Waal, J.C., de Jong, E., 2016. Avantium chemicals: The high potential for the levulinic product tree, in: *Industrial Biorenewables*. John Wiley & Sons, Ltd., pp. 97–120. https://doi.org/10.1002/9781118843796.ch4

van Heerden, C.D., Nicol, W., 2013. Continuous and batch cultures of Escherichia coli KJ134 for succinic acid fermentation: Metabolic flux distributions and production characteristics. *Microb. Cell Fact.* 12. https://doi.org/10.1186/1475-2859-12-80

Wang, B., Gebreslassie, B.H., You, F., 2013. Sustainable design and synthesis of hydrocarbon biorefinery via gasification pathway : Integrated life cycle assessment and technoeconomic analysis with multiobjective superstructure optimization. *Comput. Chem. Eng.* 52, 55–76. https://doi.org/10.1016/j.compchemeng.2012.12.008

Wang, C., Yan, D., Li, Q., Sun, W., Xing, J., 2014. Ionic liquid pretreatment to increase succinic acid production from lignocellulosic biomass. *Bioresour. Technol.* 172, 283–289. https://doi.org/10.1016/j.biortech.2014.09.045

Winters, P. (2016). Advancing the biobased economy: Renewable chemical biorefinery commercialization, progress, and market opportunities, 2016 and beyond. *Industrial Biotechnology*, 12, 290–294. https://doi.org/10.1089/ind.2016.29050.pwi

Wu, D., Li, L., Zhao, X., Peng, Y., Yang, P., Peng, X., 2019. Anaerobic digestion: A review on process monitoring. *Renew. Sustain. Energy Rev.* 103, 1–12. https://doi.org/10.1016/j.rser.2018.12.039

Xu, Y., Foster, J.L., Muir, J.P., Burson, B.L., Jessup, R.W., 2018. Succinic acid production across candidate lignocellulosic biorefinery feedstocks. *Am. J. Plant Sci.* 09, 2141–2153. https://doi.org/10.4236/ajps.2018.911155

Xu, Y., Liu, H., Du, W., Sun, Y., Ou, X., Liu, D., 2009. Integrated production for biodiesel and 1,3-propanediol with lipase-catalyzed transesterification and fermentation. *Biotechnol. Lett.* 31, 1335–1341. https://doi.org/10.1007/s10529-009-0025-2

Xue, J., Ahring, B.K., 2011. Enhancing isoprene production by genetic modification of the 1-deoxy-D-Xylulose-5-phosphate pathway in *Bacillus subtilis. Appl. Environ. Microbiol.* 77, 2399–2405. https://doi.org/10.1128/AEM.02341-10

Yang, J., Xian, M., Su, S., Zhao, G., Nie, Q., Jiang, X., Zheng, Y., Liu, W., 2012. Enhancing production of bio-isoprene using hybrid MVA pathway and isoprene synthase in E. coli. *PLoS One* 7, 1–7. https://doi.org/10.1371/journal.pone.0033509

Yang, L., Henriksen, M.M., Hansen, R.S., Lübeck, M., Vang, J., Andersen, J.E., Bille, S., Lübeck, P.S., 2020. Metabolic engineering of *Aspergillus niger* via ribonucleoprotein-based CRISPR–Cas9 system for succinic acid production from renewable biomass. *Biotechnol. Biofuels* 13, 1–12. https://doi.org/10.1186/s13068-020-01850-5

Yang, X., Lai, Z., Lai, C., Zhu, M., Li, S., Wang, J., Wang, X., 2013. Efficient production of L-lactic acid by an engineered *Thermoanaerobacterium aotearoense* with broad substrate specificity. *Biotechnol. Biofuels* 6, 1–12. https://doi.org/10.1186/1754-6834-6-124

Ye, L., Hudari, M.S. Bin, Li, Z., Wu, J.C., 2014. Simultaneous detoxification, saccharification and co-fermentation of oil palm empty fruit bunch hydrolysate for l-lactic acid production by Bacillus coagulans JI12. *Biochem. Eng. J.* 83, 16–21. https://doi.org/10.1016/j.bej.2013.12.005

Ye, L., Zhao, H., Li, Z., Wu, J.C., 2013. Improved acid tolerance of Lactobacillus pentosus by error-prone whole genome amplification. *Bioresour. Technol.* 135, 459–463. https://doi.org/10.1016/j.biortech.2012.10.042

Yu, O., Kim, K.H., 2020. Lignin to materials: A focused review on recent novel lignin applications. *Appl. Sci.* 10. https://doi.org/10.3390/app10134626

Zacher, A.H., Olarte, M. V., Santosa, D.M., Elliott, D.C., Jones, S.B., 2014. A review and perspective of recent bio-oil hydrotreating research. *Green Chem.* 16, 491–515. https://doi.org/10.1039/c3gc41382a

Zeng, A.P., Sabra, W., 2011. Microbial production of diols as platform chemicals: Recent progresses. *Curr. Opin. Biotechnol.* 22, 749–757. https://doi.org/10.1016/j.copbio.2011.05.005

Zhang, H., Lopez, P. C., Holland, C., Lunde, A., Ambye-Jensen, M., Felby, C., & Thomsen, S. T. (2018). The multi-feedstock biorefinery – Assessing the compatibility of alternative feedstocks in a 2G wheat straw biorefinery process. *GCB Bioenergy* 10(12), 946–959. https://doi.org/https://doi.org/10.1111/gcbb.12557

Zhang, Y.-H.P., 2008. Reviving the carbohydrate economy via multi-product lignocellulose biorefineries. *J. Ind. Microbiol. Biotechnol.* 35, 367–375. https://doi.org/10.1007/s10295-007-0293-6

Zhang, Yiwen, Jiang, M., Zhang, Yuqing, Cao, Q., Wang, X., Han, Y., Sun, G., Li, Y., Zhou, J., 2019. Novel lignin–chitosan–PVA composite hydrogel for wound dressing. *Mater. Sci. Eng. C* 104, 110002. https://doi.org/10.1016/j.msec.2019.110002

Zheng, X., Gu, X., Ren, Y., Zhi, Z., Lu, X., 2016. Production of 5-hydroxymethyl furfural and levulinic acid from lignocellulose in aqueous solution and different solvents. *Biofuels, Bioprod. Biorefining* 10, 917–931. https://doi.org/10.1002/bbb.1720

Zheng, Y., Zhao, J., Xu, F., Li, Y., 2014. Pretreatment of lignocellulosic biomass for enhanced biogas production. *Prog. Energy Combust. Sci.* 42, 35–53. https://doi.org/10.1016/j.pecs.2014.01.001

9 Targeted Strain Engineering to Produce Bioenergy

S. Bilal Jilani
International Centre for Genetic
Engineering and Biotechnology
Amity University

Ali Samy Abdelaal
International Centre for Genetic
Engineering and Biotechnology
Damietta University

Syed Shams Yazdani
International Centre for Genetic
Engineering and Biotechnology

CONTENTS

9.1 Introduction 241
9.2 Different Generations of Biofuels..242
    9.2.1 First Generation 242
    9.2.2 Second Generation..244
        9.2.2.1 Resistance Engineered against Inhibitors.........................245
    9.2.3 Third Generation 251
9.3 High-Energy Advanced Biofuels..253
    9.3.1 Bioalcohols 253
        9.3.1.1 n-Butanol..253
        9.3.1.2 Isobutanol...254
    9.3.2 Hydrocarbons..255
    9.3.3 Metabolic Engineering for the Production of Alkanes/Alkenes 256
References...259

9.1 INTRODUCTION

The ever increasing need for development has placed unprecedented stress on traditional energy sources, of which combustion of fossil fuels is the primary source. However, extraction and combustion of fossil fuels also result in release of by-products

DOI: 10.1201/9781003158486-9

which can be polluting as well as add to global warming. In the past decade, there has been a boom in fracking in the shale gas and oil trapped inside the sedimentary rocks of northeastern continental USA. In addition to the generation of toxic waste involved in mining fossil minerals, the abandoned wells seep out heavy compounds which pollute the water bodies and cause widespread damage to local ecology via acidification and increase in heavy metal load of the freshwaters (Akcil and Koldas, 2006). Further, cleanup of the toxins cost billions of dollar to the exchequer. Any reduction in the mining and pollution caused due to fossil fuels will not only result in the reduction of deterioration of environment, but also lead to an investment in the development of alternative fuel technologies due to diversion of cleanup funds to research and development of the same. One of the promising technologies lie in harnessing of the cellulosic carbon and channeling it toward the synthesis of biomolecules with high calorific values. The efficient harnessing of the cellulosic carbon has the potential toward the generation of biofuel with net-zero emissions. And for the fuel to gain socioeconomic acceptance it needs to be devoid of any feedstock linked to human consumption and requires least resources for cultivation. This chapter will provide an overview of the cellulosic carbon which is presently used for the generation of biofuel compounds and shed light on further development of the same.

9.2   DIFFERENT GENERATIONS OF BIOFUELS

Depending upon the starting carbon source, the process of biofuel production has been categorized into different generations. These are discussed further in this chapter.

9.2.1   FIRST GENERATION

First-generation biofuels are bioenergy compounds produced from sugar and starch derived from plant parts that are fit for human consumption. The pathway for the synthesis of ethanol by either *Escherichia coli* or *Saccharomyces cerevisiae* using glucose/xylose as carbon source(s) is described in Figure 9.1. In case of sucrose as a carbon source, an additional enzyme that can break the glycosidic bond between glucose and fructose is required. In *S. cerevisiae*, the endogenous enzyme invertase catalyzes the hydrolysis of this bond, while the absence of the same enzyme in *E. coli* results in poor utilization of sucrose. Cloning of *casAB* genes from *Klebsiella oxytoca* which encode for Enzyme II[cellobiose] and phospho-β-glucosidase has yielded *E. coli* KO11 which can also efficiently utilize sucrose as a carbon source (Da Silva et al., 2005). Using sugar cane juice as a primary carbon source, strains *E. coli* KO11 and *K. oxytoca* P2 produced maximum ethanol titers of 39.4 and 42.1 g/L, respectively. Thus, cane juice holds promise to serve as a source of carbon for biofuel production (Da Silva et al., 2005). *Pichia kudriavzevii* cells have been adapted to galactose medium and used as a biocatalyst to convert sugars in cane juice into ethanol. The rationale behind the adaptation against galactose had been to improve both the efficiency and co-utilization of sugars present in the medium by repressing the CCR effect exerted by glucose. The cane juice consisted of 14% (w/v) sucrose, 2% (w/v) glucose and 1% (w/v) fructose. At temperatures of 40°C and 45°C, the galactose-adapted yeast cells produced ethanol at titers of 71.95 and 58.53 g/L, respectively. These titers represent

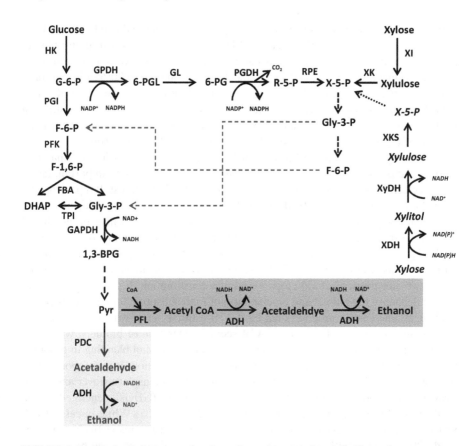

FIGURE 9.1 **Flow of glucose and xylose through central metabolic pathways to produce ethanol in *E. coli* and *S. cerevisiae*.** The ability to convert xylose to X-5-P is absent in *S. cerevisiae* and is achieved by genetic manipulation. Pathway in italics refer to the non-native gene expression in *S. cerevisiae* for xylose assimilation. Dark-shaded region highlights the fate of pyruvate in *E. coli*. Light-shaded region highlights the fate of pyruvate in *S. cerevisiae*. Abbreviations are as follows: G-6-P, glucose-6-phosphate; 6-PGL, 6-phosphogluconolactone; 6-PG, 6-phosphogluconate; R-5-P, ribulose-5-phosphate; X-5-P, xylulose-5-phosphate; Gly-3-P, glyceraldehyde-3-phosphate; F-6-P, fructose-6-phosphate; F-1,6-P, fructose-1,6-bisphosphate; DHAP, dihydroxyacetone phosphate; 1,3-BPG, 1,3-bisphosphoglycerate; 3-PG, 3-phosphoglycerate; 2-PG, 2-phosphoglycerate; PEP, phosphoenolpyruvate; Pyr, pyruvate; HK, hexokinase; PGI, phosphoglucoisomerase; PFK, phosphofructokinase; FBA, aldolase; TPI, triosephosphate isomerase; GAPDH, glyceraldehyde-3-phosphate dehydrogenase; PGK, phosphoglycerate kinase; PGM, phosphoglycerate mutase; ENO, enolase; PK, pyruvate kinase; PFL, pyruvate phosphate lyase; ADH, alcohol dehydrogenase; PDC, pyruvate decarboxylase; GPDH, glucose-6-phosphate dehydrogenase; GL, 6-phosphogluconolactonase; PGDH, 6-phosphogluconate dehydrogenase; RPE, ribulose-5-phosphate epimerase; XK, xylulose kinase; XI, xylose isomerase; XKS, xylulose kinase; XyDH, xylitol dehydrogenase; XDH, xylose reductase

32% and 30% higher ethanol levels as compared to the non-adapted cells at 40°C and 45°C, respectively (Dhaliwal et al., 2011). Another class of substrate for biofuel is crops rich in starch, such as cassava and sweet potato. Cassava is a tuberous crop mainly produced in tropical and sub-tropical regions, and up to 90% of its carbohydrate content can be starch (Jansson et al., 2009). A novel glucoamylase PoGA15A from *Penicillium oxalicum* GXU20 was expressed in *Pichia pastoris*. This strain in the presence of α-amylase fermented 150 g/L of raw cassava flour to produce ethanol titers of 57.0 g/L after 36 h with an efficiency of 93.5% (Xu et al., 2016).

The life cycle assessment of any biofuel is performed to assess its sustainability. And for the purpose, the energy return on investment (EROI) is an important factor which determines the long-term feasibility of the biofuel of interest. It represents the ratio of total energy supplied by the biofuel to the total energy utilized during its production and any value above 1 suggests net energy gains. EROI values of bioethanol sourced from sugarcane and wheat lie in the range of 3.1–9.3 and 1.6–5.8, respectively. These values represent a promising scope of these compounds for bioenergy purposes (Gasparatos et al., 2013). However, the carbon sources represented by the above two substrates are food crops for human consumption. It generates the never ending debate of the ethical and practical use of food substitute for energy purposes. And sugarcane carries an additional burden of a thirsty crop which requires copious amounts of water. However, LCA of second-generation of biofuels, which use non-food-based feedstock, revealed that the higher the ethanol blending in gasoline, the more favorable is the environmental cost of its production. It was reported that in case of E100 blend, non-food agricultural products such as corn stover and wheat straw result in the highest reduction in greenhouse gases (GHGs), generated during production of biofuel, with the values between 82% and 91%. While in case of E10 and E85 blends, the GHG reduction was <10% and >40%, respectively (Morales et al., 2015). Thus, bulk production of biofuels using lignocellulosic biomass is a feasible idea wherein lies the focus of this chapter.

9.2.2 SECOND GENERATION

Second-generation biofuels refer to biofuels produced using lignocellulosic biomass as the carbon source. However, plant biomass is a recalcitrant source of carbon and requires treatment by one of the many physical and/or chemical methods in order to make it available for microbial metabolism (Sun and Cheng, 2002). Acidic treatment is a successful method for obtaining a high yield of pentose sugars which are trapped in hemicellulose fraction. Acidic pretreatment can be performed using either concentrated or dilute acidic treatment. The acidic pretreatment results in the hydrolysis of xylan component of hemicellulose, while the glucomannan part is relatively stable and requires alkaline pH. The strong acidic treatment results in efficient solubilization of hemicellulose, and also lignin gets precipitated due to the highly acidic environment as compared to a dilute acidic pretreatment. However, strong acid treatment also leads to corrosion of industrial equipment and generation of inhibitory compounds. Under low acidic conditions, both risks are significantly reduced (Hendriks and Zeeman, 2009). However, both concentrated and dilute acidic pretreatments also result in the formation of compounds which are inhibitors of the microbial metabolism. The important inhibitory compounds are furfural, 5-hydroxymethylfurfural (5-HMF), acetic acid and aromatic compounds. Furfural is a result of breakdown of pentose sugar (e.g., xylose), while 5-HMF is a breakdown product of hexose sugar (e.g., glucose). It is important to note that furfural is one of the most potent inhibitory molecules of microbial metabolism, while 5-HMF and acetic acid are relatively less toxic.

9.2.2.1 Resistance Engineered against Inhibitors

9.2.2.1.1 Furfural/5-Hydroxymethylfurfural

As early as in 1981, it was reported that furfural inhibits growth and alcohol production by *Saccharomyces cerevisiae* over a range from 0.5 to 4 g/L. The inhibitory effect of 1 g/L furfural was maximum on triosephosphate dehydrogenase which displayed 50% inhibition, and at 2 g/L, the same was inhibited 100% (Banerjee et al., 1981).

Furfural by itself also poses a significant challenge to microbial metabolism. Furfural exerts its toxicity by withdrawing away NADPH from anabolic reactions. The transcriptomic analysis of LY180 in the presence of 0.5 g/L furfural revealed that genes involved in the biosynthesis of purines, pyrimidines, and amino acids were downregulated. In contrast, genes involved in the biosynthesis of cysteine and methionine were upregulated. Upon supplementation of AM1 minimal media with 0.1 mM amino acid and in the presence of 1 g/L furfural, it was revealed that cysteine and methionine were most pronounced in rescuing the growth of LY180. It is important to emphasize that the biosynthesis of these two sulfur-rich amino acids is a NADPH-expensive process. This fact was confirmed by the overexpression of membrane-bound transhydrogenase *pntAB*, which resulted in an increase in furfural tolerance under similar culture conditions. PntAB catalyzes the reduction of NADP$^+$ by oxidizing NADH (Miller et al., 2009b). Furfural stresses the cellular physiology by inducing the formation of reactive oxygen species (ROS) in *Saccharomyces cerevisiae* and results in fragmentation and aggregation of membranes of intracellular organelles such as mitochondria and vacuoles. The organization of nuclear chromatin is also disrupted, and cells exhibit diffuse chromatin after furfural treatment as compared to tightly packed ones in untreated control. The actin structures in furfural-treated cells also exhibited patchiness as compared to the normal cells during exponential growth (Allen et al., 2010). 5-HMF has also been implicated in increasing the ROS load of a microbial cell. And both furfural and 5-HMF have been shown to decrease GSH level, which acts as a sentinel against cellular oxidative stress. In the same study, it was shown that furfural was much more potent in depleting GSH than 5-HMF. An increase in cellular GSH level leads to an increase in tolerance against furfural. The tolerance was not observed against 5-HMF. In a similar manner, the addition of DTT to the media resulted in increased tolerance against furfural only and not to 5-HMF (Kim and Hahn, 2013). Multidrug resistance pumps and transporters have also been identified to result in conferring tolerance against furfural in *E. coli*. The overexpression of multidrug resistance genes—*sugE* and *mdtJI*—and a lactate/glycolate:H+ symporter—*lldp*—has been associated with conferring tolerance against furfural. Out of the three, the expression of *mdtJI* was most beneficial in conferring tolerance at the lowest (0.01 mM) IPTG concentration tested in the presence of 1.25 g/L furfural. At 10% xylose load, the overexpression of *mdtJI* at 0.01 mM IPTG resulted in an ethanol productivity of 0.42 g/L/h at 96 h as compared to 0.19 g/L/h for the empty vector control (Kurgan et al., 2019). Polyamines have also been implicated in conferring tolerance against furfural in *E. coli*. Four transporter genes—*potE*, *puuP*, *plaP* and *potABCD*—were reported to confer beneficial effect as compared to an empty plasmid control in media containing 5% xylose and 10 mM furfural, and induced with 0.1 mM IPTG. Of these, *potE* and *puuP* which are proton symporters involved in putrescine uptake conferred the most beneficial effect. The overexpression of *potE* was relatively more beneficial as compared to *puuP* to ferment 10% xylose in the presence of 10 mM furfural. However, both cultures were able to reach similar maximum ethanol titers of around 43 g/L at 96 h. At the same time point neither significant sugar utilization nor ethanol production could be observed in an empty plasmid control (Geddes et al., 2014). Engineering furfural tolerance in *E. coli* has also been achieved using FucO which uses NADH to oxidize furfural to furfuryl alcohol. FucO has apparent K_m values of 0.4 ± 0.2 and 0.7 ± 0.3 mM toward furfural and 5-HMF, respectively. In the presence of 10% xylose and 15 mM furfural, the overexpression of FucO in *E. coli* strain EM322 led to maximum ethanol titers of around 43 g/L at 72 h with around 90% of the maximum theoretical yield. The influence of FucO to promote lactate formation was also investigated. In the presence of 15 mM furfural, *E. coli* strain XW068 was able to ferment 10% xylose and produce lactate with approximately 85% of maximum theoretical yield at 120 h (Wang et al., 2011). The redundancy to metabolize furfural is evident from the fact that even 'cryptic gene' *ucpA* is implicated in conferring furfural tolerance. The overexpression of UcpA in the presence of 10 mM furfural led to fermentation of 10% xylose, and ethanol could be detected in the media as compared to the empty vector control where no significant amount of ethanol could be detected in the media even at 72 h (X. Wang et al., 2012). Screening of genomic libraries of *Escherichia coli* NC3, *Bacillus subtilis* YB886 and *Zymomonas mobilis* CP4 revealed a common gene component—*thyA*—to be involved in conferring tolerance against 10 mM furfural in the presence of 10% xylose as carbon source. The ethanol titers observed at 96 h in the presence of the inhibitor were comparable to that observed at 48 h in the absence of the inhibitor, which was around 33 g/L. Supplementation of the media with either 0.2 mM thymine or 0.2 mM thymidine also resulted in a significant increase in biomass, in the presence of 10 mM furfural, as compared to control without any supplements. The growth advantage observed for these two supplements were comparable to the one which involved a combination of 0.1 mM serine and 0.5 mM tetrahydrofolate. Together, these results suggest that an approach which fortifies the pyrimidine pools of cell also leads to an increase in microbial tolerance against furfural (Zheng et al., 2012).

An interesting example of convergence of stress resistance phenotypes can be demonstrated by considering the case of *Deinococcus radiodurans* and *E. coli*. The gene under consideration encodes for IrrE protein of *D. radiodurans*. IrrE is a regulator which is involved in activating the DNA repair pathways mediated by RecA and PprA in radiation-resistant *D. radiodurans*. Using error-prone PCR, a plasmid library of *E. coli irrE* mutants was generated, wherein it was observed that F2-1 clone exhibited remarkable tolerance against furfural, 5-HMF and vanillin. In separate experiments and in the presence of complex LB medium, clone F2-1 was able to increase its biomass in the presence of 2, 3.5 and 1.5 g/L of furfural, 5-HMF and vanillin, respectively. And no significant increase in biomass of the controls—host with wild type *irrE* and with empty plasmid—could be observed. Remarkably, clone F2-1 was also able to increase its biomass in the presence of 60% (v/v) of dilute acid-treated corn stover hydrolysate which had been subsequently treated with Celluclast-1.5L and Novozyme-188. The maximum OD_{600} for F2-1 was around 2.5, while in respective control, there was no remarkable growth (J. Wang et al., 2012).

Screening study involving gene disruption library in *Saccharomyces cerevisiae* revealed that the genes involved in modulating carbon flux through pentose phosphate pathway were involved in sensitivity toward furfural. The overexpression of PPP genes, viz. zwf*1*, *gnd1*, *rpe1* and *tkl1*, led to increased tolerance against furfural (Gorsich et al., 2006). The other important source set of genes responsible for conferring tolerance against the furan inhibitors are the ones involved in maintaining redox machinery. The importance of aldehyde reductases (ADs) in conferring tolerance against inhibitors can be inferred from the fact that a partially purified protein of a novel gene (Y63) isolated from *S. cerevisiae* NRRL Y12632 has demonstrated aldehyde reduction activities against 14 aldehydes which included furfural and 5-HMF, many of which are commonly generated during the pretreatment of biomass. Experimental evidence suggested that the enzyme uses NADPH as a cofactor (Liu and Moon, 2009). *S. cerevisiae* genes *adh7*, *ald4* and *gre3* exhibited strong reduction capabilities against both furfural and 5-HMF with NADH as a cofactor and not NADPH. On the other hand, *adh6* showed reduction capabilities against both aldehydes with NADPH as a cofactor (Lewis Liu et al., 2008). In case of *Saccharomyces cerevisiae* and in the presence of glucose as sole carbon source, it has been reported that the furfural reduction competes with glycerol formation which serves as a means to regenerate NAD^+. The formation of ethanol is also a strategy to regenerate NAD^+. Thus, the reduction of furfural to furfuryl alcohol and also acetaldehyde to ethanol also leads to a competition between NADH. In the presence of furfural in the media, the specific ethanol production rate decreased and acetaldehyde could be detected in the media (Palmqvist et al., 1999). A furfural reductase enzyme has also been defined from *E. coli* LYO1, which has activity against furfural with an apparent K_m and V_{max} for furfural of 1.5×10^{-4} M and 28.5 µmol/min/mg of protein, respectively. The enzyme displayed catalytic activity using NADPH as a cofactor (Gutiérrez et al., 2006). Evolutionary engineering has successfully been used to generate furan inhibitor-resistant *Saccharomyces cerevisiae* strains. The evolved strain displayed higher carbon flux via pentose phosphate pathway. The enhanced expression of *zwf1* gene drove the carbon flux via PPP which led to a consequent increase in NAD(-P)H-regenerating reactions. The activity of aldehyde-reducing enzymes was also enhanced, which contributed to cellular redox balance between NAD(P)H regeneration and consumption (Lewis Liu et al., 2009). *Saccharomyces cerevisiae* TMB3400 has been evolved in minimal media with stepwise increase in furfural concentration starting from 3 mM and going up to 20 mM. After about 300 generations, the evolved strain TMB3400-FT30-3 exhibited improved tolerance, as compared to unevolved parent strain, against furfural as well as against hydrolysate-supplemented medium. In a minimal medium with glucose as a carbon source and 17 mM furfural, the parent strain displayed a lag phase of 90 ± 5 h, while the evolved one had 16 ± 2 h. Under anaerobic conditions and 80% v/v barley straw hydrolysate, the parent strain consumed around 48 g/L glucose in 68 h, while TMB3400-FT30-3 exhausted it in around 22 h. Maximum ethanol titers for TMB3400 were around 23 g/L at 68 h, while for the evolved strain, similar ethanol titers were achieved in around 22 h (Heer and Sauer, 2008).

E. coli LY180 has been evolved in diluted hydrolysate generated by using bagasse as a source of lignocellulosic biomass. The evolution consisted of stepwise increase in concentration of hydrolysate in a minimal medium. The resulting strain MM160 was generated after more than 2 years of subculturing and had undergone greater than 2000 generations of selection pressure. The parent strain LY180 was unable to grow in 1.0 g/L furfural, while MM160 reached $OD_{550} > 3.0$ at 48 h. To test the ethanologenic traits of MM160, a hydrolysate was generated using bagasse pretreated with phosphoric acid at 180°C. At 14% w/v loadings, MM160 produced a maximum ethanol concentration of 29.0 ± 1.5 g/L with a yield of 207.1 kg/ton of biomass (Geddes et al., 2011). Genome-wide mapping of genes involved in conferring tolerance to furfural in E. coli has also been performed. The process did not involve any evolutionary adaptation rather selection of genes expressed under challenge from 0.75 g/L furfural. The strategy yielded an enrichment of around 6% of E. coli genes, which among other cellular functions were also related to cell membrane and cell wall functions. The overexpression of thyA, lpcA and groESL resulted in increased tolerance against 1.5 g/L furfural at 72 h. The first two genes are involved, respectively, in pyrimidine and lipopolysaccharide biosynthesis. The third gene encodes for chaperonin complex which is required for growth under thermal stress as well as for proper folding of proteins (Glebes et al., 2014). In a similar study as in Glebes et al. (2014), E. coli cultures were challenged with 0.75 g/L furfural which also led to the identification of yhjH and ahpC toward conferring tolerance against 1.5 g/L furfural. The former gene is involved in the regulation of flagellar motility, while the latter is a component of alkyl hydroperoxide reductase which is involved in reducing peroxides into alcohols and leads to reduction in ROS load (Glebes et al., 2015). In an adaptation study involving E. coli LY180, the strain was cultured in a bioreactor with 10% xylose as carbon source and the concentration of furfural was gradually increased from 0.5 to 1.3 g/L. The resistant strain was isolated after 54 serial transfers and designated as EMFR9. Microarray analysis revealed that two oxidoreductases yqhD and dkgA were downregulated in the evolved strain. The overexpression of both led to a decrease in furfural tolerance. The reduction of furfural to furfuryl alcohol by both enzymes was NADPH dependent. However, the apparent K_m values for furfural of YqhD (9.0 mM) and DkgA (>130 mM) were quite high, which suggests that furfural might not be the natural substrate for either of the enzymes. However, the apparent K_m values for NADPH of YqhD (8 µM) and DkgA (23 µM) were lower, which suggests that the overexpression of these enzymes leads to an increase in cellular NADPH demand which, in turn, leads to withdrawal of NADPH from biosynthetic processes and leads to a decrease in productivity. In the presence of 1 g/L furfural, EMFR9 was able to ferment 10% xylose to produce around 40 g/L ethanol, which was comparable to the unevolved LY180 strain in the absence of the inhibitor. No significant amount of ethanol could be detected in media containing strain LY180 and 10% xylose and 1 g/L furfural (Miller et al., 2009a). In a separate study, it was also shown that EMFR9 is also resistant to 5-HMF. It should be mentioned that 5-HMF is the lesser toxic cousin of furfural and thus microbes have a higher MIC against 5-HMF as compared to the same concentration of furfural. However, LY180 appeared to be sensitive to 1 g/L as well as 2.5 g/L of 5-HMF and no significant amount of ethanol could be detected in the media at 96 h. EMFR9 meanwhile reached maximum ethanol titers at 48 h in case of 1 g/L 5-HMF and at 96 h in case of 2.5 g/L 5-HMF. In both 5-HMF treatments, the maximum titers of ethanol were comparable and around 40 g/L (Miller et al., 2010).

9.2.2.1.2 Acetic Acid

Acetic acid is more of a significant problem for metabolism of *Saccharomyces cerevisiae* as compared to that of *E. coli*. Thus, in this chapter the emphasis is on the literature where tolerance against acetic acid in *S. cerevisiae* has been studied. In cell-free extracts obtained from *S. cerevisiae*, the activities of several glycolytic enzymes have been reported. Enolase and phosphoglyceromutase were most sensitive, where 50% reduction in activity was reported at acetic acid concentration <123 mM. Triosephosphate isomerase, aldolase, phosphofructokinase, glyceraldehyde-3-phosphate dehydrogenase and phosphoglycerate kinase were similarly inhibited at <358 mM, and pyruvate kinase was inhibited at <410 mM. On the contrary, hexokinase, glucose-6-phosphate isomerase, pyruvate decarboxylase and alcohol dehydrogenase were the most resistant to acetic acid with 50% inhibition occurring at a concentration >1000 mM (Pampulha and Loureiro-Dias, 1990). In *S. cerevisiae*, the presence of acetic acid leads to a decrease in specific growth rate from 0.074 to 0.061/h. As compared to the absence, the presence of 1 g/L acetic acid leads to a significant ($p \leq 0.01$) reduction in NADPH/NADP+ ratio from 4.92 ± 0.2 to 3.54 ± 0.52 (Vasserot et al., 2010). At a concentration of 170 mM, acetic acid stimulated the consumption of glucose by more than 50% and reduced the ATP yield by 70% as compared to the control in the absence of acetic acid. It was also reported that 120 mM is the maximum concentration of acetic acid where fermentation activity could be detected (Pampulha and Loureiro-Dias, 2000). ABC transporter *pdr18* gene involved in conferring pleiotropic drug resistance (PDR) has been implicated in conferring resistance to acetic acid also. At pH = 4.0 and 60 mM acetic acid, *Δpdr18* mutant displayed a lag phase of approximately 40 h, while the wild type parent BY4741 had only 10 h. The mutant also exhibited an increased plasma membrane permeability as compared to the control. The mRNA levels of *pdr18* in parent strain were three-fold higher in the face of acetic acid stress as compared to non-stressed cells (Godinho et al., 2018). The MIC of acetic acid for *S. cerevisiae* was 0.6% w/v (100 mM), and that of lactic acid was 2.5% w/v (278 mM). It was the concentration where no growth could be observed after at least 72 h after inoculation. It was also reported that the acetic acid in the presence of lactic acid also exerts synergistic inhibitory effect on microbial metabolism. In the presence of 0.5% w/v lactic acid in the media even 0.04% w/v acetic acid exerts a significant effect on the growth rate of *S. cerevisiae* ($P \leq 0.001$) (Narendranath et al., 2001). The deletion of *spi1* gene has been implicated in reducing tolerance against 60 mM acetic acid in BY4741 strain. It suggests that in response to the acetic acid challenge, the cell responds by remodeling the cell wall structure by reducing its porosity so as to limit diffusion of undissociated acid into cell cytoplasm and induce intracellular acidification (Simões et al., 2006).

Biomass hydrolysates from deciduous vegetation (alder, aspen and birch) have been shown to contain relatively high concentration of acetate (approximately 9 g/L) as compared to that of pine and spruce which contained approximately 3 g/L acetate. Without detoxification of the dilute acid-treated hydrolysates from spruce, willow, alder, pine, aspen and birch woods by *Saccharomyces cerevisiae*, it was found that the additive concentration of furfural and 5-HMF in the hydrolysates serves as a sentinel with respect to the prediction of the fermentability of the hydrolysate. A lower concentration of both inhibitors correlated with favorable fermentability (Taherzadeh et al., 1997). It has been reported that improving de novo purine biosynthesis of purines in *S. cerevisiae* results in an increased tolerance in *S. cerevisiae* against lignocellulosic inhibitors. In the presence of 100 g/L glucose and 5 g/L acetic acid, strains overexpressing *ade1*, *ade13* and *ade17* exhibited ethanol productivities of 1.21, 1.20 and 1.55 g/L/h, respectively; the same value for control strain was 1.11 g/L/h. Similarly, the strains also displayed reduced ROS load by 21.04%, 16.61% and 40.74%, respectively. The overexpression of *ade1*, *ade13* and *ade17* also increased the total adenylate pool ([ATP] + [ADP] + [AMP]) by 10.76%, 18.91% and 33.29%, respectively, as compared to the control strain under acetic acid stress (Zhang et al., 2019).

Studies involving genomic library have also been used in screening for genes involved in conferring tolerance against acetic acid. Using this approach, *whi2* has been identified as a target gene to confer tolerance against acetic acid. In the presence of 20 g/L glucose and 2.5 g/L acetic acid, strain S-*WHI2* overexpressing Whi2 was able to completely utilize the sugar within 39 h, while the control strain could only consume 3.2 g/L. The specific sugar consumption and specific ethanol production rates were >5 times higher than those of the control strain. However, in the presence of xylose, the specific sugar consumption rate was 0.245 ± 0.004 g/g cells (dry weight)/ h and no significant improvement in xylose consumption was observed in control strain. At pH = 4.0 and in a sugar mix containing either 80 g/L glucose or a mixture containing 40 g/L each of glucose and xylose, strain S-*WHI2* completely utilized glucose and more than 75% of xylose in the media and produced comparable ethanol titers (Chen et al., 2016). Another genomic library screening of *S. cerevisiae* CEN.-PK2-1D strain identified that RCK1 gene is involved in conferring tolerance against acetic acid under conditions of 35 g/L glucose and 3 g/L acetic acid with pH = 4.0. In the presence of 80 g/L glucose and 5 g/L acetic acid, the overexpression of RCK1 in strain D423R resulted in complete sugar utilization within 48 h, while the glucose consumption in the control strain D423C was only 13 g/L. In the presence of 40 g/L xylose and 5 g/L acetic acid, strain SR8R consumed 13 g/L of xylose within 96 h. It represents a consumption rate of 0.139 ± 0.001 g xylose/L/h, while the control strain SR8C did not display any significant sugar consumption. With glucose as a carbon source and in the presence of H_2O_2 and acetic acid, the ROS accumulation in D423R was 43% and 40% lower, respectively, than that of D423C strain. Similarly, ROS load was significantly lower in D423R as compared to D423C when xylose was the carbon source (Oh et al., 2019). A transcriptional profiling of *S. cerevisiae* genomic library at 70, 90 and 110 mM resulted in the identification of 648 genes whose deletion made the microbe susceptible to acetic acid. The majority of genes were grouped into transcription followed closely by carbohydrate metabolism. At a concentration of 70 mM acetic acid, $\Delta trk1$ and $\Delta arl1$ mutants displayed susceptibility toward acetic acid, which suggests that the deficiency of K$^+$ uptake impairs acetic acid tolerance and an increase in the cation concentration to 10 mM was proven to confer tolerance against 40 mM acetic acid. In $\Delta nrg1$ and $\Delta sch9$ mutants, the ability to grow in glucose was also compromised as compared to the WT control, which suggests that the signaling pathway involving Snf1 is also perturbed (Mira et al., 2010).

In *E. coli* also some work has been done to elucidate the mechanism of acetic acid tolerance. It was figured out that genes involved in biosynthesis of methionine *folM*, *metH*, *metF* and *glyA* were among the top category of genes imparting fitness upon acetic acid challenge. Another set of genes involved in the *de novo* biosynthesis of pyrimidine ribonucleotides—*pyrL*, *pyrB* and *pyrI*—were also important for imparting tolerance. In MOPS medium containing 0.2% glucose, it was found that supplementing with 10 mM methionine was beneficial in restoring 62% of the growth rate as compared to the control without 2.5 g/L acetic acid. Similarly, supplementing media with 0.4 mM of either cytosine or uracil led to 30% or 70% of the growth rate, respectively (Sandoval et al., 2011). A summary of the genetic manipulations performed in *E. coli* and *S. cerevisiae* to increase the tolerance against different inhibitors is summarized in Table 9.1.

9.2.3 THIRD GENERATION

Third-generation biofuels utilize algae or cyanobacteria for the production of compounds having bioenergy potential. The idea of third generation holds immense potential over first- and second-generation biofuels as it can be used for the generation of ethanol as well as high-density compounds such as biodiesel. The other advantage is that it does not involve cultivation on land mass. Lentic environments such as lakes and ponds and brine environments such as oceans can serve as fertile grounds for raising the third-generation biofuel 'crops' and thus relieve the stress on cultivable land for utilization toward non-food purposes. The workhorses of third-generation biofuel crops are autotrophic organisms such as cyanobacteria, microalgae and macroalgae.

Cyanobacteria have successfully been used for channeling the fixed CO_2 toward the production of ethanol. In a study involving *Synechocystis* sp. PCC 6803, the strain was used to improve both the rate of biomass and ethanol formation. The influence of four genes of Calvin–Benson–Bassham (CBB) cycle—ribulose-1,5-bisphosphate carboxylase/oxygenase (RuBisCo), fructose-1,6/sedoheptulose-1,7-bisphosphatase (FBP/SBPase), transketolase (TK) and aldolase (FBA)—was studied with the co-expression of *pdc* and *adh* genes of *Z. mobilis* which were under control of inducible promoter of *nrsB*. It was found that the strains expressing FBA, TK, FBP/SBPase and RuBisCo produced 69%, 37%, 67% and 55% higher ethanol and 10.1%, 8.8%, 15.1% and 7.7% more biomass as compared to the control *Synechocystis* strain expressing only *pdc* and *adh* genes. However, it is to be mentioned that the highest ethanol titer was only around 0.2 g/L on the seventh day (Liang et al., 2018). This finding might be beneficial to understand the contribution of different genes in

TABLE 9.1

Key Hydrolysate Inhibitors and Genes Involved in Tolerance

Inhibitor	Concentration	Organism	Target Gene	References
Furans	1 g/L furfural	*E. coli*	*pntAB*↑	Miller et al. (2009b)
	1.25 g/L furfural	*E. coli*	*mdtJI*↑	Kurgan et al. (2019)
	10 mM furfural	*E. coli*	*potE*↑, *puuP*↑	Geddes et al. (2014)
	15 mM furfural	*E. coli*	*fucO*↑	Wang et al. (2011)
	10 mM furfural	*E. coli*	*ucpA*↑	X. Wang et al. (2012)
	10 mM furfural	*E. coli*	*thyA*↑	Zheng et al. (2012)
	2 g/L furfural; 3.5 g/L 5-HMF	*E. coli*	mutant *irrE*	J. Wang et al. (2012)
	25 mM furfural	*S. cerevisiae*	*zwf1*↑, *gnd1*↑, *rpe1*↑, *tkl1*↑	Gorsich et al. (2006)
	1.5 g/L furfural	*E. coli*	*thyA*↑, *lpcA*↑, *groESL*↑	Glebes et al. (2014)
	1.5 g/L furfural	*E. coli*	*yhjH*↑, *ahpC*↑	Glebes et al. (2015)
	1 g/L furfural; 2.5 g/L 5-HMF	*E. coli*	Δ*yqhD*, Δ*dkgA*	Miller et al. (2009a) and Miller et al. (2010)
Acetic acid	60 mM	*S. cerevisiae*	Δ*pdr18*	Godinho et al. (2018)
	60 mM	*S. cerevisiae*	*spi1*↑	Simões et al. (2006)
	5 g/L	*S. cerevisiae*	*ade1*↑, *ade13*↑, *ade17*↑	Zhang et al. (2019)
	2.5 g/L	*S. cerevisiae*	*whi2*↑	Chen et al. (2016)
	5.0 g/L	*S. cerevisiae*	*rck1*↑	Oh et al. (2019)
	70 mM	*S. cerevisiae*	*trk1*↑, *arl1*↑, *nrg1*↑, *sch9*↑	Mira et al. (2010)
	2.5 g/L	*E. coli*	*folM*↑, *metH*↑, *metF*↑, *glyA*↑, *pyrL*↑, *pyrB*↑, *pyrI*↑	Sandoval et al. (2011)

ethanol production; it still falls far below the earlier reported ethanol titers and productivity. These titers are far below the values acceptable by the industry. Microalgae are also photosynthetic organisms which have minimum growth requirements and can harbor high amounts of lipids in their biomass. The highest lipid content as a percent of dry weight is found in *Chlamydomonas reinhardtii* at 21%, closely followed by *Chlorella* sp. at 19%, and the lowest lipid content has been reported in *Dunaliella salina* at 6% (Hossain et al., 2019). This lipid has the potential to be utilized as a biodiesel for bioenergy needs. Thus, efficient cultivation and extraction of lipids from *Chlamydomonas reinhardtii* and *Chlorella* sp. holds immense potential to address future bioenergy needs. Macroalgae also harbor potential for usage in bioenergy needs. The biomass generated by macroalgae type known as seaweed can be used as a substrate for conversion into ethanol. Macroalgal species are low in lipid content and not an economical source of biodiesel. However, high levels of polysaccharides and low lignin content make them an economical source of fermentation-based biofuel products. Almost 95% of the seaweed in use is cultivated and is of two types—red and brown. The majority of cultivated seaweeds belong to four genera—*Porphyra, Undaria, Laminaria* and *Gracilaria*. Brown macroalgae are abundant in carbohydrates laminarin and mannitol. Laminarin is a glucose polymer and consists of β-(1→3)-linked glucan with additional β-(1→6)-linked branches, while mannitol is a 6-carbon sugar alcohol (Ghadiryanfar et al., 2016). Since macroalgae have low lignin content, they do not require harsh treatment conditions and enzyme saccharification is sufficient to breakdown the complex polysaccharides and make the carbon amenable to microbial metabolism. The saccharified substrate was efficiently fermented by *Saccharomyces cerevisiae* IAM 4178 to produce 5.5% titers of ethanol (Yanagisawa et al., 2011). The ethanol titer was comparable with that obtained from microbes using second-generation substrates.

Overall, the third-generation biofuels hold the potential to be the most stable source of bioenergy. However, much more research is required to make them a feasible option with regard to the deployment of infrastructure and in generation of efficient cyanobacterial strains which are much more efficient in increasing their biomass so as to make the process economically feasible.

9.3 HIGH-ENERGY ADVANCED BIOFUELS

9.3.1 BIOALCOHOLS

Bioalcohols are clean and renewable fuels which can simply be defined as alcohols derived from biological resources or biomass. Presently, ethanol is the most well-established bioalcohol that has already been industrially produced, even though it has some limitations in physical properties, such as low energy density, corrosiveness and lower vapor pressure than gasoline. In addition, since ethanol is hygroscopic, it must be blended with gasoline immediately before use. This causes transportation and storage problems that make it more costly. Moreover, current vehicle engines should be adjusted in order to comply with the blending regulations mandating 5% ethanol blends in Europe and 10% in North America according to the EN228 and ASTM D5798 regulations (Das et al., 2020). These concerns have entailed the demand for higher-energy advanced bioalcohols that are similar to conventional oils, such as butanol, isobutanol, and other higher alcohols that are summarized in Figure 9.2.

9.3.1.1 n-Butanol

n-Butanol is an advanced biofuel which is considered as a sustainable alternative to gasoline. Biobutanol is naturally produced by *Clostridium* spp. via their ABE fermentation process, where acetoacetyl-CoA is dehydrated and reduced to crotonyl-CoA. Crotonyl-CoA is then reduced to butyryl-CoA, which is eventually converted into n-butanol by alcohol dehydrogenase (Atsumi et al., 2008).

Clostridium acetobutylicum produces butanol in high titers of 20 g/L via native ABE fermentation. However, it is relatively undesirable for the industrial development because of its slow growth rate and spore-forming life cycle (Atsumi and Liao, 2008).

Microbial industrial organisms such as *Escherichia coli* and *Saccharomyces cerevisiae* are primarily favored for metabolic engineering to generate and boost butanol production, as they are user-friendly and have 'omics' databases accessible. Additionally, the availability of advanced genome-editing techniques for these species often promotes their directed rewiring of cellular metabolism.

Many metabolic engineering attempts have been made to produce n-butanol from *E. coli*. The first attempt was made by the heterologous expression of n-butanol biosynthesis pathway genes, i.e., *thl, hbd, crt, bcd, etfAB* and *adhE2* from *Clostridium acetobutylicum*

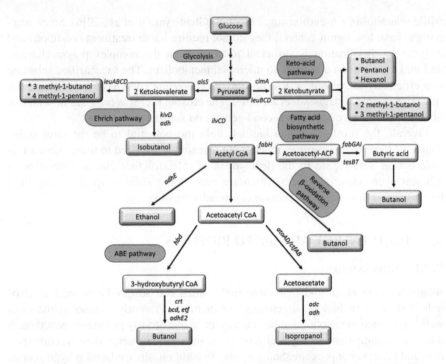

FIGURE 9.2   The metabolic pathways for bioalcohol production in *E. coli*. The biosynthetic alcohols are labeled in blue color, while the incorporated pathways are labeled in red color. Relevant reactions are represented by the name of the genes coding for the enzymes.

into *E. coli* (Atsumi et al., 2008). Further engineering was done to eliminate competing pathways in order to improve butanol production (Shen et al., 2011; Saini et al., 2016). Recently, genome editing using CRISPR/Cas9 technique has been used to integrate a butanol biosynthesis pathway into *E. coli* genome, which yielded butanol from different carbon sources (Abdelaal et al., 2019). Another attempt for butanol production was demonstrated by modifying β-oxidation pathway in the reverse biosynthetic direction to produce butanol without importing any foreign genes in *E. coli* (Dellomonaco et al., 2011). Also, n-butanol was produced by the expression of 2-ketoacid pathway in *E. coli* (Ferreira et al., 2019). Additionally, the native fatty acid biosynthesis (FASII) pathway was used to generate butanol in *E. coli* (Jawed et al., 2020).

Saccharomyces cerevisiae naturally uses the ketoacid pathway to produce n-butanol in low titers (Schadeweg and Boles, 2016). Several metabolic engineering efforts were made to enhance butanol production (Steen et al., 2008; Krivoruchko et al., 2013; Schadeweg and Boles, 2016), but further improvement is still needed to produce more butanol titers.

9.3.1.2   Isobutanol

Isobutanol is an isomer of n-butanol with a higher energy content than n-butanol (Atsumi and Liao, 2008). It is a promising biofuel candidate that has an energy density very close to that of gasoline. Also, it has lower hygroscopicity and less corrosiveness compared to ethanol (Lan and Liao, 2013). Isobutanol can be produced naturally by *Saccharomyces cerevisiae* via valine catabolism (Park and Hahn, 2019).

The modification of *S. cerevisiae* mainly includes the overexpression of valine pathway genes and/or the deletion of the competing threonine pathway (Park et al., 2014; Ida et al., 2015). The improvement in isobutanol production by *S. cerevisiae* also includes editing Entner–Doudoroff pathway genes or pyruvate circuits (Morita et al., 2017; Matsuda et al., 2013).

To engineer *E. coli* to produce isobutanol, *kivD* from *Lactobacillus lactis* and *adh2* from *S. cerevisiae* were introduced to convert ketoacids into isobutanol. For a greater accumulation of ketoacids, *ilvCD* from endogenous *E. coli* and *alsS* from *Bacillus subtilis* were overexpressed to enhance the yield by about 20 g/L isobutanol (Atsumi et al., 2008). Isobutanol-tolerant strain was developed to tolerate toxic levels of isobutanol up to 12 g/L through modifying its global transcription cAMP receptor protein (CRP) (Chong et al., 2014). Recently, Entner–Doudoroff (ED) pathway has been generated to produce isobutanol in *E. coli* along with the inhibition of the genes involved in inorganic acids, producing 15.0 g/L of isobutanol (Noda et al., 2019).

9.3.2 HYDROCARBONS

Hydrocarbons are compounds containing only of carbon and hydrogen atoms, which are abundant in the environment. Hydrocarbons can be classified into alkanes (saturated compounds), alkenes or olefins (compounds with C–C double-bonds), alkynes (compounds with C–C triple-bonds) and aromatic hydrocarbons (Heider et al., 1998). Chemical studies claimed an apparently infinite variety of hydrocarbon structures.

Bio-based production of alkanes or alkenes has been demonstrated in biological processes using several microorganisms. Methane is generated as a metabolic end product, a metabolic by-product in methanogenic bacteria. Biosynthesis of high-energy long-chain alkanes by decarbonylation of the corresponding (n + 1) aldehydes has been documented for marine algae (Dennis and Kolattukudy, 1992). The simplest alkene compound, ethylene, is produced by some bacteria and fungi (Fukuda et al., 1993). Long-chain alkenes are formed either by decarboxylation of unsaturated fatty acids (Gorgen and Boland, 1989) or by monooxygenase-catalyzed conversion of unsaturated aldehyde precursors producing alkenes and CO_2 (Reed et al., 1994). Isoprenoids are a well-known class of hydrocarbons, which are made by microorganisms (Calvin, 1980). The major sources of isoprene (2-methyl-1,3-butadiene) are Actinomyces and Bacillus species (Ladygina et al., 2006). All archaea generate significant amounts of C20 to C40 isoprenoids that vary considerably due to the degree of molecule saturation, methylation and cyclization (Koga and Morii, 2007).

Some aromatic hydrocarbons are also produced biologically. Low toluene concentrations have been found in pristine habitats, for example the anaerobic hypolimnia of lakes (Jüttner and Henatsch, 1986); they derive from phenylalanine degradation by many types of anaerobic bacteria. Phenylalanine is first oxidized by these bacteria to phenyl acetate, which is then decarboxylated (Fischer-Romero et al., 1996).

The acyl-acyl carrier protein (ACP) reductase (AAR) and aldehyde-deformylating oxygenase (ADO) constitute the alkane biosynthesis pathway in cyanobacteria. AAR converts fatty acyl-ACPs of even-numbered carbons into fatty aldehydes, which are converted by the action of ADO into an alkane/alkene with one carbon reduction (Schirmer et al., 2010). To date, the AAR–ADO pathway has showed the highest titer reported for alkane/alkene biosynthesis (Kang and Nielsen, 2016).

Alkanes/alkenes can be produced via ADs in two different routes: by the action of fatty AAR from fatty acyl-CoAs or by fatty acid reductase (FAR) from free fatty acids (FFAs) (Kang and Nielsen, 2016), followed by decarbonylation of the resulting aldehydes.

Recently, fatty acid photodecarboxylase from *Chlorella variabilis* has been identified to directly convert fatty acids into alkanes/alkenes (Bruder et al., 2019). Also, P450 fatty acid decarboxylase from *Macrococcus caseolyticus* has been identified to directly convert fatty acids to 1-alkenes (aka a-olefins) (Lee et al., 2018).

9.3.3 METABOLIC ENGINEERING FOR THE PRODUCTION OF ALKANES/ALKENES

Microbial engineering for alkane/alkene production has shown that it is feasible to engineer diverse microorganisms with AAR and ADO for the production of a diverse range of alkanes/alkenes. The first approach toward the synthesis of alkanes/alkenes was to express the cyanobacteria pathway in *E. coli* (Schirmer et al., 2010). The expression of AAR and ADO from *S. elongatus* PCC7942 produced around 25 mg/L of pentadecane and heptadecane with other alcohols and aldehydes. The inefficiency of ADOs often results in the conversion of fatty aldehydes to fatty alcohols, which is undesirable, so the expression of *N. punctiforme* PCC73102 instead of the *S. elongatus* ADO increased the yield over 300 mg/L of tridecane, pentadecene, pentadecane and heptadecane (Schirmer et al., 2010).

Depending on the *E. coli* strain, plasmids used, and culture media, the alkane/alkene production levels were variable (Y. Cao et al., 2014; Song et al., 2015). However, redirecting carbon flux toward fatty acyl-ACP synthesis is a promising approach to enhance the biosynthesis of alkanes/alkenes.

The alternative alkane/alkene synthetic pathway that uses fatty acyl-CoA or FFAs as substrates was verified. A wide range of alkanes/alkenes (tridecane, tetradecane, pentadecane, hexadecane, heptadecane and heptadecene) were produced when expression of the FAR and ADO enzymes were made from *S. elongatus* along with FabH2 from *B. subtilis* in *E. coli* with a total yield of 98.3 mg/L (Harger et al., 2012).

The production of alkanes/alkenes from FFAs was also verified by the reductase complex expression from *Photorhabdus luminescens*, and an ADO from *N. punctiforme*. The engineered strain produced a total yield of ~8 mg/L from tridecane, pentadecane, pentadecene, hexadecene, heptadecane and heptadecane (Howard et al., 2013). The composition of the synthesized alkanes can be changed through increasing the FFA pool by adding exogenous fatty acids or modifying the native pathways (Howard et al., 2013).

Escherichia coli-modified strains have also been used to find different alternatives to generate short-length alkanes of C3 (Kallio et al., 2014), C4, C5, C7 or C9 as major components of gasoline (Sheppard et al., 2016).

The highest alkane bioproduction has been reported in modified *E. coli*, which exceeded 2 g/L (Fatma et al., 2018). Different microbial genetic modifications to improve alkane/alkene production with their titers are summarized in Table 9.2.

TABLE 9.2

Genetic Modifications for Improving Alkane/Alkene Production

Strain	Modifications	Alkane/Alkene Chain	Titer	References
E. coli BL21(DE3)	Expression of reductase complex from *Photorhabdus luminescens* and aldehyde decarbonylase from *Nostoc punctiforme*	Alkanes (C13, iso-C13, C15, iso-C15, C16, C17), alkenes (C13, C15–C17)	5 mg/L	Howard et al. (2013)
E. coli BL21(DE3)	Expression of AAR and ADO from *S. elongatus* PCC7942, and overexpression of the electron transfer system Fd/FNR from *S. elongatus* PCC7942	Alkane (C15), alkene (C17)	1.31 g/L	Cao et al. (2016)
E. coli BL21(DE3)	Expression of acyl-acyl carrier protein reductase (AAR) and aldehyde-deformylating oxygenase (ADO) from *Synechococcus elongatus* PCC7942, deletion of aldehyde reductase yqhD and overexpression of transcription factor fadR	Alkanes (C15 and C17), alkene (C17)	255.6 mg/L	Song et al. (2015)
E. coli BL21(DE3)	Expression of acyl-ACP thioesterase from *Umbellularia californica*, a fatty acyl-CoA reductase from *Acinetobacter* sp. and a decarbonylase from *N. punctiforme*	Alkanes (C11, C13, C15)	8.05 mg/g CDW	Yan et al. (2016)
E. coli BL21(DE3)	Expression of AAR from *Synechococcus elongatus*, ADO from *Cyanothece* sp. pcc 7425, and a type I FAS from *Corynebacterium ammoniagenes*	Alkanes (C15 and C17)	57 mg/L C15	Coursolle et al. (2015)
E. coli BL21	Overexpression of non-heme iron (II)-dependent gene (UndA) from *Pseudomonas putida*	Alkenes (C9–C13)	6 mg/L	Rui et al. (2014)
E. coli MG1655	Reverse-β-oxidation pathway and expression of a CAR from *Nocardia iowensis* and an aldehyde reductase (AD) from *Prochlorococcus marinus*	Alkanes (C4 and C5)	1.4 mg/L	Sheppard et al. (2016)
E. coli W3110	Expression of *Clostridium acetobutylicum* fatty acyl-CoA reductase and *Arabidopsis thaliana* fatty aldehyde decarbonylase, deletions in fadE and fadR, and expression of a modified thioesterase "TesA (L109P)."	Alkanes (C9, C12–C14), alkene (C13)	580.8 mg/L	Choi and Lee (2013)

(Continued)

TABLE 9.2 (Continued)
Genetic Modifications for Improving Alkane/Alkene Production

Strain	Modifications	Alkane/Alkene Chain	Titer	References
E. coli	Expression of Cytochrome P450 OleT from Jeotgalicoccus sp. ATCC 8456	Multiple α-alkenes	97.6 mg/L	Liu et al. (2014)
S. cerevisiae BY4741	Expression OleT from Jeotgalicoccus sp. ATCC 8456 and deletion of faa1, faa4, ctt1, cta1 and ccp1 genes	Alkenes (C11, C13, C15, C17, C19)	3.7 mg/L	Chen et al. (2015)
S. cerevisiae BY4741	Expression of fatty acid α-dioxygenase (alphaDOX) from Oryza sativa, cyanobacterial aldehyde-deformylating oxygenase, and deletion of the fatty acyl-CoA synthetases FAA1 and FAA4	Alkanes (C12, C14, C16)	337.8 µg/L	Foo et al. (2017)
S. cerevisiae CEN.PK 113–11C	Construction of a chimeric citrate lyase pathway, overexpression of endogenous acetyl-CoA carboxylase and exogenous FFA synthases, mutations in the alcohol dehydrogenase and deletion in some other genes (faa1, faa4, pox1 and hfd1), and expression of CAR from Mycobacterium marinum and an ADO from Nostoc punctiforme	Alkanes (C13, C15, C17), alkenes (C15 and C17)	0.82 mg/L	Zhou et al. (2016)
N. punctiforme PCC 73102	Overexpression of acyl-acyl carrier protein reductase (AAR), aldehyde decarbonylase (ADC), and lipase genes	Alkane (C17)	12.9% of DW	Peramuna et al. (2015)
Synechocystis sp. PCC6803	Overexpression of acyl-acyl carrier protein reductase and aldehyde-deformylating oxygenase from several cyanobacteria strains and redirecting the carbon flux to acyl-ACP	Alkanes (C15 and C17), alkene (C17)	26 mg/L	Chen et al. (2015)
Yarrowia lipolytica PO1F	Overexpression of lipoxygenase from soybean (Gmlox1) and mutation of mfel	Alkane (C5)	4.98 mg/L	Blazeck et al. (2013)
Cupriavidus necator H16	Heterologous expression of AAR and ADO from S. elongatus PCC7942 and deletion of the phaCAB operon for PHB synthesis	Alkanes (C15 and C17), alkene (C17)	435 mg/L	Crépin et al. (2016)

REFERENCES

Abdelaal, A. S., K. Jawed, and S. S. Yazdani. 2019. "CRISPR/Cas9-mediated engineering of Escherichia coli for n-butanol production from xylose in defined medium." *Journal of Industrial Microbiology & Biotechnology* 46 (7): 965–975. https://doi.org/10.1007/s10295-019-02180-8.

Akcil, A., and Koldas, S. (2006). "Acid Mine Drainage (AMD): Causes, treatment and case studies." *Journal of Cleaner Production* 14 (12–13 SPEC. ISS.): 1139–1145. doi: 10.1016/j.jclepro.2004.09.006.

Allen, S. A. et al. 2010. "Furfural induces reactive oxygen species accumulation and cellular damage in Saccharomyces cerevisiae." *Biotechnology for Biofuels* 3: 1–10. doi: 10.1186/1754-6834-3-2.

Atsumi, Shota, Anthony F. Cann, Michael R. Connor, Claire R. Shen, Kevin M. Smith, Mark P. Brynildsen, Katherine J. Y. Chou, TaizoHanai, and James C. Liao. 2008. "Metabolic engineering of Escherichia coli for 1-butanol production." *Metabolic Engineering* 10 (6): 305–311. https://doi.org/10.1016/j.ymben.2007.08.003.

Atsumi, S., and J. C. Liao. 2008. "Metabolic engineering for advanced biofuels production from Escherichia coli." *Current Opinion in Biotechnology* 19 (5): 414–419. https://doi.org/10.1016/j.copbio.2008.08.008.

Atsumi, Shota, Taizo Hanai, and James C. Liao. 2008. "Non-fermentative pathways for synthesis of branched-chain higher alcohols as biofuels." *Nature* 451 (7174): 86–89. https://doi.org/10.1038/nature06450.

Banerjee, N., R. Bhatnagar, and L. Viswanathan. 1981. "Inhibition of glycolysis by furfural in Saccharomyces cerevisiae." *European Journal of Applied Microbiology and Biotechnology* 11(4): 226–228. doi: 10.1007/BF00505872.

Blazeck, John, Leqian Liu, Rebecca Knight, and Hal S. Alper. 2013. "Heterologous production of pentane in the oleaginous yeast Yarrowialipolytica." *Journal of Biotechnology* 165 (3–4): 184–194. https://doi.org/10.1016/j.jbiotec.2013.04.003.

Bruder, Stefan, Eva Johanna Moldenhauer, Robert Denis Lemke, Rodrigo Ledesma-Amaro, and Johannes Kabisch. 2019. "Drop-in biofuel production using fatty acid photodecarboxylase from Chlorella variabilis in the oleaginous yeast Yarrowialipolytica." *Biotechnology for Biofuels* 12 (1). https://doi.org/10.1186/s13068-019-1542-4.

Calvin, Melvin. 1980. "Hydrocarbons from plants: Analytical methods and observations." *Naturwissenschaften* 67 (11): 525–533. https://doi.org/10.1007/bf00450661.

Cao, Ying-Xiu, Wen-Hai Xiao, Jin-Lai Zhang, Ze-Xiong Xie, Ming-Zhu Ding, and Ying-Jin Yuan. 2016. "Heterologous biosynthesis and manipulation of alkanes in Escherichia coli." *Metabolic Engineering* 38: 19–28. https://doi.org/10.1016/j.ymben.2016.06.002.

Cao, Yujin, Wei Liu, Xin Xu, Haibo Zhang, Jiming Wang, and Mo Xian. 2014. "Production of free monounsaturated fatty acids by metabolically engineered Escherichia coli." *Biotechnology for Biofuels* 7 (1). https://doi.org/10.1186/1754-6834-7-59.

Chen, Binbin, Dong-Yup Lee, and Matthew Wook Chang. 2015. "Combinatorial metabolic engineering of Saccharomyces cerevisiae for terminal alkene production." *Metabolic Engineering* 31: 53–61. https://doi.org/10.1016/j.ymben.2015.06.009.

Chen, Y., L. Stabryla, and N. Wei. 2016. "Improved acetic acid resistance in Saccharomyces cerevisiae by overexpression of the WHI2 gene identified through inverse metabolic engineering." *Applied and Environmental Microbiology* 82 (7): 2156–2166. doi: 10.1128/AEM.03718-15.

Choi, Yong Jun, and Sang Yup Lee. 2013. "Microbial production of short-chain alkanes." *Nature* 502 (7472): 571–574. https://doi.org/10.1038/nature12536.

Chong, Huiqing, Hefang Geng, Hongfang Zhang, Hao Song, Lei Huang, and Rongrong Jiang. 2014. "EnhancingE. coliisobutanol tolerance through engineering its global transcription factor cAMP receptor protein (CRP)." *Biotechnology and Bioengineering* 111 (4): 700–708. https://doi.org/10.1002/bit.25134.

Coursolle, Dan, Jiazhang Lian, John Shanklin, and Huimin Zhao. 2015. "Production of long chain alcohols and alkanes upon coexpression of an acyl-ACP reductase and alde-hyde-deformylating oxygenase with a bacterial type-I fatty acid synthase in E. coli." *Molecular BioSystems* 11 (9): 2464–2472. https://doi.org/10.1039/c5mb00268k.

Crépin, Lucie, Eric Lombard, and Stéphane E. Guillouet. 2016. "Metabolic engineering of Cupriavidusnecator for heterotrophic and autotrophic alka(e)ne production." *Metabolic Engineering* 37: 92–101. https://doi.org/10.1016/j.ymben.2016.05.002.

Da Silva, G. P. et al. 2005. "Ethanolic fermentation of sucrose, sugarcane juice and molasses by Escherichia coli strain KO11 and Klebsiellaoxytoca strain P2." *Brazilian Journal of Microbiology* 36 (4): 395–404. doi: 10.1590/s1517-83822005000400017.

Das, Manali, PradiptaPatra, and Amit Ghosh. 2020. "Metabolic engineering for enhanc-ing microbial biosynthesis of advanced biofuels." *Renewable and Sustainable Energy Reviews* 119. https://doi.org/10.1016/j.rser.2019.109562.

Dellomonaco, Clementina, James M. Clomburg, Elliot N. Miller, and Ramon Gonzalez. 2011. "Engineered reversal of the β-oxidation cycle for the synthesis of fuels and chemicals." *Nature* 476 (7360): 355–359. https://doi.org/10.1038/nature10333.

Dennis, M., and P. E. Kolattukudy. 1992. "A cobalt-porphyrin enzyme converts a fatty alde-hyde to a hydrocarbon and CO." *Proceedings of the National Academy of Sciences* 89 (12): 5306–5310. https://doi.org/10.1073/pnas.89.12.5306.

Dhaliwal, S. S. et al. 2011. "Enhanced ethanol production from sugarcane juice by galactose adaptation of a newly isolated thermotolerant strain of Pichia kudriavzevii." *Bioresource Technology*. Elsevier Ltd., 102 (10): 5968–5975. doi: 10.1016/j.biortech.2011.02.015.

Fatma, Zia, Hassan Hartman, Mark G. Poolman, David A. Fell, Shireesh Srivastava, Tabinda Shakeel, and Syed Shams Yazdani. 2018. "Model-assisted metabolic engineering of Escherichia coli for long chain alkane and alcohol production." *Metabolic Engineering* 46: 1–12. https://doi.org/10.1016/j.ymben.2018.01.002.

Ferreira, Sofia, Rui Pereira, Filipe Liu, Paulo Vilaça, and Isabel Rocha. 2019. "Discovery and implementation of a novel pathway for n-butanol production via 2-oxoglutarate." *Biotechnology for Biofuels* 12 (1). https://doi.org/10.1186/s13068-019-1565-x.

Fischer-Romero, C., B. J. Tindall, and F. Juttner. 1996. "Tolumonasauensis gen. nov., sp. nov., a tolu-ene-producing bacterium from anoxic sediments of a freshwater lake." *International Journal of Systematic Bacteriology* 46 (1): 183–188. https://doi.org/10.1099/00207713-46-1-183.

Foo, Jee Loon, Adelia Vicanatalita Susanto, Jay D. Keasling, Susanna Su Jan Leong, and Matthew Wook Chang. 2017. "Whole-cell biocatalytic and de novo production of alkanes from free fatty acids in Saccharomyces cerevisiae." *Biotechnology and Bioengineering* 114 (1): 232–237. https://doi.org/10.1002/bit.25920.

Fukuda, H., T. Ogawa, and S. Tanase. 1993. "Ethylene production by micro-organisms." In *Advances in Microbial Physiology* edited by A.H. Rose, Volume 35, 275–306, Academic Press.

Gasparatos, A ., P. Stromberg, and K. Takeuchi. 2013. "Sustainability impacts of first-generation biofuels." *Animal Frontiers* 3 (2): 12–26. doi: 10.2527/af.2013-0011.

Geddes, C. C. et al. 2011. "Simplified process for ethanol production from sugarcane bagasse using hydrolysate-resistant *Escherichia coli* strain MM160." *Bioresource Technology*. Elsevier Ltd., 102 (3): 2702–2711. doi: 10.1016/j.biortech.2010.10.143.

Geddes, R. D. et al. 2014. "Polyamine transporters and polyamines increase furfural toler-ance during xylose fermentation with ethanologenic *Escherichia coli* strain." 80 (19): 5955–5964. doi: 10.1128/AEM.01913-14.

Ghadiryanfar, M. et al. 2016. "A review of macroalgae production, with potential applications in biofuels and bioenergy." *Renewable and Sustainable Energy Reviews*. Elsevier, 54: 473–481. doi: 10.1016/j.rser.2015.10.022.

Glebes, T. Y. et al. 2014. "Genome-wide mapping of furfural tolerance genes in Escherichia coli." *PLoS ONE* 9 (1). doi: 10.1371/journal.pone.0087540.

Glebes, T. Y. et al. 2015. "Comparison of genome-wide selection strategies to identify furfural tolerance genes in Escherichia coli." *Biotechnology and Bioengineering* 112 (1): 129–140. doi: 10.1002/bit.25325.

Godinho, C. P. et al. 2018. "Pdr18 is involved in yeast response to acetic acid stress counteracting the decrease of plasma membrane ergosterol content and order." *Scientific Reports* 8 (1): 1–13. doi: 10.1038/s41598-018-26128-7.

Gorgen, Gunther, and Wilhelm Boland. 1989. "Biosynthesis of 1-alkenes in higher plants: stereochemical implications. A model study with Carthamustinctorius (Asteraceae)." *European Journal of Biochemistry* 185 (2): 237–242. https://doi.org/10.1111/j.1432-1033.1989.tb15108.x.

Gorsich, S. W. et al. 2006. "Tolerance to furfural-induced stress is associated with pentose phosphate pathway genes ZWF1, GND1, RPE1, and TKL1 in Saccharomyces cerevisiae." *Applied Microbiology and Biotechnology* 71 (3): 339–349. doi: 10.1007/s00253-005-0142-3.

Gutiérrez, T., L. O. Ingram, and J. F. Preston. 2006. "Purification and characterization of a furfural reductase (FFR) from Escherichia coli strain LYO1- An enzyme important in the detoxification of furfural during ethanol production." *Journal of Biotechnology* 121(2): 154–164. doi: 10.1016/j.jbiotec.2005.07.003.

Harger, Matthew, Lei Zheng, Austin Moon, Casey Ager, JuHye An, Chris Choe, Yi-Ling Lai, Benjamin Mo, David Zong, Matthew D. Smith, Robert G. Egbert, Jeremy H. Mills, David Baker, Ingrid Swanson Pultz, and Justin B. Siegel. 2012. "Expanding the product profile of a microbial alkane biosynthetic pathway." *ACS Synthetic Biology* 2 (1): 59–62. https://doi.org/10.1021/sb300061x.

Heer, D., and U. Sauer. 2008. "Identification of furfural as a key toxin in lignocellulosic hydrolysates and evolution of a tolerant yeast strain." *Microbial Biotechnology* 1 (6): 497–506. doi: 10.1111/j.1751-7915.2008.00050.x.

Heider, Johann, Alfred M. Spormann, Harry R. Beller, and Friedrich Widdel. 1998. "Anaerobic bacterial metabolism of hydrocarbons." *FEMS Microbiology Reviews* 22 (5): 459–473. https://doi.org/10.1111/j.1574-6976.1998.tb00381.x.

Hendriks, A. T. W. M., and G. Zeeman. 2009. "Pretreatments to enhance the digestibility of lignocellulosic biomass." *Bioresource Technology* 100 (1): 10–18. doi: 10.1016/j.biortech.2008.05.027.

Hossain, N., T. M. I. Mahlia, and R. Saidur. 2019. "Latest development in microalgae-biofuel production with nano-additives." *Biotechnology for Biofuels*. BioMed Central Ltd., 12 (1): 1–16. doi: 10.1186/s13068-019-1465-0.

Howard, T. P., S. Middelhaufe, K. Moore, C. Edner, D. M. Kolak, G. N. Taylor, D. A. Parker, R. Lee, N. Smirnoff, S. J. Aves, and J. Love. 2013. "Synthesis of customized petroleum-replica fuel molecules by targeted modification of free fatty acid pools in Escherichia coli." *Proceedings of the National Academy of Sciences* 110 (19): 7636–7641. https://doi.org/10.1073/pnas.1215966110.

Ida, Kengo, Jun Ishii, Fumio Matsuda, Takashi Kondo, and Akihiko Kondo. 2015. "Eliminating the isoleucine biosynthetic pathway to reduce competitive carbon outflow during isobutanol production by Saccharomyces cerevisiae." *Microbial Cell Factories* 14 (1). https://doi.org/10.1186/s12934-015-0240-6.

Jansson, C. et al. 2009. "Cassava, a potential biofuel crop in China." *Applied Energy*. Elsevier Ltd., 86 (SUPPL. 1): S95–S99. doi: 10.1016/j.apenergy.2009.05.011.

Jawed, Kamran, Ali SamyAbdelaal, Mattheos A. G. Koffas, and Syed Shams Yazdani. 2020. "Improved butanol production using FASII pathway in E. coli." *ACS Synthetic Biology* 9 (9): 2390–2398. https://doi.org/10.1021/acssynbio.0c00154.

Jüttner, F., and J. J. Henatsch. 1986. "Anoxic hypolimnion is a significant source of biogenic toluene." *Nature* 323 (6091): 797–798. https://doi.org/10.1038/323797a0.

Kallio, Pauli, AndrásPásztor, Kati Thiel, M. Kalim Akhtar, and Patrik R. Jones. 2014. "An engineered pathway for the biosynthesis of renewable propane." *Nature Communications* 5 (1). https://doi.org/10.1038/ncomms5731.

Kang, Min-Kyoung, and Jens Nielsen. 2016. "Biobased production of alkanes and alkenes through metabolic engineering of microorganisms." *Journal of Industrial Microbiology & Biotechnology* 44 (4–5): 613–622. https://doi.org/10.1007/s10295-016-1814-y.

Kim, D., and J. S. Hahn. 2013. "Roles of the Yap1 transcription factor and antioxidants in Saccharomyces cerevisiae's tolerance to furfural and 5-Hydroxymethylfurfural, which function as Thiol-Reactive electrophiles generating oxidative stress." *Applied and Environmental Microbiology* 79 (16): 5069–5077. doi: 10.1128/AEM.00643-13.

Koga, Yosuke, and Hiroyuki Morii. 2007. "Biosynthesis of ether-type polar lipids in archaea and evolutionary considerations." *Microbiology and Molecular Biology Reviews* 71 (1): 97–120. https://doi.org/10.1128/mmbr.00033-06.

Krivoruchko, Anastasia, Cristina Serrano-Amatriain, Yun Chen, VerenaSiewers, and Jens Nielsen. 2013. "Improving biobutanol production in engineered Saccharomyces cerevisiae by manipulation of acetyl-CoA metabolism." *Journal of Industrial Microbiology & Biotechnology* 40 (9): 1051–1056. https://doi.org/10.1007/s10295-013-1296-0.

Kurgan, G. et al. 2019. "Bioprospecting of native efflux pumps to enhance furfural tolerance in ethanologenic Escherichia coli." *Applied and Environmental Microbiology* 85 (6): 1–11. doi: 10.1128/AEM.02985-18.

Ladygina, N., E. G. Dedyukhina, and M. B. Vainshtein. 2006. "A review on microbial synthesis of hydrocarbons." *Process Biochemistry* 41 (5): 1001–1014. https://doi.org/10.1016/j.procbio.2005.12.007.

Lan, Ethan I., and James C. Liao. 2013. "Microbial synthesis of n-butanol, isobutanol, and other higher alcohols from diverse resources." *Bioresource Technology* 135: 339–349. https://doi.org/10.1016/j.biortech.2012.09.104.

Lee, Jong-Won, Narayan P. Niraula, and Cong T. Trinh. 2018. "Harnessing a P450 fatty acid decarboxylase from Macrococcuscaseolyticus for microbial biosynthesis of odd chain terminal alkenes." *Metabolic Engineering Communications* 7. https://doi.org/10.1016/j.mec.2018.e00076.

Lewis Liu, Z. et al. 2008. "Multiple gene-mediated NAD(P)H-dependent aldehyde reduction is a mechanism of in situ detoxification of furfural and 5-hydroxymethylfurfural by Saccharomyces cerevisiae." *Applied Microbiology and Biotechnology* 81 (4): 743–753. doi: 10.1007/s00253-008-1702-0.

Lewis Liu, Z., M. Ma, and M. Song. 2009. "Evolutionarily engineered ethanologenic yeast detoxifies lignocellulosic biomass conversion inhibitors by reprogrammed pathways." *Molecular Genetics and Genomics* 282 (3): 233–244. doi: 10.1007/s00438-009-0461-7.

Liang, F. et al. 2018. "Engineered cyanobacteria with enhanced growth show increased ethanol production and higher biofuel to biomass ratio." *Metabolic Engineering.* Elsevier Inc., 46 (December 2017): 51–59. doi: 10.1016/j.ymben.2018.02.006.

Liu, Yi, Cong Wang, Jinyong Yan, Wei Zhang, Wenna Guan, Xuefeng Lu, and Shengying Li. 2014. "Hydrogen peroxide-independent production of α-alkenes by OleTJE P450 fatty acid decarboxylase." *Biotechnology for Biofuels* 7 (1). https://doi.org/10.1186/1754-6834-7-28.

Liu, Z. L., and J. Moon. 2009. "A novel NADPH-dependent aldehyde reductase gene from Saccharomyces cerevisiae NRRL Y-12632 involved in the detoxification of aldehyde inhibitors derived from lignocellulosic biomass conversion." *Gene.* Elsevier B.V., 446 (1): 1–10. doi: 10.1016/j.gene.2009.06.018.

Matsuda, Fumio, Jun Ishii, Takashi Kondo, Kengo Ida, Hironori Tezuka, and Akihiko Kondo. 2013. "Increased isobutanol production in Saccharomyces cerevisiae by eliminating competing pathways and resolving cofactor imbalance." *Microbial Cell Factories* 12 (1). https://doi.org/10.1186/1475-2859-12-119.

Miller, E. N. et al. 2009a. "Silencing of NADPH-dependent oxidoreductase genes (yqhD and dkgA) in furfural-resistant ethanologenic Escherichia coli." *Applied and Environmental Microbiology* 75 (13): 4315–4323. doi: 10.1128/AEM.00567-09.

Miller, Elliot N. et al. 2009b. "Furfural inhibits growth by limiting sulfur assimilation in ethanologenic Escherichia coli strain LY180." *Applied and Environmental Microbiology* 75 (19): 6132–6141. doi: 10.1128/AEM.01187-09.

Miller, E. N. et al. 2010. "Genetic changes that increase 5-hydroxymethyl furfural resistance in ethanol-producing Escherichia coli LY180." *Biotechnology Letters* 32 (5): 661–667. doi: 10.1007/s10529-010-0209-9.

Mira, N. P. et al. 2010. "Genome-wide identification of Saccharomyces cerevisiae genes required for tolerance to acetic acid." *Microbial Cell Factories*. BioMed Central Ltd., 9 (1): 79. doi: 10.1186/1475-2859-9-79.

Morales, M. et al. 2015. "Life cycle assessment of lignocellulosic bioethanol: Environmental impacts and energy balance." *Renewable and Sustainable Energy Reviews*. Elsevier, 42: 1349–1361. doi: 10.1016/j.rser.2014.10.097.

Morita, Keisuke, Yuta Nomura, Jun Ishii, Fumio Matsuda, Akihiko Kondo, and Hiroshi Shimizu. 2017. "Heterologous expression of bacterial phosphoenol pyruvate carboxylase and Entner–Doudoroff pathway in Saccharomyces cerevisiae for improvement of isobutanol production." *Journal of Bioscience and Bioengineering* 124 (3): 263–270. https://doi.org/10.1016/j.jbiosc.2017.04.005.

Narendranath, N. V., K. C. Thomas, and W. M. Ingledew. 2001. "Effects of acetic acid and lactic acid on the growth of Saccharomyces cerevisiae in a minimal medium." *Journal of Industrial Microbiology and Biotechnology* 26 (3): 171–177. doi: 10.1038/sj.jim. 7000090.

Noda, Shuhei, Yutaro Mori, Sachiko Oyama, Akihiko Kondo, Michihiro Araki, and Tomokazu Shirai. 2019. "Reconstruction of metabolic pathway for isobutanol production in Escherichia coli." *Microbial Cell Factories* 18 (1). https://doi.org/10.1186/s12934-019-1171-4.

Oh, E. J. et al. 2019. "Overexpression of RCK1 improves acetic acid tolerance in Saccharomyces cerevisiae." *Journal of Biotechnology*. Elsevier, 292 (January): 1–4. doi: 10.1016/j.jbiotec.2018.12.013.

Palmqvist, E., J. S. Almeida, and B. Hahn-Hägerdal. 1999. "Influence of furfural on anaerobic glycolytic kinetics of saccharomyces cerevisiae in batch culture." *Biotechnology and Bioengineering* 62 (4): 447–454. doi: 10.1002/(SICI)1097-0290(19990220)62:4<447::AID-BIT7>3.0.CO;2-0.

Pampulha, M. E., and M. C. Loureiro-Dias. 1990. "Activity of glycolytic enzymes of Saccharomyces cerevisiae in the presence of acetic acid." *Applied Microbiology and Biotechnology* 34 (3): 375–380. doi: 10.1007/BF00170063.

Pampulha, M. E., and M. C. Loureiro-Dias. 2000. "Energetics of the effect of acetic acid on growth of Saccharomyces cerevisiae." *FEMS Microbiology Letters* 184 (1): 69–72. doi: 10.1016/S0378-1097(00)00022-7.

Park, Seong-Hee, and Ji-Sook Hahn. 2019. "Development of an efficient cytosolic isobutanol production pathway in Saccharomyces cerevisiae by optimizing copy numbers and expression of the pathway genes based on the toxic effect of α-acetolactate." *Scientific Reports* 9 (1). https://doi.org/10.1038/s41598-019-40631-5.

Park, Seong-Hee, Sujin Kim, and Ji-Sook Hahn. 2014. "Metabolic engineering of Saccharomyces cerevisiae for the production of isobutanol and 3-methyl-1-butanol." *Applied Microbiology and Biotechnology* 98 (21): 9139–9147. https://doi.org/10.1007/s00253-014-6081-0.

Peramuna, Anantha, Ray Morton, and Michael Summers. 2015. "Enhancing alkane production in cyanobacterial lipid droplets: A model platform for industrially relevant compound production." *Life* 5 (2): 1111–1126. https://doi.org/10.3390/life5021111.

Reed, J. R., D. Vanderwel, S. Choi, J. G. Pomonis, R. C. Reitz, and G. J. Blomquist. 1994. "Unusual mechanism of hydrocarbon formation in the housefly: Cytochrome P450 converts aldehyde to the sex pheromone component (Z)-9-tricosene and CO_2." *Proceedings of the National Academy of Sciences* 91 (21): 10000–10004. https://doi.org/10.1073/pnas.91.21.10000.

Rui, Zhe, Xin Li, Xuejun Zhu, Joyce Liu, Bonnie Domigan, Ian Barr, Jamie H. D. Cate, and Wenjun Zhang. 2014. "Microbial biosynthesis of medium-chain 1-alkenes by a nonheme iron oxidase." *Proceedings of the National Academy of Sciences* 111 (51): 18237–18242. https://doi.org/10.1073/pnas.1419701112.

Saini, M., S. Y. Li, Z. W. Wang, C. J. Chiang, and Y. P. Chao. 2016. "Systematic engineering of the central metabolism in Escherichia coli for effective production of n-butanol." *Biotechnology for Biofuels* 9: 69. https://doi.org/10.1186/s13068-016-0467-4. https://www.ncbi.nlm.nih.gov/pubmed/26997975.

Sandoval, N. R. et al. 2011. "Elucidating acetate tolerance in E. coli using a genome-wide approach." *Metabolic Engineering*. Elsevier, 13 (2): 214–224. doi: 10.1016/j.ymben.2010.12.001.

Schadeweg, Virginia, and Eckhard Boles. 2016. "n-Butanol production in Saccharomyces cerevisiae is limited by the availability of coenzyme A and cytosolic acetyl-CoA." *Biotechnology for Biofuels* 9 (1). https://doi.org/10.1186/s13068-016-0456-7.

Schirmer, A., M. A. Rude, X. Li, E. Popova, and S. B. delCardayre. 2010. "Microbial biosynthesis of alkanes." *Science* 329 (5991): 559–562. https://doi.org/10.1126/science.1187936.

Shen, C. R., E. I. Lan, Y. Dekishima, A. Baez, K. M. Cho, and J. C. Liao. 2011. "Driving forces enable high-titer anaerobic 1-butanol synthesis in Escherichia coli." *Applied and Environmental Microbiology* 77 (9): 2905–2915. https://doi.org/10.1128/AEM.03034-10. https://www.ncbi.nlm.nih.gov/pubmed/21398484.

Sheppard, Micah J., Aditya M. Kunjapur, and Kristala L. J. Prather. 2016. "Modular and selective biosynthesis of gasoline-range alkanes." *Metabolic Engineering* 33: 28–40. https://doi.org/10.1016/j.ymben.2015.10.010.

Simões, T. et al. 2006. "The SPI1 gene, encoding a glycosylphosphatidylinositol-anchored cell wall protein, plays a prominent role in the development of yeast resistance to lipophilic weak-acid food preservatives." *Applied and Environmental Microbiology* 72 (11): 7168–7175. doi: 10.1128/AEM.01476-06.

Song, Xuejiao, Haiying Yu, and Kun Zhu. 2015. "Improving alkane synthesis in Escherichia coli via metabolic engineering." *Applied Microbiology and Biotechnology* 100 (2): 757–767. https://doi.org/10.1007/s00253-015-7026-y.

Steen, E. J., R. Chan, N. Prasad, S. Myers, C. J. Petzold, A. Redding, M. Ouellet, and J. D. Keasling. 2008. "Metabolic engineering of Saccharomyces cerevisiae for the production of n-butanol." *Microbial Cell Factories* 7: 36. https://doi.org/10.1186/1475-2859-7-36. https://www.ncbi.nlm.nih.gov/pubmed/19055772.

Steen, Eric J., Yisheng Kang, Gregory Bokinsky, Zhihao Hu, Andreas Schirmer, Amy McClure, Stephen B. del Cardayre, and Jay D. Keasling. 2010. "Microbial production of fatty-acid-derived fuels and chemicals from plant biomass." *Nature* 463 (7280): 559–562. https://doi.org/10.1038/nature08721.

Sun, Y. and J. Cheng. 2002. "Hydrolysis of lignocellulosic materials for ethanol production: A review." *Bioresource Technology* 83 (1): 1–11. doi: 10.1016/S0960-8524(01)00212-7.

Taherzadeh, M. J. et al. 1997. "Characterization and fermentation of dilute-acid hydrolyzates from wood." *Industrial and Engineering Chemistry Research* 36 (11): 4659–4665. doi: 10.1021/ie9700831.

Vasserot, Y., F. Mornet, and P. Jeandet. 2010. "Acetic acid removal by Saccharomyces cerevisiae during fermentation in oenological conditions. Metabolic consequences." *Food Chemistry*. Elsevier Ltd., 119 (3): 1220–1223. doi: 10.1016/j.foodchem.2009.08.008.

Wang, J. et al. 2012. "Global regulator engineering significantly improved Escherichia coli tolerances toward inhibitors of lignocellulosic hydrolysates." *Biotechnology and Bioengineering* 109 (12): 3133–3142. doi: 10.1002/bit.24574.

Wang, X. et al. 2011. "Increased furfural tolerance due to overexpression of NADH-dependent oxidoreductase FucO in Escherichia coli strains engineered for the production of ethanol and lactate." *Applied and Environmental Microbiology* 77 (15): 5132–5140. doi: 10.1128/AEM.05008-11.

Wang, X. et al. 2012. "Increased furan tolerance in Escherichia coli due to a Cryptic ucpA gene." *Applied and Environmental Microbiology* 78 (7): 2452–2455. doi: 10.1128/AEM.07783-11.

Xu, Q. S., Y. S. Yan, and J. X. Feng. 2016. "Efficient hydrolysis of raw starch and ethanol fermentation: A novel raw starch-digesting glucoamylase from Penicilliumoxalicum." *Biotechnology for Biofuels*. BioMed Central Ltd., 9 (1): 1–18. doi: 10.1186/s13068-016-0636-5.

Yan, Hong, Zheng Wang, Fang Wang, Tianwei Tan, and Luo Liu. 2016. "Biosynthesis of chain-specific alkanes by metabolic engineering in Escherichia coli." *Engineering in Life Sciences* 16 (1): 53–59. https://doi.org/10.1002/elsc.201500057.

Yanagisawa, M. et al. 2011. "Production of high concentrations of bioethanol from seaweeds that contain easily hydrolyzable polysaccharides." *Process Biochemistry*. Elsevier Ltd., 46 (11): 2111–2116. doi: 10.1016/j.procbio.2011.08.001.

Zhang, M. M. et al. 2019. "Enhanced acetic acid stress tolerance and ethanol production in Saccharomyces cerevisiae by modulating expression of the de novo purine biosynthesis genes." *Biotechnology for Biofuels*. BioMed Central Ltd., 12 (1): 1–13. doi: 10.1186/s13068-019-1456-1.

Zheng, H. et al. 2012. "Increase in furfural tolerance in ethanologenic Escherichia coli LY180 by plasmid-based expression of ThyA." *Applied and Environmental Microbiology* 78 (12): 4346–4352. doi: 10.1128/AEM.00356-12.

Zhou, Yongjin J., Nicolaas A. Buijs, Zhiwei Zhu, Jiufu Qin, VerenaSiewers, and Jens Nielsen. 2016. "Production of fatty acid-derived oleochemicals and biofuels by synthetic yeast cell factories." *Nature Communications* 7 (1). https://doi.org/10.1038/ncomms11709.

10 Saccharide to Biodiesel

Farha Deeba, Kukkala Kiran Kumar,
and Naseem A. Gaur
Yeast Biofuel Group, International Centre for
Genetic Engineering and Biotechnology (ICGEB)

CONTENTS

10.1   Introduction 267
10.2   Physicochemical Properties and Chemical Constitution of Saccharides ...268
10.3   Source of Saccharides 270
10.4   Oleaginous Microbial Platform for Biodiesel Production 271
        10.4.1   Oleaginous Microalgae 271
        10.4.2   Oleaginous Yeast and Filamentous Fungi 273
        10.4.3   Oleaginous Bacteria 273
        10.4.4   Co-Cultivation of Microorganisms 274
10.5   Sugar Conversion into SCO 274
10.6   Lipid Extraction Methods 275
10.7   Transesterification 276
10.8   Purification of Biodiesel 278
10.9   Fatty Acid Profiles and Biodiesel Properties 279
10.10 Conclusions 280
References 280

10.1   INTRODUCTION

Rapid population growth, unbalanced food supply and diminishing fossil fuel resources have caused the world energy threats (Patel et al., 2020). Biodiesel is found to be a sustainable alternative for fossil diesel and emits low level of greenhouse gases (Mahlia et al., 2020). The low CO_2 emission levels deprived of sulphur and other harmful elements are the key factors that make it green and environment-friendly (Hill et al., 2006).

Biodiesel production and consumption has increased by 14% from 2016 to 2020 globally, driven by biofuel policies in the USA, Argentina, Brazil, Indonesia and EU by Food and Agriculture Organization (FAO) and Organisation for Economic Co-operation and Development (OECD). This accounts for an increase in production from 33.2 billion litres in 2016 to 37.9 billion litres in 2020. Waste-derived biodiesel was expected to grow to 4.4 billion litres, which is 53% rise in production (*OECD/FAO | S&P Global Platts*, 2020). In one study, it was claimed that vegetable oil cost accounts for up to 77% of total expenses for biodiesel production at a smaller scale (Skarlis et al., 2018). It is also clear that as operating costs get higher, there is a decline in entrepreneur and government interests (Apostolakou et al., 2009). This study also mentions that production at 10–50 k tonne per year is also not feasible economically.

The edible oil-derived biodiesel is not sustainable due to the high demands in food sector. So there is a need for abundant, non-edible and low-cost feedstocks to meet the fuel demand of the present transportation (Patel et al., 2019). This led to the exploration of saccharide-based biofuels as an alternative renewable source of energy. An important characteristic for the implementation of saccharide-based biofuels is sustainability, which should be assessed during the production process. To achieve sustainable productivity, single-cell oils (SCOs) are known as a better substitute for biodiesel production (Papanikolaou, 2012). SCOs have several advantages in terms of productivity, easy genetic modifications and ability to grow under controlled environments irrespective of climatic conditions (Sitepu et al., 2014). These microbes can utilize saccharides for growth and lipid agglomeration. Several species of microbes which can surpass 20% w/w of lipid production per dry cell weight (DCW) are generally termed as oleaginous (Kumar et al., 2020). However, certain microbes can reach >70% (w/w) lipid under high C/N conditions (Papanikolaou and Aggelis, 2011). For the production of SCO, conventional or pure sugars, (preferentially glucose), were used as the main source of saccharide by microbial cell factories. The utilization of other sugars such as xylose, arabinose, sucrose and fructose present in agro-industrial wastes can reduce the overall operational costs (Do et al., 2019). To access the sugars from lignocellulosic materials, different pre-treatment methods are necessary to degrade the lignin (Baruah et al., 2018). The common steps in the conversion of agro-industrial wastes into biodiesel by microbes are as follows: hydrolysis of agro-industrial wastes into fermentable sugars and microbial conversion of the sugars released for biodiesel production. However, pre-treatment methods not only release sugars, but also generate inhibitors that inhibit microbial growth and their lipid production ability. Hence, microbes with inhibitor tolerance or detoxification procedures are used to enable effective fermentation of sugars into lipids (Zhang et al., 2018). The lipids obtained from microbial cells are then transesterified to produce fatty acid methyl esters (FAMEs), i.e. biodiesel. Figure 10.1 presents an overview of biodiesel production using saccharides by different microorganisms.

This chapter focuses on the application of saccharides for biodiesel production using microbes. The fatty acid composition and important biodiesel properties to determine fuel quality have also been discussed. This chapter provides inspiring information on saccharides utilization, which should facilitate more efficient production of biodiesel from lignocellulosic biomass using oleaginous microorganisms.

10.2 PHYSICOCHEMICAL PROPERTIES AND CHEMICAL CONSTITUTION OF SACCHARIDES

Sugars are categorized into carbohydrates as they are structurally made of carbon, hydrogen and oxygen. Based on the number of carbons, simple sugars are classified as trioses (3-carbon base), tetroses (4-carbon base), pentoses (5-carbon base) and hexoses (6-carbon base). Glucose is the most abundant monosaccharide. It has a chiral structure ['d-' (dextrogyre, dextrorotatory) and 'l-' (levogyre, laevorotatory)];

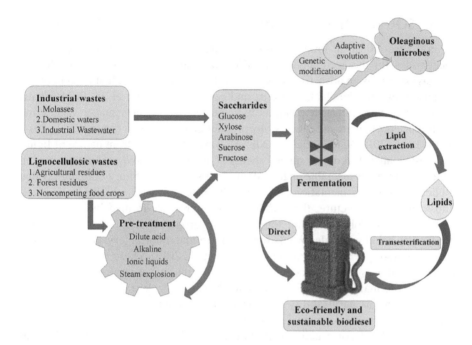

FIGURE 10.1 An overview of biodiesel production from saccharides by microorganisms.

TABLE 10.1

Physical and Chemical Properties of Different Monosaccharides (Pardo et al., 2018)

	Saccharide	d(+)-Glucose	l(+)-Arabinose	d(+)-Xylose
Appearance	White translucent crystals	Crystals or colourless to white crystalline powder	White powder	White powder
Sugar derivative	–	Sorbitol	–	Xylitol
Molecular formula	–	$C_6H_{12}O_6$	$C_5H_{10}O_5$	$C_5H_{10}O_5$
Water density at 20°C (kg m^{-3})	1587	1540	1585	1525
Water solubility (g/100 ml at 298 K)	203.9	91	83.4	55
Molar mass (g/mol)	342.3	180.1	150.1	150.1
Fusion point (K)	459	419	431–436	419–425

all carbons in positions 2–5 with various substitute radicals create different stereo-isomers with mirror symmetry; and chemically, their behaviour is similar. The more common C6 and C5 sugars with these structures are d(+)-glucose and l(+)-arabinose, and d(+)-xylose. The physicochemical properties of these sugars are depicted in Table 10.1. Except for water solubility and molecular mass, there is very little variation in their properties (Pardo et al., 2018).

10.3   SOURCE OF SACCHARIDES

Several agro-industrial low-cost substrates are being explored as a repository of saccharides for the accumulation of SCOs (Anwar et al., 2014). Lignocellulosic biomass stands as an attractive substrate for the production of valuable compounds because they are abundant, low cost and renewable (Anwar et al., 2014; Fatma et al., 2018; Vaz, 2017). It contains high percentage of fermentable sugars and micronutrients that are used for biofuel production by microbes (Anwar et al., 2014; Fatma et al., 2018; Valdés et al., 2020b). However, the recalcitrant nature of lignocellulosic materials requires special chemical, physical or biological pre-treatment, which produces fermentable sugars along with several inhibitory compounds that hinder microbial fermentation (Jin et al., 2015; Kumar et al., 2020). In lignocellulosic biomass structure, the cellulose microfibres are formed by cross-linked molecules of cellobiose bound together by hydrogen bonds. These microfibres are further held by Van der Waals force to form macrofibrils (crystalline regions), which are highly recalcitrant regions (Brandt et al., 2013). These macrofibrils bind to hemicelluloses and hemicelluloses to lignin, making a very stable structure. Lignin is the main contributing factor for biomass recalcitrance; at the molecular level, controls permeability and mechanical properties of the cells; and also gives protection from chemical degradation and microorganisms. Certain degree of flexibility for deconstruction is provided by the branches of the hemicelluloses and the amorphous regions of the cellulose. Various pre-treatment methods are employed to achieve the deconstruction of biomass (Lorenci Woiciechowski et al., 2020). Steam explosion, dilute acid, ammonia fibre expansion and organosolv are some of the pre-treatment technologies that are of industrial application. On the other hand, enzymatic hydrolysis involves the application of enzyme cocktails containing hemicellulases and cellulases to hydrolyse the major polysaccharides (cellulose and hemicellulose) into fermentable sugars (Van Dyk and Pletschke, 2012). Enzymatic hydrolysis can be operated as simultaneous saccharification and fermentation (SSF) or as separate hydrolysis and fermentation (SHF) process. SSF facilitates the conversion of simple sugars immediately, with an improved yield of enzymatic hydrolysis (Jin et al., 2012). Consolidated bioprocessing (CBP) involves the integration of enzyme production into the SSF system and is considered as the efficient low-cost process for biofuel production (Jin et al., 2012). Oleaginous microbes that can consume major sugars such as glucose and xylose along with the less available sugars such as arabinose, mannose or galactose are preferable (Huang et al., 2013). The pre-treatment methods that produce less inhibitory compounds were employed, or several detoxification processes were considered to diminish inhibitors concentration (Jin et al., 2012; Kumar et al., 2020). Genetic and metabolic engineering techniques (Tsai et al., 2019), evolutionary adaptation (Daskalaki et al., 2019) and the use of specialized microbial consortia (Anwar et al., 2014) can be considered for biotechnological valorization of lignocellulosic biomass. Sub-merged culture has been reported as the most preferred culture method for lipid production, along with solid-state fermentation as viable option (Jin et al., 2012). However, at industrial level the technological advancement still needs to be achieved for cost-effective SCO production.

Apart from lignocellulosic wastes, industrial wastes such as molasses which contain saccharides such as sucrose, fructose, glucose and glycerol (Singh et al., 2020; Worland et al., 2020) are also being widely used for SCO production.

10.4 OLEAGINOUS MICROBIAL PLATFORM FOR BIODIESEL PRODUCTION

Bacteria, filamentous fungi, yeast and microalgae that can store high lipid (>20% w/w) content inside their cell are considered for biofuel production (Patel et al., 2020), as depicted in Table 10.2. The lipids of these oleaginous microbes generally comprise C16–C18 fatty acids that are favourable to be utilized as biodiesel (Knothe, 2009). The metabolism of these microorganisms is adapted under specific growth environment to transform the carbon source into storage lipids inside the cell (Zhang and Liu, 2019).

10.4.1 OLEAGINOUS MICROALGAE

Oleaginous microalgae are the important storehouse of renewable biofuels (Elrayies, 2018). Microalgae utilize organic and inorganic carbon sources by four diverse types

TABLE 10.2
Various Oleaginous Microorganisms Grown on Different Carbon Sources

Organism	Carbon Source	% w/w Lipids	References
	Bacteria		
R. opacus PD630	Dairy wastewater	14	S. Kumar et al. (2015)
	Dextrose	70	
R. opacus PD630	Kraft hardwood pulp	46	Kurosawa et al. (2013)
Bacillus subtilis	Hydrolysate of cotton stalk	39.8	Q. Zhang et al. (2014)
Gordonia sp.	Sugar cane molasses	96	Gouda et al. (2008)
	Yeast and Fungi		
Cryptococcus sp.	Banana peel hydrolysate	34	Han et al. (2019)
R. kratochvilovae HIMPA1	Hydrolysate of Cassia fistula fruit pulp	53.18	Patel et al. (2015b)
	Hemp seed aqueous extract	55.56	Patel et al. (2014)
	Phenol + glucose	64.92	Patel et al. (2017b)
R. toruloides	Brewers' spent grain	56	Patel et al. (2018b)
L. starkeyi	Xylose and glucose	48	Bonturi et al. (2015)
R. glutinis	Monosodium glutamate with glucose	20	Xue et al. (2008)
C. curvatus	Waste cooking oil	70	Patel and Matsakas
C. curvatus	Glucose	53	(2019)
L. starkeyi CBS 1807	Sweet sorghum stalks juice	30	Matsakas et al. (2014)
Rhodotorula pacifica INDKK	Pongamia shell hydrolysate	55.89	Kumar et al. (2020)

(Continued)

TABLE 10.2 (*Continued*)
Various Oleaginous Microorganisms Grown on Different Carbon Sources

Organism	Carbon Source	% w/w Lipids	References
Microalgae			
Chlorella protothecoides	Glucose	49	Li et al. (2007)
Chlorella protothecoides	Sugarcane bagasse hydrolysate	34	Mu et al. (2015)
Chlorella vulgaris	Cyperus esculentus hydrolysate	34.4	Wang et al. (2013)
Chlorella sp.	Piggery wastewater	22	Kuo et al. (2015)
Chlorella protothecoides	Brewer fermentation waste + crude glycerol	52	Feng et al. (2014)
Scenedesmus bijuga	Anaerobically digested food wastewater	30.7	Feng et al. (2014)
Chlorella pyrenoidosa	Food waste hydrolysate	20	Shin et al. (2015)
Scenedesmus sp.	Domestic wastewater	32.2	Valdés et al. (2020a)
C. vulgaris NIES-227	Glucose	89	Shen et al. (2015)
Auxenochlorella protothecoides	Birch biomass hydrolysate	66	Patel et al. (2018a)
	Spruce biomass hydrolysate	63	
Fusarium oxysporum	Glucose, fructose and sucrose mixture	53	Matsakas et al. (2017)
	Fructose	26	
	Sucrose	49	
	Glucose	42	
Fusarium equiseti UMN-1	Glucose	56	Yang and Hu (2019)
Sarocladium kiliense ADH17	Glucose and glycerol	33	Nouri et al. (2019)
Mortierella alpina LP M 301	Glucose + potassium nitrate	31	Eroshin et al. (2000)
Microsphaeropsis sp.	Corncob waste liquor	22	Venkata Subhash and Venkata Mohan (2011)
Mixed Cultures			
Chlorella vulgaris + Mesorhizobium sangaii	BG11-N medium	51.2	Wei et al. (2020)
Chlorella vulgaris + bacteria	Sea food wastewater	32.15	Nguyen et al. (2019)
R. toruloides & C. vulgaris	Food waste hydrolysate	58	Zeng et al. (2018)
Indigenous microalgae + bacteria	Effluent from sewage sludge fermentation	17	Cho et al. (2017)
C. pyrenoidosa + bacteria	Landfill leachate	20.8	Zhao et al. (2014)
Chlorella sp. + R. toruloides	Wine distillery wastewater	63.4	Ling et al. (2014)
Chlorella sp. + Rhodotorula glutinis	Crude glycerol	39.5	Cheirsilp et al. (2012)
Chlorella sp. + Rhodotorula glutinis	Effluent from seafood processing plant	62.2	Cheirsilp et al. (2011)

of cultivation, namely photoheterotrophic, heterotrophic, mixotrophic and autotrophic (Guldhe et al., 2017). However, in the past decades, heterotrophic cultivation has been preferred as it offers advantages such as low cost and low daily maintenance to cultivate (Fan et al., 2011). Glucose is the most preferred substrate for the heterotrophic mode, but it should be derived from low-cost substrates to cut down the production cost (Chen and Jiang, 2017). Various low-cost raw materials such as birch, spruce, beech, rice straw, molasses, corn stover, sugarcane bagasse, wheat straw and industrial wastewater have been employed for heterotrophic cultivation (Arora et al., 2017; Patel et al., 2018a). Microalgae strains such as *Botryococcus, Chlamydomonas, Scenedesmus, Chlorococcum, Nannochloropsis, Chlorella, Isochrysis, Auxenochlorella* are shown to have excellent lipid production ability (Deshmukh et al., 2019; Dourou et al., 2018; Finco et al., 2017; Malibari et al., 2018; Patel et al., 2020).

The technical problems in microalgae cultivation can be solved by preferring oleaginous yeast for lipid production due to their short lipid production time with high titres in contrast to microalgae.

10.4.2 OLEAGINOUS YEAST AND FILAMENTOUS FUNGI

Yeast species such as *Trichosporon, Lipomyces, Yarrowia, Rhodosporidium, Candida, Rhodotorula* and *Cryptococcus* can produce lipids up to 60%–80% w/w (Patel et al., 2016a). To reach economic feasibility, lipid production by oleaginous yeast strains on various lignocellulosic materials has been demonstrated (Matsakas et al., 2015). Filamentous fungi are also considered for making biofuel as they produce unique fatty acids such as linolenic acid (Shu and Tsai, 2016). They can utilize various low-cost feedstocks such as glycerol, sewage sludge, monosodium glutamate wastewater, waste molasses and lignocellulosic materials (Fakas et al., 2009; Fan et al., 2012). *Cunninghamella echinulata*, grown on orange peels and glucose, accumulated linolenic acid (14.1%) out of 46.6% total lipids (Gema et al., 2002). Also, the fungus *Mortierella alpina* LPM 301 produced lipids (31.1%) mainly consisting of arachidonic acid (60.4%) when provided with glucose and potassium nitrate (Eroshin et al., 2000). *Aspergillus niger* was grown on sugarcane distillery wastewater or vinasse for biodiesel production (Du et al., 2018). In one study, maximum lipid (2.27 g/L) was produced by *Aspergillus awamori* when grown on pure vinasse (Du et al., 2018).

10.4.3 OLEAGINOUS BACTERIA

Though oleaginous bacteria can produce TAGs which are inadequate for biodiesel manufacturing relative to yeast and microalgae (Cho and Park, 2018), they are used for biodiesel production. Some of the well-known oleaginous bacteria are *Arthrobacter, Acinetobacter* sp., *Gordonia* sp. and *Rhodococcus* sp. Out of them, *Rhodococcus* sp. has widely been used as it can grow on different substrates (Wells et al., 2015) and it has the ability to use lignin for lipid production (Kosa and Ragauskas, 2013, 2012). In a study, *R. opacus* produced 26.8% (w/w) lipid content on pre-treated pine and lignocellulosic effluents (Wells et al., 2015). This microbe has also been used for the production of lipids from pre-treated kraft lignin (Wei et al., 2015).

10.4.4 Co-Cultivation of Microorganisms

To overcome the limitations faced by single type of microbe, co-cultivation of different types of microbes for lipid production has been studied. Several researches have shown that microalgae growth can be promoted by bacterial metabolites such as vitamin B12 (Xie et al., 2013) and indole-3-acetic acid (De-Bashan et al., 2008), which therefore contributes strongly to enhanced lipid and biomass productivity. Indigenous bacteria and microalga *Chlorella pyrenoidosa* consortium was used in a landfill leachate and municipal wastewater mixture to obtain a lipid productivity of 24.1 mg/L/d along with the removal of 95% ammonium nitrogen (Zhao et al., 2014). Microalgae easily assimilated the various nitrogen-containing compounds decomposed by bacteria, which leads to enhanced productivity. Other studies have shown that the co-cultivation of cyanobacteria and microalgae can increase lipid productivity. Cyanobacterium *Synechocystis salina* and microalga *Pseudokirchneriella subcapitata* together gave 51.6% more lipid productivity in comparison with single culture (Gonçalves et al., 2016). Mixed cultures of yeasts and microalgae have been reported to enhance lipid yields (Reyna-Martínez et al., 2015). Microalgae can utilize nutrients and carbon dioxide and generate oxygen required by oleaginous yeast. The yeasts generate sufficient amount of carbon dioxide required for microalgae growth. Hence, during co-cultivation a symbiosis for lipid production happens between these two taxa. *Saccharomyces cerevisiae* and *Chlorella* sp. mixed culture leads to an increased lipid production by 65.2% and biomass by 28.1% in comparison with a single culture of *Chlorella* sp. (Shu et al., 2013). A mixed culture of microalga *C. pyrenoidosa* and yeast *R. toruloides* (Ling et al., 2014) utilized a mixture of domestic wastewater and distillery wastewater for lipid production (Saayman and Viljoen-Bloom, 2017). A higher lipid yield (4.6 g/L) was obtained as compared to the monocultures of yeast and algae. Co-cultivation of *C. vulgaris* and *R. glutinis* on crude glycerol revealed an enhanced production of lipid (Cheirsilp et al., 2012). Co-cultivation of *R. glutinis* and *C. vulgaris* on seafood processing plant effluent showed a greater production of lipid than each strain culture individually (Cheirsilp et al., 2011). More studies are required during co-cultivation to tweak the ratios of each strain to enlighten the detailed mechanism of lipid production for improved lipid yields.

10.5 SUGAR CONVERSION INTO SCO

The conversion of sugar into SCO occurs via de novo biosynthesis, which includes three physiological phases: (a) growth phase, (b) oleaginous phase and (c) lipid turnover phase. During the growth phase, through glycolytic pathway and the pentose phosphate pathway (PPP), carbon is converted into cell mass which is rich in polysaccharides and proteins, while limited synthesis of polar lipids necessary for the cell membranes construction occurred (Dourou et al., 2017). During stationary phase, the depletion of one essential nutrient (e.g., magnesium, phosphate, sulphate or nitrogen) induces the accumulation of oil (Bellou et al., 2016). Finally, during lipid turnover phase, the degradation of TAGs is done to meet the energy for maintenance of cell (Beopoulus and Nicaud, 2012). The studies related to metabolism of hexoses and pentoses have extensively been done (Valdés et al., 2020a). The sugars conversion of molasses and cellobiose into storage lipids has also been reported (Sagia et al., 2020; Y. Yu et al., 2020). The yields obtained by cultivating oleaginous microbes on individual sugars are not sufficient to estimate the effective conversion of lignocellulosic biomass into SCO. As an alternative, the simultaneous assimilation of pentoses and hexoses present in substrates is essential for obtaining sustainable production of SCO (Kim et al., 2010). Various studies on oleaginous microbes (especially yeasts) cultivated on mixed sugars have revealed the sequential assimilation of individual sugars demonstrating diauxic growth (Poontawee et al., 2018; Yamada et al., 2017).

Furthermore, some strains are also reported to consume simultaneously more than two sugars. For instance, *Rhodosporidium* and *Pseudozyma* strains have shown simultaneous consumption of glucose, xylose and fructose (Patel et al., 2015a) and glucose, xylose and arabinose, respectively (Tanimura et al., 2016). Similarly, *Y. lipolytica* and *R. kratochvilovae* presented a simultaneous consumption of different hexoses and pentoses on a mixed sugar medium (Patel et al., 2015b; Tsigie et al., 2011). Recently, strains of *Sugiyamaella paludigena*, *Scheffersomyces coipomensis* and *Meyerozyma guilliermondii* from decaying wood can utilize xylose, mannose and glucose simultaneously, which are the most abundant sugars present in lignocellulosic biomass (Valdés et al., 2020b).

Nutrient (usually nitrogen) deficiency usually induces lipid accumulation. This is due to a reduction in the concentration of adenosine monophosphate (AMP) which led to the inhibition of NAD^+-dependent mitochondrial isocitrate dehydrogenase (NAD^+-ICDH) (Ratledge and Wynn, 2002). This inhibition causes Krebs cycle deregulation causing an increment in citric acid inside mitochondria which is then secreted out to cytosol by exchanging malate. Later, citric acid by the action of ATP-dependent citrate lyase (ACL) is converted to acetyl-CoA and oxaloacetate. The high activity of ACL and the null or low activity of ICDH in the cytoplasm are the key factors for lipids accumulation (Arous et al., 2016; Dourou et al., 2017; Valdés et al., 2020b).

Genetic engineering that facilitates targeting different stages of de novo lipid production in various oleaginous microbes has been developed (Lazar et al., 2014). A previous study has shown that the overexpression of TAG, G3P and FA synthesis genes significantly increases lipid agglomeration (Aguilar et al., 2017; Dulermo and Nicaud, 2011). Other enzymes that have been reported to affect the accumulation of lipid are ACL (H. Zhang et al., 2014), ACC (Fakas, 2017), ME (Zhou et al., 2014) (Xue et al., 2015), fatty acid synthase (FASI), FASII (Runguphan and Keasling, 2014) and some desaturases (Chuang et al., 2010; Xie et al., 2015).

10.6 LIPID EXTRACTION METHODS

The conventional method of lipids quantification involves their extraction from the cells using solvents. Several extraction techniques such as bead milling, microwave, ultrasound and detergent-assisted methods were used (Meullemiestre et al., 2016; Yellapu et al., 2016) to improve the lipid recovery. The efficiency of lipid extraction from microbial cells determines the sustainable production of biofuels. The robust and thick cell wall of microbes hinders lipid recovery methods and increases the overall production cost, which discourages the usage of microbial biomass as feedstock for industrial-scale biodiesel production (Kapoore et al., 2018). Hence, to improve the lipid extraction several pre-treatment methods were employed, which are more environmental friendly and sustainable for large-scale production (Dvoretsky et al., 2016). The two different routes, wet and dry, are employed for lipid extraction from microorganisms. Due to low energy demand and reduced cost, the wet route is advantageous as compared to the dry route (Yu et al., 2011). The Bligh & Dyer and Folch methods are the widely used methods for lipid extraction, wherein a 2:1 ratio by volume mixtures of chloroform and methanol are used as solvents (Bligh and Dyer, 1959; Folch et al., 1957; Halim et al., 2012). Though the Folch method takes less time, it also has low sensitivity relative to other processes (R. R. Kumar et al., 2015). The more precise method to extract lipids is the Bligh & Dyer method, as during extraction proteins precipitation occurs in the interface of the two liquid layers. Hence, this method is more appropriate at large scale and pilot scale. Modified lipid extraction methods were also used by several researchers to further improve the lipid recovery. The usage of methyl tert-butyl ether for lipid extraction with a better suitability and recovery for the lipidome was also reported (Matyash et al., 2008). Acidic treatment (HCl) of biomass was also employed to improve the polyunsaturated fatty acids recovery (Matyash et al., 2008). However, these two methods include the usage of chloroform and methanol that are toxic, which is harmful to human health and environment. In this context, solvents such as 2-ethoxyethanol (2-EE) have been shown to be environmentally safer and more effective for the recovery of lipids than hexane, methanol or chloroform (Jones et al., 2012). Another alternative for the extraction of lipid is supercritical fluid extraction. It offers a highly effective extraction, but has not reached commercial scale (Kitada et al., 2009). Supercritical fluids such as water, CO_2, ethanol, methanol, ethane, benzene, toluene and ethylene are employed for this method (Hernández et al., 2014).

10.7 TRANSESTERIFICATION

During transesterification process, the oils are transformed into fatty acid ethyl esters (FAEEs) or FAMEs with the by-product glycerol in the presence of a catalyst and ethanol/methanol (Patel et al., 2016b). The efficiency of transesterification reaction depends on many different factors such as reaction time and temperature, solvents used, and the percentage and type of catalyst. The transesterification approaches for SCOs can be divided into direct and conventional processes. The first method (conventional method) involves multiple steps such as microbial biomass drying, cell breakage by biological, chemical or mechanical methods, extraction, purification and finally transesterification of the obtained oil. This method operates at high temperature, with a longer reaction time and a large number of solvents, which increases the overall biodiesel production cost. In the second method, i.e. direct or in situ transesterification, several steps are not needed. Based on the catalyst used, the transesterification process can be alkali/acid homogenous or alkali/acid heterogeneous, or enzymatic (Yellapu et al., 2018). The catalyst selection for each reaction is an essential step to improve the transesterification efficiency (Yellapu et al., 2018).The utilization of alkali catalysts with microalgal cells leads to the formation of soap, which affects the downstream process. Therefore, acids and, in particular, inorganic acids are preferred (Bharti et al., 2019). During homogenous catalysis, the reaction yield decreases and the cost increases, making it unsuitable for industrial-scale usage (Yellapu et al., 2018). The heterogeneous catalysts have great advantage over homogenous catalysts as they can be reused several times, requires fewer units and has low cost (Bharti et al., 2019; Yellapu et al., 2018). But they require higher pressures and temperatures and take a longer time (Yellapu et al., 2018). About 30%–40% of the total downstream expenses accounted for by the transesterification reaction is due to the catalyst (Tran et al., 2012). The transesterification process catalysed by chemical (acid or alkali) means has been considered as a good choice and hence implemented by biodiesel industries as they have a high rate of conversion in less time. But they are energy-intensive and non-environment-friendly. However, enzymatic transesterification has the knack of utilizing biocatalysts that solves the soap formation issue (Tran et al., 2012). The high cost at the end of this method makes it incompatible for utilization at an industrial scale. Enzyme immobilization on a support material or a carrier can partially solve this problem. In this case, the catalyst attains stability and can be reused with minimum by-products (Selvakumar and Sivashanmugam, 2019; Tran et al., 2012). Immobilization can be done depending on the enzyme selection, the reaction environment and the solvents used (Taher et al., 2011). Moreover, the selection of suitable enzyme and reaction-determining parameters, such as concentration, temperature, pH and time, plays an important role in the entire process (Selvakumar and Sivashanmugam, 2019). Since they have high activity and are produced by waste products, these can enhance the efficiency of transesterification reaction while decreasing the cost (Christopher et al., 2014; Selvakumar and Sivashanmugam, 2019; Taher et al., 2011). Also, substrate specificity, catalytic activity and stability under room temperature conditions are provided by enzymatic transesterification (Selvakumar and Sivashanmugam, 2019; Tran et al., 2012). Moreover, enzymes aid in the recovery and separation of the products except for catalytic action (Taher et al., 2011; Tran et al., 2012). Thereafter, enzyme catalyst consists of an eco-friendly and sustainable transesterification process (Selvakumar and Sivashanmugam, 2019). Most recently, the use of nanomaterials for the transesterification reaction has been considered as it overcomes the problems associated with heterogeneous catalysts, such as mass transfer limitations, the enzyme deactivation and the prolonged reaction time. Also, nano-catalysts have a high catalytic activity and offer large specific areas that increase catalyst and substrate interaction, which, in turn, increases the efficiency of the reaction (Bharti et al., 2019). For example, microwave method and enhanced ultrasound method can increase the efficiency by lowering cost and are appropriate for large-scale production (Martinez-Guerra et al., 2014).

The alternative strategy is to directly produce FAMEs/FAEEs by microbes to eliminate the additional transesterification step. Several genetic modifications in different microbes including bacteria and yeast have been reported. In one study, 16 mg/L FAMEs was produced by the deletion of *metJ* (methionine regulator) and the overexpression of methionine adenosyltransferase in *E.coli* (Nawabi et al., 2011). In another study, *S*-adenosyl-*L*-methionine-dependent methylation of free fatty acids was enhanced by introducing *Drosophila melanogaster* juvenile hormone acid

O-methyltransferase (*Dm*JHAMT) in *E. coli* producing medium-chain FAMEs at titres of 0.56 g/L (Sherkhanov et al., 2016). FAEEs production was studied by heterologous expression of wax ester synthases (Lian and Zhao, 2014; Shi et al., 2012; Thompson and Trinh, 2014). Among them, five wax ester synthases, *Psychrobacter arcticus* 273-4, *Mus musculus* C57BL/6, *Acinetobacter baylyi* ADP1, *Rhodococcus opacus* PD630 and *Marinobacter hydrocarbonoclasticus* DSM 8798, were expressed in yeast, and among them, the highest amount of FAEEs (6.3 mg/L) was produced by *M. hydrocarbonoclasticus* DSM 8798 (ws2) (Shi et al., 2012). Other metabolic engineering approaches applied for FAEEs production in yeast improved FAEEs titres (15.8 mg/L) by stopping acetyl-CoA carboxylase (*ACC1*) through ser659 and ser157 mutations (Shi et al., 2014a). The removal of fatty acid catabolic pathways by the deletion of acyl-CoA:sterol acyltransferases (*ARE1* and *ARE2*), diacylglycerol acyltransferases (*DGA1* and *LRO1*) and fatty acyl-CoA oxidase (*POX1*) produced 17.2 mg/L of FAEEs (Runguphan and Keasling, 2014). In one study, increased FAEEs production (34 mg/L) was obtained by integrating six copies of ws2 expression cassette, while FAEEs production of 48 mg/L was obtained by increasing the overexpression of glyceraldehyde-3-phosphate dehydrogenase (*gapN*) and acyl-CoA-binding protein (encoded by *ACB1*) (Shi et al., 2014b). In a recent study, *Yarrowia lipolytica* produced a yield of 360.8 mg/L by heterologous expression of wax ester synthase gene *MaAtfA* or *MhAtfA* (A. Yu et al., 2020). Though several studies have reported FAMEs/FAEEs production, the yield was considerably low (Kalscheuer et al., 2006; Shi et al., 2014b) and more efforts are required in terms of genetic modification and process development to improve yields that become sustainable.

10.8 PURIFICATION OF BIODIESEL

The final product quality is crucial for biodiesel production. The crude biodiesel obtained after the transesterification step is not pure and still has non-desirable lipids, base or acid solvent, water, metal ions, enzymes and soap which have to be removed to attain pure biodiesel (Jin et al., 2015). Several approaches have been reported for this purpose, such as filtration, sodium sulphate and the usage of solvents such as hexane in combination with vacuum (Yellapu et al., 2018). After the transesterification reaction of microalgal biomass, separation of lipids from biomass is the first step of purification, which can be performed by filtration or centrifugation (Yellapu et al., 2018). But, when biodiesel (non-polar) is the key product, glycerol (polar) formed as a major by-product is removed by gravitational or centrifugation techniques (Jin et al., 2015; Yellapu et al., 2018).

In most instances, the previous steps and the by-products in the final mixture decide the method to be selected. Membrane separation, wet washing and dry washing are the most common techniques (Yellapu et al., 2018). The most known, conventional and traditional method for biodiesel production is wet washing. It is also suitable for removing excess of chemicals and contaminants from the previous steps. But it requires higher quantities of water for washing which should be completely removed from the end product, necessitating additional water treatment for its disposal. So, water removal and disposal increases the overall production cost and time (Okumuş et al., 2019).

The dry washing method is an alternative to the above process, which offers the benefit of choosing the utmost adsorbent. Silica, starch, ion exchange resin and cellulolytic derivatives are the most efficient compounds for dry washing (Gomes et al., 2018). A quest for economical, eco-friendly constituents to purify biodiesel without interference with the main product is necessary. Some materials such as chamotte have been proposed, which are cheap and have a high efficacy for biodiesel purification (Saengprachum and Pengprecha, 2016). Some novel methods such as membrane technology are gaining attention (Saengprachum and Pengprecha, 2016). These membranes are made up of coating and support materials, and this whole process depends on rejection coefficients. Polydimethylsiloxane and polyvinylidene fluoride membranes are two commonly used membranes (Yellapu et al., 2018), which along with ceramic materials become suitable for organic solvents (Stojković et al., 2014). The advantage of using membranes is their thermal and chemical stability for a wide range of required solvents, temperatures, and pH which reduces the corrosion and degradation rates (Stojković et al., 2014; Yellapu et al., 2018). Although it is a low-cost operation, its high purchase cost makes it inappropriate for usage at industrial level. Fowling is the basic drawback of membranes. This process can be hindered by glycerol, soap, particles and solutes which clog the membrane, and specific solvents are needed to solve this issue (Wang et al., 2009). However, low energy consumption of membranes eases the environmental footprint and makes it suitable for future perspectives (Shuit et al., 2012; Wang et al., 2009). In another example, a two-step method of wet washing and then dry washing leads to an improved biodiesel quality (Stojković et al., 2014).

Apparently, biodiesel is the end product and should be pure to meet the quality standards set by the European Union (EU) and the American Society of Testing and Materials (ASTM) organizations. There should be a balance between the purification process and, the efficiency and operation, environmental and purchase costs. The ultimate research target is to make the production of biodiesel more profitable with a reduced amount of environmental footprint.

10.9 FATTY ACID PROFILES AND BIODIESEL PROPERTIES

Biodiesel is obtained by transesterifying vegetable oils or other feedstocks chiefly consisting of TAG, with monohydric alcohols to produce the corresponding mono-alkyl esters. The biodiesel quality is influenced by numerous parameters such as viscosity, density, heating value, cold flow properties, cetane number and flash point. As a diesel fuel substitute, these properties are necessary to determine the biodiesel potential. The biodiesel properties are influenced by the type of fatty acid present. The fatty acids mainly comprise of C18 and C16 acids, namely stearic (octadecanoic), palmitic (hexadecanoic), linolenic (octadecatrienoic), linoleic (octadecadienoic) and oleic (octadecenoic). Few oils have the exception of containing high quantities of saturated acids in the C12–C16 range, such as coconut oil. The presence of high saturated fatty acids (SFAs) in FAMEs decreases the chances of oxidation, increasing its shelf life; however, unsaturated fatty acids (UFAs) control the cold flow properties. Thus, fuel properties are regulated by tweaking the SFA-to-UFA ratio (Patel et al., 2017a). The standards set by ASTM D67513 (the USA) and EN 142144 (Europe) have proposed limits which serve in the progress of biodiesel standards around the globe.

10.10 CONCLUSIONS

The implementation of renewable and low-cost carbon sources can mitigate the problem of expensive media required for biofuel production. Fatty acid-derived biofuels, i.e. biodiesel, are gaining attention as a suitable alternative for diesel fuel. Different microbial hosts such as bacteria, yeast, fungi and microalgae are being explored intensively for biodiesel production. These microorganisms can utilize saccharides of waste lignocellulosic biomass and transform them into storage lipid form. Metabolic engineering techniques improve fermentation properties of microbes which would certainly be helpful in making sustainable biodiesel production.

So, there is a need for systemic, strategic and collective endeavours to overcome the hurdles mentioned in this chapter, such as the robustness of microbes in terms of utilizing different saccharides along with tolerance to several stresses and inhibitors, reaching industrial scale-up and commercialization of biodiesel in a sustainable way. This will, in turn, cut down the heavy dependency on fossil fuels and provide green energy for transportation.

REFERENCES

Aguilar, L.R., Pardo, J.P., Lomelí, M.M., Bocardo, O.I.L., Juárez Oropeza, M.A., Guerra Sánchez, G., 2017. Lipid droplets accumulation and other biochemical changes induced in the fungal pathogen *Ustilago maydis* under nitrogen-starvation. *Arch. Microbiol.* 199, 1195–1209. https://doi.org/10.1007/s00203-017-1388-8

Anwar, Z., Gulfraz, M., Irshad, M., 2014. Agro-industrial lignocellulosic biomass a key to unlock the future bio-energy: A brief review. *J. Radiat. Res. Appl. Sci.* 7, 163–173. https://doi.org/10.1016/j.jrras.2014.02.003

Apostolakou, A.A., Kookos, I.K., Marazioti, C., Angelopoulos, K.C., 2009. Techno-economic analysis of a biodiesel production process from vegetable oils. *Fuel Process. Technol.* 90, 1023–1031. https://doi.org/10.1016/j.fuproc.2009.04.017

Arora, N., Gulati, K., Patel, A., Pruthi, P.A., Poluri, K.M., Pruthi, V., 2017. A hybrid approach integrating arsenic detoxification with biodiesel production using oleaginous microalgae. *Algal Res.* 24, 29–39. https://doi.org/10.1016/j.algal.2017.03.012

Arous, F., Azabou, S., Jaouani, A., Zouari-Mechichi, H., Nasri, M., Mechichi, T., 2016. Biosynthesis of single-cell biomass from olive mill wastewater by newly isolated yeasts. *Environ. Sci. Pollut. Res.* 23, 6783–6792. https://doi.org/10.1007/s11356-015-5924-2

Baruah, J., Nath, B.K., Sharma, R., Kumar, S., Deka, R.C., Baruah, D.C., Kalita, E., 2018. Recent trends in the pretreatment of lignocellulosic biomass for value-added products. *Front. Energy Res.* https://doi.org/10.3389/fenrg.2018.00141

Bellou, S., Triantaphyllidou, I.E., Mizerakis, P., Aggelis, G., 2016. High lipid accumulation in *Yarrowia lipolytica* cultivated under double limitation of nitrogen and magnesium. *J. Biotechnol.* https://doi.org/10.1016/j.jbiotec.2016.08.001

Beopoulus, A., Nicaud, J.-M., 2012. Yeast: A new oil producer? *OCL* 19, 22–28. https://doi.org/10.1684/ocl.2012.0426

Bharti, R.K., Katiyar, R., Dhar, D.W., Prasanna, R., Tyagi, R., 2019. *In situ* transesterification and prediction of fuel quality parameters of biodiesel produced from *Botryococcus* sp. MCC31. *Biofuels* 1–10. https://doi.org/10.1080/17597269.2019.1594592

Bligh, E.G., Dyer, W.J., 1959. A rapid method of total lipid extraction and purification. *Can. J. Biochem. Physiol.* 37, 911–917.

Bonturi, N., Matsakas, L., Nilsson, R., Christakopoulos, P., Miranda, E., Berglund, K., Rova, U., 2015. Single cell oil producing yeasts *Lipomyces starkeyi* and *Rhodosporidium toruloides*: Selection of extraction strategies and biodiesel property prediction. Energies 8, 5040–5052. https://doi.org/10.3390/en8065040

Brandt, A., Gräsvik, J., Hallett, J.P., Welton, T., 2013. Deconstruction of lignocellulosic biomass with ionic liquids. Green Chem. https://doi.org/10.1039/c2gc36364j

Cheirsilp, B., Kitcha, S., Torpee, S., 2012. Co-culture of an oleaginous yeast *Rhodotorula glutinis* and a microalga *Chlorella vulgaris* for biomass and lipid production using pure and crude glycerol as a sole carbon source. Ann. Microbiol. 62, 987–993. https://doi.org/10.1007/s13213-011-0338-y

Cheirsilp, B., Suwannarat, W., Niyomdecha, R., 2011. Mixed culture of oleaginous yeast *Rhodotorula glutinis* and microalga *Chlorella vulgaris* for lipid production from industrial wastes and its use as biodiesel feedstock. N. Biotechnol. 28, 362–368. https://doi.org/10.1016/j.nbt.2011.01.004

Chen, H.H., Jiang, J.G., 2017. Lipid accumulation mechanisms in auto- and heterotrophic microalgae. J. Agric. Food Chem. https://doi.org/10.1021/acs.jafc.7b03495

Cho, H.U., Cho, H.U., Park, J.M., Park, J.M., Kim, Y.M., 2017. Enhanced microalgal biomass and lipid production from a consortium of indigenous microalgae and bacteria present in municipal wastewater under gradually mixotrophic culture conditions. Bioresour. Technol. 228, 290–297. https://doi.org/10.1016/j.biortech.2016.12.094

Cho, H.U., Park, J.M., 2018. Biodiesel production by various oleaginous microorganisms from organic wastes. Bioresour. Technol. https://doi.org/10.1016/j.biortech.2018.02.010

Christopher, L.P., Hemanathan Kumar, Zambare, V.P., 2014. Enzymatic biodiesel: Challenges and opportunities. Appl. Energy. https://doi.org/10.1016/j.apenergy.2014.01.017

Chuang, L. Te, Chen, D.C., Nicaud, J.M., Madzak, C., Chen, Y.H., Huang, Y.S., 2010. Co-expression of heterologous desaturase genes in Yarrowia lipolytica. N. Biotechnol. 27, 277–282. https://doi.org/10.1016/j.nbt.2010.02.006

Daskalaki, A., Perdikouli, N., Aggeli, D., Aggelis, G., 2019. Laboratory evolution strategies for improving lipid accumulation in *Yarrowia lipolytica. Appl. Microbiol. Biotechnol.* 103, 8585–8596. https://doi.org/10.1007/s00253-019-10088-7

De-Bashan, L.E., Antoun, H., Bashan, Y., 2008. Involvement of indole-3-acetic acid produced by the growth-promoting bacterium *Azospirillum* spp. in promoting growth of *Chlorella vulgaris.* J. Phycol. 44, 938–947. https://doi.org/10.1111/j.1529-8817.2008.00533.x

Deshmukh, S., Kumar, R., Bala, K., 2019. Microalgae biodiesel: A review on oil extraction, fatty acid composition, properties and effect on engine performance and emissions. *Fuel Process. Technol.* https://doi.org/10.1016/j.fuproc.2019.03.013

Do, D.T.H., Theron, C.W., Fickers, P., 2019. Organic wastes as feedstocks for non-conventional yeast-based bioprocesses. *Microorganisms* 7, 229. https://doi.org/10.3390/microorganisms7080229

Dourou, M., Aggeli, D., Papanikolaou, S., Aggelis, G., 2018. Critical steps in carbon metabolism affecting lipid accumulation and their regulation in oleaginous microorganisms. *Appl. Microbiol. Biotechnol.* https://doi.org/10.1007/s00253-018-8813-z

Dourou, M., Mizerakis, P., Papanikolaou, S., Aggelis, G., 2017. Storage lipid and polysaccharide metabolism in *Yarrowia lipolytica* and *Umbelopsis isabellina. Appl. Microbiol. Biotechnol.* 101, 7213–7226. https://doi.org/10.1007/s00253-017-8455-6

Du, Z.Y., Alvaro, J., Hyden, B., Zienkiewicz, K., Benning, N., Zienkiewicz, A., Bonito, G., Benning, C., 2018. Enhancing oil production and harvest by combining the marine alga *Nannochloropsis oceanica* and the oleaginous fungus *Mortierella elongata. Biotechnol. Biofuels* 11. https://doi.org/10.1186/s13068-018-1172-2

Dulermo, T., Nicaud, J.M., 2011. Involvement of the G3P shuttle and B-oxidation pathway in the control of TAG synthesis and lipid accumulation in *Yarrowia lipolytica. Metab. Eng.* 13, 482–491. https://doi.org/10.1016/j.ymben.2011.05.002

Dvoretsky, D., Dvoretsky, S., Temnov, M., Akulinin, E., Peshkova, E., 2016. Enhanced lipid extraction from microalgae *chlorella vulgaris* biomass: Experiments, modelling, optimization. *Chem. Eng. Trans.* 49, 175–180. https://doi.org/10.3303/CET1649030

Elrayies, G.M., 2018. Microalgae: Prospects for greener future buildings. *Renew. Sustain. Energy Rev.* https://doi.org/10.1016/j.rser.2017.08.032

Eroshin, V.K., Satroutdinov, A.D., Dedyukhina, E.G., Chistyakova, T.I., 2000. Arachidonic acid production by *Mortierella alpina* with growth-coupled lipid synthesis. *Process Biochem.* 35, 1171–1175. https://doi.org/10.1016/S0032-9592(00)00151-5

Fakas, S., 2017. Lipid biosynthesis in yeasts: A comparison of the lipid biosynthetic pathway between the model nonoleaginous yeast *Saccharomyces cerevisiae* and the model oleaginous yeast *Yarrowia lipolytica. Eng. Life Sci.* 17, 292–302. https://doi.org/10.1002/elsc.201600040

Fakas, S., Papanikolaou, S., Batsos, A., Galiotou-Panayotou, M., Mallouchos, A., Aggelis, G., 2009. Evaluating renewable carbon sources as substrates for single cell oil production by *Cunninghamella echinulata* and *Mortierella isabellina. Biomass and Bioenergy* 33, 573–580. https://doi.org/10.1016/j.biombioe.2008.09.006

Fan, J., Andre, C., Xu, C., 2011. A chloroplast pathway for the de novo biosynthesis of triacylglycerol in *Chlamydomonas reinhardtii. FEBS Lett.* 585, 1985–1991. https://doi.org/-10.1016/j.febslet.2011.05.018

Fan, J., Yan, C., Andre, C., Shanklin, J., Schwender, J., Xu, C., 2012. Oil accumulation is controlled by carbon precursor supply for fatty acid synthesis in *Chlamydomonas reinhardtii. Plant Cell Physiol.* 53, 1380–1390. https://doi.org/10.1093/pcp/pcs082

Fatma, S., Hameed, A., Noman, M., Ahmed, T., Shahid, M., Tariq, M., Sohail, I., Tabassum, R., 2018. Lignocellulosic biomass: A sustainable bioenergy source for the future. *Protein Pept. Lett.* 25, 148–163. https://doi.org/10.2174/0929866525666180122144504

Feng, X., Walker, T.H., Bridges, W.C., Thornton, C., Gopalakrishnan, K., 2014. Biomass and lipid production of *Chlorella protothecoides* under heterotrophic cultivation on a mixed waste substrate of brewer fermentation and crude glycerol. *Bioresour. Technol.* 166, 17–23. https://doi.org/10.1016/j.biortech.2014.03.120

Finco, A.M. de O., Mamani, L.D.G., Carvalho, J.C. de, de Melo Pereira, G.V., Thomaz-Soccol, V., Soccol, C.R., 2017. Technological trends and market perspectives for production of microbial oils rich in omega-3. *Crit. Rev. Biotechnol.* 37, 656–671. https://doi.org/10.1080/07388551.2016.1213221

Folch, J., Lees, M., Sloane Stanley, G.H., 1957. A simple method for the isolation and purification of total lipides from animal tissues. *J. Biol. Chem.* 226, 497–509. https://doi.org/10.3989/scimar.2005.69n187

Gema, H., Kavadia, A., Dimou, D., Tsagou, V., Komaitis, M., Aggelis, G., 2002. Production of γ-linolenic acid by *Cunninghamella echinulata* cultivated on glucose and orange peel. *Appl. Microbiol. Biotechnol.* 58, 303–307. https://doi.org/10.1007/s00253-001-0910-7

Gomes, M.G., Santos, D.Q., Morais, L.C. de, Pasquini, D., 2018. Purification of biodiesel by dry washing and the use of starch and cellulose as natural adsorbents: Part II – study of purification times. *Biofuels* 1–9. https://doi.org/10.1080/17597269.2018.1510721

Gonçalves, A.L., Pires, J.C.M., Simões, M., 2016. Biotechnological potential of *Synechocystis salina* co-cultures with selected microalgae and cyanobacteria: Nutrients removal, biomass and lipid production. *Bioresour. Technol.* 200, 279–286. https://doi.org/10.1016/j.biortech.2015.10.023

Gouda, M.K., Omar, S.H., Aouad, L.M., 2008. Single cell oil production by *Gordonia* sp. DG using agro-industrial wastes. *World J. Microbiol. Biotechnol.* 24, 1703–1711. https://doi.org/10.1007/s11274-008-9664-z

Guldhe, A., Ansari, F.A., Singh, P., Bux, F., 2017. Heterotrophic cultivation of microalgae using aquaculture wastewater: A biorefinery concept for biomass production and nutrient remediation. *Ecol. Eng.* 99, 47–53. https://doi.org/10.1016/j.ecoleng.2016.11.013

Halim, R., Danquah, M.K., Webley, P.A., 2012. Extraction of oil from microalgae for biodiesel production: A review. *Biotechnol. Adv.* https://doi.org/10.1016/j.biotechadv.2012.01.001

Han, S., Kim, G.Y., Han, J.I., 2019. Biodiesel production from oleaginous yeast, *Cryptococcus* sp. by using banana peel as carbon source. *Energy Reports* 5, 1077–1081. https://doi. org/10.1016/j.egyr.2019.07.012

Hernández, D., Solana, M., Riaño, B., García-González, M.C., Bertucco, A., 2014. Biofuels from microalgae: Lipid extraction and methane production from the residual biomass in a biorefinery approach. *Bioresour. Technol.* 170, 370–378. https://doi.org/10.1016/j. biortech.2014.07.109

Hill, J., Nelson, E., Tilman, D., Polasky, S., Tiffany, D., 2006. Environmental, economic, and energetic costs and benefits of biodiesel and ethanol biofuels. *Proc. Natl. Acad. Sci. U. S. A.* 103, 11206–11210. https://doi.org/10.1073/pnas.0604600103

Huang, C., Chen, Xue fang, Xiong, L., Chen, Xin de, Ma, L. long, Chen, Y., 2013. Single cell oil production from low-cost substrates: The possibility and potential of its industrialization. *Biotechnol. Adv.* https://doi.org/10.1016/j.biotechadv.2012.08.010

Jin, M., Gunawan, C., Balan, V., Lau, M.W., Dale, B.E., 2012. Simultaneous saccharification and co-fermentation (SSCF) of AFEX™ pretreated corn stover for ethanol production using commercial enzymes and *Saccharomyces cerevisiae* 424A(LNH-ST). *Bioresour. Technol.* 110, 587–594. https://doi.org/10.1016/j.biortech.2012.01.150

Jin, M., Slininger, P.J., Dien, B.S., Waghmode, S., Moser, B.R., Orjuela, A., Sousa, L. da C., Balan, V., 2015. Microbial lipid-based lignocellulosic biorefinery: Feasibility and challenges. *Trends Biotechnol.* 33, 43–54. https://doi.org/10.1016/J.tibtech.2014.11.005

Jones, J., Manning, S., Montoya, M., Keller, K., Poenie, M., 2012. Extraction of algal lipids and their analysis by HPLC and mass spectrometry. *J. Am. Oil Chem. Soc.* 89, 1371–1381. https://doi.org/10.1007/s11746-012-2044-8

Kalscheuer, R., Stölting, T., Steinbüchel, A., 2006. Microdiesel: *Escherichia coli* engineered for fuel production. *Microbiology* 152, 2529–2536. https://doi.org/10.1099/mic.0.29028-0

Kapoore, R., Butler, T., Pandhal, J., Vaidyanathan, S., 2018. Microwave-assisted extraction for microalgae: From biofuels to biorefinery. *Biology (Basel).* 7, 18. https://doi. org/10.3390/biology7010018

Kim, J.H., Block, D.E., Mills, D.A., 2010. Simultaneous consumption of pentose and hexose sugars: An optimal microbial phenotype for efficient fermentation of lignocellulosic biomass. *Appl. Microbiol. Biotechnol.* 88, 1077–1085. https://doi.org/10.1007/s00253-010-2839-1

Kitada, K., Machmudah, S., Sasaki, M., Goto, M., Nakashima, Y., Kumamoto, S., Hasegawa, T., 2009. Supercritical CO_2 extraction of pigment components with pharmaceutical importance from *Chlorella vulgaris*. *J. Chem. Technol. Biotechnol.* 84, 657–661. https://doi.org/10.1002/jctb.2096

Knothe, G., 2009. Improving biodiesel fuel properties by modifying fatty ester composition. *Energy Environ. Sci.* 2, 759–766. https://doi.org/10.1039/b903941d

Kosa, M., Ragauskas, A.J., 2013. Lignin to lipid bioconversion by oleaginous *Rhodococci*. *Green Chem.* 15, 2070–2074. https://doi.org/10.1039/c3gc40434j

Kosa, M., Ragauskas, A.J., 2012. Bioconversion of lignin model compounds with oleaginous *Rhodococci*. *Appl. Microbiol. Biotechnol.* 93, 891–900. https://doi.org/10.1007/s00253-011-3743-z

Kumar, K.K., Deeba, F., Sauraj, Negi, Y.S., Gaur, N.A., 2020. Harnessing pongamia shell hydrolysate for triacylglycerol agglomeration by novel oleaginous yeast *Rhodotorula pacifica* INDKK. *Biotechnol. Biofuels* 13, 175. https://doi.org/10.1186/s13068-020-01814-9

Kumar, R.R., Rao, P.H., Arumugam, M., 2015. Lipid extraction methods from microalgae: A comprehensive review. *Front. Energy Res.* https://doi.org/10.3389/fenrg.2014.00061

Kumar, S., Gupta, N., Pakshirajan, K., 2015. Simultaneous lipid production and dairy wastewater treatment using *Rhodococcus opacus* in a batch bioreactor for potential biodiesel application. *J. Environ. Chem. Eng.* 3, 1630–1636. https://doi.org/10.1016/j.jece.2015.05.030

Kuo, C.M., Chen, T.Y., Lin, T.H., Kao, C.Y., Lai, J.T., Chang, J.S., Lin, C.S., 2015. Cultivation of *Chlorella* sp. GD using piggery wastewater for biomass and lipid production. *Bioresour. Technol.* 194, 326–333. https://doi.org/10.1016/j.biortech.2015.07.026

Kurosawa, K., Wewetzer, S.J., Sinskey, A.J., 2013. Engineering xylose metabolism in triacylg-lycerol-producing *Rhodococcus opacus* for lignocellulosic fuel production. *Biotechnol. Biofuels* 6, 134. https://doi.org/10.1186/1754-6834-6-134

Lazar, Z., Dulermo, T., Neuvéglise, C., Crutz-Le Coq, A.M., Nicaud, J.M., 2014. Hexokinase-A limiting factor in lipid production from fructose in *Yarrowia lipolytica*. *Metab. Eng.* 26, 89–99. https://doi.org/10.1016/j.ymben.2014.09.008

Li, X., Xu, H., Wu, Q., 2007. Large-scale biodiesel production from microalga *Chlorella pro-tothecoides* through heterotrophic cultivation in bioreactors. *Biotechnol. Bioeng.* 98, 764–771. https://doi.org/10.1002/bit.21489

Lian, J., Zhao, H., 2014. Recent advances in biosynthesis of fatty acids derived products in *Saccharomyces cerevisiae* via enhanced supply of precursor metabolites. *J. Ind. Microbiol. Biotechnol.* https://doi.org/10.1007/s10295-014-1518-0

Ling, J., Nip, S., Cheok, W.L., de Toledo, R.A., Shim, H., 2014. Lipid production by a mixed culture of oleaginous yeast and microalga from distillery and domestic mixed waste-water. *Bioresour. Technol.* 173, 132–139. https://doi.org/10.1016/j.biortech.2014.09.047

Lorenci Woiciechowski, A., Dalmas Neto, C.J., Porto de Souza Vandenberghe, L., de Carvalho Neto, D.P., Novak Sydney, A.C., Letti, L.A.J., Karp, S.G., Zevallos Torres, L.A., Soccol, C.R., 2020. Lignocellulosic biomass: Acid and alkaline pretreatments and their effects on biomass recalcitrance – Conventional processing and recent advances. *Bioresour. Technol.* https://doi.org/10.1016/j.biortech.2020.122848

Mahlia, T.M.I., Syazmi, Z.A.H.S., Mofijur, M., Abas, A.E.P., Bilad, M.R., Ong, H.C., Silitonga, A.S., 2020. Patent landscape review on biodiesel production: Technology updates. *Renew. Sustain. Energy Rev.* https://doi.org/10.1016/j.rser.2019.109526

Malibari, R., Sayegh, F., Elazzazy, A.M., Baeshen, M.N., Dourou, M., Aggelis, G., 2018. Reuse of shrimp farm wastewater as growth medium for marine microalgae iso-lated from Red Sea – Jeddah. *J. Clean. Prod.* 198, 160–169. https://doi.org/10.1016/j.jclepro.2018.07.037

Martinez-Guerra, E., Gude, V.G., Mondala, A., Holmes, W., Hernandez, R., 2014. Microwave and ultrasound enhanced extractive-transesterification of algal lipids. *Appl. Energy* 129, 354–363. https://doi.org/10.1016/j.apenergy.2014.04.112

Matsakas, L., Bonturi, N., Miranda, E.A., Rova, U., Christakopoulos, P., 2015. High con-centrations of dried sorghum stalks as a biomass feedstock for single cell oil produc-tion by *Rhodosporidium toruloides*. *Biotechnol. Biofuels* 8. https://doi.org/10.1186/s13068-014-0190-y

Matsakas, L., Giannakou, M., Vörös, D., 2017. Effect of synthetic and natural media on lipid production from *Fusarium oxysporum*. *Electron. J. Biotechnol.* 30, 95–102. https://doi.org/10.1016/j.ejbt.2017.10.003

Matsakas, L., Sterioti, A.A., Rova, U., Christakopoulos, P., 2014. Use of dried sweet sorghum for the efficient production of lipids by the yeast *Lipomyces starkeyi* CBS 1807. *Ind. Crops Prod.* 62, 367–372. https://doi.org/10.1016/j.indcrop.2014.09.011

Matyash, V., Liebisch, G., Kurzchalia, T. V., Shevchenko, A., Schwudke, D., 2008. Lipid extraction by methyl tert-butyl ether for high-throughput lipidomics, in: *Journal of Lipid Research*. American Society for Biochemistry and Molecular Biology, pp. 1137–1146. https://doi.org/10.1194/jlr.D700041-JLR200

Meullemiestre, A., Breil, C., Abert-Vian, M., Chemat, F., 2016. Microwave, ultrasound, ther-mal treatments, and bead milling as intensification techniques for extraction of lip-ids from oleaginous *Yarrowia lipolytica* yeast for a biojetfuel application. *Bioresour. Technol.* 211, 190–199. https://doi.org/10.1016/j.biortech.2016.03.040

Mu, J., Li, S., Chen, D., Xu, H., Han, F., Feng, B., Li, Y., 2015. Enhanced biomass and oil production from sugarcane bagasse hydrolysate (SBH) by heterotrophic oleaginous microalga *Chlorella protothecoides*. *Bioresour. Technol.* 185, 99–105. https://doi.org/-10.1016/j.biortech.2015.02.082

Nawabi, P., Bauer, S., Kyrpides, N., Lykidis, A., 2011. Engineering *Escherichia coli* for biodiesel production utilizing a bacterial fatty acid methyltransferase. *Appl. Environ. Microbiol.* 77, 8052–8061. https://doi.org/10.1128/AEM.05046-11

Nguyen, T.D.P., Nguyen, D.H., Lim, J.W., Chang, C.-K., Leong, H.Y., Tran, T.N.T., Vu, T.B.H., Nguyen, T.T.C., Show, P.L., 2019. Investigation of the relationship between bacteria growth and lipid production cultivating of microalgae *Chlorella Vulgaris* in seafood wastewater. *Energies* 12, 2282. https://doi.org/10.3390/en12122282

Nouri, H., Moghimi, H., Nikbakht Rad, M., Ostovar, M., Farazandeh Mehr, S.S., Ghanaatian, F., Talebi, A.F., 2019. Enhanced growth and lipid production in oleaginous fungus, *Sarocladium kiliense* ADH17: Study on fatty acid profiling and prediction of biodiesel properties. *Renew. Energy* 135, 10–20. https://doi.org/10.1016/j.renene.2018.11.104

OECD/FAO I S&P Global Platts, 2020.

Okumuş, Z.Ç., Doğan, T.H., Temur, H., 2019. Removal of water by using cationic resin during biodiesel purification. *Renew. Energy* 143, 47–51. https://doi.org/10.1016/j.renene.2019.04.161

Papanikolaou, S., 2012. Oleaginous yeasts: Biochemical events related with lipid synthesis and potential biotechnological applications. *Ferment. Technol.* 01, 1–3. https://doi.org/10.4172/2167-7972.1000e103

Papanikolaou, S., Aggelis, G., 2011. Lipids of oleaginous yeasts. Part I: Biochemistry of single cell oil production. *Eur. J. Lipid Sci. Technol.* 113, 1031–1051. https://doi.org/10.1002/ejlt.201100014

Pardo, L.M.F., Galán, J.E.L., Ramírez, T.L.L., 2018. Saccharide biomass for biofuels, biomaterials, and chemicals. *Biomass Green Chem.* 1, 11–30.

Patel, A., Arora, N., Mehtani, J., Pruthi, V., Pruthi, P.A., 2017a. Assessment of fuel properties on the basis of fatty acid profiles of oleaginous yeast for potential biodiesel production. *Renew. Sustain. Energy Rev.* 77, 604–616. https://doi.org/10.1016/j.rser.2017.04.016

Patel, A., Arora, N., Sartaj, K., Pruthi, V., Pruthi, P.A., 2016a. Sustainable biodiesel production from oleaginous yeasts utilizing hydrolysates of various non-edible lignocellulosic biomasses. *Renew. Sustain. Energy Rev.* 62, 836–855. https://doi.org/10.1016/j.rser.2016.05.014

Patel, A., Karageorgou, D., Rova, E., Katapodis, P., Rova, U., Christakopoulos, P., Matsakas, L., 2020. An overview of potential oleaginous microorganisms and their role in biodiesel and omega-3 fatty acid-based industries. *Microorganisms* 8, 434. https://doi.org/10.3390/microorganisms8030434

Patel, A., Matsakas, L., 2019. A comparative study on de novo and ex novo lipid fermentation by oleaginous yeast using glucose and sonicated waste cooking oil. *Ultrason. Sonochem.* 52, 364–374. https://doi.org/10.1016/j.ultsonch.2018.12.010

Patel, A., Matsakas, L., Rova, U., Christakopoulos, P., 2018a. Heterotrophic cultivation of *Auxenochlorella protothecoides* using forest biomass as a feedstock for sustainable biodiesel production. *Biotechnol. Biofuels* 11, 169. https://doi.org/10.1186/s13068-018-1173-1

Patel, A., Mikes, F., Bühler, S., Matsakas, L., 2018b. Valorization of Brewers' spent grain for the production of lipids by oleaginous yeast. *Molecules* 23, 3052. https://doi.org/10.3390/molecules23123052

Patel, A., Pravez, M., Deeba, F., Pruthi, V., Singh, R.P., Pruthi, P.A., 2014. Boosting accumulation of neutral lipids in *Rhodosporidium kratochvilovae* HIMPA1 grown on hemp (*Cannabis sativa* Linn) seed aqueous extract as feedstock for biodiesel production. *Bioresour. Technol.* 165, 214–222. https://doi.org/10.1016/j.biortech.2014.03.142

Patel, A., Pruthi, P.A., Pruthi, V., 2016b. Oleaginous yeast-a promising candidatea for high quality biodiesel production, in: *Advances in Biofeedstocks and Biofuels: Production Technologies for Biofuels.* Wiley, pp. 107–128. https://doi.org/10.1002/9781119117551.ch4

Patel, A., Pruthi, V., Singh, R.P., Pruthi, P.A., 2015a. Synergistic effect of fermentable and non-fermentable carbon sources enhances TAG accumulation in oleaginous yeast *Rhodosporidium kratochvilovae* HIMPA1. *Bioresour. Technol.* 188, 136–144. https://doi.org/10.1016/j.biortech.2015.02.062

Patel, A., Sartaj, K., Arora, N., Pruthi, V., Pruthi, P.A., 2017b. Biodegradation of phenol via meta cleavage pathway triggers de novo TAG biosynthesis pathway in oleaginous yeast. *J. Hazard. Mater.* 340, 47–56. https://doi.org/10.1016/j.jhazmat.2017.07.013

Patel, A., Sartaj, K., Pruthi, P.A., Pruthi, V., Matsakas, L., 2019. Utilization of clarified butter sediment waste as a feedstock for cost-effective production of biodiesel. *Foods* 8, 234. https://doi.org/10.3390/foods8070234

Patel, A., Sindhu, D.K., Arora, N., Singh, R.P., Pruthi, V., Pruthi, P.A., 2015b. Biodiesel production from non-edible lignocellulosic biomass of *Cassia fistula* L. fruit pulp using oleaginous yeast *Rhodosporidium kratochvilovae* HIMPA1. *Bioresour. Technol.* 197, 91–98. https://doi.org/10.1016/j.biortech.2015.08.039

Poontawee, R., Yongmanitchai, W., Limtong, S., 2018. Lipid production from a mixture of sugarcane top hydrolysate and biodiesel-derived crude glycerol by the oleaginous red yeast, *Rhodosporidiobolus fluvialis. Process Biochem.* 66, 150–161. https://doi.org/10.1016/j.procbio.2017.11.020

Ratledge, C., Wynn, J.P., 2002. The biochemistry and molecular biology of lipid accumulation in oleaginous microorganisms. *Adv. Appl. Microbiol.* 51, 1–51.

Reyna-Martínez, R., Gomez-Flores, R., López-Chuken, U.J., González-González, R., Fernández-Delgadillo, S., Balderas-Rentería, I., 2015. Lipid production by pure and mixed cultures of *Chlorella pyrenoidosa* and *Rhodotorula mucilaginosa* isolated in Nuevo Leon, Mexico. *Appl. Biochem. Biotechnol.* 175, 354–359. https://doi.org/10.1007/s12010-014-1275-6

Runguphan, Weerawat, Keasling, J.D., 2014. Metabolic engineering of *Saccharomyces cerevisiae* for production of fatty acid-derived biofuels and chemicals. *Metab. Eng.* 21, 103–113. https://doi.org/10.1016/j.ymben.2013.07.003

Saayman, M., Viljoen-Bloom, M., 2017. The biochemistry of malic acid metabolism by wine yeasts – A review. *South African J. Enol. Vitic.* 27. https://doi.org/10.21548/27-2-1612

Saengprachum, N., Pengprecha, S., 2016. Preparation and characterization of aluminum oxide coated extracted silica from rice husk ash for monoglyceride removal in crude biodiesel production. *J. Taiwan Inst. Chem. Eng.* 58, 441–450. https://doi.org/10.1016/j.jtice.2015.06.037

Sagia, S., Sharma, A., Singh, S., Chaturvedi, S., Nain, P.K.S., Nain, L., 2020. Single cell oil production by a novel yeast *Trichosporon mycotoxinivorans* for complete and eco-friendly valorization of paddy straw. *Electron. J. Biotechnol.* 44, 60–68. https://doi.org/10.1016/j.ejbt.2020.01.009

Selvakumar, P., Sivashanmugam, P., 2019. Ultrasound assisted oleaginous yeast lipid extraction and garbage lipase catalyzed transesterification for enhanced biodiesel production. *Energy Convers. Manag.* 179, 141–151. https://doi.org/10.1016/j.enconman.2018.10.051

Shen, X.F., Chu, F.F., Lam, P.K.S., Zeng, R.J., 2015. Biosynthesis of high yield fatty acids from *Chlorella vulgaris* NIES-227 under nitrogen starvation stress during heterotrophic cultivation. *Water Res.* 81, 294–300. https://doi.org/10.1016/j.watres.2015.06.003

Sherkhanov, S., Korman, T.P., Clarke, S.G., Bowie, J.U., 2016. Production of FAME biodiesel in *E. coli* by direct methylation with an insect enzyme. *Sci. Rep.* 6. https://doi.org/10.1038/srep24239

Shi, S., Chen, Y., Siewers, V., Nielsen, J., 2014a. Improving production of malonyl coenzyme A-derived metabolites by abolishing Snf1-dependent regulation of Acc1. *MBio* 5. https://doi.org/10.1128/mBio.01130-14

Shi, S., Valle-Rodríguez, J.O., Khoomrung, S., Siewers, V., Nielsen, J., 2012. Functional expression and characterization of five wax ester synthases in *Saccharomyces cerevisiae* and their utility for biodiesel production. *Biotechnol. Biofuels* 5. https://doi.org/10.1186/1754-6834-5-7

Shi, S., Valle-Rodríguez, J.O., Siewers, V., Nielsen, J., 2014b. Engineering of chromosomal wax ester synthase integrated *Saccharomyces cerevisiae* mutants for improved biosynthesis of fatty acid ethyl esters. *Biotechnol. Bioeng.* 111, 1740–1747. https://doi.org/10.1002/bit.25234

Shin, D.Y., Cho, H.U., Utomo, J.C., Choi, Y.N., Xu, X., Park, J.M., 2015. Biodiesel production from *Scenedesmus bijuga* grown in anaerobically digested food wastewater effluent. *Bioresour. Technol.* 184, 215–221. https://doi.org/10.1016/j.biortech.2014.10.090

Shu, C.H., Tsai, C.C., 2016. Enhancing oil accumulation of a mixed culture of *Chlorella* sp. and *Saccharomyces cerevisiae* using fish waste hydrolysate. *J. Taiwan Inst. Chem. Eng.* 67, 377–384. https://doi.org/10.1016/j.jtice.2016.08.022

Shu, C.H., Tsai, C.C., Chen, K.Y., Liao, W.H., Huang, H.C., 2013. Enhancing high quality oil accumulation and carbon dioxide fixation by a mixed culture of Chlorella sp. and *Saccharomyces cerevisiae. J. Taiwan Inst. Chem. Eng.* 44, 936–942. https://doi.org/10.1016/j.jtice.2013.04.001

Shuit, S.H., Ong, Y.T., Lee, K.T., Subhash, B., Tan, S.H., 2012. Membrane technology as a promising alternative in biodiesel production: A review. *Biotechnol. Adv.* https://doi.org/10.1016/j.biotechadv.2012.02.009

Singh, G., Sinha, S., Kumar, K.K., Gaur, N.A., Bandyopadhyay, K.K., Paul, D., 2020. High density cultivation of oleaginous yeast isolates in 'mandi" waste for enhanced lipid production using sugarcane molasses as feed.' *Fuel* 276, 118073. https://doi.org/10.1016/j.fuel.2020.118073

Sitepu, I.R., Jin, M., Fernandez, J.E., Boundy-mills, K.L., 2014. Identification of oleaginous yeast strains able to accumulate high intracellular lipids when cultivated in alkaline pretreated corn stover. *Appl. Microbiol. Biotechnol.* 98, 7645–7657. https://doi.org/10.1007/s00253-014-5944-8

Skarlis, S., Kondili, E., Kaldellis, J.K., 2012. Small-scale biodiesel production economics: A case study focus on Crete Island. *J. Clean. Prod.* 20, 20–26. https://doi.org/10.1016/j.jclepro.2011.08.011.

Stojković, I.J., Stamenković, O.S., Povrenović, D.S., Veljković, V.B., 2014. Purification technologies for crude biodiesel obtained by alkali-catalyzed transesterification. *Renew. Sustain. Energy Rev.* https://doi.org/10.1016/j.rser.2014.01.005

Taher, H., Al-Zuhair, S., Al-Marzouqi, A.H., Haik, Y., Farid, M.M., 2011. A review of enzymatic transesterification of microalgal oil-based biodiesel using supercritical technology. *Enzyme Res.* https://doi.org/10.4061/2011/468292

Tanimura, A., Takashima, M., Sugita, T., Endoh, R., Ohkuma, M., Kishino, S., Ogawa, J., Shima, J., 2016. Lipid production through simultaneous utilization of glucose, xylose, and l-arabinose by *Pseudozyma hubeiensis*: A comparative screening study. *AMB Express* 6, 58. https://doi.org/10.1186/s13568-016-0236-6

Thompson, R.A., Trinh, C.T., 2014. Enhancing fatty acid ethyl ester production in *Saccharomyces cerevisiae* through metabolic engineering and medium optimization. *Biotechnol. Bioeng.* 111, 2200–2208. https://doi.org/10.1002/bit.25292

Tran, D.T., Yeh, K.L., Chen, C.L., Chang, J.S., 2012. Enzymatic transesterification of microalgal oil from *Chlorella vulgaris* ESP-31 for biodiesel synthesis using immobilized Burkholderia lipase. *Bioresour. Technol.* 108, 119–127. https://doi.org/10.1016/j.biortech.2011.12.145

Tsai, Y.Y., Ohashi, T., Wu, C.C., Bataa, D., Misaki, R., Limtong, S., Fujiyama, K., 2019. Delta-9 fatty acid desaturase overexpression enhanced lipid production and oleic acid content in *Rhodosporidium toruloides* for preferable yeast lipid production. *J. Biosci. Bioeng.* 127, 430–440. https://doi.org/10.1016/j.jbiosc.2018.09.005

Tsigie, Y.A., Wang, C.-Y., Truong, C.-T., Ju, Y.-H., 2011. Lipid production from *Yarrowia lipolytica* Po1g grown in sugarcane bagasse hydrolysate. *Bioresour. Technol.* 102, 9216–9222. https://doi.org/10.1016/J.BIORTECH.2011.06.047

Valdés, G., Mendonça, R.T., Aggelis, G., 2020a. Lignocellulosic biomass as a substrate for oleaginous microorganisms: A review. *Appl. Sci.* 10, 1–43. https://doi.org/10.3390/app10217698

Valdés, G., Mendonça, R.T., Parra, C., Aggelis, G., 2020b. Patterns of lignocellulosic sugar assimilation and lipid production by newly isolated yeast strains from Chilean Valdivian forest. *Appl. Biochem. Biotechnol.* 1–23. https://doi.org/10.1007/s12010-020-03398-4

Van Dyk, J.S., Pletschke, B.I., 2012. A review of lignocellulose bioconversion using enzymatic hydrolysis and synergistic cooperation between enzymes-Factors affecting enzymes, conversion and synergy. *Biotechnol. Adv.* https://doi.org/10.1016/j.biotechadv.2012.03.002

Vaz, S., 2017. Biomass and the green chemistry principles, in: *Biomass and Green Chemistry: Building a Renewable Pathway.* Springer International Publishing, pp. 1–9. https://doi.org/10.1007/978-3-319-66736-2_1

Venkata Subhash, G., Venkata Mohan, S., 2011. Biodiesel production from isolated oleaginous fungi *Aspergillus* sp. using corncob waste liquor as a substrate. *Bioresour. Technol.* 102, 9286–9290. https://doi.org/10.1016/j.biortech.2011.06.084

Wang, W., Zhou, W., Liu, J., Li, Y., Zhang, Y., 2013. Biodiesel production from hydrolysate of *Cyperus esculentus* waste by *Chlorella vulgaris*. *Bioresour. Technol.* 136, 24–29. https://doi.org/10.1016/j.biortech.2013.03.075

Wang, Y., Wang, X., Liu, Y., Ou, S., Tan, Y., Tang, S., 2009. Refining of biodiesel by ceramic membrane separation. *Fuel Process. Technol.* 90, 422–427. https://doi.org/10.1016/j.fuproc.2008.11.004

Wei, Z., Wang, H., Li, X., Zhao, Q., Yin, Y., Xi, L., Ge, B., Qin, S., 2020. Enhanced biomass and lipid production by co-cultivation of *Chlorella vulgaris* with *Mesorhizobium sangaii* under nitrogen limitation. *J. Appl. Phycol.* 32, 233–242. https://doi.org/10.1007/s10811-019-01924-4

Wei, Z., Zeng, G., Huang, F., Kosa, M., Huang, D., Ragauskas, A.J., 2015. Bioconversion of oxygen-pretreated Kraft lignin to microbial lipid with oleaginous *Rhodococcus opacus* DSM 1069. *Green Chem.* 17, 2784–2789. https://doi.org/10.1039/c5gc00422e

Wells, T., Wei, Z., Ragauskas, A., 2015. Bioconversion of lignocellulosic pretreatment effluent via oleaginous *Rhodococcus opacus* DSM 1069. *Biomass Bioenergy* 72, 200–205. https://doi.org/10.1016/j.biombioe.2014.11.004

Worland, A.M., Czajka, J.J., Xing, Y., Harper, W.F., Moore, A., Xiao, Z., Han, Z., Wang, Y., Su, W.W., Tang, Y.J., 2020. Metabolic analysis, terpenoid biosynthesis, and morphology of *Yarrowia lipolytica* during utilization of lipid-derived feedstock. *Metab. Eng. Commun.* e00130. https://doi.org/10.1016/J.MEC.2020.E00130

Xie, B., Bishop, S., Stessman, D., Wright, D., Spalding, M.H., Halverson, L.J., 2013. *Chlamydomonas reinhardtii* thermal tolerance enhancement mediated by a mutualistic interaction with vitamin B12-producing bacteria. *ISME J.* 7, 1544–1555. https://doi.org/10.1038/ismej.2013.43

Xie, D., Jackson, E.N., Zhu, Q., 2015. Sustainable source of omega-3 eicosapentaenoic acid from metabolically engineered *Yarrowia lipolytica*: from fundamental research to commercial production. *Appl. Microbiol. Biotechnol.* https://doi.org/10.1007/s00253-014-6318-y

Xue, F., Miao, J., Zhang, X., Luo, H., Tan, T., 2008. Studies on lipid production by *Rhodotorula glutinis* fermentation using monosodium glutamate wastewater as culture medium. *Bioresour. Technol.* 99, 5923–5927. https://doi.org/10.1016/j.biortech.2007.04.046

Xue, J., Niu, Y.F., Huang, T., Yang, W.D., Liu, J.S., Li, H.Y., 2015. Genetic improvement of the microalga *Phaeodactylum tricornutum* for boosting neutral lipid accumulation. *Metab. Eng.* 27, 1–9. https://doi.org/10.1016/j.ymben.2014.10.002

Yamada, R., Yamauchi, A., Kashihara, T., Ogino, H., 2017. Evaluation of lipid production from xylose and glucose/xylose mixed sugar in various oleaginous yeasts and improvement of lipid production by UV mutagenesis. *Biochem. Eng. J.* 128, 76–82. https://doi.org/10.1016/j.bej.2017.09.010

Yang, Y., Hu, B., 2019. Investigation on the cultivation conditions of a newly isolated *Fusarium* fungal strain for enhanced lipid production. *Appl. Biochem. Biotechnol.* 187, 1220–1237. https://doi.org/10.1007/s12010-018-2870-8

Yellapu, S.K., Bezawada, J., Kaur, R., Kuttiraja, M., Tyagi, R.D., 2016. Detergent assisted lipid extraction from wet yeast biomass for biodiesel: A response surface methodology approach. *Bioresour. Technol.* 218, 667–673. https://doi.org/10.1016/j.biortech.2016.07.011

Yellapu, S.K., Bharti, Kaur, R., Kumar, L.R., Tiwari, B., Zhang, X., Tyagi, R.D., 2018. Recent developments of downstream processing for microbial lipids and conversion to biodiesel. *Bioresour. Technol.* https://doi.org/10.1016/j.biortech.2018.01.129

Yu, A., Zhao, Yu, Li, J., Li, S., Pang, Y., Zhao, Yakun, Zhang, C., Xiao, D., 2020. Sustainable production of FAEE biodiesel using the oleaginous yeast *Yarrowia lipolytica*. *Microbiologyopen* 9. https://doi.org/10.1002/mbo3.1051

Yu, W.L., Ansari, W., Schoepp, N.G., Hannon, M.J., Mayfield, S.P., Burkart, M.D., 2011. Modifications of the metabolic pathways of lipid and triacylglycerol production in microalgae. *Microb. Cell Fact.* https://doi.org/10.1186/1475-2859-10-91

Yu, Y., Xu, Z., Chen, S., Jin, M., 2020. Microbial lipid production from dilute acid and dilute alkali pretreated corn stover via *Trichosporon dermatis*. *Bioresour. Technol.* 295. https://doi.org/10.1016/j.biortech.2019.122253

Zeng, Y., Xie, T., Li, P., Jian, B., Li, X., Xie, Y., Zhang, Y., 2018. Enhanced lipid production and nutrient utilization of food waste hydrolysate by mixed culture of oleaginous yeast *Rhodosporidium toruloides* and oleaginous microalgae *Chlorella vulgaris*. *Renew. Energy* 126, 915–923. https://doi.org/10.1016/j.renene.2018.04.020

Zhang, C., Liu, P., 2019. The new face of the lipid droplet: Lipid droplet proteins. *Proteomics* 19, 1700223. https://doi.org/10.1002/pmic.201700223

Zhang, H., Zhang, L., Chen, H., Chen, Y.Q., Chen, W., Song, Y., Ratledge, C., 2014. Enhanced lipid accumulation in the yeast *Yarrowia lipolytica* by over-expression of ATP: Citrate lyase from *Mus musculus*. *J. Biotechnol.* 192, 78–84. https://doi.org/10.1016/j.jbiotec.2014.10.004

Zhang, Q., Li, Y., Xia, L., 2014. An oleaginous endophyte *Bacillus subtilis* HB1310 isolated from thin-shelled walnut and its utilization of cotton stalk hydrolysate for lipid production. *Biotechnol Biofuels.* 7, 1–13. https://doi.org/10.1186/s13068-014-0152-4

Zhang, Y., Xia, C., Lu, M., Tu, M., 2018. Effect of overliming and activated carbon detoxification on inhibitors removal and butanol fermentation of poplar prehydrolysates. *Biotechnol. Biofuels* 11, 178. https://doi.org/10.1186/s13068-018-1182-0

Zhao, X., Zhou, Y., Huang, S., Qiu, D., Schideman, L., Chai, X., Zhao, Y., 2014. Characterization of microalgae-bacteria consortium cultured in landfill leachate for carbon fixation and lipid production. *Bioresour. Technol.* 156, 322–328. https://doi.org/10.1016/j.biortech.2013.12.112

Zhou, Y.J., Buijs, N.A., Siewers, V., Nielsen, J., 2014. Fatty acid-derived biofuels and chemicals production in *Saccharomyces cerevisiae*. *Front. Bioeng. Biotechnol.* 2, 32. https://doi.org/10.3389/fbioe.2014.00032

11 Second-Generation Bioethanol and Biobutanol – Methods and Prospects

Guruprasad K, Anurag Singh,
Bhawna Madan, and Mohan Yama
Bharat Petroleum Corporation Ltd.

CONTENTS

11.1 Introduction 292
11.2 Feedstocks and Their Composition 293
11.3 Conversion of Biomass to Biofuels/Biochemicals.....................................293
11.4 Pretreatment Strategies..294
    11.4.1 Physical Pretreatment Methods 295
        11.4.1.1 Mechanical Pretreatment...295
        11.4.1.2 Irradiation Pretreatment...295
    11.4.2 Chemical Pretreatment Methods 296
        11.4.2.1 Acid Pretreatment 296
        11.4.2.2 Alkaline Pretreatment...297
        11.4.2.3 Organic Solvent Pretreatment 297
    11.4.3 Physicochemical Pretreatment Methods...297
        11.4.3.1 Steam Explosion 297
        11.4.3.2 Liquid Hot Water 298
        11.4.3.3 Ammonia Fiber Explosion 298
    11.4.4 Biological Pretreatment Methods 299
        11.4.4.1 Microbes-Based Pretreatment.......................................299
        11.4.4.2 Enzyme-Based Pretreatment 299
11.5 Enzymatic Hydrolysis..299
    11.5.1 Enzymes for Hydrolysis of Lignocellulosic Biomass 300
        11.5.1.1 Cellulases 300
        11.5.1.2 Xylanases 300
    11.5.2 Factors Affecting Enzymatic Hydrolysis...300
        11.5.2.1 Enzyme-Related Factors 301
        11.5.2.2 Substrate-Related Factors 301

DOI: 10.1201/9781003158486-11

11.6 Bioethanol..302
       11.6.1   Fermentative Microorganisms...302
       11.6.2   Thermophilic Fermentative Microorganisms for
                Lignocellulosic Ethanol Production 303
       11.6.3   Downstream Processing of Ethanol 304
                11.6.3.1   Distillation 304
                11.6.3.2   Alternative Recovery Techniques 305
                11.6.3.3   Pervaporation..305
                11.6.3.4   Gas Stripping 307
                11.6.3.5   Vacuum Fermentation..307
                11.6.3.6   Adsorption...308
                11.6.3.7   Solvent Extraction...308
       11.6.4   Improvements in Ethanol Production 308
       11.6.5   Opportunities and Challenges 309
11.7 Biobutanol...310
       11.7.1   ABE/IBE Fermentation—Role of Clostridia............................310
       11.7.2   Amelioration of Butanol Fermentation.....................................311
       11.7.3   Consolidated Bioprocessing 311
       11.7.4   Downstream Processing of Butanol 313
                11.7.4.1   Pervaporation..314
                11.7.4.2   Hybrid Technologies 315
       11.7.5   Opportunities and Challenges 315
11.8 Comparison of Fuel Characteristics of Gasoline and Bioalcohols............315
11.9 Conclusions...316
References...317

11.1 INTRODUCTION

Biofuels (bio-based fuels) produced from renewable sources are envisaged to be promising alternate transportation fuels in view of the anticipated depletion of fossil fuels and the environmental concerns associated with them. Renewable biofuels are considered to provide energy security, economic stability, and also assistance in rural development. Therefore, biofuels such as bioethanol and biobutanol produced from natural feedstocks have emerged as promising transportation fuels due to their sustainability and environmental benefits. Further, they can also reduce the dependency on crude oil reserves. Bioethanol is widely used as a transportation fuel or additive to gasoline in spark-ignition (SI) engine due to its attractive properties of high octane number and reduced exhaust emissions. The next promising and competitive biofuel is biobutanol, which has superior properties to be used in SI engine without engine modification (Yusoff et al., 2015).

Second-generation biofuels are basically derived from non-food crops or crop residues. These include agricultural and forest residues, and grasses which are cheaper sources of sugars for bioethanol production. Agricultural residues include cereal straws, stovers, bagasse, etc., whereas the forest residues include softwood, hard wood, sawdust, pruning and bark thinning residues, etc. (Zabed et al., 2017).

However, the sugar content of these feedstocks vary from plant to plant and species to species.

11.2 FEEDSTOCKS AND THEIR COMPOSITION

The composition of the feedstocks is complex in nature with cross-linked polysaccharide networks, lignin, and glycosylated proteins. These polysaccharides are mainly cellulose and hemicellulose with lignin acting as a cementing material (Vohra et al., 2014). Cellulose is a crystalline insoluble polymer with repeating units of β-D-glucopyranose and is a chief constituent of the cell wall. Hemicelluloses are amorphous heterogeneous polymers which may contain pentose sugars (β-D-xylose and α-L-arabinose), hexose sugars (β-D-mannose, β-D-glucose, and α-D-galactose), and/-or uronic acids (α-D-glucuronic, α-D-4-O-methylgalacturonic, and α-D-galacturonic acids). Small amounts of other sugars such as α-L-rhamnose and α-L-fucose may also be present. Further, the hydroxyl groups of sugars can be partially substituted with acetyl groups (Gírio et al., 2010). Lignin is a complex heteropolymeric material composed of phenylpropanoid subunits which are made up of coniferyl alcohol (guaiacyl propanol), p-coumaryl alcohol (p-hydroxyphenyl propanol), and sinapyl alcohol (syringyl alcohol) and linked by alkyl–aryl, alkyl–alkyl, and aryl–aryl ether bonds (Rajesh Banu et al., 2019; Zhao et al., 2020). The biomass also contains several other components such as solvent extractives, proteins, and ashes. These components are present in minor quantities and are not significant in the perspective of biomass refining and ethanol production.

11.3 CONVERSION OF BIOMASS TO BIOFUELS/BIOCHEMICALS

Lignocellulosic materials are abundant natural resources and have enormous potential for biofuels/biochemicals production due to the presence of large amounts of sugars in the form of cellulose and hemicellulose polymers. However, their conversion into biofuels/biochemicals is relatively difficult than that of sugar-rich or starch-rich materials owing to the recalcitrant nature of these substrates. The conversion typically consists of four processes, viz. pretreatment, enzymatic hydrolysis, fermentation, and downstream processing. There could be additional steps based on the process configuration (Figure 11.1). These include feedstock handling, milling the biomass to small and homogeneous particles, detoxification of the hydrolysate, etc. (Vohra et al., 2014).

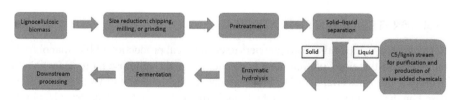

FIGURE 11.1 General process overview of second-generation ethanol/butanol production from lignocellulosic biomass.

Pretreatment is the first and one of the most important steps in biorefining of lignocellulosic biomass. The principal objective of pretreatment process is to expose the cellulose fibers for subsequent enzymatic hydrolysis (Sharma et al., 2020). This mainly involves either delignification or hydrolysis of hemicellulose, which then exposes the cellulose portion for enzymatic hydrolysis. However, decreasing the crystallinity of cellulose and increasing the surface area are also important aspects to be considered as these can have considerable effects on the recovery of cellulosic sugars and then the final ethanol yield (Galbe and Zacchi, 2007).

In the hydrolysis step, polymeric cellulose and hemicellulose are converted to monomeric sugars. Enzymatic hydrolysis can be carried out in two ways: acid-catalyzed hydrolysis or enzyme-catalyzed hydrolysis. Acid-catalyzed hydrolysis usually occurs at high temperature ($100°C$–$240°C$), leading to the formation of inhibitors such as 5-hydroxymethylfurfural (HMF) and furfural (Sharma et al., 2020). However, enzymatic hydrolysis offers numerous advantages over acid-catalyzed hydrolysis as it can be operated at mild conditions ($50°C$–$60°C$, pH 4.8–5.0). Moreover, high sugar recoveries could be achieved with no inhibitor formation (Balat, 2011). Hence, enzymatic hydrolysis is the most widely used process for the saccharification reactions in a biorefinery concept.

Fermentation is the step where biofuels/biochemicals are directly produced due to the metabolic activity of the fermentation organisms. Hydrolysate produced after the saccharification is used as the substrate where the organisms (yeast or bacteria) convert monomeric sugars into acids, gases, solvents, and/or other products. In case of ethanol fermentation, *Saccharomyces cerevisiae* (commonly known as baker's yeast) is the most widely used organism because of its high ethanol productivity and tolerance. *Pichia stipitis, Kluyveromyces fragilis*, and *Candida shehatae* are some of the other reported yeast strains for the production of ethanol from different sugars (Sharma et al., 2020). Theoretically, based on stoichiometric equations, 1 kg of sugar (glucose or xylose) can produce 0.51 kg of ethanol and 0.49 kg of carbon dioxide during fermentation (Aditiya et al., 2016; Sharma et al., 2020).

The downstream processes that are involved vary depending on the product being recovered. In case of ethanol, recovery step involves distillation wherein the broth from fermentation process is further processed to remove water content, yielding high-quality ethanol product (anhydrous ethanol). Distillation process exploits the difference in boiling points of the mixtures in a solution. When the mixture is heated to the ethanol boiling point of $78.2°C$, ethanol will be vaporized and separated from the other components. For fuel applications, the purity of the ethanol has to be minimum of 99.5% by volume (Aditiya et al., 2016).

11.4   PRETREATMENT STRATEGIES

Recalcitrance is one of the major barriers to economical production of bioethanol from biomass, and therefore, pretreatment is essential to make cellulose more accessible to hydrolytic enzymes (Balat, 2011). Pretreatment can be divided into various categories based on the mechanism. They can be broadly divided into four categories, viz. physical, chemical, physicochemical, and biological. The pretreatment parameters employed for different substrates would vary considerably based on the relative proportion and composition of cellulose, hemicellulose, lignin, and the types of bonds present. Hence, it is not possible to apply the same pretreatment technology to different types of biomass or different fractions of the same biomass (Seidl and Goulart, 2016).

11.4.1 Physical Pretreatment Methods

Physical pretreatment increases the accessible surface area and pore size of lignocellulosic biomass and decreases the crystallinity and degree of polymerization of the cellulose. The strategies include mechanical operations, different types of irradiations, and ultrasonic pretreatment techniques. These methods are briefly discussed in the following section.

11.4.1.1 Mechanical Pretreatment

This type of pretreatment employs chipping, milling, and grinding the lignocellulosic biomass into fine powder to increase the available surface area of the cellulose. It can also increase the bulk density and improve flow properties. The type of the physical pretreatment used depends on the final particle size achieved; for example, the particle size reduces to 10–30 mm after chipping and 0.2–2 mm after milling and grinding (Kumari and Singh, 2018). Mills are commonly used for this purpose, and different variants include knife, hammer, pin, ball, vibratory, colloid, attrition, extruder, and centrifugal mills (Bhutto et al., 2017). Among these, hammer mills are commonly used for size reduction as they are easy to operate, relatively inexpensive, and produce a wide range of particle sizes. The shear and impact actions of the mills result in the reduction of the particle size. However, the power requirements depend on the desired final particle size and biomass characteristics and they increase rapidly with decreasing particle size (Balat, 2011).

11.4.1.2 Irradiation Pretreatment

Irradiation of feedstock using gamma rays, electron beams, microwaves, and ultrasound is some of the widely studied tools for the modification of polymeric materials through degradation, decrystallization, and cross-linking.

Gamma irradiation: This ionizing radiation penetrates the biomass and transfers its energy to atoms of biomass components causing localized energy absorptions within the macromolecules resulting in long- and short-lived radicals. These free radicals cause structural modifications which help in polysaccharide degradation and cell wall deconstruction by scission of glyosidic bond (Kumar et al., 2020). Other effects also include lignin modification and decrease in cellulose crystallinity which also offers favorable conditions for subsequent enzymatic hydrolysis.

Electron beam irradiation: Electron beams are highly charged ionizing radiations obtained from linear electron accelerator. When these beams are directed on biomass, it causes energy transfer to the atoms of biomass disrupting the cell wall polymers, generating free radicals. This results in decrystallization of the cellulose chains by breaking of hydrogen bonds, decreasing the degree of polymerization and inducing chain scissions (Ma et al., 2014).

Microwave irradiation: Microwaves are low-frequency non-ionizing waves whose effect on biomass can be explained by thermal and non-thermal effects. Thermal effect due to the increase in temperature and pressure induces disruption of cell wall components. In contrast, the non-thermal forces are generated due to relaxation and polarization of dielectric substances under the influence of electromagnetic field. This realignment of polar molecules results in the breakdown of cellulose crystallinity due to disruption of hydrogen bonds, thereby enhancing enzymatic hydrolysis (Kumar et al., 2020).

Ultrasound pretreatment: Sound waves with frequencies higher than 20 kHz cause biomass rupture, floc fragmentation, or its destruction. These ultrasound waves break β-O-4 and α-O-4 linkages of lignin, which leads to the formation of cavitation bubbles by splitting the polysaccharides and lignin fraction of biomass. The bubbles grow to a critical size and collapse destructively causing the creation of localized hot spots, for a very small fraction of time, with a pressure of 500 atm and temperature of 5000°C (Bhutto et al., 2017). This causes intense turbulence, shear, and shock to the substrate, leading to morphological changes and thereby increasing the rate of enzymatic hydrolysis (Kumar et al., 2020).

11.4.2 CHEMICAL PRETREATMENT METHODS

Chemical pretreatment is considered to be one of the most promising methods to improve the fractionation of lignocellulosic biomass by removing lignin and/or hemicelluloses, thereby decreasing the crystallinity and degree of polymerization of the cellulosic component. Various chemicals ranging from oxidizing agents, alkalis, acids, organic solvents, and salts have extensively been studied for their efficacy on pretreatment. Many of these chemicals have been used alone or in combination with other chemicals to achieve the desired results during pretreatment. Each chemical acts differently based on the process conditions employed. In general, acids are used to solubilize hemicellulose component, whereas alkali and organic solvent pretreatments are used to solubilize and/or degrade lignin portion in the lignocellulosic biomass. Chemical pretreatment methods are believed to be the most suitable methods for the commercial-scale applications (Aditiya et al., 2016). The main advantages with these methods are that most of these chemicals are cheap, easy to procure, less hassle in storage, and of long shelf life.

11.4.2.1 Acid Pretreatment

Hydrochloric acid, phosphoric acid, nitric acid, and sulfuric acid are the most common acids used in the deconstruction of lignocellulosic biomass. Several organic acids such as peracetic acid, maleic acid, lactic acid, and acetic acid have also been used for pretreatment. The purpose of acid pretreatment is to hydrolyze the hemicellulose portion of the biomass and expose cellulose fibers for enzymatic digestion. Acid pretreatment can operate either under a high temperature and low acid concentration (dilute acid pretreatment) or under a low temperature and high acid concentration (concentrated acid pretreatment). Both these methods work on agricultural feedstocks, such as corn stover, rice straw, and wheat straw (Balat, 2011). However, dilute acid pretreatment has widely been studied because of the lower chemical consumption, process cost, and effectiveness. In spite of these benefits, there are some important disadvantages associated with this method. These acids cause corrosion problems that necessitate expensive materials of construction, stream neutralization before fermentation, and formation of degradation products such as 5-hydroxymethyl-2-furaldehyde (HMF) and furfural inhibitory to fermentation process.

11.4.2.2 Alkaline Pretreatment

Alkaline pretreatment technologies use the application of various alkaline reagents for the purpose of improving digestibility of lignocellulosic biomass. Sodium hydroxide, ammonia, and calcium hydroxide are the commonly used reagents for this purpose. Alkaline pretreatments are carried out under milder conditions, some of them even at ambient temperature, as evidenced by soaking in aqueous ammonia (Kim et al., 2008). Such methods eliminate the need for expensive materials and special designs to cope with corrosion and severe reaction conditions. It is also possible to recover and reuse chemical reagents in some of the alkaline pretreatment methods. The major reactions include dissolution of lignin and hemicellulose, and saponification of intermolecular ester bonds. These also alter the degree of polymerization of these components and bring about changes in the physical properties of treated solids such as changes in surface area, porosity, and crystallinity (Kim et al., 2016).

11.4.2.3 Organic Solvent Pretreatment

Solvents such as ethanol, methanol, ethylene glycol, and glycerol can be used for effective fractionation of lignocellulosic biomass into cellulose, lignin, and hemicellulose components. With only minor degradation, this pretreatment allows separation of high-purity cellulose in solid form. The mechanism involves dissolution of lignin and hemicellulose into organic phase. This results in reduced lignin recalcitrance and increased surface area of cellulose, thereby enhancing enzymatic accessibility for hydrolysis and subsequent fermentation yields. The mild pretreatment temperature and neutral pH conditions also reduce carbohydrate degradation into undesired compounds such as furfural and HMF (Zhang et al., 2016). This process can also be performed using a catalytic agent, which includes inorganic acids such as hydrochloric and sulfuric acids and bases such as sodium hydroxide, ammonia, and lime. The main drawbacks of this method are the usage of organic solvents with low boiling point, which leads to the risk of high-pressure operation and flammability of the solvents. However, the main advantages are recyclability of the solvents, which lowers the operation cost, and the recovery of high-quality lignin.

11.4.3 Physicochemical Pretreatment Methods

This category involves pretreatment methods that combine physical and chemical processes for dissolving hemicellulose and/or the alteration of lignin structure. It includes pretreatment methods such as steam explosion, liquid hot water, and ammonia fiber explosion. These pretreatments depend on solvents used and process conditions that affect the physical and chemical properties of the biomass.

11.4.3.1 Steam Explosion

Steam explosion is the widely used physicochemical pretreatment method for treating lignocellulosic biomass. During the process, biomass is treated for few seconds to few minutes with pressurized steam at high temperature (20–50 bar, 160°C–270°C) followed by sudden pressure release. This facilitates the breakdown of structural components, shearing (sudden decompression and evaporation of moisture), lignin transformation, and auto-hydrolysis of glyosidic bonds (Haghighi Mood et al., 2013), resulting in hydrolysis of hemicellulose. Water at high temperature possesses acidic properties and in combination with acetic acid released from hemicellulose, causes the breakdown of hemicellulose. The biomass can also be impregnated with acid prior to steam explosion to decrease the time and temperature of reaction, improve hydrolysis rate, and decrease the production of inhibitory compounds (Behera et al., 2014). Lignin is removed to a small extent, but it is redistributed on the fiber surface as a result of melting and depolymerization/repolymerization reactions (Li et al., 2007). The main advantage of this pretreatment is the low energy requirement and no recycling or environmental costs. Hence, it is considered to be one of the most cost-effective pretreatment processes for hardwoods and agricultural residues. Steam pretreatment using a catalyst has been claimed to be the closest to *commercialization*.

11.4.3.2 Liquid Hot Water

In liquid hot water pretreatment, biomass is subjected to treatment with water at high temperature (160°C–220°C) and pressure. High pressure is employed to keep the water in liquid state. No rapid expansion or decompression is required unlike steam explosion. A catalyst such as an acid can also be added, which would make the method similar to dilute acid pretreatment. Since water content is much higher than in steam explosion pretreatment, the resulting sugar solution is in dilute form, thus requiring more energy-intensive downstream processes (Galbe and Zacchi, 2007). The liquid hot water cleaves hemiacetal linkages in biomass liberating acetic and uronic acids into the liquid phase. These acids then catalyze the removal of oligosaccharides and further hydrolyze hemicellulose to monomeric sugars. The lignin is also hydrolyzed, rendering the cellulose accessible to the enzymatic hydrolysis. Complete delignification is not possible because of the recondensation of lignin (Behera et al. (2014)). The advantages of this pretreatment include (a) no requirement of any chemicals except water, (b) no issues related to corrosion of the equipment, and (c) formation of very low concentration of inhibitors (Zhuang et al. (2016)).

11.4.3.3 Ammonia Fiber Explosion

Ammonia fiber explosion is similar to steam explosion process in which the biomass is exposed to liquid ammonia at moderate temperature (below 100°C) and high pressure (above 30 bar) for a certain period of time followed by rapid pressure release. Rapid expansion of ammonia causes swelling of biomass fibers, partial decrystallization of cellulose, and cleavage of lignin–carbohydrate complex (Haghighi Mood et al., 2013). It does not cause solubilization of hemicellulose or lignin fraction, but opens up the structure of lignocellulosic biomass, thereby increasing the water-holding capacity and digestibility of biomass. The advantages of this method are recyclability of ammonia after pretreatment (as ammonia is very volatile at atmospheric pressure), no requirement of particle size reduction, and residual ammonia serving as a nitrogen source for fermentation (Behera et al., 2014).

11.4.4 Biological Pretreatment Methods

Biological pretreatment employs microorganisms or enzymes that selectively degrade the lignocellulosic components of the feedstocks. Biological pretreatment is favorable due to its sustainability to the environment as the process requires mild conditions and no additional chemicals are required.

11.4.4.1 Microbes-Based Pretreatment

Various species of fungi such as brown, white, and soft rot fungi possess the capability to degrade the biomass. Brown rot fungi is known to degrade cellulose, while white and soft rot fungi are known to degrade cellulose and lignin components (Aditiya et al., 2016). Out of these, white rot fungi have extensively been studied as these organisms possess good hydrolytic and ligninolytic systems for breaking polysaccharides and lignin structures, respectively. The major factors affecting the pretreatment are particle size, pH, temperature, moisture content, and nutrient requirements (Kumar et al., 2020). The advantages associated are low cost, low energy, less water utilization, reduced waste, low inhibitor generation, and easy downstream processing. However, the drawback of this process is *long incubation time, making it unfavorable for industrial production.*

11.4.4.2 Enzyme-Based Pretreatment

The application of enzymes for selectively removing biomass components is also one of the approaches for pretreatment. The most widely used enzymes for this purpose include xylanases. Numerous studies have focused on the production of xylanases from *Aspergillus oryzae* and their application in hydrolyzing the agricultural residues into xylooligosaccharides (Bhardwaj et al., 2017; Bhardwaj et al., 2019). Partially purified xylanase enzymes can also be used for such purposes, thereby decreasing the cost associated with downstream purification of enzymes. However, limitations still exist with cost of enzyme production, stability, shelf life, and reusability (Kumar et al., 2020).

11.5 ENZYMATIC HYDROLYSIS

The polymeric forms of cellulose and hemicellulose components are converted into their monomeric form during enzymatic hydrolysis. Yeast or other fermenting organisms cannot directly utilize the sugars in polymeric form. Hence, these polymers are converted into simple sugars to be effectively metabolized to produce the product of interest. This is performed using either acid-based or enzyme-based methods. However, since acid-based methods could lead to sugar degradation and production of inhibitory compounds, enzyme-based methods are widely utilized to achieve high sugar recovery with no production of toxic compounds. Moreover, enzymatic hydrolysis requires moderate process conditions to operate, whereas acid hydrolysis requires high temperature (>120°C) to achieve desired sugar recovery. These reasons make the enzyme-based hydrolysis methods more preferable for commercialization of the second-generation ethanol/butanol technologies.

11.5.1 Enzymes for Hydrolysis of Lignocellulosic Biomass

The complex nature of lignocellulosic biomass restricts the usage of a single enzyme to perform all the reactions involved in the hydrolysis of polymeric molecules into monomers. There is a need for enzyme cocktails which consist of several different enzymes which can perform the reactions in a sequential manner (Binod et al., 2019). The major enzymes involved in the hydrolysis reactions of lignocellulosic biomass are discussed below.

11.5.1.1 Cellulases

These are the enzymes involved in the hydrolysis of cellulose to glucose. Complete degradation takes place by the synergistic action of three enzymes—endoglucanases, exoglucanases, and β-glucosidases. The function of endoglucanases is to hydrolyze the internal β-1,4-glucosidic linkages randomly at amorphous sites in the cellulose chain. Exoglucanases, also known as cellobiohydrolases, cleave the long-chain oligosaccharides produced by the action of endoglucanases to short-chain oligosaccharides and cellobioses. Finally, β-glucosidases, also known as cellobiases, hydrolyze the glycosidic bonds in cellobiose and produce glucose as the final product.

Fungi are the main source for these cellulolytic enzymes. Enzymes derived from *Trichoderma reesei* have widely been used in research as well as commercial production. The main advantages are the full complement production of cellulases, stability of the produced cellulases under the enzymatic hydrolysis conditions, and resistance to chemical inhibitors (Taherzadeh and Karimi, 2007).

11.5.1.2 Xylanases

These enzymes catalyze the hydrolysis of hemicellulose component in lignocellulosic biomass. Unlike cellulose, hemicellulose is loosely packed in the cell wall and hence it is relatively more susceptible to enzymatic hydrolysis than cellulose. Most of the commercially available xylanase enzymes are produced from *Trichoderma reesei, Humicola insolens, or Bacillus* sp. The complete breakdown requires the action of several enzymes that include endo-1,4-β-xylanase, β-xylosidases, α-arabinofuranosidases, α-glucuronidases, and esterases. Endo-1,4-β-xylanase is the most important xylan-degrading enzyme which hydrolyzes the glycosidic bonds in the xylan backbone releasing xylooligosaccharides and β-D-xylopyranosyl oligomers. At the later stage of hydrolysis, mono-, di-, and trisaccharides of β-D-xylopyranosyl are produced. β-Xylosidase enzymes hydrolyze the xylooligosaccharides and xylobiose to xylose. α-Arabinofuranosidases cleave the arabinose and 4-O-methyl glucuronic acid substituents from the xylan backbone. α-Glucuronidase cleaves α-1,2-glycosidic linkages between xylose and D-glucuronic acid. Esterases are enzymes which act on the ester linkages between xylose and acetic acid. This enzyme removes O-acetyl groups from β-D-pyranosyl residues of acetyl xylan (Binod et al., 2019).

11.5.2 Factors Affecting Enzymatic Hydrolysis

Several factors affect the enzymatic saccharification of lignocellulosic biomass. These can be divided into enzyme-related and substrate-related factors. They are briefly described in the following sections.

11.5.2.1 Enzyme-Related Factors

There are many aspects with respect to enzyme itself which can determine the success of hydrolysis. They include enzyme adsorption onto lignin, end-product inhibition, and mechanical and thermal inactivation. Temperature and pH also play a critical role as each enzyme cocktail has its own temperature and pH optimum at which maximum hydrolytic efficiency is observed. Most of the cellulases have their optimum temperature at 50°C, and any deviation can lead to a significant decrease in the performance of the enzymes. Likewise, most enzyme cocktails have their optimum pH in the range of 4.5–5.0 and any deviations could also lead to poor enzyme performance (Binod et al., 2019).

The products of enzymatic hydrolysis are also known to cause inhibition of specific enzymes in the cocktail. Cellobiose and glucose are formed during the hydrolysis process. Cellobiose is known to directly inhibit both cellobiohydrolases and endoglucanases, whereas glucose directly inhibits β-glucosidase (Andrić et al., 2010). Hence, enzyme cocktails are supplemented with excess β-glucosidase in order to prevent the buildup of cellobiose, thereby mitigating the problems of product inhibition.

Enzymes are also known to exhibit non-specific binding to lignin on the surface of the lignocellulose substrate. Hydrophobic moieties of cellulase enzyme complexes can form irreversible binding with the hydrophobic lignin molecules. To overcome this issue, surfactants such as Tween and Triton X are commonly employed (Eriksson et al., 2002). They exhibit a positive effect with biomass containing high lignin content due to their hydrophobic interaction with lignin causing steric repulsion of enzyme from the lignin surface. This leads to availability of more enzyme molecules for hydrolysis, leading to an increase in the sugar yield (Kristensen et al., 2007).

11.5.2.2 Substrate-Related Factors

Substrate-related factors such as structural features and substrate loading play an important role affecting the rate of enzymatic hydrolysis. Structural features include crystallinity of cellulose, degree of polymerization, available surface area, particle size, etc. Many of these factors can be regulated by choosing an appropriate pretreatment technique. Moreover, conversion technologies of lignocellulosic biomass will become economically sustainable only if enzymatic hydrolysis is carried out at high biomass loading so that a concentrated sugar solution is obtained, which will lead to less effluent generation and less energy and costs associated with downstream processing. The main challenge with high biomass loading is the lack of available free water in the reactor. Water is essential in hydrolysis for mass transfer and lubricity. Less free water increases the viscosity and affects the mixing of biomass. Increased biomass loading can also lead to end-product inhibition of cellulolytic enzymes. These limitations can be overcome by adopting fed-batch strategies through which higher biomass loading can be achieved with lower inhibition. Fed-batch addition of biomass and fresh enzyme will increase the sugar yield by replacing the enzyme non-productively bound to lignin (Binod et al., 2019).

Though a variety of potential fermentation products are possible through biochemical means, this chapter will only focus on ethanol and n-butanol fermentations.

11.6 BIOETHANOL

11.6.1 FERMENTATIVE MICROORGANISMS

Saccharomyces cerevisiae is widely employed for ethanol production owing to several factors such as high fermentation rate, high ethanol tolerance, tolerance to fermentation inhibitors growth under strictly anaerobic conditions, ability to withstand low pH, and insensitivity to bacteriophage infection. Furthermore, it is non-pathogenic yeast, having well-characterized genome and proteome, and therefore, has been used commercially used for the production of biopharmaceuticals and industrial enzymes (Jansen et al. (2017)). Lignocellulosic biomass contains both C6 and C5 sugars. However, *S. cerevisiae* does not have C5 sugar metabolism and lignocellulosic biomass contains both C6 and C5 sugars. Given this, the co-fermentation approach using co-culture of *S. cerevisiae* (having capability to ferment C6 sugars) and *Scheffersomyces (Pichia) stipitis (having capability to ferment both C6 and C5 sugars)* has been employed for the fermentation of lignocellulosic biomass (Santosh et al., 2017). Another approach is genetic engineering of *Saccharomyces cerevisiae* for xylose metabolism. Two xylose-fermenting pathways have successfully been integrated into *S. cerevisiae* (Figure 11.2), namely (a) xylose reductase (XR), xylitol dehydrogenase (XDH) pathway prevailing in fungi (Ha et al., 2010; Wohlbach et al., 2011) and (b) xylose isomerase (XI) pathway found in bacteria (Kuyper et al., 2004). In xylose metabolism, xylose is first isomerized to xylulose, which is then phosphorylated to xylulose 5-phosphate, which, in turn, enters pentose phosphate pathway (PPP) and subsequently to glycolysis pathway and ethanol fermentation. PPP plays a crucial role in bioconversion of xylose to ethanol, where the overexpression of four rate-limiting enzymes of PPP, i.e., transaldolase (TAL1), transketolase (TKL1), ribose-5-phosphate ketol-isomerase (RKI1), and d-ribulose 5-phosphate 3-epimerase (RPE1), resulted in an increased fermentation of xylulose (Johansson and Hahn-Hägerdal, 2002; Karhumaa et al., 2007; Kobayashi et al., 2018; Kuyper et al., 2005; Lee et al., 2013; Lindorfer et al., 2019; Qi et al., 2015). However, engineered *S. cerevisiae* with integrated xylose metabolic pathway and overexpressed PPP genes have the limited capability to metabolize xylose in the presence of glucose because native *S. cerevisiae* sugar transporters belonging to the Hxt family have a low affinity for xylose. Therefore, heterologous expression of pentose transporters in

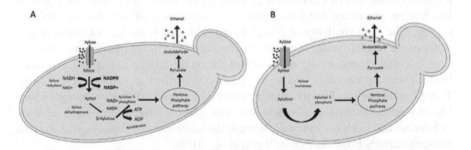

FIGURE 11.2 Engineered yeast with xylose metabolic pathway: (a) XR, XDH pathway present in fungi. (b). XI pathway present in bacteria.

Saccharomyces cerevisiae and engineering of endogenous Hxt transporters for the high affinity of xylose is crucial for co-processing of C6 and C5 sugars (Gonçalves et al., 2014; Nijland et al., 2018; Nijland and Driessen, 2020; Reznicek et al., 2015). Additionally, in the lignocellulosic ethanol process, there is an accumulation of acetic acid and fermentation inhibitors; therefore, robust strains of *S. cerevisiae* are isolated, which have the tolerance to fermentation inhibitors.

With the advancement in advanced molecular biology techniques such as clustered regularly interspaced short palindromic repeats/Cas9 (CRISPR/Cas9), it has become relatively easier to create robust genetically engineered industrial *S. cerevisiae*. Genetic engineering of *S. cerevisiae* strain for xylose metabolic pathway, overexpressing PPP enzymes, and aldose reductase deletion has been carried out by CRISPR/Cas9 technology, and the engineered strain was capable of growing anaerobically and ferment xylose (Bracher et al., 2019). CRISPR/Cas9 technology has also been used to create homogenous *S. cerevisiae* strains having deletion of alcohol dehydrogenase that has resulted in 74.7% improved ethanol yield (Xue et al., 2018). To obtain *S. cerevisiae* having the capability to metabolize cellulose directly without the supplementation of cellulases, *S. cerevisiae* has been engineered for the expression of cellulase enzymes with CRISPR/Cas9 technology (Yang et al., 2018). Furthermore, molecular biology technologies will create robust *S. cerevisiae* strains with higher ethanol yield and the capability to metabolize cellulose.

11.6.2 Thermophilic Fermentative Microorganisms for Lignocellulosic Ethanol Production

In the ethanologenic process, thermophiles have several advantages (Taylor et al., 2009) such as the ability to carry out simultaneous saccharification and fermentation (SSF) (Takagi et al., 1977); the ability to ferment cellulose (Sommer et al., 2004; Zaldivar et al., 2001); tolerance to pH, temperature, and inhibitors (Hartley FRS and Shama, 1987; Takami et al., 2004); being a commercially viable process (Dien et al., 2004; Zaldivar et al., 2001); and the possibility of product recovery along with fermentation and reduction in microbial contamination (Skinner and Leathers, 2004). The SSF process initially demonstrated by Takagi et al. (1977) combines cellulose hydrolysis/saccharification with fermentation in the same reactor for ethanol production. This process has several benefits such as reduction in the accumulation of sugars overcoming feedback inhibition by sugars, thereby increasing ethanol yield and saccharification rate (Wyman and Hinman, 1990). The use of a single fermenter for both saccharification and fermentation has the additional advantage of reduced CapEx cost. Moreover, ethanol production at high temperatures offers several other advantages, including reduced cooling costs, and the presence of ethanol during fermentation results in reduced microbial contamination (Abdel-Banat et al., 2010; Wyman, 1994).

In view of the above, thermostable cellulolytic enzymes and thermotolerant yeast, owing to higher yields, have tremendous potential in commercial bioconversion processes. Therefore, several thermotolerant yeast strains have been isolated based on their growth at higher temperatures and high ethanol tolerance. The thermotolerant yeast *Kluyveromyces marxianus* isolates have been employed for SSF at 42°C–45°C in the presence of cellulases (Ballesteros et al. (2004); Gough et al. (1996); Hari Krishna et al. (2001); Lark et al. (1997); Oliva et al. (2004)). Moreover, *K. marxianus* strain has genetically been engineered for cellulase and has been shown to ferment cellulose to ethanol at 45°C and has great potential to substitute *S. cerevisiae* (Hong et al., 2007). Various *Pichia kudriavzevii* have been reported to be thermostable up to 45°C, resulting in good ethanol yield (Techaparin et al., 2017; Yuangsaard et al., 2013). The thermotolerant and ethanol-tolerant *Pichia kudriavzevii* NUNS-4, NUNS-5, and NUNS-6 have been isolated from soil sugarcane fields in Uttar Pradesh, India. Among these isolated strains, *Pichia kudriavzevii* NUNS-4 resulted in ethanol production of 88.60 g/L and 54.30 g/L at 40°C and 45°C, respectively (Pongcharoen et al., 2018). The thermotolerant yeast *Issatchenkia orientalis* IPE 100 has shown growth at 42°C and has an ethanol productivity of 0.91 g/L/h using corn stalk (Kitagawa et al., 2010; Kwon et al., 2011). Furthermore, the commercial viability of genetically engineered thermophiles has been demonstrated by various bioenergy-focused industries and has therefore the potential to make the lignocellulosic ethanol process cost-effective (Taylor et al., 2009).

11.6.3 Downstream Processing of Ethanol

In case of second-generation ethanol, the fermentation broth typically contains 4–5 wt.% of ethanol. The broth contains both solid and liquid. Thus, solid–liquid separation unit operation is required before the liquid can be processed to recover ethanol. For fuel-grade ethanol, a purity of >99.5 wt.% is mandated. Anhydrous ethanol can be produced by conventional (distillation) and non-conventional (non-distillation) routes. Both strategies have been discussed in the following section.

11.6.3.1 Distillation

Ethanol and water are the major components obtained after fermentation. The ethanol content in the broth is very low (around 5%) (Kanchanalai et al., 2013). The ethanol separation and purification to fuel-grade ethanol from fermentation broth involves two energy-intensive steps, viz. distillation and dehydration.

Distillation is the dominant purification technology in the industry utilized for the concentration of ethanol from dilute fermentation broth. Distillation is the preferred process due to high ethanol recovery (99+%) (Zentou et al., 2019). However, the process has a limitation for achieving fuel-grade ethanol (>99.5 wt.%) due to azeotrope formation. Thus, distillation is coupled with dehydration techniques in order to obtain anhydrous ethanol suitable for fuel application. The process is energy-intensive due to low concentration of ethanol and involves high operational cost.

A typical distillation column for ethanol recovery consists of 'trays,' also referred to as stages or contactors. The vapor rises from the bottom of the tower which is at a higher temperature, and simultaneously, the condensed liquid flows down the tower. Trays provide the surface area to facilitate the contact between the rising vapor and the descending liquid in order to allow mass transfer between the two phases and therefore achieve effective separation.

Since the broth contains both solid and liquid components, their separation occurs in a distillation column commonly known as mash or beer column. In this column, 40–45 wt.% of ethanol in water mixture is obtained from the top, while the solid residues are obtained from the bottom. The top stream is further concentrated to 70 wt.% and subsequently to about 95% using distillation. This is the primary energy-consuming step and thus translates to almost 60%–80% of the total separation cost of bioethanol from water. The second step is the dehydration of the mixture to obtain anhydrous ethanol (>99.5 wt.%), which involves any of the complex processes such as pressure-swing adsorption by molecular sieves, azeotropic distillation, extractive distillation, or a combination of these methods.

In extractive distillation, a liquid solvent extracts water from the ethanol–water mixture in a column and produces fuel-grade ethanol. The liquid solvent is recycled by regenerating it in a separate column. The solvent generally used is ethylene glycol. Similar to extractive distillation system, azeotropic distillation system comprises of azeotropic distillation column and solvent regeneration column. Brazil and the United States, the largest ethanol producers, use ethylene glycol in extractive distillation and cyclohexane in azeotropic distillation as solvents for the production of ethanol (Singh and Rangaiah, 2017).

Another conventional ethanol separation process involves three columns: distillation column for pre-concentration of ethanol, followed by dehydration using an extractive distillation system which has two columns. The first column which is an extractive distillation column produces fuel-grade ethanol, and the second column known as the solvent regeneration column is used for recycling of solvent.

11.6.3.2 Alternative Recovery Techniques

Currently, integrated fermentation and separation processes are gaining popularity as they can limit the fermentation inhibition due to the toxic effect of the product. Different separation methods that can be integrated into fermentation include pervaporation, adsorption, gas stripping, vacuum fermentation, and solvent extraction (Zentou et al., 2019). Although many of these processes have been realized in pilot scale, their integration in industrial scale is yet to be achieved.

11.6.3.3 Pervaporation

Although ethanol separation process is mature, research is focused on the development of energy-efficient and economic processes. In this respect, membrane technologies have gained attention due to their high separation efficiency, low energy demand and operating costs, and no waste streams generation, and they can be used in the separation of temperature-sensitive materials. Among the membrane technologies, pervaporation is widely used for the separation of organic–organic, organic–aqueous, and azeotropic mixtures (Ong et al., 2016; Vane, 2005). The technique is also economically viable and was first put to industrial use in 1985 in Karlsruhe-Maxau (Germany) with another plant coming up in Bétheniville (France) for dehydration of 94% ethanol (Zentou et al., 2019).

Pervaporation technique is based on selective adsorption and diffusion. It has high selectivity and low energy demand/requirement compared to other conventional separation processes such as liquid–liquid extraction, distillation, gas stripping, and adsorption. Pervaporation uses a non-porous membrane that enables separation based on differential interaction of molecular interaction of feed components with

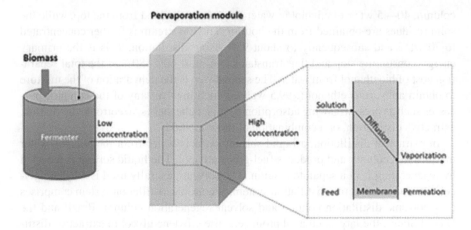

FIGURE 11.3   Schematic representation of the pervaporation process.

the membrane. Consequently, separation depends on the membrane characteristics and chemical composition of the mixture. Atmospheric pressure is applied on the upstream side of the membrane, and a vacuum is applied on the downstream. The concentration gradient resulting from the vacuum pressure is the driving force for separation. Therefore, the transport of molecules across the membrane involves three steps: (a) selective adsorption on the membrane, (b) diffusion across the membrane, and (c) release of molecules on the retentate side as the gas phase. The technique selectively concentrates the fermented solvent on the permeate side, while the nutrients and microbes are retained in the fermentation broth (Figure 11.3).

Pervaporation system consists of a feed vessel and pump, heating device, membrane module, vacuum pump, and cold trap. The membrane selection depends on the nature of the molecules to be separated. Hydrophobic membranes preferentially allow select organic compounds as compared to water. Therefore, organic compounds are recovered in permeate, while, if a hydrophilic membrane is used, the mixture is dehydrated, and water is released in the permeate side.

The biggest limitation of the pervaporation process is membrane fouling, leading to loss of productivity over time. Extensive research has been conducted to determine the effect of several factors on membrane fouling, such as size and shape of the module, type of membrane, and the operating conditions (Hassan et al., 2013; Gaykawad et al., 2013). Recently, Kamelian et al. (2020) have reported superhydrophobic pervaporation membrane for addressing the biofouling issue.

Numerous studies have been carried out to evaluate the performance of pervaporation process for ethanol dehydration and ethanol recovery using hydrophilic and ethanol-selective membranes, respectively. However, a majority of these studies have been conducted using ethanol/water standard mixture and are of little significance when integrating the process with fermentation broth due to the different chemical composition of the fermentation broth than ethanol/water mixture. Moreover, hydrophilic membranes used for ethanol purification cannot be applied for ethanol recovery from fermentation broth due to the very low concentration of ethanol in the broth. Several significant studies have been reported on ethanol recovery from the lignocellulosic fermentation broth. However, the number of membranes employed for ethanol recovery is limited, and there is scope for the development of new materials for membrane (Gaykawad et al., 2013; Cai et al., 2016; Chen et al., 2014). Further, research work is needed to address the fouling issue and design of membrane module for integration at the industrial scale.

11.6.3.4 Gas Stripping

Gas stripping is one of the methods that can easily be integrated with fermentation for *in situ* removal of ethanol and does not require complex equipment or any plant modification. In this process, gas is sparged through the fermenter to evaporate the bioethanol, which is then recovered from the gas phase using a condenser. The technique is based on the principle of difference in volatilities of the compounds, and the separation is governed by Henry's law. The compound of low boiling points is thus stripped easily, such as acetaldehyde, which is generally observed as an impurity in ethanol.

$$H = \frac{P_{vap}}{C_{sat}}$$

where H = Henry's constant (moles/L atm), P_{vap} = the partial pressure of a pure compound (atm), and C_{sat} = the saturation concentration of the pure compound in liquid phase (moles or mg/L).

Zhang et al. (2005) compared the effect of common stripping gases such as CO_2, N_2, and air on cell growth and ethanol production in gas stripping ethanol fermentation (GSEF). Although, with air as stripping gas, cell growth was stimulated, ethanol productivity decreased. CO_2 was observed to be inhibitory to yeast growth. N_2 was claimed to be the most preferred gas leading to two times increase in ethanol productivity. The operating conditions such as feed temperature, stripping temperature and flow rate, liquid height to column diameter were also found to affect the ethanol productivity. Several gas stripping operations were developed to reduce the cost of ethanol recovery. Taylor et al. (2010) performed stripping experiment in continuous fermentation process without recycling of the stripping gas. Productivity was increased on stripping the reaction products (ethanol and CO_2). After recovering the ethanol, CO_2 and the gaseous stream were purified by water wash scrubbing. The cost can be reduced by replacing part of N_2 with CO_2 since the mass of N_2 is two-thirds of that of CO_2 for the same volume.

11.6.3.5 Vacuum Fermentation

The fermentation is carried out under vacuum condition, which allows for the evaporation of ethanol at fermentation temperature (usually around 30°C–35°C) which is well below the boiling point of the ethanol–water mixture (78.3°C–100°C). The ethanol is recovered from the vapor phase by condensation. The approach effectively minimizes the inhibitory effect of ethanol by its continuous removal, thereby showing improved productivity (Abdullah and Ariyanti, 2012; Shihadeh et al., 2014;

Ghose et al., 1984). However, the practical application is limited due to high energy cost incurred for maintaining the vacuum. The reports on vacuum fermentation technique are relatively scarce in the literature.

11.6.3.6   Adsorption

This technique utilizes a porous adsorbent material having a pore size corresponding to the molecular size of ethanol for selective removal of ethanol. The fermentation broth is passed through a bed of packed adsorbent material, and effluent is recycled back into the fermentation broth. The adsorbed ethanol is desorbed by desorption gas, and the adsorbent material is regenerated for the next cycle.

Activated charcoal is one of the earliest and most common adsorbent materials used for the separation of ethanol from the mixture (Silvestre-Albero et al., 2009). Several other adsorbents explored include zeolites (SiO_2/Al_2O_3) and ion-exchange resins (Wee et al., 2005). Although several adsorbent materials reported in the literature exhibited high affinity for ethanol, the majority of the studies were performed using the model mixture of ethanol–water, ethanol–glucose, or ABE (acetone–butanol–ethanol) during the investigation. Therefore, the studies do not confirm the suitability of these adsorbent materials for integration into fermentation as the affinity toward ethanol will greatly be affected by the presence of other molecules in the fermentation broth. Jones et al. reported F-600 activated carbon for *in situ* ethanol adsorption from fermentation broth and achieved an ethanol production of 45 g/L compared to 28 g/L in control (Jones et al., 2011). Seo et al. reported an ethanol adsorption of 0.163 g/g using molecular-sieving carbon (MSC) in pilot scale for bioethanol production (Seo et al., 2018).

11.6.3.7   Solvent Extraction

During the solvent extraction process, the fermentation broth is passed through an extraction unit containing extraction solvent wherein the solvent extracts the ethanol. The concentrated product is recovered in a column, and both the fermentation broth and solvent are transferred back to the fermenter. The contact between fermentation broth and the extractant solvent is necessary for effective mass transfer. It can either be direct by using an agitator or packed column, or be indirect, *via* a non-wetted porous membrane. The solvent used for extraction of ethanol should have the properties such as high separation factor for ethanol, chemical stability, low solubility in water, and density difference from aqueous phase for fast separation, must not form an emulsion with water, should be non-toxic, and should not have an inhibitory effect on microbes (Offeman et al., 2005). The most commonly used solvents are ketones, esters, and alcohols because of their low reactivity and high distribution coefficients.

11.6.4   Improvements in Ethanol Production

Research and development aims for making the 2G ethanol production cost-effective by optimizing all unit operations, i.e., pretreatment, hydrolysis, and fermentation (Robak and Balcerek, 2018). The parameters of pretreatment are optimized to minimize energy-intensive processes, obtain higher sugar recovery, reduce the degradation of sugars, and increase the purity of recovered lignin. The steam explosion process saves energy and is environmentally benign compared to other methods. In case of chemical pretreatment, recycling of chemical compounds is carried out. Effective pretreatment results in substrate with high surface area, minimizes the enzyme deactivation, and increases enzymatic hydrolysis. The high cost of cellulolytic enzymes and low yield of hydrolysis reaction are major factors for high operating cost of lignocellulosic biorefinery (Ellilä et al., 2017). Therefore, novel enzymes with high hydrolytic efficiency have been developed by mutagenesis and protein engineering approaches. In addition, strategies for enzyme reuse and on-site production of enzymes have been developed. Further, the use of additives such as surfactants (Tween and polyethylene glycol) during hydrolysis has shown improvements in enzymatic hydrolysis (Brondi et al., 2020). The use of SSF has the advantage of less enzyme loading with increased rate of hydrolysis. Therefore, *Saccharomyces cerevisiae* with high temperature tolerance is being developed. Also, *S. cerevisiae* is being genetically engineered to co-ferment both glucose and xylose efficiently. Additionally, ethanol fermentation carried out at high temperature is more economical as temperature control during fermentation is not required, resulting in reduced cooling costs of fermenter, especially in hotter regions. Moreover, it has the advantage of higher ethanol productivity and reduced risks of contamination by undesirable microorganisms (Arora et al., 2015; Auesukaree et al., 2012; Siedlarz et al., 2016; Taylor et al., 2009).

11.6.5 OPPORTUNITIES AND CHALLENGES

Global climatic concerns about rising CO_2 emission, increasing worldwide energy demand, and the security of fuel supply have encouraged the development of sustainable fuels such as biofuels (Acharya and Perez-Pena, 2020; Lee and Lavoie, 2013). Because of the above factors, several policies have been implemented in both the developed and developing countries for the promotion of biofuels. These policies involve mandatory biofuel blending, providing tax subsidies, and the introduction of flex-fuel vehicle programs. Moreover, governments have also provided various funding schemes to develop advanced technologies for the production of biofuels and to encourage academia–industry collaborations (Hoekman, 2009). As a result of biofuel policies, regulations, and incentives, worldwide bioethanol production has increased from 17 billion liters in 2000 to 108 billion liters in 2018 (https://ethanolrfa. org/). In 2018, 90% of the total global bioethanol supply is majorly contributed by the United States (56%), Brazil (28%), and the European Union (5%) (Acharya and Perez-Pena, 2020). In Asian countries, according to International Energy Agency (-IEA) (Karatzos et al., 2014), World Energy Balances and Statistics 2014, China has the highest bioethanol production followed by Thailand, Viet Nam, and Australia ((Yamaguchi et al.2013)). The bioethanol supply is expected to grow tremendously in the future; however, the major bottleneck of lignocellulosic ethanol technology is two to three times higher cost of second-generation ethanol compared to gasoline on an energy equivalent basis (Carriquiry et al. (2011)). Thus, bringing down the cost of lignocellulosic biorefinery is a major challenge.

11.7 BIOBUTANOL

The demand for renewable fuels in transportation sector is anticipated to grow, and biobutanol has the potential to drive this growth in future. Recently, with the depleting fossil fuel reserves, biological production of butanol as a superior biofuel candidate has gained much attention among researchers. The petroleum industry now looks very committed to the use of butanol as fuel, and there have been worldwide efforts for its production through renewable sources. Fuel properties of biobutanol are better than bioethanol due to its higher energy content and relatively better air-to-fuel ratio (Yusoff et al., 2015). Also, biobutanol has a higher flash point, has a lower vapor pressure, is less volatile, and is therefore relatively safer to handle.

Though the biological conversion of lignocellulosic wastes looks to be a potential source for the production of butanol, the cost of pretreatment, cost of enzymes capable of converting cellulose into monomeric sugars, and low sugar and butanol yields have been identified to be hindering the commercialization prospects of biobutanol.

11.7.1 ABE/IBE FERMENTATION—ROLE OF CLOSTRIDIA

Regardless of the starting lignocellulosic feedstock, the degradation of cellulose or hemicellulose yields hexoses and pentoses that need to be fermented to solvents. Over the past few decades, butanol fermentation has made significant advances and breakthroughs in the bioproduction from various alternative feedstocks (Algayyim et al., 2018). Butanol is produced naturally by bacteria of the genus *Clostridia*. They are obligate, anaerobic, rod-shaped, gram-positive, and spore-forming bacteria. Butanol-producing *Clostridia* include a variety of species, including *acetobutylicum, beijerinckii, sporogenes, saccharoperbutylacetonicum, saccharoacetobutylicum, aurantibutyricum, pasteurianum*, and *tetanomorphum* (Jones and Woods, 1986; Kumar and Gayen, 2011). *Clostridia* can metabolize a variety of carbon sources including glucose, xylose, arabinose, mannose, fructose, sucrose, lactose, cellobiose, and starch. This feature of *Clostridia* broadens the substrate pool and makes it possible to utilize lignocellulosic biomass as a feedstock.

A typical acetone–butanol–ethanol (ABE) fermentation by *Clostridia* consists of two phases, namely acidogenesis and solventogenesis. In acidogenesis, the bacteria grow and produce volatile fatty acids such as acetate and butyrate. During solventogenesis, the growth of the bacteria stops and the metabolism changes. The volatile fatty acids are then converted into acetone, butanol, and ethanol with a typical ratio of 3:6:1, respectively (Veza et al., 2021). Acetone is not a preferred metabolite in ABE fermentation due to its corrosive nature. Therefore, attempts were made to reduce the acetone production through metabolic engineering of *Clostridium* sp. (Papoutsakis, 2008). Some of the *Clostridia* species produce isopropanol instead of acetone in IBE fermentation. These mixed alcohols can directly be used as a biofuel mixture (Papoutsakis, 2008), and hence, the isopropanol–butanol–ethanol (IBE) mixture is considered better than ABE for fuel applications. Further, the energy density of isopropanol is higher than acetone (23.9 MJ/L vs 22.6 MJ/L) (Peralta-Yahya and Keasling, 2010; Rassadin et al., 2006). In nature, several

Clostridia sp. produce isopropanol along with butanol and ethanol (Chen and Hiu, 1986; George et al., 1983).

For economical production of biofuels, research is primarily focused on utilizing renewable feedstocks such as agriculture residues as substrates (Green, 2011; Qureshi et al., 2007, 2008a), which are abundant and sustainable (Kumar et al., 2012). Lignocellulosic biomass sources that have been used for solvent production are wheat straw, corn stover, barley straw, corn fiber, switchgrass, dried distillers grains with solubles (DDGS), corncob, bagasse, and rice straw (Dalal et al. (2019)). Recent reports (Table 11.1) on solvent production from various sources of lignocellulosic biomass are tabulated in Table 11.1.

11.7.2 Amelioration of Butanol Fermentation

Nakayama et al. (2011) showed butanol production from crystalline cellulose using a co-culture of *C. thermocellum* and *C. saccharoperbutylacetonicum*. Tran et al. (-2010) reported the use of a mixed culture consisting of *B. subtilis* and *C. butylicum* without anaerobic treatment. They observed a 6.5-fold increase in ABE production in this study from cassava starch compared to a pure culture. Genetic engineering of *Clostridia* was also tried to enhance the solvent production, butanol tolerance, and also the ratio of butanol in the total solvent. Higher butanol production (172 mM) was seen in *C. acetobutylicum* ATCC 824 strains overexpressed with the *SpoOA* gene compared to the strains in which the gene was inactivated (13 mM butanol). *SpoOA* is postulated to be a transcriptional regulator that positively controls sporulation and solvent production (Harris et al., 2002).

Relatively higher butanol production has been observed using a fed-batch strategy compared to batch production (Darmayanti et al., 2018; Niglio et al., 2019). Continuous multi-stage (two or more) fermentation systems were also shown to improve the butanol production due to the increased volumetric productivity compared to batch fermentation (Qureshi et al., 2000). Simultaneous saccharification and fermentation of acid-pretreated rice straw using *C. beijerinckii* P260 was reported (Qureshi et al., 2008b), wherein both the hydrolysis and fermentation processes were shown to operate simultaneously in the same reactor.

11.7.3 Consolidated Bioprocessing

Consolidated bioprocessing (CBP) is basically the integration of enzyme production, subsequent saccharification, and fermentation in one step without external supply of hydrolytic enzymes. It is considered as a cost-effective alternative for the production of biofuels from lignocellulosic biomass (Olguin-Maciel et al., 2020). Compared to classical CBP performed by a genetically engineered strain with hydrolytic and solventogenic properties, CBP strategy using an enriched consortium provides a number of advantages (Gaida et al., 2016). Hydrolytic enzymes secreted by diverse microorganisms ensure degradation of extensive biomass substrates. Further, complementary metabolic activities of different microbes in the consortium can accomplish CBP more efficiently than a single strain genetically modified for multiple functions (Peng et al., 2016).

TABLE 11.1
Reports on Solvent Production from Various Sources of Lignocellulosic Biomass

Organism	Substrate	Pretreatment Process	Yield (g/L) Butanol	Total Solvents	References
C. beijerinckii C-01	Rice straw	Ammonia	7.0	9.5	Dalal et al. (2019)
C. beijerinckii NCIMB 4110	Corn stalks	NaOH and ethanol	9.9	12.8	Tang et al. (2017)
C. saccharoperbutylacetonicum N1-4 (ATCC 27021)	Synthetic rice straw hydrolysate (sterile)	NA	5.1 ± 0.3 (sterile)	8.1 ± 0.8 (sterile)	Chen et al. (2013)
	Synthetic rice straw hydrolysate (non-sterile)		2.8 ± 2.1 (non-sterile)	4.6 ± 3.3 (non-sterile)	
C. sporogenes BE01	Rice straw	Dilute acid	3.43	5.32	Gottumukkala et al. (2013)
C. beijerinckii P260	Corn stover hydrolysate	Dilute acid	10.4	16.0	Qureshi et al. (2010b)
	Switchgrass hydrolysate + glucose suppl.		5.8	8.9	
C. beijerinckii P260	Barley straw hydrolysate (BSH)	Dilute acid	~4.5	7.1	Qureshi et al. (2010a)
	BSH + wheat straw hydrolysate (1:1)		13.6	17.2	
C. beijerinckii BA101	Corn fiber	Dilute acid	9.3	9.3	Qureshi et al. (2008a)
C. beijerinckii P260	Wheat straw	Dilute acid	12.0	25.0	Qureshi et al. (2007)
C. beijerinckii NCIMB 8052	Corn steep water (CSW) medium (1.6% solids + 6% glucose)	–	6.0	10.7	Parekh et al. (1998)

11.7.4 DOWNSTREAM PROCESSING OF BUTANOL

The ABE fermentation process yields aqueous mixtures of acetone, 1-butanol, and ethanol, which is a complex ternary system and can result in the formation of azeotropic mixtures during distillation. However, owing to the heterogeneous nature of water–n-butanol azeotrope, the mixture can be separated by a two-column distillation system without requiring any additional compound (Luyben, 2008).

The acetone, butanol, and ethanol mixture is separated on the basis of their boiling point difference. In a conventional distillation process, the mixture containing the products from the batch fermenter is heated to 100°C and the products are removed from the broth by a stream of vapors (Roffler et al., 1987). The vapors obtained comprise 30 wt.% of acetone, butanol, and ethanol, and the rest of 70 wt.% is water. Vapors are passed through a series of four distillation columns for separation. The first column which operates at 0.7 atm pressure removes about 99.5 wt.% of acetone. The residual bottoms fractions of the first column are sent to the ethanol column which operates at 0.3 atm pressure. From the top of this column, 95 wt.% ethanol is obtained. The bottoms fractions of the ethanol column are sent to a decanter for separating water and n-butanol mixture. The aqueous phase containing 9.5 wt.% of n-butanol is redirected to a water stripper. The butanol-rich phase, with 23 wt.% water content, is sent to an n-butanol stripper, wherein 99.7 wt.% n-butanol is recovered.

Although distillation is a well-established process for butanol recovery from ABE fermentation broth, it is energy-intensive due to high energy consumption involved in evaporation of water. The energy required for producing 99.5 wt.% n-butanol by the traditional distillation-decanter method is 14.5 MJ/kg (Vane et al., 2013). Hence, alternate energy-efficient non-distillation processes are being explored for butanol recovery (Green, 2011; Kujawska et al., 2015).

The critical problem in the fermentation process of biobutanol production is the toxic effect of the product on microbial strains, which severely affects the productivity of the process. Hence, *in situ* product recovery (ISPR) is significant in biobutanol production to ensure continuous removal of butanol from the fermentation broth. Several biobutanol separation techniques that can be integrated with the fermentation process are the same as described for ethanol in the previous section, which includes adsorption, liquid–liquid extraction, gas stripping, pervaporation, and reverse osmosis.

Liquid–liquid extraction is a prime option among others in terms of butanol yield and carbon consumption. However, the *in situ* butanol extraction requires a high volume ratio of extractant to fermentation broth (Groot et al., 1990). In gas stripping method, nitrogen or biogas (CO_2 and H_2) is used as a carrier gas for removing solvents from the fermentation broth. The gaseous solvents are later recovered by cold traps, and carrier gas is circulated back to the fermenter. The gas stripping method is relatively energy-intensive due to its low selectivity toward butanol. Thus, a secondary separation process is required to further purify butanol, which leads to an increase in the overall energy consumption for the entire process.

11.7.4.1 Pervaporation

Pervaporation (PV) technique is looked upon as one of the potential techniques for *in situ* butanol separation from fermentation broth because it is environmentally friendly and energy-saving and has no harmful effect on microorganisms, and no medium ingredients are removed from the fermentation broth. In an integrated fermentation and PV process, fermentation process and n-butanol separation are performed simultaneously to avoid butanol toxicity to microorganisms.

The membrane is a part of the process, and over the years, research has been directed toward the development of membranes with high permeability (flux) and selectivity (separation factor) with improved stability (Zhu et al., 2020). Often, there is an issue with membrane biofouling due to the formation of biofilm in fermentation broth, which adheres to the membrane surface. This leads to deterioration of membrane over time. Broth pretreatment can reduce this problem to a certain extent, and additionally, the development of anti-fouling pervaporation membrane can also be explored to address the issue of membrane fouling.

Polydimethylsiloxane (PDMS) is the most widely used membrane because of its good thermal, chemical, and mechanical stability and high separation performance in addition to ease of fabrication. Many commercially available PDMS membranes such as Pervap 1060, Pervap 2200, Pervap 4060, and Pervatech PDMS have been applied for butanol separation (Kujawski et al., 2014). High permeate flux can be achieved by hollow-fiber-supported composite membranes owing to the higher surface area (Dong et al., 2014).

Mixture matrix membranes which are prepared by the addition of fillers in the polymeric matrix are promising for overcoming the trade-off between permeability and selectivity of polymeric membranes. ZIF-8 having excellent superhydrophobic properties is used by many researchers for organophilic pervaporation, as fillers in the polymeric matrix (Fan et al., 2014).

Various other materials have been explored as a membrane for butanol recovery, such as polyether block amide (PEBA) composed of different segments of rigid polyamide (PA) and flexible polyether (PE) blocks, wherein PA segments result in good mechanical strength of the membrane and PE segments ensure high affinity for organic solvents. Poly[1-(trimethylsilyl)-1-propyne] (PTMSP) polymers which have a high free volume fraction (34%) exhibits extremely high permeability or ultra-high flux in pervaporation (Zhu et al., 2020). Polymers of intrinsic microporosity (PIMs) show superior separation performance due to their spider-like structure which prevents polymer chains from packing and rotating freely. They can be combined with other polymers such as polyvinylidene fluoride (PVDF) or PDMS for composite membrane, or as mixed matrix membranes (MMMs) for better separation performance. Inorganic membranes such as MFI zeolites and mesoporous silica have also been explored owing to their superior chemical and thermal properties compared to polymeric membranes for biobutanol separation. Although inorganic membranes have suitably been applied for ethanol–water separation, n-butanol recovery from a dilute aqueous solution remains a challenge.

Ionic liquid membranes have also been investigated for biobutanol separation owing to their low volatility and high thermal stability (Cascon and Choudhari,

2013). However, they suffer from the drawback of toxic effect to microbes. The practical application is limited by their instability.

11.7.4.2 Hybrid Technologies

Hybrid ISPR (*in situ* product removal) is a technique that combines various ISPR techniques for n-butanol recovery and, thus, possesses the advantage of each method. Gas stripping and pervaporation hybrid process has been reported, where gas stripping method is employed for the *in situ* recovery of n-butanol and pervaporation operated *ex situ* removes n-butanol from stripped gas (Xue et al., 2016).

Lu et al. reported integrated extraction and gas stripping (Lu and Li, 2014). Biocompatible solvent oleyl alcohol was used for the extraction of butanol *in situ*, and butanol is then removed by gas stripping from the oleyl alcohol phase. The advantage of this process is that since butanol is continuously removed from the solvent, the saturation level of butanol in the solvent is not reached and, thus, the amount of oleyl alcohol required for the process is small. The integrated extraction–gas stripping process can be applied for *in situ* product removal in a packed bed, where simultaneous production and extraction of n-butanol can take place inside the packed bed. Gas stripping method can be used for regenerating oleyl alcohol containing n-butanol and then recycled back into the packed bed.

11.7.5 Opportunities and Challenges

The major constraints of biobutanol fermentation are low titers and therefore large process volumes, low rate of production, and toxicity of butanol to the microorganism. It is therefore essential to develop robust microbial strains which can overcome these challenges. Further, cost-effective production needs to be achieved by employing integrated fermentation and simultaneous butanol recovery techniques.

11.8 COMPARISON OF FUEL CHARACTERISTICS OF GASOLINE AND BIOALCOHOLS

Gasoline or motor spirit (MS) is the lighter fraction (C_4–C_{12}) of crude oil and used in internal combustion (IC) engines. The fuel characteristics of gasoline are measured by parameters such as octane number, calorific value, olefin content, aromatic content, sulfur content, and oxygenate content (Table 11.2). Gasoline quality is mainly determined by its octane number, which is the indication of its anti-knocking property. Knocking is the result of undesirable auto-ignition. A high octane number is desirable in fuel, which ensures that the fuel will burn in a controlled manner.

Alcohols derived from biosources, viz. methanol, ethanol, and butanol, are combustible and hence can be used as fuel. The octane number of bioalcohols is higher than that of gasoline, and as a result, engines can be operated at a higher compression ratio using bioalcohols (Mamat et al., 2019). The efficiency of an IC engine is determined by its compression ratio, and hence, a higher thermal efficiency can be achieved with bioalcohols. Also, since bioalcohols are oxygenated fuels, their combustion leads to better oxidation of CO, reduced unburnt hydrocarbons, and reduced

TABLE 11.2

Fuel Characteristics of Bioalcohols vis-à-vis Gasoline

Fuel Property	Units	Gasoline	Methanol	Ethanol	Butanol
Oxygen	% wt	0	49.93	34.73	21.6
Boiling point	°C	25–215	65	79	117
Density	kg/m³	746	791.3	789.4	
Viscosity at 40°C	mm²/s	0.4–0.8	0.59	1.13	2.22
Lower calorific value	MJ/kg	43.4	19.9	26.7	34.4
Higher calorific value	MJ/kg	46.4	23	29.7	37.3
Motor octane number (MON)	-	81	88.6	92	85
Research octane number (RON)	-	91	108.7	108	98
Latent heat of vaporization	kJ/kg	350–400	1109	924	582
Aromatic content	% volume	Max. 35	0	0	0
Sulfur content	ppm	Max. 10	0	0	0

Source: Mamat et al. (2019) and Chupka et al. (2015).

particulate matter compared to base gasoline. In this respect, bioalcohols are cleaner fuels than gasoline.

Sulfur is a natural component of crude oil along with aromatics, and both must be removed during the refining process. Burning of sulfur-rich fuel leads to SO_x emissions, which is an environmental pollutant, and aromatics are a significant source of unburnt hydrocarbons or particulate matter. Benzene which is present in crude oil is a known carcinogen. In India, the government has implemented BS norms which dictate the maximum permissible amount of these components in the fuel. Since the past three revisions in BS norms, sulfur has been reduced to 10 ppm from 150 ppm. In stark contrast, bioalcohols do not contain sulfur and aromatics and, thus, are environmentally friendly as compared to gasoline.

However, the only drawback of bioalcohols is their lower energy density as compared to gasoline. For instance, the calorific value of methanol is half that of gasoline, so twice the amount of fuel needs to be injected for achieving the same power output in an IC engine. Butanol is less volatile compared to ethanol and gasoline and, thus, requires higher injection pressure or cold-start device for improving fuel–air mixture (Lapuerta et al., 2017). But, butanol has the highest energy content compared to methanol and ethanol.

11.9 CONCLUSIONS

Tremendous research efforts are being put to develop cost-effective second-generation biofuel production processes. However, numerous challenges have to be overcome before commercializing such technologies at industrial scale (Vohra et al., 2014). On an energy equivalent basis, it is estimated that the production cost of second-generation biofuels is 2–3 times higher than petroleum fuels. In order to cut down the production cost, several challenges need to be addressed for the conversion of lignocellulosic biomass into biofuels and chemicals using biochemical platforms. The major challenges are in the areas of (a) feedstock production and storage, (b) feedstock supply chain, (c) the development of energy-efficient technologies (pretreatment, enzymatic hydrolysis, and microbial fermentation) to reduce both capital and revenue expenditure, (d) simultaneous production of value-added chemicals along with bioethanol, (e) establishment of biofuel/biochemical standards, (f) biofuel distribution, (g) societal acceptance, and (h) environmental impact minimization (Hoekman, 2009; Menon and Rao, 2012; Luo et al., 2010).

REFERENCES

Abdel-Banat B M.A., Hoshida, H., Ano, A., Nonklang, S., Akada, R., 2010. High-temperature fermentation: How can processes for ethanol production at high temperatures become superior to the traditional process using mesophilic yeast? *Appl. Microbiol. Biotechnol.* 85, 861–867. https://doi.org/10.1007/s00253-009-2248-5

Abdullah, A., Ariyanti, D., 2012. Enhancing ethanol production by fermentation using Saccharomyces cereviseae under vacuum condition in batch operation. *Int. J. Renew. Energy Dev.* 1, 6–9.

Acharya, R.N., Perez-Pena, R., 2020. Role of comparative advantage in biofuel policy adoption in Latin America. *Sustainability* 12, 15–21. https://doi.org/10.3390/su12041411

Aditiya, H.B., Mahlia, T.M.I., Chong, W.T., Nur, H., Sebayang, A.H., 2016. Second generation bioethanol production: A critical review. *Renew. Sustain. Energy Rev.* 66, 631–653. https://doi.org/10.1016/j.rser.2016.07.015

Algayyim, S.J.M., Wandel, A.P., Yusaf, T., Hamawand, I., 2018. Production and application of ABE as a biofuel. *Renew. Sustain. Energy Rev.* 82, 1195–1214. https://doi.org/10.1016/j.rser.2017.09.082

Andrić, P., Meyer, A.S., Jensen, P.A., Dam-Johansen, K., 2010. Reactor design for minimizing product inhibition during enzymatic lignocellulose hydrolysis: I. Significance and mechanism of cellobiose and glucose inhibition on cellulolytic enzymes. *Biotechnol. Adv.* 28, 308–324. https://doi.org/10.1016/j.biotechadv.2010.01.003

Arora, R., Behera, S., Kumar, S., 2015. Bioprospecting thermophilic / thermotolerant microbes for production of lignocellulosic ethanol : A future perspective. *Renew. Sustain. Energy Rev.* 51, 699–717. https://doi.org/10.1016/j.rser.2015.06.050

Auesukaree, C., Koedrith, P., Saenpayavai, P., Asvarak, T., Benjaphokee, S., Sugiyama, M., Kaneko, Y., Harashima, S., Boonchird, C., 2012. Characterization and gene expression profiles of thermotolerant Saccharomyces cerevisiae isolates from Thai fruits. *J. Biosci. Bioeng.* 114, 144–149. https://doi.org/10.1016/j.jbiosc.2012.03.012

Balat, M., 2011. Production of bioethanol from lignocellulosic materials via the biochemical pathway: A review. *Energy Convers. Manag.* 52, 858–875. https://doi.org/10.1016/j.enconman.2010.08.013

Ballesteros, M., Oliva, J.M., Negro, M.J., Manzanares, P., Ballesteros, I., 2004. Ethanol from lignocellulosic materials by a simultaneous saccharification and fermentation process (SFS) with Kluyveromyces marxianus CECT 10875. *Process Biochem.* 39, 1843–1848. https://doi.org/10.1016/j.procbio.2003.09.011

Behera, S., Arora, R., Nandhagopal, N., Kumar, S., 2014. Importance of chemical pretreatment for bioconversion of lignocellulosic biomass. *Renew. Sustain. Energy Rev.* 36, 91–106. https://doi.org/10.1016/j.rser.2014.04.047

Bhardwaj, N., Chanda, K., Kumar, B., Prasad, H., Sharma, G.D., Verma, P., 2017. Statistical optimization of nutritional and physical parameters for xylanase production from newly isolated Aspergillus oryzae LC1 and its application in the hydrolysis of lignocellulosic agro-residues. *BioResources* 12, 8519–8538. https://doi.org/10.15376/biores.12.4.8519-8538

Bhardwaj, N., Kumar, B., Agarwal, K., Chaturvedi, V., Verma, P., 2019. Purification and characterization of a thermo-acid/alkali stable xylanases from Aspergillus oryzae LC1 and its application in Xylo-oligosaccharides production from lignocellulosic agricultural wastes. *Int. J. Biol. Macromol.* 122, 1191–1202. https://doi.org/10.1016/j.ijbiomac.2018.09.070

Bhutto, A.W., Qureshi, K., Harijan, K., Abro, R., Abbas, T., Bazmi, A.A., Karim, S., Yu, G., 2017. Insight into progress in pre-treatment of lignocellulosic biomass. *Energy* 122, 724–745. https://doi.org/10.1016/j.energy.2017.01.005

Binod, P., Gnansounou, E., Sindhu, R., Pandey, A., 2019. Enzymes for second generation biofuels: Recent developments and future perspectives. *Bioresour. Technol. Reports* 5, 317–325. https://doi.org/10.1016/j.biteb.2018.06.005 https://doi.org/10.1038/s41598-020-64316-6

Bracher, J.M., Verhoeven, M.D., Wisselink, H.W., Crimi, B., Nijland, J.G., Driessen, A.J.M., Klaassen, P., Maris, A.J.A. Van, Daran, J.M.G., Pronk, J.T., 2018. Biotechnology for biofuels the penicillium chrysogenum transporter Pc AraT enables high - affinity , glucose - insensitive l - arabinose transport in Saccharomyces cerevisiae. *Biotechnol. Biofuels* 1–16. https://doi.org/10.1186/s13068-018-1047-6

Brondi, M.G., Elias, A.M., Furlan, F.F., Giordano, R.C., Farinas, C.S., 2020. Performance targets defined by retro-techno-economic analysis for the use of soybean protein as saccharification additive in an integrated biorefinery. *Sci. Rep.* 10, 1–13. https://doi.org/10.1038/s41598-020-64316-6

Cai, D., Hu, S., Chen, C., Wang, Y., Zhang, C., Miao, Q., Qin, P., Tan, T., 2016. Immobilized ethanol fermentation coupled to pervaporation with silicalite-1/polydimethylsiloxane/-polyvinylidene fluoride composite membrane. *Bioresour. Technol.* 220, 124–131. https://doi.org/https://doi.org/10.1016/j.biortech.2016.08.036

Carriquiry, M.A., Du, X., Timilsina, G.R., 2011. Second generation biofuels: Economics and policies. *Energy Policy* 39, 4222–4234. https://doi.org/10.1016/j.enpol.2011.04.036

Cascon, H.R., Choudhari, S.K., 2013. 1-Butanol pervaporation performance and intrinsic stability of phosphonium and ammonium ionic liquid-based supported liquid membranes. *J. Memb. Sci.* 429, 214–224. https://doi.org/https://doi.org/10.1016/j.memsci.2012.11.028

Chen, J., Zhang, H., Wei, P., Zhang, L., Huang, H., 2014. Pervaporation behavior and integrated process for concentrating lignocellulosic ethanol through polydimethylsiloxane (PDMS) membrane. *Bioprocess Biosyst. Eng.* 37, 183–191. https://doi.org/10.1007/s00449-013-0984-5

Chen, J.-S., Hiu, S.F., 1986. Acetone-butanol-isopropanol production byClostridiumbeijerinckii (synonym, Clostridiumbutylicum). *Biotechnol. Lett.* 8, 371–376. https://doi.org/10.1007/BF01040869

Chen, W.H., Chen, Y.C., Lin, J.G., 2013. Evaluation of biobutanol production from non-pretreated rice straw hydrolysate under non-sterile environmental conditions. *Bioresour. Technol.* 135, 262–268. https://doi.org/10.1016/j.biortech.2012.10.140

Chupka, G.M., Christensen, E., Fouts, L., Alleman, T.L., Ratcliff, M.A., McCormick, R.L., 2015. Heat of vaporization measurements for ethanol blends up to 50 volume percent in several hydrocarbon blendstocks and implications for knock in SI engines. *SAE Int. J. Fuels Lubr.* 8, 251–263.

Dalal, J., Das, M., Joy, S., Yama, M., Rawat, J., 2019. Efficient isopropanol-butanol (IB) fermentation of rice straw hydrolysate by a newly isolated Clostridium beijerinckii strain C-01. *Biomass and Bioenergy* 127, 105292. https://doi.org/10.1016/j.biombioe.2019.105292

Darmayanti, R.F., Tashiro, Y., Noguchi, T., Gao, M., Sakai, K., Sonomoto, K., 2018. Novel biobutanol fermentation at a large extractant volume ratio using immobilized Clostridium saccharoperbutylacetonicum N1–4. *J. Biosci. Bioeng.* 126, 750–757. https://doi.org/10.1016/j.jbiosc.2018.06.006

Dien, B.S., Cotta, M., Jeffries, T., 2004. Bacteria engineered for fuel ethanol production : Current Status Bacteria engineered for fuel ethanol production : Current status. *Appl. Microbiol. Biotechnol.* 63, 258–266. https://doi.org/10.1007/s00253-003-1444-y

Dong, Z., Liu, G., Liu, S., Liu, Z., Jin, W., 2014. High performance ceramic hollow fiber supported PDMS composite pervaporation membrane for bio-butanol recovery. *J. Memb. Sci.* 450, 38–47. https://doi.org/https://doi.org/10.1016/j.memsci.2013.08.039

Ellilä, S., Fonseca, L., Uchima, C., Cota, J., Goldman, G.H., Saloheimo, M., Sacon, V., Siika-Aho, M., 2017. Development of a low-cost cellulase production process using Trichoderma reesei for Brazilian biorefineries. *Biotechnol. Biofuels* 10, 1–17. https://doi.org/10.1186/s13068-017-0717-0

Eriksson, T., Börjesson, J., Tjerneld, F., 2002. Mechanism of surfactant effect in enzymatic hydrolysis of lignocellulose. *Enzyme Microb. Technol.* 31, 353–364. https://doi.org/10.1016/S0141-0229(02)00134-5

Fan, H., Shi, Q., Yan, H., Ji, S., Dong, J., Zhang, G., 2014. Simultaneous spray self-assembly of highly loaded ZIF-8–PDMS nanohybrid membranes exhibiting exceptionally high biobutanol-permselective pervaporation. *Angew. Chemie Int. Ed.* 53, 5578–5582. https://doi.org/https://doi.org/10.1002/anie.201309534

Gaida, S.M., Liedtke, A., Jentges, A.H.W., Engels, B., Jennewein, S., 2016. Metabolic engineering of Clostridium cellulolyticum for the production of n-butanol from crystalline cellulose. *Microb. Cell Fact.* 15, 1–11. https://doi.org/10.1186/s12934-015-0406-2

Galbe, M., Zacchi, G., 2007. Pretreatment of lignocellulosic materials for efficient bioethanol production. *Adv. Biochem. Eng. Biotechnol.* 108, 41–65. doi: 10.1007/10_2007_070. PMID: 17646946.

Gaykawad, S.S., Zha, Y., Punt, P.J., van Groenestijn, J.W., van der Wielen, L.A.M., Straathof, A.J.J., 2013. Pervaporation of ethanol from lignocellulosic fermentation broth. *Bioresour. Technol.* 129, 469–476. https://doi.org/https://doi.org/10.1016/j.biortech.2012.11.104

George, H.A., Johnson, J.L., Moore, W.E.C., 1983. Acetone, isopropanol, and butanol production by Clostridium beijerinckii (syn. Clostridium butylicum) and Clostridium aurantibutyricum. *Appl. Environ. Microbiol.* 45, 1160–1163. https://doi.org/10.1128/aem.45.3.1160-1163.1983

Ghose, T.K., Roychoudhury, P.K., Ghosh, P., 1984. Simultaneous saccharification and fermentation (SSF) of lignocellulosics to ethanol under vacuum cycling and step feeding. *Biotechnol. Bioeng.* 26, 377–381. https://doi.org/10.1002/bit.260260414

Gírio, F.M., Fonseca, C., Carvalheiro, F., Duarte, L.C., Marques, S., Bogel-Łukasik, R., 2010. Hemicelluloses for fuel ethanol: A review. *Bioresour. Technol.* 101, 4775–4800. https://doi.org/10.1016/j.biortech.2010.01.088

Gonçalves, D.L., Matsushika, A., de Sales, B.B., Goshima, T., Bon, E.P.S., Stambuk, B.U., 2014. Xylose and xylose/glucose co-fermentation by recombinant Saccharomyces cerevisiae strains expressing individual hexose transporters. *Enzyme Microb. Technol.* 63, 13–20. https://doi.org/10.1016/j.enzmictec.2014.05.003

Gottumukkala, L.D., Parameswaran, B., Valappil, S.K., Mathiyazhakan, K., Pandey, A., Sukumaran, R.K., 2013. Biobutanol production from rice straw by a non acetone producing Clostridium sporogenes BE01. *Bioresour. Technol.* 145, 182–187. https://doi.org/10.1016/j.biortech.2013.01.046

Gough, S., Flynn, O., Hack, C.J., Marchant, R., 1996. Fermentation of molasses using a thermotolerant yeast, Kluyveromyces marxianus IMB3: Simplex optimisation of media supplements. *Appl. Microbiol. Biotechnol.* 46, 187–190. https://doi.org/10.1007/s002530050803

Green, Edward M., 2011. Fermentative production of butanol-the industrial perspective. *Curr. Opin. Biotechnol.* 22, 337–343. https://doi.org/10.1016/j.copbio.2011.02.004

Groot, W.J., Soedjak, H.S., Donck, P.B., van der Lans, R.G.J.M., Luyben, K.C.A.M., Timmer, J.M.K., 1990. Butanol recovery from fermentations by liquid-liquid

extraction and membrane solvent extraction. *Bioprocess Eng.* 5, 203–216. https://doi.
org/10.1007/BF00376227

Ha, S., Galazka, J.M., Rin, S., Choi, J., Yang, X., Seo, J., 2010. Engineered Saccharomyces
cerevisiae capable of simultaneous cellobiose and xylose fermentation 1–6. https://
doi.org/10.1073/pnas.1010456108/-/DCSupplemental.www.pnas.org/cgi/doi/10.1073/
pnas.1010456108

Haghighi Mood, S., Hossein Golfeshan, A., Tabatabaei, M., Salehi Jouzani, G., Najafi, G.H.,
Gholami, M., Ardjmand, M., 2013. Lignocellulosic biomass to bioethanol, a comprehensive review with a focus on pretreatment. *Renew. Sustain. Energy Rev.* 27, 77–93.
https://doi.org/10.1016/j.rser.2013.06.033

Hari Krishna, S., Janardhan Reddy, T., Chowdary, G. V., 2001. Simultaneous saccharification and fermentation of lignocellulosic wastes to ethanol using a thermotolerant yeast.
Bioresour. Technol. 77, 193–196. https://doi.org/10.1016/S0960-8524(00)00151-6

Harris, L.M., Welker, N.E., Papoutsakis, E.T., 2002. Northern, morphological, and fermentation
analysis of spo0A inactivation and overexpression in Clostridium acetobutylicum ATCC
824. *J. Bacteriol.* 184, 3586–3597. https://doi.org/10.1128/JB.184.13.3586-3597.2002

Hartley, F.R.S., BS and Shama, G., 1987. The Royal Society is collaborating with JSTOR to
digitize, preserve, and extend access to Philosophical Transactions of the Royal Society
of London. Series A, Mathematical and Physical Sciences. ® www.jstor.org. *Phil.
Trans. R Soc. L. A* 321, 555–568.

Hassan, I. Ben, Ennouri, M., Lafforgue, C., Schmitz, P., Ayadi, A., 2013. Experimental study
of membrane fouling during crossflow microfiltration of yeast and bacteria suspensions:
Towards an analysis at the microscopic level. membranes (basel). 3, 44–68. https://doi.
org/10.3390/membranes3020044

Hoekman, S.K., 2009. Biofuels in the U. S. – Challenges and opportunities 34, 14–22. https://-
doi.org/10.1016/j.renene.2008.04.030

Hong, J., Wang, Y., Kumagai, H., Tamaki, H., 2007. Construction of thermotolerant yeast
expressing thermostable cellulase genes. *J. Biotechnol.* 130, 114–123. https://doi.org/-
https://doi.org/10.1016/j.jbiotec.2007.03.008

Jansen, M.L.A., Bracher, J.M., Papapetridis, I., Verhoeven, M.D., de Bruijn, H., de Waal, P.P.,
van Maris, A.J.A., Klaassen, P., Pronk, J.T., 2017. *Saccharomyces cerevisiae* strains for
second-generation ethanol production: From academic exploration to industrial implementation. *FEMS Yeast Research* 17(5). https://doi.org/10.1093/femsyr/fox044

Johansson, B., Hahn-Hägerdal, B., 2002. The non-oxidative pentose phosphate pathway controls the fermentation rate of xylulose but not of xylose in *Saccharomyces cerevisiae*
TMB3001. *FEMS Yeast Res.* 2, 277–282. https://doi.org/10.1016/S1567-1356(02)00114-9

Jones, D.T., Woods, D.R., 1986. Acetone-butanol fermentation revisited. *Microbiol. Rev.* 50,
484–524. https://doi.org/10.1128/mr.50.4.484-524.1986

Jones, R.A., Gandier, J.A., Thibault, J., Tezel, F.H., 2011. Enhanced ethanol production
through selective adsorption on bacterial fermentation. *Biotechnol. Bioprocess Eng.* 16,
531–541. https://doi.org/10.1007/s12257-010-0299-1

Kamelian, F.S., Mohammadi, T., Naeimpoor, F., Sillanpää, M., 2020. One-step and low-cost designing of two-layered active-layer superhydrophobic silicalite-1/PDMS membrane for simultaneously achieving superior bioethanol pervaporation and fouling/biofouling resistance.
ACS Appl. Mater. Interfaces 12, 56587–56603. https://doi.org/10.1021/acsami.0c17046

Kanchanalai, P., Lively, R.P., Realff, M.J., Kawajiri, Y., 2013. Cost and energy savings using
an optimal design of reverse osmosis membrane pretreatment for dilute bioethanol purification. *Ind. Eng. Chem. Res.* 52, 11132–11141. https://doi.org/10.1021/ie302952p

Karatzos, S., Mcmillan, J.D., **Saddler**, J.N., 2014. The potential and challenges of drop - in
biofuels. *IEA Bioenergy | Task 39 Rep. Biorefinery.*

Karhumaa, K., Garcia Sanchez, R., Hahn-Hägerdal, B., Gorwa-Grauslund, M.-F., 2007.
Comparison of the xylose reductase-xylitol dehydrogenase and the xylose isomerase

pathways for xylose fermentation by recombinant Saccharomyces cerevisiae. *Microb. Cell Fact.* 6, 5. https://doi.org/10.1186/1475-2859-6-5

Kim, J.S., Lee, Y.Y., Kim, T.H., 2016. A review on alkaline pretreatment technology for bioconversion of lignocellulosic biomass. *Bioresour. Technol.* 199, 42–48. https://doi.org/10.1016/j.biortech.2015.08.085

Kim, T.H., Taylor, F., Hicks, K.B., 2008. Bioethanol production from barley hull using SAA (soaking in aqueous ammonia) pretreatment. *Bioresour. Technol.* 99, 5694–5702. https://doi.org/10.1016/j.biortech.2007.10.055

Kitagawa, T., Tokuhiro, K., Sugiyama, H., Kohda, K., Isono, N., Hisamatsu, M., Takahashi, H., Imaeda, T., 2010. Construction of a β-glucosidase expression system using the multistress-tolerant yeast Issatchenkia orientalis. *Appl. Microbiol. Biotechnol.* 87, 1841–1853. https://doi.org/10.1007/s00253-010-2629-9

Kobayashi, Y., Sahara, T., Ohgiya, S., Kamagata, Y., Fujimori, K.E., 2018. Systematic optimization of gene expression of pentose phosphate pathway enhances ethanol production from a glucose / xylose mixed medium in a recombinant Saccharomyces cerevisiae. *AMB Express* 1–11. https://doi.org/10.1186/s13568-018-0670-8

Kristensen, J.B., Börjesson, J., Bruun, M.H., Tjerneld, F., Jørgensen, H., 2007. Use of surface active additives in enzymatic hydrolysis of wheat straw lignocellulose. *Enzyme Microb. Technol.* 40, 888–895. https://doi.org/10.1016/j.enzmictec.2006.07.014

Kujawska, A., Kujawski, J., Bryjak, M., Kujawski, W., 2015. ABE fermentation products recovery methods—A review. *Renew. Sustain. Energy Rev.* 48, 648–661. https://doi.org/https://doi.org/10.1016/j.rser.2015.04.028

Kujawski, J., Kujawska, A., Bryjak, M., Kujawski, W., 2014. Pervaporative removal of acetone, butanol and ethanol from binary and multicomponent aqueous mixtures. *Sep. Purif. Technol.* 132, 422–429. https://doi.org/10.1016/j.seppur.2014.05.047

Kumar, B., Bhardwaj, N., Agrawal, K., Chaturvedi, V., Verma, P., 2020. Current perspective on pretreatment technologies using lignocellulosic biomass: An emerging biorefinery concept. *Fuel Process. Technol.* 199. https://doi.org/10.1016/j.fuproc.2019.106244

Kumar, M., Gayen, K., 2011. Developments in biobutanol production: New insights. *Appl. Energy* 88, 1999–2012. https://doi.org/10.1016/j.apenergy.2010.12.055

Kumar, M., Goyal, Y., Sarkar, A., Gayen, K., 2012. Comparative economic assessment of ABE fermentation based on cellulosic and non-cellulosic feedstocks. *Appl. Energy* 93, 193–204. https://doi.org/10.1016/j.apenergy.2011.12.079

Kumari, D., Singh, R., 2018. Pretreatment of lignocellulosic wastes for biofuel production: A critical review. *Renew. Sustain. Energy Rev.* 90, 877–891. https://doi.org/10.1016/j.rser.2018.03.111

Kuyper, M., Hartog, M.M.P., Toirkens, M.J., Almering, M.J.H., Winkler, A.A., Dijken, J.P. Van, Pronk, J.T., 2005. Metabolic engineering of a xylose-isomerase-expressing Saccharomyces cerevisiae strain for rapid anaerobic xylose fermentation 5, 399–409. https://doi.org/10.1016/j.femsyr.2004.09.010

Kuyper, M., Winkler, A.A., Dijken, J.P. Van, Pronk, J.T., 2004. Minimal metabolic engineering of Saccharomyces cerevisiae for efficient anaerobic xylose fermentation : A proof of principle 4, 655–664. https://doi.org/10.1016/j.femsyr.2004.01.003

Kwon, Y., Ma, A., Li, Q., Wang, F., Zhuang, G., Liu, C., 2011. Bioresource technology effect of lignocellulosic inhibitory compounds on growth and ethanol fermentation of newly-isolated thermotolerant Issatchenkia orientalis. *Bioresour. Technol.* 102, 8099–8104. https://doi.org/10.1016/j.biortech.2011.06.035

Lapuerta, M., Ballesteros, R., Barba, J., 2017. Strategies to introduce n-butanol in gasoline blends. *Sustainability.* https://doi.org/10.3390/su9040589

Lark, N., Xia, Y., Qin, C.G., Gong, C.S., Tsao, G.T., 1997. Production of ethanol from recycled paper sludge using cellulase and yeast, Kluyveromyces marxianus. *Biomass and Bioenergy* 12, 135–143. https://doi.org/10.1016/S0961-9534(96)00069-4

Lee, R.A., Lavoie, J.M., 2013. From first- to third-generation biofuels: Challenges of producing a commodity from a biomass of increasing complexity. *Anim. Front.* 3, 6–11. https://doi.org/10.2527/af.2013-0010

Lee, W., Nan, H., Kim, H.J., Jin, Y., 2013. Simultaneous saccharification and fermentation by engineered Saccharomyces cerevisiae without supplementing extracellular. *J. Biotechnol.* 167, 316–322. https://doi.org/10.1016/j.jbiotec.2013.06.016

Li, J., Henriksson, G., Gellerstedt, G., 2007. Lignin depolymerization/repolymerization and its critical role for delignification of aspen wood by steam explosion. *Bioresour. Technol.* 98, 3061–3068. https://doi.org/10.1016/j.biortech.2006.10.018

Lu, K.-M., Li, S.-Y., 2014. An integrated in situ extraction-gas stripping process for Acetone–Butanol–Ethanol (ABE) fermentation. *J. Taiwan Inst. Chem. Eng.* 45, 2106–2110. https://doi.org/https://doi.org/10.1016/j.jtice.2014.06.023

Luo, L., Van der Voet, E., Huppes, G., 2010. Biorefining of lignocellulosic feedstock - Technical, economic and environmental considerations. *Bioresour. Technol.* 101, 5023–5032. https://doi.org/10.1016/j.biortech.2009.12.109

Luyben, W.L., 2008. Control of the heterogeneous azeotropic n-butanol/water distillation system. *Energy & Fuels* 22, 4249–4258. https://doi.org/10.1021/ef8004064

Ma, X., Zheng, X., Zhang, M., Yang, X., Chen, L., Huang, L., Cao, S., 2014. Electron beam irradiation of bamboo chips: Degradation of cellulose and hemicelluloses. *Cellulose* 21, 3865–3870. https://doi.org/10.1007/s10570-014-0402-4

Mamat, R., Sani, S., Kadarohman, A., Sardjono, R., 2019. An overview of higher alcohol and biodiesel as alternative fuels in engines. *Energy Reports* 5, 467–479. https://doi.org/10.1016/j.egyr.2019.04.009

Menon, V., Rao, M., 2012. Trends in bioconversion of lignocellulose: Biofuels, platform chemicals & biorefinery concept. *Prog. Energy Combust. Sci.* 38, 522–550. https://doi.org/10.1016/j.pecs.2012.02.002

Nakayama, S., Kiyoshi, K., Kadokura, T., Nakazato, A., 2011. Butanol production from crystalline cellulose by Cocultured Clostridium thermocellum and Clostridium saccharoperbutylacetonicum N1-4. *Appl. Environ. Microbiol.* 77, 6470–6475. https://doi.org/10.1128/AEM.00706-11

Niglio, S., Marzocchella, A., Rehmann, L., 2019. Clostridial conversion of corn syrup to Acetone-Butanol-Ethanol (ABE) via batch and fed-batch fermentation. *Heliyon* 5, e01401. https://doi.org/10.1016/j.heliyon.2019.e01401

Nijland, J.G., Driessen, A.J.M., 2020. Engineering of pentose transport in Saccharomyces cerevisiae for biotechnological applications 7, 1–13. https://doi.org/10.3389/fbioe.2019.00464

Nijland, J.G., Shin, H.Y., de Waal, P.P., Klaassen, P., Driessen, A.J.M., 2018. Increased xylose affinity of Hxt2 through gene shuffling of hexose transporters in Saccharomyces cerevisiae. *J. Appl. Microbiol.* 124, 503–510. https://doi.org/10.1111/jam.13670

Offeman, R.D., Stephenson, S.K., Robertson, G.H., Orts, W.J., 2005. Solvent extraction of ethanol from aqueous solutions. I. Screening methodology for solvents. *Ind. Eng. Chem. Res.* 44, 6789–6796. https://doi.org/10.1021/ie0500319

Olguin-Maciel, E., Singh, A., Chable-Villacis, R., Tapia-Tussell, R., Ruiz, H.A., 2020. Consolidated bioprocessing, an innovative strategy towards sustainability for biofuels production from crop residues: An overview. *Agronomy* 10, 1834. https://doi.org/10.3390/agronomy10111834

Oliva, J.M., Ballesteros, I., Negro, M.J., Manzanares, P., Cabañas, A., Ballesteros, M., 2004. Effect of binary combinations of selected toxic compounds on growth and fermentation of Kluyveromyces marxianus. *Biotechnol. Prog.* 20, 715–720. https://doi.org/10.1021/bp034317p

Ong, Y.K., Shi, G.M., Le, N.L., Tang, Y.P., Zuo, J., Nunes, S.P., Chung, T.-S., 2016. Recent membrane development for pervaporation processes. *Prog. Polym. Sci.* 57, 1–31. https://doi.org/10.1016/j.progpolymsci.2016.02.003

Papoutsakis, E.T., 2008. Engineering solventogenic clostridia. *Curr. Opin. Biotechnol.* 19, 420–429. https://doi.org/10.1016/j.copbio.2008.08.003

Parekh, M., Formanek, J., Blaschek, H.P., 1998. Development of a cost-effective glucose-corn steep medium for production of butanol by Clostridium beijerinckii. *J. Ind. Microbiol. Biotechnol.* 21, 187–191. https://doi.org/10.1038/sj.jim.2900569

Peng, X. "Nick," Gilmore, S.P., O'Malley, M.A., 2016. Microbial communities for biopro-cessing: Lessons learned from nature. *Curr. Opin. Chem. Eng.* 14, 103–109. https://doi.org/10.1016/j.coche.2016.09.003

Peralta-Yahya, P.P., Keasling, J.D., 2010. Advanced biofuel production in microbes. *Biotechnol. J.* 5, 147–162. https://doi.org/10.1002/biot.200900220

Pongcharoen, P., Chawneua, J., Tawong, W., 2018. High temperature alcoholic fermentation by new thermotolerant yeast strains Pichia kudriavzevii isolated from sugarcane field soil. *Agric. Nat. Resour.* 52, 511–518. https://doi.org/10.1016/j.anres.2018.11.017

Qi, X., Zha, J., Liu, G.-G., Zhang, W., Li, B.-Z., Yuan, Y.-J., 2015. Heterologous xylose isomer-ase pathway and evolutionary engineering improve xylose utilization in Saccharomyces cerevisiae. *Front. Microbiol.* 6, 1165. https://doi.org/10.3389/fmicb.2015.01165

Qureshi, N., Ezeji, T.C., Ebener, J., Dien, B.S., Cotta, M.A., Blaschek, H.P., 2008a. Butanol production by Clostridium beijerinckii. Part I: Use of acid and enzyme hydrolyzed corn fiber. *Bioresour. Technol.* 99, 5915–5922. https://doi.org/10.1016/j.biortech.2007.09.087

Qureshi, N., Saha, B.C., Cotta, M.A., 2007. Butanol production from wheat straw hydroly-sate using Clostridium beijerinckii. *Bioprocess Biosyst. Eng.* 30, 419–427. https://doi.org/10.1007/s00449-007-0137-9

Qureshi, N., Saha, B.C., Cotta, M.A., 2008b. Butanol production from wheat straw by simul-taneous saccharification and fermentation using Clostridium beijerinckii: Part II-Fed-batch fermentation. *Biomass and Bioenergy* 32, 176–183. https://doi.org/10.1016/j.biombioe.2007.07.005

Qureshi, N., Saha, B.C., Dien, B., Hector, R.E., Cotta, M.A., 2010a. Production of butanol (a biofuel) from agricultural residues: Part I - Use of barley straw hydrolysate. *Biomass and Bioenergy* 34, 559–565. https://doi.org/10.1016/j.biombioe.2009.12.024

Qureshi, N., Saha, B.C., Hector, R.E., Dien, B., Hughes, S., Liu, S., Iten, L., Bowman, M.J., Sarath, G., Cotta, M.A., 2010b. Production of butanol (a biofuel) from agricultural resi-dues: Part II - Use of corn stover and switchgrass hydrolysates. *Biomass and Bioenergy* 34, 566–571. https://doi.org/10.1016/j.biombioe.2009.12.023

Qureshi, N., Schripsema, J., Lienhardt, J., Blaschek, H.P., 2000. Continuous solvent produc-tion by Clostridium beijerinckii BA101 immobilized by adsorption onto brick. *World J. Microbiol. Biotechnol.* 16, 377–382. https://doi.org/10.1023/A:1008984509404

Rajesh Banu, J., Kavitha, S., Yukesh Kannah, R., Poornima Devi, T., Gunasekaran, M., Kim, S.H., Kumar, G., 2019. A review on biopolymer production via lignin valorization. *Bioresour. Technol.* 290, 121790. https://doi.org/10.1016/j.biortech.2019.121790

Rassadin, V.G., Shlygin, O.Y., Likhterova, N.M., Slavin, V.N., Zharov, A.V., 2006. Problems in production of high-octane, unleaded automotive gasolines. *Chem. Technol. Fuels Oils* 42, 235–242. https://doi.org/10.1007/s10553-006-0064-5

Reznicek, O., Facey, S.J., de Waal, P.P., Teunissen, A.W.R.H., de Bont, J.A.M., Nijland, J.G., Driessen, A.J.M., Hauer, B., 2015. Improved xylose uptake in Saccharomyces cerevi-siae due to directed evolution of galactose permease Gal2 for sugar co-consumption. *J. Appl. Microbiol.* 119, 99–111. https://doi.org/10.1111/jam.12825

Robak, K., Balcerek, M., 2018. Review of second generation bioethanol production from residual biomass. *Food Technol. Biotechnol.* 56, 174–187. https://doi.org/10.17113/ftb.56.02.18.5428

Roffler, S., Blanch, H.W., Wilke, C.R., 1987. Extractive fermentation of acetone and butanol: Process design and economic evaluation. *Biotechnol. Prog.* 3, 131–140. https://doi.org/https://doi.org/10.1002/btpr.5420030304

Santosh, I., Ashtavinayak, P., Amol, D., Sanjay, P., 2017. Enhanced bioethanol production from different sugarcane bagasse cultivars using co-culture of Saccharomyces cerevisiae and Scheffersomyces (Pichia) stipitis. https://doi.org/10.1016/j.jece.2017.05.045

Seidl, P.R., Goulart, A.K., 2016. Pretreatment processes for lignocellulosic biomass conversion to biofuels and bioproducts. *Curr. Opin. Green Sustain. Chem.* 2, 48–53. https://doi.org/10.1016/j.cogsc.2016.09.003

Seo, D.-J., Takenaka, A., Fujita, H., Mochidzuki, K., Sakoda, A., 2018. Practical considerations for a simple ethanol concentration from a fermentation broth via a single adsorptive process using molecular-sieving carbon. *Renew. Energy* 118, 257–264. https://doi.org/https://doi.org/10.1016/j.renene.2017.11.019

Sharma, B., Larroche, C., Dussap, C.G., 2020. Comprehensive assessment of 2G bioethanol production. *Bioresour. Technol.* 313, 123630. https://doi.org/10.1016/j.biortech.2020.123630

Shihadeh, J., Huang, H., Rausch, K., Tumbleson, M., Singh, V., 2014. Vacuum stripping of ethanol during high solids fermentation of corn. *Appl. Biochem. Biotechnol.* 173. https://doi.org/10.1007/s12010-014-0855-9

Siedlarz, P., Sroka, M., Dylag, M., Nawrot, U., Gonchar, M., Kus-Liśkiewicz, M., 2016. Preliminary physiological characteristics of thermotolerant Saccharomyces cerevisiae clinical isolates identified by molecular biology techniques. *Lett. Appl. Microbiol.* 62, 277–282. https://doi.org/10.1111/lam.12542

Silvestre-Albero, A., Silvestre-Albero, J., Sepúlveda-Escribano, A., Rodríguez-Reinoso, F., 2009. Ethanol removal using activated carbon: Effect of porous structure and surface chemistry. *Microporous Mesoporous Mater.* 120, 62–68. https://doi.org/https://doi.org/10.1016/j.micromeso.2008.10.012

Singh, A., Rangaiah, G.P., 2017. Review of technological advances in bioethanol recovery and dehydration. *Ind. Eng. Chem. Res.* 56, 5147–5163. https://doi.org/10.1021/acs.iecr.7b00273

Skinner, K.A., Leathers, T.D., 2004. Bacterial contaminants of fuel ethanol production 401–408. https://doi.org/10.1007/s10295-004-0159-0

Sommer, P., Georgieva, T., Ahring, B.K., 2004. Potential for using thermophilic anaerobic bacteria for bioethanol production from hemicellulose. Biochem Soc Trans. 32(Pt 2):283-9. doi: 10.1042/bst0320283. PMID: 15046590.

Taherzadeh, M.J., Karimi, K., 2007. Enzyme-based hydrolysis processes for ethanol from lignocellulosic materials: A review, *BioResources.* https://doi.org/10.15376/biores.2.4.707-738

Takami, H., Takaki, Y., Chee, G., Nishi, S., Shimamura, S., 2004. Thermoadaptation trait revealed by the genome sequence of thermophilic Geobacillus kaustophilus. 32, 6292–6303. https://doi.org/10.1093/nar/gkh970

Tang, C., Chen, Y., Liu, J., Shen, T., Cao, Z., Shan, J., Zhu, C., Ying, H., 2017. Sustainable biobutanol production using alkali-catalyzed organosolv pretreated cornstalks. *Ind. Crops Prod.* 95, 383–392. https://doi.org/10.1016/j.indcrop.2016.10.048

Taylor, F., Marquez, M.A., Johnston, D.B., Goldberg, N.M., Hicks, K.B., 2010. Continuous high-solids corn liquefaction and fermentation with stripping of ethanol. *Bioresour. Technol.* 101, 4403–4408. https://doi.org/10.1016/j.biortech.2010.01.092

Taylor, M.P., Eley, K.L., Martin, S., Tuffin, M.I., Burton, S.G., Cowan, D.A., 2009. Thermophilic ethanologenesis : Future prospects for second-generation bioethanol production. https://doi.org/10.1016/j.tibtech.2009.03.006

Techaparin, A., Thanonkeo, P., Klanrit, P., 2017. High-temperature ethanol production using thermotolerant yeast newly isolated from Greater Mekong Subregion. *Brazilian J. Microbiol.* [publication Brazilian Soc. Microbiol.] 48, 461–475. https://doi.org/10.1016/j.bjm.2017.01.006

Tran, H.T.M., Cheirsilp, B., Hodgson, B., Umsakul, K., 2010. Potential use of Bacillus subtilis in a co-culture with Clostridium butylicum for acetone-butanol-ethanol production from cassava starch. *Biochem. Eng. J.* 48, 260–267. https://doi.org/10.1016/j.bej.2009.11.001

Vane, L.M., 2005. A review of pervaporation for product recovery from biomass fermentation processes. *J. Chem. Technol. Biotechnol.* 80, 603–629.

Vane, L.M., Alvarez, F.R., Rosenblum, L., Govindaswamy, S., 2013. Hybrid vapor stripping–vapor permeation process for recovery and dehydration of 1-butanol and acetone/butanol/ethanol from dilute aqueous solutions. Part 2. Experimental validation with simple mixtures and actual fermentation broth. *J. Chem. Technol. Biotechnol.* 88, 1448–1458. https://doi.org/https://doi.org/10.1002/jctb.4086

Veza, I., Muhamad Said, M.F., Latiff, Z.A., 2021. Recent advances in butanol production by acetone-butanol-ethanol (ABE) fermentation. *Biomass and Bioenergy* 144, 105919. https://doi.org/10.1016/j.biombioe.2020.105919

Vohra, M., Manwar, J., Manmode, R., Padgilwar, S., Patil, S., 2014. Bioethanol production: Feedstock and current technologies. *J. Environ. Chem. Eng.* 2, 573–584. https://doi.org/10.1016/j.jece.2013.10.013

Wee, Y.-J., Yun, J.-S., Lee, Y.Y., Zeng, A.-P., Ryu, H.-W., 2005. Recovery of lactic acid by repeated batch electrodialysis and lactic acid production using electrodialysis wastewater. *J. Biosci. Bioeng.* 99, 104–108. https://doi.org/10.1263/jbb.99.104

Wohlbach, D.J., Kuo, A., Sato, T.K., Potts, K.M., Salamov, A.A., LaButti, K.M., Sun, H., Clum, A., Pangilinan, J.L., Lindquist, E.A., Lucas, S., Lapidus, A., Jin, M., Gunawan, C., Balan, V., Dale, B.E., Jeffries, T.W., Zinkel, R., Barry, K.W., Grigoriev, I. V., Gasch, A.P., 2011. Comparative genomics of xylose-fermenting fungi for enhanced biofuel production. *Proc. Natl. Acad. Sci. U. S. A.* 108, 13212–13217. https://doi.org/10.1073/pnas.1103039108

Wyman, C.E., 1994. Ethanol from lignocellulosic biomass: Technology, economics, and opportunities. *Bioresour. Technol.* 50, 3–15. https://doi.org/10.1016/0960-8524(94)90214-3

Wyman, C.E., Hinman, N., 1990. Fundamentals of production from renewable feedstocks and use as a transportation fuel. *Appl. Biochem. Biotechnol.* 24/25, 735–753.

Xue, C., Liu, F., Xu, M., Zhao, J., Chen, L., Ren, J., Bai, F., Yang, S.-T., 2016. A novel in situ gas stripping-pervaporation process integrated with acetone-butanol-ethanol fermentation for hyper n-butanol production. *Biotechnol. Bioeng.* 113, 120–129. https://doi.org/10.1002/bit.25666

Xue, T., Liu, K., Chen, D., Yuan, X., Fang, J., Yan, H., Huang, L., Chen, Y., He, W., 2018. Improved bioethanol production using CRISPR/Cas9 to disrupt the ADH2 gene in Saccharomyces cerevisiae. *World J. Microbiol. Biotechnol.* 34, 154. https://doi.org/10.1007/s11274-018-2518-4

Yamaguchi, K., Matsumoto, M., Kusdiana, D., Ikeda, T., Hoshi, H., Kan, S., Chew, C.S., Omar, N., Nang, H.L.L., Masigan, M.C., Guzman, R.B.D., Petai, K., Raungkraikonkit, D., 2013. Study on ASIAN potential of biofuel market, ERIA Research Project Report No. 25, Economic Research Institute for ASEAN and East Asia, 20, 1–231.

Yang, P., Wu, Y., Zheng, Z., Cao, L., Zhu, X., Mu, D., Jiang, S., 2018. CRISPR-Cas9 Approach Constructing Cellulase sestc-Engineered Saccharomyces cerevisiae for the Production of Orange Peel Ethanol. *Front. Microbiol.* 9, 2436. https://doi.org/10.3389/fmicb.2018.02436

Yuangsaard, N., Yongmanitchai, W., Yamada, M., Limtong, S., 2013. Selection and characterization of a newly isolated thermotolerant Pichia kudriavzevii strain for ethanol production at high temperature from cassava starch hydrolysate. *Antonie van Leeuwenhoek, Int. J. Gen. Mol. Microbiol.* 103, 577–588. https://doi.org/10.1007/s10482-012-9842-8

Yusoff, M.N.A.M., Zulkifli, N.W.M., Masum, B.M., Masjuki, H.H., 2015. Feasibility of bioethanol and biobutanol as transportation fuel in spark-ignition engine: A review. *RSC Adv.* 5, 100184–100211. https://doi.org/10.1039/c5ra12735a

Zabed, H., Sahu, J.N., Suely, A., Boyce, A.N., Faruq, G., 2017. Bioethanol production from renewable sources: Current perspectives and technological progress. *Renew. Sustain. Energy Rev.* 71, 475–501. https://doi.org/10.1016/j.rser.2016.12.076

Zaldivar, J., Nielsen, J., Olsson, L., 2001. Fuel ethanol production from lignocellulose: A challenge for metabolic engineering and process integration. *Appl. Microbiol. Biotechnol.* 56, 17–34. https://doi.org/10.1007/s002530100624

Zentou, H., Abidin, Z.Z., Yunus, R., Awang Biak, D.R., Korelskiy, D., 2019. Overview of alternative ethanol removal techniques for enhancing bioethanol recovery from fermentation broth. *Processess.* https://doi.org/10.3390/pr7070458

Zhang, J., Liu, H.-J., Liu, D., 2005. Effect of different types of gas in gas stripping ethanol fermentation (GSEF). *Guocheng Gongcheng Xuebao/The Chinese J. Process Eng.* 15, 349–352.

Zhang, K., Pei, Z., Wang, D., 2016. Organic solvent pretreatment of lignocellulosic biomass for biofuels and biochemicals: A review. *Bioresour. Technol.* 199, 21–33. https://doi.org/10.1016/j.biortech.2015.08.102

Zhao, Y., Shakeel, U., Saif Ur Rehman, M., Li, H., Xu, X., Xu, J., 2020. Lignin-carbohydrate complexes (LCCs) and its role in biorefinery. *J. Clean. Prod.* 253, 120076. https://doi.org/10.1016/j.jclepro.2020.120076

Zhu, H., Liu, G., Jin, W., 2020. Recent progress in separation membranes and their fermentation coupled processes for biobutanol recovery. *Energy & Fuels* 34, 11962–11975. https://doi.org/10.1021/acs.energyfuels.0c02680

Zhuang, X., Wang, W., Yu, Q., Qi, W., Wang, Q., Tan, X., Zhou, G., Yuan, Z., 2016. Liquid hot water pretreatment of lignocellulosic biomass for bioethanol production accompanying with high valuable products. *Bioresour. Technol.* 199, 68–75. https://doi.org/10.1016/j.biortech.2015.08.051

12 Biological Production of Diols – Current Perspective

Koel Saha, Divya Mudgil, and Sanjukta Subudhi
The Energy and Resources Institute

Aishwarya Srivastava and Nidhi Adlakha
NCR Biotech Science Cluster

CONTENTS

12.1 Introduction: Overview on the Importance and Production of Diols 327
12.2 Biological Production of Various Diols 329
     12.2.1 1,3-Propanediol (1,3-PDO) 329
     12.2.2 1,2-Propanediol (1,2-PDO) 330
     12.2.3 2,3-Butanediol (2,3-BDO) 331
     12.2.4 1,4-Butanediol (1,4-BDO).. 332
     12.2.5 2,4-Pentanediol 332
12.3 A Brief Outline on the Various Aspects and Applications of Diols 333
12.4 Production of 2,3-Butanediol from Lignocellulosic Biomass – A Brief 334
     12.4.1 Feedstock Used for 2,3-BDO Production 335
     12.4.2 Potential Microorganism 335
     12.4.3 Fermentation Strategies 336
     12.4.4 Effect of Substrate Concentration.. 338
     12.4.5 Downstream Processing 339
References 340

12.1 INTRODUCTION: OVERVIEW ON THE IMPORTANCE AND PRODUCTION OF DIOLS

Lignocellulosic biomass is the copious renewable resource being replenished continuously by the photosynthesis. It consists of crop residues, energy crops, animal manures and forest residues. It can serve as a prospective low-cost substrate in the bioenergy sector, the chemical industry, the pulp and paper industry, etc. The incessant and increasing energy demands, fossil fuel depletion and rising environmental disquiet have cumulatively fetched the attention of the scientists to work on the utilization of lignocellulose biotechnology. Zhang (2008) reported the production of almost 200 million tons of lignocellulosic biomass worldwide every year. According to the US Department of Energy, the viable sources of lignocellulosic biomass are agricultural residues, branches of trees, different grasses and municipal garbage wastes. Proper handling and genuine utilization of these substances result in the reduction in waste generation as well as greenhouse gas emission. Also, the conversion of these into valuable products can build a good revenue for the industry (Roy et al. 2020). For this purpose, a proper understanding of the chemical nature, crystallinity and intermolecular interactions of the biomass is very important. This complex biopolymer contains cellulose, hemicellulose, lignin and few inorganic components. Cellulose, hemicellulose and lignin predominantly comprise approximately 98% dry weight of lignocellulose (Mosier et al. 2005). Cellulose, the primary structural component of natural fibre, is defined as the most abundant biopolymer and is present in grass, wood, stalks, etc. (Roy et al. 2020). The linear polymer structure of cellulose consists of both amorphous and crystalline regions (Jonoobi et al. 2009). The homogeneous polymer is made of repeating units of glucose monomers which are linked by β-1,4-glycosidic bond. The glucose units are linked by single oxygen atom between C_1 of one unit to the C_4 of the next unit. As one water molecule is eliminated in this reaction, the glucose units become anhydroglucose units (Kalia et al. 2011). Hemicellulose, a heterogeneous polymer, consists of different sugars such as galactose, xylose, arabinose, glucose, mannose and sugar acids. The monomer units bind to each other by glycosidic bond and form the branched polymer structure (Saha 2003). Due to its applications in various fields such as bioplastics, carbon fibre, biofuel, nanoparticles, agriculture, adsorbent in solution, dispersants and electrochemistry, it is defined as a value-added product (Norgren and Edlund 2014). In the complex lignocellulose matrix, lignin is cross-linked to polysaccharide cellulose and hemicellulose. It forms ester linkage with the hemicellulose and hydrogen bond with cellulose (Harmsen et al. 2010). As the structure of lignocellulose is recalcitrant for conversion, proper treatment of the biomass is necessary to disrupt the strong intermolecular bond to make all the constituent accessible for transformation. The production of diols from lignocellulose requires biomass pretreatment, hydrolysis of complex polymers, fermentation of hexoses, separation and effluent treatment (Ojeda et al. 2009). Diols are indispensable platform chemicals with applications in innumerable industrial sectors including polymer, rubber, lubricant, cosmetic, and personal care. Among the diols, 1,2-propanediol, 1,3-propanediol, 2,3-butanediol, 1,4-butanediol and 2,4-pentanediol have received much attention owing to the ease of their biological production and extensive applications. The importance of this dihydroxy hydrocarbon is inevitable from its ever growing compound annual growth rate (CAGR), with the recent being 12.2%; in fact, a market survey of diols indicates 490 million USD market size of mere 1,3-propanediol. This lucrative market size prompted chemists to innovate the chemical production of diols, and thus the 20th century witnessed the production of diols and other simple hydrocarbons mainly from crude oil fractionation (Alper et al. 1999). However, the depletion of conventional sources and dependency on foreign reserves impelled the paradigm shift to renewable and self-sustainable technologies. There had been many attempts concerning the chemical conversion of biomass to 1,2-propanediol (Cortright et al. 2002; Xiao et al. 2013). But the chemical transformation method suffers from two major drawbacks: (a) poor recovery and (b) non-environment-friendliness. To combat the challenges, scientists have developed efficient technologies for biological conversion of lignocellulose-based sugar into dihydroxy compounds. A group in University of Wisconsin established the fermentation approach for the conversion of wood- or corn-derived glucose to 1,2-propanediol (Altaras et al. 2001); however, the yield of this process is very low, as metabolically glycerol is the preferred substrate for propanediol production in *Thermosaccharolyticum* sp. Till date, only the production of butanediol from cellulose and hemicellulose has successfully been achieved at an industrial scale.

Considering the importance of other diols, efforts are being laid on the development of effective biomass-based biological methods for the production of dihydroxy hydrocarbons.

12.2 BIOLOGICAL PRODUCTION OF VARIOUS DIOLS

12.2.1 1,3-PROPANEDIOL (1,3-PDO)

In 1881, August Freund demonstrated the production of 1,3-PDO by fermenting glycerol using *Clostridium pasteurianum* (Freund 1881), making it one of the oldest fermentation products. Its use in various synthetic reactions has surged the interest of researchers towards the exploration of alternative hyper-producing strain, which led to the identification of *Klebsiella* sps., *Lactobacillus* sps., etc. All of these strains anaerobically ferment glycerol or glucose to 1,3-PDO; however, the fate of pyruvate is diverse, resulting in differential yields of 1,3-PDO (Table 12.1). The typical metabolic route from glycerol to 1,3-PDO involves two enzymes: glycerol dehydratase and 1,3-PDO dehydrogenase. The maximum theoretical yield of 1,3-PDO from glucose is 0.61 g/g, whereas that from glycerol is 0.72 g/g. In general, the yield of 1,3-PDO ranges between 0.11 and 0.72 mol/mol of glycerol, depending on the pathway and the by-product produced (Figure 12.1). As a case example, in most of the *Clostridium* sp., 1,3-PDO production is coupled with the formation of acetic acid, butyric acid or butanol (Dabrock et al. 1992) that preferentially consume reducing equivalents, generated by the anaerobic fermentation of glycerol to pyruvate and thus resulting in minimal 1,3-PDO production.

Another approach involves the use of *Enterobacterium* or *Klebsiella* sp., which favourably produces 1,3-propanediol, and acetic acid is produced as a by-product. (The conversion of pyruvate to acetic acid does not utilize NADH.) An improvement in these strains, using chemical mutagenesis approach, led to an increased titre of 1,3-PDO of up to 108 g/L. But the use of *Enterobacteria* and *Klebsiella* in industrial sector has been restricted as these strains are classified as opportunistic pathogens. Therefore, the focus has been diverted to native non-pathogenic *Clostridium* species which demonstrated a 1,3-PDO titre of 94 g/L. However, *Clostridium* sp. produces other by-products alongside.

To overcome the challenges associated with the by-product formation, DuPont and Genencor have metabolically engineered *Escherichia coli* for 1,3-PDO production from glucose (Nakamura and Whited 2003) and eventually commercialized this process. However, the economics of 1,3-PDO from ethylene oxide (chemical process developed by Shell) is not a sustainable process and thus not comparable with the

TABLE 12.1

Diol Yield Efficiency of Potential Microbes from Simple Substrates – Glucose and Glycerol

Strains	Yield (g/g glucose)	Yield (g/g glycerol)	References
1,2-Propanediol			
Clostridium thermosaccharolyticum	0.27	0.20	Cameron et al. (1986)
E. coli		0.21	Clomburg and Gonzalez (2011)
Saccharomyces cerevisiae			Jung et al. (2011)
1,3-Propanediol			
Clostridium butyricum CNCM1211		0.52	Himmi et al. (1999)
K. pneumoniae AC 15		0.53	Zheng (2008)
C. freundii DSM 30040		0.51	Boenigk et al. (1993)
Lactobacillus diolivorans		0.65	Pflügl et al. (2012)
Klebsiella pneumoniae ATCC 15380		0.46	Homann et al. (1990)
Clostridium butyricum VPI3266		0.57	Saint-Amans et al. (1994)
Engineered *Klebsiella pneumoniae*		0.62	Jung et al. (2015) and Lee et al. (2018)
2,3-Butanediol			
Bacillus licheniformis	0.45		Wang et al. (2012)
Engineered *E. coli*	0.31	0.21	Lee et al. (2012)
Klebsiella pneumoniae		0.36	Petrov and Petrova (2009)
Klebsiella oxytoca	0.48		Ji et al. (2010)
Paenibacillus polymyxa	0.41		Nakashimada et al. (2000)
Enterobacter aerogenes	0.48		Petrov and Petrova (2009)
1,4-Butanediol			
Engineered *E.coli*	0.13		Yim et al. (2011)

1,3-PDO production through biological route, emphasizing the need for continuous innovation to improve microbial production of 1,3-PDO from saccharide substrates.

12.2.2   1,2-Propanediol (1,2-PDO)

1,2-Propanediol (1,2-PDO) is a dihydroxy compound whose central carbon atom is a stereocentre, and its importance in the industrial sector is clearly evident from its huge annual production in the United States. The chemical route for 1,2-PDO synthesis involves catalytic hydrogenation of lactic acid ester (Goodlove et al. 1989) or bioreduction of acetol (Levene et al. 1943); all of these approaches are expensive, challenging and dependent on crude oil, demanding an alternate and sustainable route.

Several organisms have been discovered with the ability to ferment glucose to 1,-2-PDO, e.g. *Thermoanaerobacterium* sp. (Altaras et al. 2001), *Clostridium thermosaccharolyticum* (Tran-Din and Gottschalk 1985), etc.; however, the reported yield is

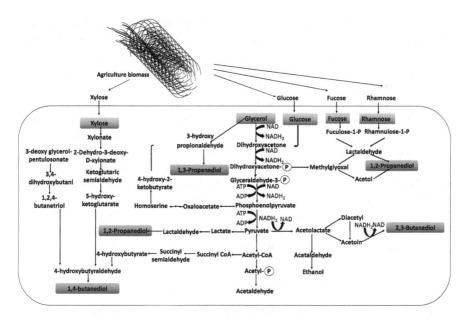

FIGURE 12.1 Microbial conversion of lignocellulosic biomass into diols.

very low. Another native producer, *E.coli*, also has preponderance to convert glycerol to 1,2-PDO as its genome harbours *mgs* and *gldA* genes (Afschar et al. 1993), and it has been reported that the insertion of these genes into industrial strain *S. cerevisiae* genome enables it to produce this bioproduct with a titre of 0.32 g/L. Jeon et al. (2009) further manipulated *Saccharomyces* genome by incorporating a combination of *mgs* gene, which transforms dihydroxyacetone phosphate to methylglyoxal, and *Citrobacter freundii dha*D gene that converts acetol to 1,2-PDO, and this has led the titre to increase to 0.45 g/L. A further improvement in yield was achieved by knocking out side pathway, i.e. lactate dehydrogenase, resulting in 0.21 g of 1,2-PDO per gram of glycerol (Bennett and San 2001). The maximum theoretical yield of 1,2-PDO from glycerol is 0.67 g/g. Efforts are ongoing to increase the productivity and yield of 1,2-PDO using co-fermentation approach as it has been observed that using both glucose and glycerol as substrate results in a shift from NADH consumption to NADH-producing reaction – a promising approach towards improving the yield of 1,2-PDO.

12.2.3 2,3-BUTANEDIOL (2,3-BDO)

2,3-Butanediol (2,3-BDO) is a four-carbon compound that has wide applications as an industry platform chemical as well as a liquid fuel. The unique stereoisomerism in 2,3-BDO accounts for its wide applications; for example, the levo isomer has a very low freezing point ($-60°C$), making it a suitable antifreeze, whereas dextro is majorly employed for the synthesis of agrochemicals and in food industries (Ma et al. 2018). 2,3-BDO can be converted into methyl ethyl ketone (MEK), which is used as a flavouring agent in food and dairy industries. Microbial strains involved in 2,3-BDO production include *Klebsiella pneumoniae* (Ma et al. 2009), *Enterobacter, Bacillus polymyxa* (Hespell 1996) and *Bacillus licheniformis* (Wang et al. 2012). Along with native producers, research studies focused on the genetic manipulation of robust microbial strains for improved 2,3-BDO production, including the genetic manipulation of *Lactobacillus plantarum* that has been reported to produce 2,3-BDO with 98% yield efficiency (Paul et al. 2010).

E.coli is a preferred heterologous host owing to the ease of its manipulation, and thus, efforts are being laid to produce 2,3-BDO by engineering *E.coli* to be used as a cell factory. These manipulations involved the integration of two BDO biosynthetic genes – acetolactate decarboxylase and alcohol dehydrogenase from *Klebsiella pneumoniae*. Table 12.1 illustrates the 2,3-BDO yield details from glucose and glycerol. The diverse applications of 2,3-BDO urged the need to search for inexpensive substrate(s) as the cost of the feed is one of the major challenges for driving this process towards commercialization. In this regard, *Bacillus licheniformis* and *Bacillus polymyxa* have intensively been studied with respect to the use of low-cost lignocellulose biomass as the feed for 2,3-BDO production (Adlakha et al. 2015).

12.2.4   1,4-Butanediol (1,4-BDO)

1,4-Butanediol (1,4-BDO) has widely been used for the production of plastics, polyesters and spandex fibres (Zeng and Sabra 2011; Forte et al. 2016). Its annual demand is approximately 1 million metric ton (Liu and Lu 2015). Till date, no native producer of 1,4-BDO has been reported and thus genetic engineering approaches are being employed for the biological production of this important hydrocarbon. First, Yim et al. (2011) reported the biological production of 1,4-BDO in *E.coli* by overexpressing a series of six enzymes for the conversion of succinate into 1,4-BDO. However, the inefficient alcohol dehydrogenase limited the 1,4-BDO yield. Therefore, Hwang et al. (2014) replaced this intrinsic gene with butanol dehydrogenase and butyraldehyde dehydrogenase from *Clostridium saccharoperbutylacetonicum* that has resulted in a fourfold higher titre of 1,4-BDO. An enhancement in 1,4-butanediol yield efficiency was observed with the modification of the pathway through TCA cycle bypass. However, this has resulted in concurrent accumulation of 1,2,4-butanetriol (1,2,4-BTO) (Tai et al. 2016). Further, Jia et al. (2017) adopted a rational protein engineering approach to engineer the diol dehydratase and thereby enabled the expansion of the 1,2,4-BTO biosynthesis pathway to divert towards 1,4-BDO production. Further progress in this domain was made by Meng et al. (2020) and Wu et al. (2017), through the implementation of CRISPR-based mutation of few genes involved in the 1,4-BDO biosynthesis pathway. The overall manipulation studies resulted in enhancing the titre to 0.9 g/L in 48 h. Further focused research explorations are required to screen for more active homologues from other potential species, which might serve as a promising approach for the optimized production of 1,4-BDO from different saccharide substrates.

12.2.5   2,4-Pentanediol

2,4-Pentanediol is a five-carbon compound which is important for chiral synthons for naturally occurring bioactive substances. Biological production of 2,4-pentanediol is reported in *Candida boidinii* (Matsumura et al. 1994). However, reports concerning the microbial production of 2,4-pentanediol is scarce. This could be attributed to (a) the negative impact of longer dihydroxy hydrocarbon on microbial growth and/or (b) complex and unknown pathway(s) involved and (c) the lack of reports concerning native producer. Nevertheless, efforts are ongoing in searching for promising biological alternative(s).

Of all the diols, scientific community has achieved success in transforming plant-based polysaccharides to 2,3-butanediol, and the platform developed serves as the basis for biological production of other diols.

12.3 A BRIEF OUTLINE ON THE VARIOUS ASPECTS AND APPLICATIONS OF DIOLS

Compounds containing two hydroxyl groups have extensive applications as fuels and value-added industry platform chemicals. In view of this, the large-scale production of these diols has intensively been explored in the recent past (Sabra et al. 2015). The literature survey illustrates the use of broad-spectrum renewable substrates as feed for the biological production of 1,2-propanediol, 1,3-propanediol, 2,3-butanediol and 1,4-butanediol (Zeng and Sabra 2011). Fermentative production of 1,3-PDO, which is produced through anaerobic metabolism of glycerol, has extensively been explored through the use of microbe(s) that could feed on crude glycerol. Crude glycerol is a major by-product of biodiesel industry. Nevertheless, the glycerol-utilizing microorganisms have the potential to tolerate the crude glycerol impurities such as methanol, fatty acids and salts (Chatzifragkou et al. 2010). Few of the species, *Enterobacter*, *Clostridia*, *Klebsiella*, are reported to utilize crude glycerol as feed for 1,3-PDO production. Among these microbes, *Klebsiella pneumoniae* is being widely explored for 1,3-PDO synthesis owing to its high substrate tolerance and high titre (Sabra et al. 2015). 1,3-PDO is used as a monomer for the synthesis of polyether, polyesters and polyurethanes (Wang et al. 2014).

Another important propane diol is 1,2-PDO, commonly named as propylene glycol. It appears as an uncoloured hygroscopic liquid that has a high boiling point. It is mostly used as de-icer, antifreeze, heat transfer fluid and solvent (Clomburg and Gonzalez 2011). This has got profound applications in cosmetic, food and pharmaceutical industries (Rode et al. 2010). Being a less toxic compound, it is considered as a potential alternate to ethylene glycol. 1,2-PDO is recognized as safe by US Food and Drug Administration for use in medicine, food and cosmetics (Jung et al. 2011). Not only propane diols, but butane diols also serve as platform chemicals for several downstream applications. The most useful products of 1,4-BDO are tetrahydrofuran (THF) and polybutylene terephthalate (PBT). THF is widely used in spandex fibres production that is popular in apparel industry, whereas PBT has extensive applications in electronics and automotive industries (Sabra et al. 2015). 1,4-BDO finds applications in cosmetic formulation, cleaning agent, printing ink, agricultural chemicals, as an adhesive in plastic and footwear industry (Werawattanachai et al. 2007), tissue engineering, drug delivery system, hydrogel production (Díaz et al. 2014). No specific metabolic pathway has been identified to date for the production of 1,4-BDO.

TABLE 12.2

Industrial Applications of Short-Chain Diols

Diols	Applications	References
1,3-Propanediol	Precursor for polyester (PTT), polyether and polyurethane synthesis	Biebl et al. (1999)
	Used in solvents, adhesives, resins, detergents and cosmetics	
1,2-Propanediol	Non-toxic replacement of ethylene glycol in automobiles	Zeng and Sabra (2011)
2,3-Butanediol	Liquid fuel	Adlakha et al. (2015)
	Rubber	
	Flavouring agent	
1,4-Butanediol	Polybutylene terephthalate synthesis	Haas et al. (2005) and
	Pyrrolidones and solvent	
	Manufacture of polytetramethylene ether glycol – a component of spandex fibres	
2,4-Pentanediol	Chiral synthon for naturally occurring bioactive substance	Matsumura et al. (1994)

Thus, either different solvents serve as the major source of 1,4-BDO, or genetically modified microorganisms serve as a cell factory for 1,4-BDO biosynthesis using commercial sugar as feed (Yim et al. 2011). 2,3-Butanediol is widely acknowledged as the most promising chemical owing to its prospective applications in aeronautical, food and fuel industries (Sabra et al. 2015), as a starter material for the manufacture of gamma-butyrolactone, methyl ethyl ketone, 1,3-butadiene, etc. (Li et al. 2015). The annual production of 2,3-BDO has been increased up to 4%–7% to meet its rising market demand (Li et al. 2012; 2013). Because of expensive chemical synthesis and limitations in obtaining the downstream derivatives, synthetic 2,3-BDO could not achieve high market demand. In this context, biologically synthesized 2,3-BDO is gaining attention due to its potential to meet the rising market demand. Various applications of diols as platform chemical for the production of solvents, adhesives, resins, detergents, cosmetics, polyester resins for film, etc., have enlisted diols as one of the industrially important biochemicals (Table 12.2).

Surveying the importance of all the diols, this chapter focuses on the different steps involved in the biological production of 2,3-butanediol from lignocellulosic biomass.

12.4 PRODUCTION OF 2,3-BUTANEDIOL FROM LIGNOCELLULOSIC BIOMASS – A BRIEF

The native form of lignocellulose is not accessible to enzyme for hydrolysis due to its complex structure. Pretreatment is the process by which this biopolymer can be converted into simplified form so that it can easily be hydrolysed by the enzyme(s) (Hazeena et al. 2020). Major challenges of disrupting the recalcitrant structure of lignocellulose are irregular structural sequence of lignin, low degradability of hemicellulose and highly crystalline structure of cellulose. Thus, the major goal of the pretreatment process is efficient separation of cellulose, hemicellulose and lignin from each other through breakage of the intermolecular cross-linkages. 2,3-BDO production from lignocellulosic biomass involves multiple steps: pretreatment of biomass followed by enzymatic hydrolysis of complex sugar, microbial conversion of simple sugar(s) to 2,3-BDO and subsequent downstream extraction of 2,3-BDO from the fermentation broth.

12.4.1 FEEDSTOCK USED FOR 2,3-BDO PRODUCTION

Feedstock accounts for the major cost of the biological production of diol. Glucose has widely been used for the fermentation process, but the high cost of glucose limits its use in the large-scale production of 2,3-BDO from the same. Thus, using traditional starch or sugar for 2,3-BDO production in the industrial scale is not economically feasible. Considering this, recently, sugars obtained from low-cost biomass are being explored for the production of diol. As an alternate, glycerol, the major by-product of biodiesel and bioethanol industry, can be used as a potential substrate for diol synthesis (Jiang et al. 2014). Various types of substrates employed for the synthesis of 2,3-butanediol were thoroughly reviewed by Ji et al. (2011).

Cellulosic feedstocks are commonly described as lignocellulose-derived substrates. Lignocellulose is a potential renewable source for the production of value-added chemicals and biofuels and has immense potential for the reduction in greenhouse gas emission (Naik et al. 2010). Different lignocellulose biomass sources including agricultural residues such as sugarcane bagasse, corncob, kenaf core powder have been explored for the production of 2,3-butanediol (Table 12.3).

The native form of lignocellulosic biomass is not accessible to enzyme(s) for hydrolysis due to its complex structure. Therefore, lignocellulose biomass needs pretreatment for further use as feed for diol production.

12.4.2 POTENTIAL MICROORGANISM

Fermentative production of 2,3-BDO has intensively been investigated with diverse groups of bacteria. The industrially important bacterial species for 2,3-BDO production are *Klebsiella pneumoniae, Bacillus polymyxa, Bacillus subtilis, Pseudomonas hydrophila, Enterobacter, Bacillus* and *Serratia* (Ji et al. 2011). Several investigations (Table 12.4) have been carried out using native producers, such as *Klebsiella* sps., *Serratia marcescens, Enterobacter aerogenes, L. lactis, Lactobacillus brevis, Pediococcus pentosaceus, Raoultella terrigena, Raoultella planticola, Bacillus licheniformis, Bacillus amyloliquefaciens* and *Aerobacter indologenes* (Jiang et al. 2014). Though several microbes have been reported to produce 2,3-butanediol with a significant yield efficiency, only few are known to produce 2,3-BDO with a high titre and high yield efficiency. Few of the yeast strains and marine algal strains have also been reported for 2,3-butanediol production and have been explored for industrial-scale production. This is mainly due to the broad-spectrum substrate utilization efficiency of these host organisms (Yang et al. 2011). The microbial conversion of biomass into 2,3-BDO is an environmentally sustainable process.

TABLE 12.3

Production of 2,3-BDO from Various Cellulosic Feedstocks

Cellulosic Biomass	Microorganism	BDO Conc. (g/L)	Yield	References
Wheat straw	Paenibacillus polymyxa DSM 365	32.5	0.33	Okonkwo et al. (2021)
Empty palm fruit bunch	Genetically modified Escherichia coli	11	0.48 g/g	Sathesh-Prabu et al. (2020)
Non-detoxified corncob hydrolysate	Enterobacter cloacae M22	24.32	–	Wu et al. (2019)
Soy hull (Acid hydrolysate) (Enzyme hydrolysate)	K. pneumoniae	21.9±1.9 20.1±0.3	0.40 g/g 0.50 g/g	Cortivo et al. (2019)
Kenaf core powder	K. pneumoniae KMK05	10.42	0.385 g/g	Saratale et al. (2018)
Corn stover	Paenibacillus polymyxa ATCC 12321	18.80	0.313 g/g	Ma et al. (2018)
Sugarcane bagasse	Metabolically engineered Enterobacter aerogenes	–	0.395 g/g	Um et al. (2017)
Corncob	E. cloacae CICC 10011	52.5	0.42 g/g	Ling et al. (2017)
Oil palm frond	E. Cloacae SG1	7.67	25.56%	Hazeena et al. (2016)
Yellow poplar	Enterobacter aerogenes KCTC 2190	14.27	79.5%	Joo et al. (2016)
Larix		12.44	68.6%	
Rice hull		10.24	59.12%	
Rice waste biomass	K. pneumoniae KMK-05	11.44±0.55	0.381±0.015 g/g	Saratale et al. (2016)
Rice straw	Klebsiella sp. Zmd30	24.6	62%	Wong et al. (2012)
Jatropha hulls	K. oxytoca	31.41 (2,3-BDO + acetoin)	80.4%	Jiang et al. (2012)
Water hyacinth	Klebsiella oxytoca NRRL B 199	7.5	15 g/100 g delignified water hyacinth	Motwani et al. (1993)

12.4.3 FERMENTATION STRATEGIES

Fermentation process is one of the most important and complicated aspects for the product formation in industrial scale. It is the strategy of translating the process from a small-scale to large-scale production. Hence, it is also extremely important to understand the underlying challenges for scale-up as the microbes rarely behave in the same way in a large-scale fermentor as they behave in small laboratory scale. Such problems arise because mixing and aeration can easily be accomplished in the small-scale fermentation rather than in the large industrial fermentor (Wang et al.

TABLE 12.4

Different Microorganisms Reported for the Production of 2,3-Butanediol

Microorganism	Substrate and Conditions	Method	BDO Conc. (g/L)	BDO productivity [g/(L h)]	Product Yield (g/g)	References
Klebsiella oxytoca M3	pH 7, 30°C, 200 rpm, glycerol 35–40 g/L 160 h	Fed-batch	131.5	–	–	Cho et al. (2015)
Klebsiella pneumoniae DSM 2026	pH 7, 37°C, 200 rpm, glycerol 40 g/L, 32 h	Fed-batch anaerobic	5.3	–	–	Rahman et al. (2015)
Bacillus amyloliquefaciens B10-127	pH 6.5, 37°C, 180 rpm, 20–50 g/L, 88 h	Fed batch aerobic	102	–	–	Yang et al. (2015)
Enterobacter cloacae	Corn stover hydrolysate	Fed-batch	119.4	2.3	0.475	Li et al. (2015)
Enterobacter cloacae	Corncob residue	Fed-batch SSF	43.2	0.55	0.42	Zhang et al. (2016)
Klebsiella oxytoca	Cellulose hydrolysate	Batch	32.4	0.54	0.41	Jiang et al. (2013)
Clostridium acetobutylicum	Glucose/syngas	Fed batch	17	–	0.27	Köpke et al. (2011)
Klebsiella oxytoca	Corncob hydrolysate	Fed-batch	35.7	0.59	0.50	Cheng et al. (2010)
Klebsiella oxytoca	Glucose	Fed-batch	131.4	1.64	0.48	Ji et al. (2010)
K. variicola SRP3	pH 5, 35°C, 200 rpm, glycerol 50 g/L, 96 h	Batch aerobic	25.2 29.8	–	–	Petrov and Petrova (2010)
Klebsiella pneumoniae	Corncob molasses	Fed-batch	82.5	1.35	0.41	Wang et al. (2010)
Serratia marcescens	Sucrose	Fed-batch	152	2.67	0.46	Zhang et al. (2010)
Klebsiella pneumoniae G31	pH 8, 37°C, 200 rpm, glycerol 30 g/L, 280 h	Fed-batch	49.2	–	–	Petrov and Petrova (2009)
Klebsiella pneumoniae G31	37°C, 200 rpm, glycerol 30 g/L 150 h	Fed-batch	70	–	–	Cheng et al. (2004)
Klebsiella oxytoca	Corncob cellulose	Batch SSF	25.0	0.36	0.31	Cao et al. (1997)
Klebsiella pneumoniae	Wood hydrolysate	Batch	13.6	0.28	0.29	Grover et al. (1990)
Klebsiella oxytoca	Xylose	Batch	29.6	1.35	0.295	Jansen et al. (1984)
Bacillus polymyxa	Xylose	Batch	4.2	–	–	Laube et al. (1984)

2016). Oxygen transfer and mixing are the most important parameters that impact the metabolic pathway leading to 2,3-BDO production (Zeng et al. 1990). It depends more on the surface exposed than on the bioreactor volume. Oxygen transfer is much difficult to obtain in a large-scale production system, because 2,3-BDO formation is dependent on oxygen transfer rate which can reduce the metabolite production during the fermentation. Fermentation process parameters are sequentially optimized for batch, fed-batch and continuous-mode processes (Perego et al. 2003; Jiayang et al. 2006). Batch fermentation involves close culture system containing initial limited amount of nutrients. The starter culture microorganisms pass through a number of phases, such as lag phase, log phase, stationary phase and death phase. Fed-batch fermentation is a batch process, which is fed continuously or sequentially with the growth medium and/or feed (Syu 2001). In this process, the culture is regularly maintained under active mode. This process is implemented for the large-scale production of the product. The development of appropriate fermentation strategy depends largely on the operating parameters, and thus, it is essential to optimize the operating parameters to enhance the process efficiency to get high titre and yield efficiency.

12.4.4 Effect of Substrate Concentration

Most of the 2,3-BDO fermentation studies are being explored with 5%–10% substrate (sugar) concentrations (Garg and Jain 1995). 2,3-BDO yield and production rate mainly depend on the type of substrate and its concentration. It has been reported that with increased substrate concentrations, the toxicity level also increases, resulting in poor substrate utilization by the host organism (Jansen and Tsao 1983). Hence, in industrial-scale fermentations, substrates are frequently diluted to lower the sugar concentrations in the fermentation broth (Voloch et al. 1985). *E. aerogenes,* a facultative anaerobe, has intensively been explored for 2,3-BDO production with varying glucose concentrations ranging from 9 to 72 g/L (Converti et al. 2003). The maximum 2,3-BDO titre was observed with 3.5% of sugar. 2,3-BDO yield efficiency was high with lower sugar concentration. The specific growth rate of *K. oxytoca* has been found to decrease with an increase in the xylose concentration (Jansen et al. 1984). The maximum 2,3-BDO titre was observed with 10% xylose. *K. pneumoniae* has also been reported to utilize xylose for 2,3-BDO production (Jansen and Tsao 1983). Research studies carried out with *B. amyloliquefaciens* revealed a maximum 2,3-BDO productivity with 12% glucose (120 g/L). This study has shown that with lower feed concentration, the fermentation rate was faster. Similar results were observed with *B. polymyxa* when glucose was used as the substrate (Laube et al. 1984). 2,-3-BDO productivity of *B. licheniformis* was maximum with 2% glucose-based fermentation broth supplemented with 1% peptone and 1% beef extract. 2,3-BDO yield was 94% (Nilegaonkar et al. 1992). A thermophilic *B. licheniformis* strain has been explored with different concentrations of glucose, 64–180 g/L. With high glucose concentrations (>152 g/L), feed conversion rate got inhibited and 2,3-BDO production was not observed (Li et al. 2013). A maximum 2,3-BDO titre of *B. licheniformis DSM 8785* was observed with a high sugar concentration (18%) when fermentation was carried out in a 3.5–L bioreactor at 30°C, 400 rpm with an aeration rate of 1.2 L/min (Jurchescu et al. 2013).

12.4.5 Downstream Processing

Cost is one of the major challenges for the fermentative production of 2,3-BDO in large scale, and this is attributed mainly to the expensive downstream purification of 2,3-BDO (Haider et al. 2020). Due to the high boiling point (180°C) at atmospheric pressure, 2,3-BDO does not form azeotrope (Harvianto et al. 2018) and thus it can be purified by the traditional distillation process. However, the drawback of this process is the requirement of high energy input. In contrast to this, different energy-efficient separation processes such as liquid–liquid extraction, aqueous two-phase extraction and membrane-based separation approaches have been developed. Advanced membrane technology has shown promising results and is considered significant due to its ease of integration with other processes in a biorefinery approach. The primary

TABLE 12.5
Recovery of 2,3-Butanediol by Different Downstream Processing Methods

Method	Materials	2,3-Butanediol Recovery	References
Salting-out extraction	Ionic liquid [C$_2$mim] [CF$_3$SO$_3$] 25% and K$_2$HPO$_4$ 30%	95.7% recovery at top phase	Dai et al. (2018)
Extraction-assisted distillation (simulation study)	Isobutanol and 1-butanol	99 wt.% recovery	Haider et al. (2018)
Alcohol precipitation and vacuum distillation	Isopropanol	76.2% recovery and 96.1% purity	Jeon et al. (2014)
Nanofiltration and reverse osmosis	Polyamide membrane Cellulose acetate membrane	5.62 g/L in retentate 5.58 g/L in retentate	Davey et al. (2016)
Sugaring out	t-Butanol/glucose/water	76.3% recovery at top phase	Dai et al. (2015)
Pervaporation	Polydimethylsiloxane membrane	70.6% (w/w) concentration in third recycled permeate	Shao and Kumar (2011)
Solvent extraction and salting out	20% K$_2$HPO$_4$/19% ethanol (mass fraction)	72.2 g/L obtained at top phase formed in 400 g scale operation	Jianying et al. (2011)
Aqueous two-phase extraction	32% ethanol and 16% ammonium sulphate	91.7% recovery	Li et al. (2010)
Aqueous two-phase extraction	24% w/w ethanol and 25% w/w K$_2$HPO$_4$	98.13% recovery	Jiang et al. (2009)
Liquid–liquid extraction	Oleyl alcohol	68% recovery. Concentration increased from 17.9 g/L to 23.01 g/L	Anvari and Khayati (2009)
Aqueous two-phase extraction	2-Propanol/ammonium sulphate	93.7% recovery	Sun et al. (2009)
Extraction and pervaporation	1-Butanol, polydimethylsiloxane membrane	98 wt.% purity	Shao and Kumar (2009)
Vacuum membrane distillation	Polytetrafluoroethylene (PTFE) membrane	Concentrated from 40 g/L to about 650 g/L	Qureshi et al. (1994)

benefits include energy efficiency, limited consumption of hazardous chemicals, eco-friendly feature, reusability and separation efficiency (He et al. 2012; Curcio et al. 2016; Saha et al. 2017; Lipnizki et al. 2020). Table 12.5 illustrates different downstream processes employed for the recovery of 2,3-BDO from the fermentation broth.

REFERENCES

Adlakha, N., Pfau, T., Ebenhöh, O. and Yazdani, S.S., 2015. Insight into metabolic pathways of the potential biofuel producer, *Paenibacillus polymyxa* ICGEB2008. *Biotechnology for Biofuels*, *8*(1), pp.1–10.

Afschar, A.S., Rossell, C.V., Jonas, R., Chanto, A.Q. and Schaller, K., 1993. Microbial production and downstream processing of 2, 3-butanediol. *Journal of Biotechnology*, *27*(3), pp.317–329.

Alper, K.R., Lotsof, H.S., Frenken, G.M., Luciano, D.J. and Bastiaans, J., 1999. Treatment of acute opioid withdrawal with ibogaine. *American Journal on Addictions*, *8*(3), pp.234–242.

Altaras, N.E. and Cameron, D.C., 1999. Metabolic engineering of a 1, 2-propanediol pathway in Escherichia coli. *Applied and Environmental Microbiology*, *65*(3), pp.1180–1185.

Altaras, N.E., Etzel, M.R. and Cameron, D.C., 2001. Conversion of sugars to 1, 2-propanediol by Thermoanaerobacterium thermosaccharolyticum HG-8. *Biotechnology Progress*, *17*(1), pp.52–56.

Anvari, M. and Khayati, G., 2009. In situ recovery of 2, 3-butanediol from fermentation by liquid–liquid extraction. *Journal of Industrial Microbiology and Biotechnology*, *36*(2), pp.313–317.

Bennett, G.N. and San, K.Y., 2001. Microbial formation, biotechnological production and applications of 1, 2-propanediol. *Applied microbiology and biotechnology*, *55*(1), pp.1–9.

Biebl, H., Menzel, K., Zeng, A.P. and Deckwer, W.D., 1999. Microbial production of 1, 3-propanediol. *Applied Microbiology and Biotechnology*, *52*(3), pp.289–297.

Boenigk, R., Bowien, S., Gottschalk, G., 1993. Fermentation of glycerol to 1,3-propanediol in continuous cultures of *Citrobacter freundii*. *Applied Microbiology Biotechnology*, 38, pp.453–457.

Cameron, D., Cooney, C., 1986. A Novel Fermentation: The production of R(–)–1,2–propanediol and acetol by clostridium thermosaccharolyticum. *Nature Biotechnology*, 4, pp.651–654.

Cao, N., Xia, Y., Gong, C.S. and Tsao, G.T., 1997. Production of 2, 3-butanediol from pretreated corn cob by *Klebsiella oxytoca* in the presence of fungal cellulase. *Applied Biochemistry and Biotechnology*, pp. 63–65 (pp. 129–139). Humana Press Inc.

Chatzifragkou, A., Dietz, D., Komaitis, M., Zeng, A.P. and Papanikolaou, S., 2010. Effect of biodiesel-derived waste glycerol impurities on biomass and 1, 3-propanediol production of *Clostridium butyricum* VPI 1718. *Biotechnology and Bioengineering*, *107*(1), pp.76–84.

Cheng, K.K., Liu, D.H., Sun, Y. and Liu, W.B., 2004. 1, 3-Propanediol production by *Klebsiella pneumoniae* under different aeration strategies. *Biotechnology Letters*, 26 (11), pp.911–915.

Cheng, K.K., Liu, Q., Zhang, J.A., Li, J.P., Xu, J.M. and Wang, G.H., 2010. Improved 2, 3-butanediol production from corncob acid hydrolysate by fed-batch fermentation using *Klebsiella oxytoca*. *Process Biochemistry*, *45*(4), pp.613–616.

Cho, S., Kim, T., Woo, H.M., Kim, Y., Lee, J. and Um, Y., 2015. High production of 2, 3-butanediol from biodiesel-derived crude glycerol by metabolically engineered *Klebsiella oxytoca* M1. *Biotechnology for Biofuels*, *8*(1), pp.1–12.

Clomburg, J.M. and Gonzalez, R., 2011. Metabolic engineering of Escherichia coli for the production of 1, 2-propanediol from glycerol. *Biotechnology and Bioengineering, 108* (4), pp.867–879.

Converti, A., Perego, P. and Del Borghi, M., 2003. Effect of specific oxygen uptake rate on Enterobacter aerogenes energetics: Carbon and reduction degree balances in batch cultivations. *Biotechnology and Bioengineering, 82*(3), pp.370–377.

Cortivo, P.R., Machado, J., Hickert, L.R., Rossi, D.M. and Ayub, M.A., 2019. Production of 2, 3-butanediol by *Klebsiella pneumoniae* BLh-1 and *Pantoea agglomerans* BL1 cultivated in acid and enzymatic hydrolysates of soybean hull. *Biotechnology Progress, 35*(3), p.e2793.

Cortright, R.D., Sanchez-Castillo, M. and Dumesic, J.A., 2002. Conversion of biomass to 1, 2-propanediol by selective catalytic hydrogenation of lactic acid over silica-supported copper. *Applied Catalysis B: Environmental, 39*(4), pp.353–359.

Curcio, S., De Luca, G., Saha, K. and Chakraborty, S., 2016. Advance membrane separation processes for biorefineries. In *Membrane Technologies for Biorefining* (pp. 3–28). Woodhead Publishing.

Dabrock, B., Bahl, H. and Gottschalk, G., 1992. Parameters affecting solvent production by *Clostridium pasteurianum*. *Applied and Environmental Microbiology, 58*(4), pp.1233–1239.

Dai, J., Wang, H., Li, Y. and Xiu, Z.L., 2018. Imidazolium ionic liquids-based salting-out extraction of 2, 3-butanediol from fermentation broths. *Process Biochemistry, 71*, pp.175–181.

Dai, J.Y., Liu, C.J. and Xiu, Z.L., 2015. Sugaring-out extraction of 2, 3-butanediol from fermentation broths. *Process Biochemistry, 50*(11), pp.1951–1957.

Davey, C.J., Havill, A., Leak, D. and Patterson, D.A., 2016. Nanofiltration and reverse osmosis membranes for purification and concentration of a 2, 3-butanediol producing gas fermentation broth. *Journal of Membrane Science, 518*, pp.150–158.

Díaz, A., Katsarava, R. and Puiggalí, J., 2014. Synthesis, properties and applications of biodegradable polymers derived from diols and dicarboxylic acids: from polyesters to poly (ester amide) s. *International Journal of Molecular Sciences, 15*(5), pp.7064–7123.

Forte, A., Zucaro, A., Basosi, R. and Fierro, A., 2016. LCA of 1, 4-butanediol produced via direct fermentation of sugars from wheat straw feedstock within a territorial biorefinery. *Materials, 9*(7), p.563.

Freund, A., 1881. About the formation and preparation of trimethylene alcohol from glycerine. *Monthly Books for Chemistry and Related Parts of Other Sciences, 2*(1), pp.636–641.

Garg, S.K. and Jain, A., 1995. Fermentative production of 2, 3-butanediol: A review. *Bioresource Technology, 51*(2–3), pp.103–109.

Goodlove, P.E., Cunningham, P.R., Parker, J. and Clark, D.P., 1989. Cloning and sequence analysis of the fermentative alcohol-dehydrogenase-encoding gene of Escherichia coli. *Gene, 85*(1), pp.209–214.

Grover, B.P., Garg, S.K. and Verma, J., 1990. Production of 2, 3-butanediol from wood hydrolysate by *Klebsiella pneumoniae*. *World Journal of Microbiology and Biotechnology, 6*(3), pp.328–332.

Haas, T., Jaeger, B., Weber, R., Mitchell, S.F. and King, C.F., 2005. New diol processes: 1, 3-propanediol and 1, 4-butanediol. *Applied Catalysis A: General, 280*(1), pp.83–88.

Haider, J., Harvianto, G.R., Qyyum, M.A. and Lee, M., 2018. Cost-and energy-efficient butanol-based extraction-assisted distillation designs for purification of 2, 3-butanediol for use as a drop-in fuel. *ACS Sustainable Chemistry & Engineering, 6*(11), pp.14901–14910.

Haider, J., Qyyum, M.A. and Lee, M., 2020. Purification step enhancement of the 2, 3-butanediol production process through minimization of high pressure steam consumption. *Chemical Engineering Research and Design, 153*, pp.697–708.

342   Biomass for Bioenergy and Biomaterials

Harmsen, P.F.H., Huijgen, W.J.J., Bermudez Lopez, L.M. and Bakker, R.R.C., 2010. *Literature Review of Physical and Chemical Pretreatment Processes for Lignocellulosic Biomass.* Wageningen, The Netherlands: Wagenigen UR Food and Biobased Research; 1, Biosynergy Report No. 1184.

Harvianto, G.R., Haider, J., Hong, J., Long, N.V.D., Shim, J.J., Cho, M.H., Kim, W.K. and Lee, M., 2018. Purification of 2, 3-butanediol from fermentation broth: Process development and techno-economic analysis. *Biotechnology for Biofuels, 11*(1), pp.1–16.

Hazeena, S.H., Pandey, A. and Binod, P., 2016. Evaluation of oil palm front hydrolysate as a novel substrate for 2, 3-butanediol production using a novel isolate *Enterobacter cloacae* SG1. *Renewable Energy, 98*, pp.216–220.

Hazeena, S.H., Sindhu, R., Pandey, A. and Binod, P., 2020. Lignocellulosic bio-refinery approach for microbial 2, 3-Butanediol production. *Bioresource Technology, 302*, p.122873.

He, Y., Bagley, D.M., Leung, K.T., Liss, S.N. and Liao, B.Q., 2012. Recent advances in membrane technologies for biorefining and bioenergy production. *Biotechnology Advances, 30*(4), pp.817–858.

Hespell, R.B., 1996. Fermentation of xylan, corn fiber, or sugars to acetoin and butanediol by *Bacillus polymyxa* strains. *Current Microbiology, 32*(5), pp.291–296.

Himmi, E, L., Bories, A., Barbirato, F., 1999. Nutrient requirements for glycerol conversion to 1,3 -propanediol by *Clostridium butyricum. Bioresource Technology, 67*, pp.123-128.

Homann, T., Tag, C., Biebl, H., Deckwer, W, D., Schink, B., 1990. Fermentation of glycerol to 1,3-propanediol by klebsiella and Citrobacter strains. *Applied Microbiology Biotechnology, 33*, pp.121–126.

Hwang, H.J., Park, J.H., Kim, J.H., Kong, M.K., Kim, J.W., Park, J.W., Cho, K.M. and Lee, P.C., 2014. Engineering of a butyraldehyde dehydrogenase of *Clostridium saccharoperbutylacetonicum* to fit an engineered 1, 4-butanediol pathway in *Escherichia coli. Biotechnology and Bioengineering, 111*(7), pp.1374–1384.

Jansen, N.B., Flickinger, M.C. and Tsao, G.T., 1984. Production of 2, 3-butanediol from D-xylose by *Klebsiella oxytoca* ATCC 8724. *Biotechnology and Bioengineering, 26*(4), pp.362–369.

Jansen, N.B. and Tsao, G.T., 1983. Bioconversion of pentoses to 2, 3-butanediol by Klebsiella pneumoniae. *Advances in Biochemical Engineering/Biotechnology, 27*, pp.85–99.

Jeon, E., Lee, S., Kim, D., Yoon, H., Oh, M., Park, C., Lee, J., 2009. Development of a Saccharomyces cerevisiae strain for the production of 1,2-propanediol by gene manipulation. *Enzyme and Microbial Technology, 45*(1), pp.42-47.

Jeon, S., Kim, D.K., Song, H., Lee, H.J., Park, S., Seung, D. and Chang, Y.K., 2014. 2, 3-Butanediol recovery from fermentation broth by alcohol precipitation and vacuum distillation. *Journal of Bioscience and Bioengineering, 117*(4), pp.464–470.

Ji, X.J., Huang, H. and Ouyang, P.K., 2011. Microbial 2, 3-butanediol production: A state-of-the-art review. *Biotechnology Advances, 29*(3), pp.351–364.

Ji, X.J., Huang, H., Zhu, J.G., Ren, L.J., Nie, Z.K., Du, J. and Li, S., 2010. Engineering *Klebsiella oxytoca* for efficient 2, 3-butanediol production through insertional inactivation of acetaldehyde dehydrogenase gene. *Applied Microbiology and Biotechnology, 85*(6), pp.1751–1758.

Jia, W., Jain, R., Shen, X., Sun, X., Cheng, M., Liao, J.C., Yuan, Q., Yan, Y., 2017. Rational engineering of diol dehydratase enables 1,4-butanediol biosynthesis from xylose. *Metabolic Engineering, 40*, pp.148–156.

Jiang, B., Li, Z.G., Dai, J.Y., Zhang, D.J. and Xiu, Z.L., 2009. Aqueous two-phase extraction of 2, 3-butanediol from fermentation broths using an ethanol/phosphate system. *Process Biochemistry, 44*(1), pp.112–117.

Jiang, L.Q., Fang, Z., Guo, F. and Yang, L.B., 2012. Production of 2, 3-butanediol from acid hydrolysates of Jatropha hulls with *Klebsiella oxytoca. Bioresource Technology, 107*, pp.405–410.

Jiang, L.Q., Fang, Z., Li, X.K. and Luo, J., 2013. Production of 2, 3-butanediol from cellulose and Jatropha hulls after ionic liquid pretreatment and dilute-acid hydrolysis. *AMB Express*, *3*(1), pp.1–8.

Jiang, Y., Liu, W., Zou, H., Cheng, T., Tian, N. and Xian, M., 2014. Microbial production of short chain diols. *Microbial Cell Factories*, *13*(1), pp.1–17.

Jianying, D.A.I., Zhang, Y. and Zhilong, X.I.U., 2011. Salting-out extraction of 2, 3-butanediol from Jerusalem artichoke-based fermentation broth. *Chinese Journal of Chemical Engineering*, *19*(4), pp.682–686.

Jiayang, Q.I.N., Zijun, X., Cuiqing, M.A., Nengzhong, X.I.E., Peihai, L.I.U. and Ping, X.U., 2006. Production of 2, 3-butanediol by *Klebsiella pneumoniae* using glucose and ammonium phosphate. *Chinese Journal of Chemical Engineering*, *14*(1), pp.132–136.

Jonoobi, M., Harun, J., Mishra, M. and Oksman, K., 2009. Chemical composition, crystallinity and thermal degradation of bleached and unbleached kenaf bast (*Hibiscus cannabinus*) pulp and nanofiber. *BioResources*, *4*(2), pp.626–639.

Joo, J., Lee, S.J., Yoo, H.Y., Kim, Y., Jang, M., Lee, J., Han, S.O., Kim, S.W. and Park, C., 2016. Improved fermentation of lignocellulosic hydrolysates to 2, 3-butanediol through investigation of effects of inhibitory compounds by *Enterobacter aerogenes*. *Chemical Engineering Journal*, *306*, pp.916–924.

Jung, J.Y., Yun, H.S., Lee, J.W. and Oh, M.K., 2011. Production of 1, 2-propanediol from glycerol in *Saccharomyces cerevisiae*. *Journal of Microbiology and Biotechnology*, *21*(8), pp.846–853.

Jung, M.Y., Jung, H.M., Lee, J. and Oh, M.K., 2015. Alleviation of carbon catabolite repression in *Enterobacter aerogenes* for efficient utilization of sugarcane molasses for 2, 3-butanediol production. *Biotechnology for Biofuels*, *8*(1), pp.1–12.

Jurchescu, I.M., Hamann, J., Zhou, X., Ortmann, T., Kuenz, A., Prüße, U. and Lang, S., 2013. Enhanced 2, 3-butanediol production in fed-batch cultures of free and immobilized *Bacillus licheniformis* DSM 8785. *Applied Microbiology and Biotechnology*, *97*(15), pp.6715–6723.

Kalia, S., Dufresne, A., Cherian, B.M., Kaith, B.S., Avérous, L., Njuguna, J. and Nassiopoulos, E., 2011. Cellulose-based bio-and nanocomposites: A review. *International Journal of Polymer Science*, *2011*, pp.1–35.

Köpke, M., Mihalcea, C., Liew, F., Tizard, J.H., Ali, M.S., Conolly, J.J., Al-Sinawi, B. and Simpson, S.D., 2011. 2, 3-Butanediol production by acetogenic bacteria, an alternative route to chemical synthesis, using industrial waste gas. *Applied and Environmental Microbiology*, *77*(15), pp.5467–5475.

Laube, V.M., Groleau, D. and Martin, S.M., 1984. 2, 3-Butanediol production from xylose and other hemicellulosic components by *Bacillus polymyxa*. *Biotechnology Letters*, *6*(4), pp.257–262.

Lee, J, H., Jung, M, Y., Oh, M, K., 2018. High-yield production of 1,3-propanediol from glycerol by metabolically engineered *Klebsiella pneumoniae*. *Biotechnology Biofuels*, 11, pp.104.

Lee, J., Na, D., Park, J., Lee, J., Choi, S., Lee, S, Y., 2012. Systems metabolic engineering of microorganisms for natural and non-natural chemicals. *Nature Chemical Biology*, 8, pp.536–546.

Levene, P, A., Walti, A., 1943. 1-Propylene glycol. In: Blatt A H, editor. *Organic Syntheses Collective*. 2. J (pp.545–547). New York: Wiley & Sons, Inc.

Li, L., Li, K., Wang, Y., Chen, C., Xu, Y., Zhang, L., Han, B., Gao, C., Tao, F., Ma, C. and Xu, P., 2015. Metabolic engineering of *Enterobacter cloacae* for high-yield production of enantiopure (2R, 3R)-2, 3-butanediol from lignocellulose-derived sugars. *Metabolic Engineering*, *28*, pp.19–27.

Li, L., Su, F., Wang, Y., Zhang, L., Liu, C., Li, J., Ma, C. and Xu, P., 2012. Genome sequences of two thermophilic *Bacillus licheniformis* strains, efficient producers of platform chemical 2, 3-butanediol. *Journal of Bacteriology*, *194*(15), pp.4133–4134.

Li, Z., Teng, H. and Xiu, Z., 2010. Aqueous two-phase extraction of 2, 3-butanediol from fermentation broths using an ethanol/ammonium sulfate system. *Process Biochemistry*, 45(5), pp.731–737.

Li, L., Zhang, L., Li, K., Wang, Y., Gao, C., Han, B., Ma, C. and Xu, P., 2013. A newly isolated *Bacillus licheniformis* strain thermophilically produces 2, 3-butanediol, a platform and fuel bio-chemical. *Biotechnology for Biofuels*, 6(1), pp.1–13.

Ling, H., Cheng, K., Ge, J. and Ping, W., 2017. Corncob mild alkaline pretreatment for high 2, 3-butanediol production by spent liquor recycle process. *BioEnergy Research*, 10(2), pp.566–574.

Lipnizki, F., Thuvander, J. and Rudolph, G., 2020. Membrane processes and applications for biorefineries. In: Figoli, A., Li, Y., and Basile, A. (eds.) *Current Trends and Future Developments on (Bio-) Membranes* (pp. 283–301). Elsevier.

Liu, H. and Lu, T., 2015. Autonomous production of 1, 4-butanediol via a de novo biosynthesis pathway in engineered *Escherichia coli*. *Metabolic Engineering*, 29, pp.135–141.

Ma, C., Wang, A., Qin, J., Li, L., Ai, X., Jiang, T., Tang, H. and Xu, P., 2009. Enhanced 2, 3-butanediol production by *Klebsiella pneumoniae* SDM. *Applied Microbiology and Biotechnology*, 82(1), pp.49–57.

Ma, K., He, M., You, H., Pan, L., Wang, Z., Wang, Y., Hu, G., Cui, Y. and Maeda, T., 2018. Improvement of (R, R)-2, 3-butanediol production from corn stover hydrolysate by cell recycling continuous fermentation. *Chemical Engineering Journal*, 332, pp.361–369.

Matsumura, S., Kawai, Y., Takahashi, Y. and Toshima, K., 1994. Microbial production of (2 R, 4 R)-2, 4-pentanediol by enatioselective reduction of acetylacetone and stereoinversion of 2, 4-pentanediol. *Biotechnology Letters*, 16(5), pp.485–490.

Meng, W., Zhang, Y., Cao, M., Zhang, W., Lü, C., Yang, C., Gao, C., Xu, P. and Ma, C., 2020. Efficient 2, 3-butanediol production from whey powder using metabolically engineered. *Klebsiella Oxytoca. Microbial Cell Factories*, 19(1), pp.1–10.

Mosier, N., Wyman, C., Dale, B., Elander, R., Lee, Y.Y., Holtzapple, M. and Ladisch, M., 2005. Features of promising technologies for pretreatment of lignocellulosic biomass. *Bioresource Technology*, 96(6), pp.673–686.

Motwani, M., Seth, R., Daginawala, H.F. and Khanna, P., 1993. Microbial production of 2, 3-butanediol from water hyacinth. *Bioresource Technology*, 44(3), pp.187–195.

Naik, S.N., Goud, V.V., Rout, P.K. and Dalai, A.K., 2010. Production of first and second generation biofuels: A comprehensive review. *Renewable and Sustainable Energy Reviews*, 14(2), pp.578–597.

Nakamura, C.E. and Whited, G.M., 2003. Metabolic engineering for the microbial production of 1, 3-propanediol. *Current Opinion in Biotechnology*, 14(5), pp.454–459.

Nakashimada, Y., Marwoto, B., Kashiwamura, T., Kakizono, T., Nishio, N., 2000. Enhanced 2,3-butanediol production by addition of acetic acid in *Paenibacillus polymyxa*. *Bioscience Bioengineering*, 90(6), pp.661–664.

Nilegaonkar, S., Bhosale, S.B., Kshirsagar, D.C. and Kapadi, A.H., 1992. Production of 2, 3-butanediol from glucose by *Bacillus licheniformis*. *World Journal of Microbiology and Biotechnology*, 8(4), pp.378–381.

Norgren, M. and Edlund, H., 2014. Lignin: Recent advances and emerging applications. *Current Opinion in Colloid & Interface Science*, 19(5), pp.409–416.

Ojeda, T.F., Dalmolin, E., Forte, M.M., Jacques, R.J., Bento, F.M. and Camargo, F.A., 2009. Abiotic and biotic degradation of oxo-biodegradable polyethylenes. *Polymer Degradation and Stability*, 94(6), pp.965–970.

Okonkwo, C.C., Ujor, V. and Ezeji, T.C., 2021. Production of 2, 3-Butanediol from non-detoxified wheat straw hydrolysate: Impact of microbial inhibitors on *Paenibacillus polymyxa* DSM 365. *Industrial Crops and Products*, 159, p.113047.

Paul, B., 2010. Enhanced pyruvate to 2,3-butanediol conversion in lactic acid bacteria.

Perego, P., Converti, A. and Del Borghi, M., 2003. Effects of temperature, inoculum size and starch hydrolyzate concentration on butanediol production by *Bacillus licheniformis*. *Bioresource Technology*, 89(2), pp.125–131.

Petrov, K. and Petrova, P., 2010. Enhanced production of 2, 3-butanediol from glycerol by forced pH fluctuations. *Applied Microbiology and Biotechnology*, 87(3), pp.943–949.

Petrov, K. and Petrova, P., 2009. High production of 2, 3-butanediol from glycerol by *Klebsiella pneumoniae* G31. *Applied Microbiology and Biotechnology*, 84(4), pp.659–665.

Pflügl, S., Marx, H., Mattanovich, D. and Sauer, M., 2012. 1, 3-Propanediol production from glycerol with *Lactobacillus diolivorans*. *Bioresource Technology*, 119, pp.133–140.

Qureshi, N., Meagher, M.M. and Hutkins, R.W., 1994. Recovery of 2, 3-butanediol by vacuum membrane distillation. *Separation Science and Technology*, 29(13), pp.1733–1748.

Rahman, M.S., Yuan, Z., Ma, K., Xu, C.C. and Qin, W., 2015. Aerobic conversion of glycerol to 2, 3-butanediol by a novel *Klebsiella variicola* SRP3 strain. *Journal of Microbial and Biochemistry Technology*, 7, pp.299–304.

Rode, C.V., Ghalwadkar, A.A., Mane, R.B., Hengne, A.M., Jadkar, S.T. and Biradar, N.S., 2010. Selective hydrogenolysis of glycerol to 1, 2-propanediol: Comparison of batch and continuous process operations. *Organic Process Research & Development*, 14(6), pp.1385–1392.

Roy, R., Rahman, M.S. and Raynie, D.E., 2020. Recent advances of greener pretreatment technologies of lignocellulose. *Current Research in Green and Sustainable Chemistry*, p.100035.

Sabra, W., Groeger, C. and Zeng, A.P., 2015. Microbial cell factories for diol production. In: *Advances in Biochemical Engineering/Biotechnology*, Springer-Verlag, Berlin Heidelberg, pp.165–197.

Saha, B.C., 2003. Hemicellulose bioconversion. *Journal of Industrial Microbiology and Biotechnology*, 30(5), pp.279–291.

Saha, K., Sikder, J., Chakraborty, S., da Silva, S.S. and dos Santos, J.C., 2017. Membranes as a tool to support biorefineries: Applications in enzymatic hydrolysis, fermentation and dehydration for bioethanol production. *Renewable and Sustainable Energy Reviews*, 74, pp.873–890.

Saint-Amans, S., Perlot, P., Goma, G., Soucaille, P., 1994. High production of 1,3-propanediol from glycerol by *Clostridium butyricum* in a simply controlled fed-batch system. *Biotechnology Letters*, 17, pp.211–216.

Saratale, G.D., Jung, M.Y. and Oh, M.K., 2016. Reutilization of green liquor chemicals for pretreatment of whole rice waste biomass and its application to 2, 3-butanediol production. *Bioresource Technology*, 205, pp.90–96.

Saratale, R.G., Shin, H.S., Ghodake, G.S., Kumar, G., Oh, M.K. and Saratale, G.D., 2018. Combined effect of inorganic salts with calcium peroxide pretreatment for kenaf core biomass and their utilization for 2, 3-butanediol production. *Bioresource Technology*, 258, pp.26–32.

Sathesh-Prabu, C., Kim, D. and Lee, S.K., 2020. Metabolic engineering of *Escherichia coli* for 2, 3-butanediol production from cellulosic biomass by using glucose-inducible gene expression system. *Bioresource Technology*, 309, p.123361.

Shao, P. and Kumar, A., 2011. Process energy efficiency in pervaporative and vacuum membrane distillation separation of 2, 3-butanediol. *The Canadian Journal of Chemical Engineering*, 89(5), pp.1255–1265.

Shao, P. and Kumar, A., 2009. Separation of 1-butanol/2, 3-butanediol using ZSM-5 zeolite-filled polydimethylsiloxane membranes. *Journal of Membrane Science*, 339(1–2), pp.143–150.

Sun, L.H., Jiang, B. and Xiu, Z.L., 2009. Aqueous two-phase extraction of 2, 3-butanediol from fermentation broths by isopropanol/ammonium sulfate system. *Biotechnology Letters*, 31(3), pp.371–376.

Syu, M.J., 2001. Biological production of 2, 3-butanediol. *Applied Microbiology and Biotechnology*, 55(1), pp.10–18.

Tai, Y.S., Xiong, M., Jambunathan, P., Wang, J., Wang, J., Stapleton, C. and Zhang, K., 2016. Engineering nonphosphorylative metabolism to generate lignocellulose-derived products. *Nature Chemical Biology*, 12(4), pp.247–253.

Tran-Din, K. and Gottschalk, G., 1985. Formation of D (-)-1, 2-propanediol and D (-)-lactate from glucose by *Clostridium sphenoides* under phosphate limitation. *Archives of Microbiology*, 142(1), pp.87–92.

Um, J., Kim, D.G., Jung, M.Y., Saratale, G.D. and Oh, M.K., 2017. Metabolic engineering of *Enterobacter aerogenes* for 2, 3-butanediol production from sugarcane bagasse hydrolysate. *Bioresource Technology*, 245, pp.1567–1574.

Voloch, M., Jansen, N.B., Ladisch, M.R., Tsao, G.T., Narayan, R. and Rodwell, V., 1985. 2, 3-Butanediol. In: Moo-Young, M., Blanch, H.W., Drew, S. and Wang, D.I.C. (eds.) *Comprehensive Biotechnology*. Pergamon, Oxford, pp. 933–947.

Wang, A., Wang, Y., Jiang, T., Li, L., Ma, C. and Xu, P., 2010. Production of 2, 3-butanediol from corncob molasses, a waste by-product in xylitol production. *Applied Microbiology and Biotechnology*, 87(3), pp.965–970.

Wang, H., Zhang, N., Qiu, T., Zhao, J., He, X. and Chen, B., 2014. Optimization of a continuous fermentation process producing 1, 3-propane diol with Hopf singularity and unstable operating points as constraints. *Chemical Engineering Science*, 116, pp.668–681.

Wang, Q., Chen, T., Zhao, X. and Chamu, J., 2012. Metabolic engineering of thermophilic *Bacillus licheniformis* for chiral pure D-2, 3-butanediol production. *Biotechnology and Bioengineering*, 109(7), pp.1610–1621.

Wang, R., Unrean, P. and Franzén, C.J., 2016. Model-based optimization and scale-up of multi-feed simultaneous saccharification and co-fermentation of steam pre-treated lignocellulose enables high gravity ethanol production. *Biotechnology for Biofuels*, 9(1), pp.1–13.

Werawattanachai, N., Towiwat, P., Unchern, S. and Maher, T.J., 2007. Neuropharmacological profile of tetrahydrofuran in mice. *Life Sciences*, 80(18), pp.1656–1663.

Wong, C.L., Huang, C.C., Lu, W.B., Chen, W.M. and Chang, J.S., 2012. Producing 2, 3-butanediol from agricultural waste using an indigenous *Klebsiella sp.* Zmd30 strain. *Biochemical Engineering Journal*, 69, pp.32–40.

Wu, J., Zhou, Y.J., Zhang, W., Cheng, K.K., Liu, H.J. and Zhang, J.A., 2019. Screening of a highly inhibitor-tolerant bacterial strain for 2, 3-BDO and organic acid production from non-detoxified corncob acid hydrolysate. *AMB Express*, 9(1), pp.1–9.

Wu, M.Y., Sung, L.Y., Li, H., Huang, C.H. and Hu, Y.C., 2017. Combining CRISPR and CRISPRi systems for metabolic engineering of E. coli and 1, 4-BDO biosynthesis. *ACS Synthetic Biology*, 6(12), pp.2350–2361.

Xiao, Z., Jin, S., Pang, M. and Liang, C., 2013. Conversion of highly concentrated cellulose to 1, 2-propanediol and ethylene glycol over highly efficient CuCr catalysts. *Green Chemistry*, 15(4), pp.891–895.

Yang, T., Rao, Z., Zhang, X., Lin, Q., Xia, H., Xu, Z. and Yang, S., 2011. Production of 2, 3-butanediol from glucose by GRAS microorganism *Bacillus amyloliquefaciens*. *Journal of Basic Microbiology*, 51(6), pp.650–658.

Yang, T., Rao, Z., Zhang, X., Xu, M., Xu, Z. and Yang, S.T., 2015. Enhanced 2, 3-butanediol production from biodiesel-derived glycerol by engineering of cofactor regeneration and manipulating carbon flux in *Bacillus amyloliquefaciens*. *Microbial Cell Factories*, 14(-1), pp.1–11.

Yim, H., Haselbeck, R., Niu, W., Pujol-Baxley, C., Burgard, A., Boldt, J., Khandurina, J., Trawick, J.D., Osterhout, R.E., Stephen, R. and Estadilla, J., 2011. Metabolic engineering of Escherichia coli for direct production of 1, 4-butanediol. *Nature Chemical Biology*, 7(7), pp.445–452.

Zeng, A.P., Biebl, H. and Deckwer, W.D., 1990. Effect of pH and acetic acid on growth and 2, 3-butanediol production of *Enterobacter aerogenes* in continuous culture. *Applied Microbiology and Biotechnology*, *33*(5), pp.485–489.

Zeng, A.P. and Sabra, W., 2011. Microbial production of diols as platform chemicals: Recent progresses. *Current Opinion in Biotechnology*, *22*(6), pp.749–757.

Zhang, C., Li, W., Wang, D., Guo, X., Ma, L. and Xiao, D., 2016. Production of 2, 3-butanediol by *Enterobacter cloacae* from corncob-derived xylose. *Turkish Journal of Biology*, *40*(4), pp.856–865.

Zhang, L., Yang, Y., Sun, J.A., Shen, Y., Wei, D., Zhu, J. and Chu, J., 2010. Microbial production of 2, 3-butanediol by a mutagenized strain of *Serratia marcescens* H30. *Bioresource Technology*, *101*(6), pp.1961–1967.

Zhang, Y.P., 2008. Reviving the carbohydrate economy via multi-product lignocellulose biorefineries. *Journal of Industrial Microbiology and Biotechnology*, *35*(5), pp.367–375.

13 Market Analysis of Biomass for Biofuels and Biomaterials

Brajesh Barse
International Centre for Genetic Engineering
and Biotechnology (ICGEB)

Navin Tamrakar
Innovation Consultant for Specialty Chemicals and
Advanced Composites Industry at Chemical Market Intel

Syed Shams Yazdani
International Centre for Genetic Engineering
and Biotechnology (ICGEB)

CONTENTS

13.1 Introduction 350
13.2 Commercially Available Biomass 351
    13.2.1 Classification of Agricultural Biomass......................... 351
13.3 Opportunities, Growth Drivers and Challenges in Biomass-Derived
    Products 352
    13.3.1 Affordable and Clean Energy.. 352
    13.3.2 Climate Action... 352
    13.3.3 Life Below Water.. 352
    13.3.4 Good Health and Well-Being 352
13.4 Global Biomass Supply Analysis in the Energy and Fuels Segment.......... 353
13.5 Market Analysis of Biomass-Based Fuel Production 357
13.6 Global Market Analysis of Biomaterials 359
    13.6.1 Bio-based Polymers – Supply–Demand Analysis 359
    13.6.2 Analysis of Biomass Consumption to Produce Biopolymers
    in 2019.. 360
13.7 Market and Industry Analysis of Bio-based Chemicals.............. 360
    13.7.1 Biorefinery 360
        13.7.1.1 Syngas Platform.. 361
        13.7.1.2 Biogas Platform.. 361
        13.7.1.3 C6 and C6/C5 Sugar Platform 361

DOI: 10.1201/9781003158486-13

             13.7.1.4   Selective Dehydration, Hydrogenation and

                                  Oxidation Processes 362

             13.7.1.5   Organic Solutions Platform 362

             13.7.1.6   Lignin-Based Platform 362

     13.7.2   Bio-based Chemicals Industry 362

             13.7.2.1   C1-Containing Compounds 362

             13.7.2.2   C2-Containing Compounds 363

             13.7.2.3   C3-Containing Compounds 363

             13.7.2.4   Bio-based Propylene Glycol 364

             13.7.2.5   C4-Containing Compounds 364

             13.7.2.6   C5-Containing Compounds 365

             13.7.2.7   C6-Containing Compounds 365

             13.7.2.8   Others 365

13.8   Conclusions 366

References 366

13.1   INTRODUCTION

Biomass characterises plentiful carbon-neutral renewable resource. The enhanced use of biomass addresses the climate issues and provides renewable energy and chemicals for societal needs. The biological matter of living organisms is defined as biomass (Tackling Increasing Plastic Waste, 2019). It can be categorised as agricultural biomass, forest biomass and waste streams. Biomass-based energy, chemicals and materials are becoming popular due to sustainability and availability in abundance of biomass. It also helps in curbing pollution across air, water and land. Biomass-based industries need to overcome certain challenges such as biomass storage and transportation, high capital and operating costs, and the lack availability of technology in developing countries.

In energy segment, the supply of biomass is defined in energy equivalent. Major sources of biomass for the energy industry are municipal and industrial wastes, solid biofuels, liquid biofuels and biogases. Global biomass supply for energy increased from 50.5 to 55.5 EJ with a CAGR of 1.2% from 2010 to 2018. Solid biofuels accounted for more than 85% of the total supply, which is 47.6 EJ. Historically, biogas achieved the highest CAGR of 6.1% from 2010 to 2018.

In biopolymers segment, biogenic by-products, starch, sugar, cellulose and non-edible plant oil are the sources of biomass. Globally, 5000 kilotonnes of biomass feedstock were consumed to produce 3900 kilotonnes of bio-based polymers. It is expected that the capacity of bio-based polymers will grow with a CAGR of 2.6% from 2019 to 2024. Biogenic by-products are the largest source of feedstock, which accounted for 47% of total biomass consumption in the year 2019.

Chemicals and petrochemicals industry defines the products based on carbon chains, e.g. C1, C2, C3, C4, etc.

Bio-syngas and bio-methanol are the key products in the C1 chain that is fully commercialised. SODRA, Carbon Recycling International, Shell, BASF and Veolia are the key players that produce bio-methanol commercially.

Bioethanol (generally considered under liquid biofuels), bio-based acetic acid, bio-ethylene and its derivatives such as bio-based ethylene glycol are the fully commercialised products in the C2 chain. India Glycols and Greencol Taiwan Corporation are the key players producing bioethanol commercially. India's Godavari Biorefineries is the only commercial producer of bio-based acetic acid.

Bio-propane, bio-based polypropylene, bio-based propylene glycols, glycerol, bio-based epichlorohydrin and bio-based lactic acid are the fully commercialised products in the C3 chain, and the capacity of glycerol is the highest among all the products in C3 chain.

Biobutanol, bio-based 1,4-butanediol, bio-based ethyl acetate and bio-based succinic acid are the fully commercialised products in the C4 chain, and the capacity of bio-based succinic acid is the highest among all the products in the C4 chain.

Levulinic acid, xylitol, furfural and itaconic acid are the fully commercialised products in the C5 chain, and the capacity of furfural is the highest among all the products in the C5 chain. TransFurans Chemicals, Pennakem and Silvateam are the key producers of these products at commercial scale.

Lysine, sorbitol and citric acid are the fully commercialised products in the C6 chain, and the capacity of citric acid is the highest among all the products in the C6 chain. Cargill, DSM, BBCA and Ensign are the key producers of these products at commercial scale.

13.2 COMMERCIALLY AVAILABLE BIOMASS

The biological matter of living organisms is defined as biomass (Tackling Increasing Plastic waste, 2019). Biomass offers other significant environmental and consumer benefits, including improving forest health, protecting air quality and offering the most dependable renewable energy source. Biomass is the biodegradable part which is generated from the following sources (Felix Colmorgen et al., 2020).

13.2.1 CLASSIFICATION OF AGRICULTURAL BIOMASS

Agriculture biomass can be classified into two segments - Dedicated crops are the crops for biofuels (corn, sugarcane, rapeseed, oil palm, Jatropha, sorghum, cassava, etc.) and by-products/residues are from the herbaceous plants: straw from cereals, rice, corn, bagasse, empty fruit bunch from oil palm, pruning from stover, empty corn cobs, etc. (WBA Fact Sheet, 2012).

- *Forest biomass*: Forest and its related industries such as fisheries and aquaculture have been included in this category. The extraction of stem for pulp and timber industry, thinning of plantations and cutting of old strands for pulp and timber industry generate lots of forest residues (https://www.altenergymag.com/article/2009/08/biomass-wastes/530/). These operations generate lots of residues which can further be converted into energy and chemicals.
- *Waste streams*: This is the biodegradable section of municipal and industrial wastes. The key biomass types from municipal wastes are food waste, woody waste materials, non-recycled paper, etc. (Felix Colmorgen

et al., 2020). Industrial waste mainly comprises of wastes generated from the woodworking industry, food industry, pulp and paper industry, etc. (Waste-to-Energy from Municipal Solid Wastes Report, 2019).

13.3 OPPORTUNITIES, GROWTH DRIVERS AND CHALLENGES IN BIOMASS-DERIVED PRODUCTS

The use of biomass in the field of bioenergy, bio-based chemicals and materials is highly aligned with some of the United Nations Sustainable Development Goals (Press Release – Trash to treasure, August 25, 2017).

13.3.1 AFFORDABLE AND CLEAN ENERGY

The number of people using polluting and unhealthy fuels is 2.8 billion across the globe. Energy segment contributes to 73% of human-caused greenhouse gas emissions. The use of biomass-based power generation will reduce greenhouse gas emissions drastically.

13.3.2 CLIMATE ACTION

The net CO_2 emissions must reduce by 45% from baseline, and it should reduce to zero by 2050 to limit the global warming to 1.5°C. The use of biomass across different industries will reduce greenhouse gas emissions drastically.

13.3.3 LIFE BELOW WATER

Reduction in ocean acidification and underwater pollution is a key goal. Approximately 8.5 million tonnes of non-biodegradable plastics go into the ocean heavily impacting the marine life. Bio-based biodegradable polymers help in reducing plastic wastes in both land and sea. (Islas et al. (2019); The Guardian News Article from Environment Section (2020), https://www.theguardian.com/environment/2020/oct/06/more-than--14m-tonnes-of-plastic-believed-to-be-at-the-bottom-of-the-ocean)

13.3.4 GOOD HEALTH AND WELL-BEING

Due to pollution, 7 million people die every year. The use of biofuels will help in the reduction of pollution which in turn will bring reduction in deaths caused due to pollution. Initiatives such as ethanol blending and use of other biofuels in transportation segment have already been started in different countries.

To put it in a nutshell, biomass-based energy, chemicals and materials are becoming one of the most sustainable alternatives to fossil-based industries and the key growth drivers of biomass as biofuels and feedstocks for bio-based chemicals and materials (Figure 13.1).

The biomass-based industry has come a long way in recent years, but there are certain challenges which the industry needs to overcome. Figure 13.2 shows some of the key challenges which the industry needs to overcome to grow at a faster pace.

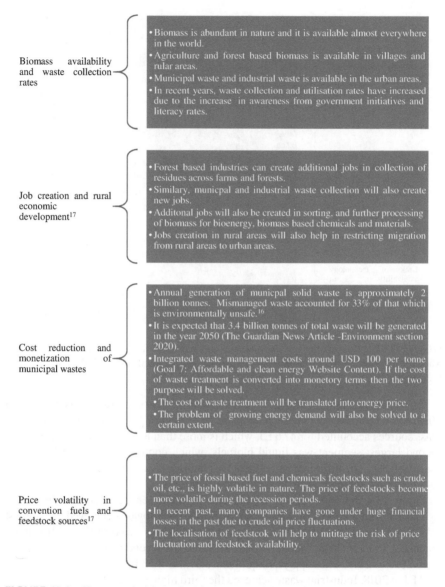

Biomass availability and waste collection rates

- Biomass is abundant in nature and it is available almost everywhere in the world.
- Agriculture and forest based biomass is available in villages and rular areas.
- Municipal waste and industrial waste is available in the urban areas.
- In recent years, waste collection and utilisation rates have increased due to the increase in awareness from government initiatives and literacy rates.

Job creation and rural economic development[17]

- Forest based industries can create additional jobs in collection of residues across farms and forests.
- Similary, municpal and industrial waste collection will also create new jobs.
- Additonal jobs will also be created in sorting, and further processing of biomass for bioenergy, biomass based chemicals and materials.
- Jobs creation in rural areas will also help in restricting migration from rural areas to urban areas.

Cost reduction and monetization of municipal wastes

- Annual generation of municpal solid waste is approximately 2 billion tonnes. Mismanaged waste accounted for 33% of that which is environmentally unsafe.[16]
- It is expected that 3.4 billion tonnes of total waste will be generated in the year 2050 (The Guardian News Article -Environment section 2020).
- Integrated waste management costs around USD 100 per tonne (Goal 7: Affordable and clean energy Website Content). If the cost of waste treatment is converted into monetory terms then the two purpose will be solved.
- The cost of waste treatment will be translated into energy price.
- The problem of growing energy demand will also be solved to a certain extent.

Price volatility in convention fuels and feedstock sources[17]

- The price of fossil based fuel and chemicals feedstocks such as crude oil, etc., is highly volatile in nature. The price of feedstocks become more volatile during the recession periods.
- In recent past, many companies have gone under huge financial losses in the past due to crude oil price fluctuations.
- The localisation of feedstcok will help to mititage the risk of price fluctuation and feedstock availability.

FIGURE 13.1 Key growth drivers for biomass and biobased feedstocks. Biomass for heating & cooling Vision Document – Executive Summary (July 2010), Trends in Solid Waste Management, The World Bank Group Web Article (2020).

13.4 GLOBAL BIOMASS SUPPLY ANALYSIS IN THE ENERGY AND FUELS SEGMENT

Biomass can be converted into electricity, thermal energy and transportation fuels. The consumption of biomass in terms of energy equivalent is provided in this section (WBA Global Bioenergy Statistics, 2020).

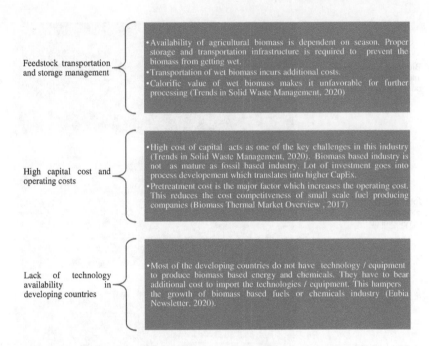

Feedstock transportation and storage management
- Availability of agricultural biomass is dependent on season. Proper storage and transportation infrastructure is required to prevent the biomass from getting wet.
- Transportation of wet biomass incurs additional costs.
- Calorific value of wet biomass makes it unfavorable for further processing (Trends in Solid Waste Management, 2020)

High capital cost and operating costs
- High cost of capital acts as one of the key challenges in this industry (Trends in Solid Waste Management, 2020). Biomass based industry is not as mature as fossil based industry. Lot of investment goes into process developement which translates into higher CapEx.
- Pretreatment cost is the major factor which increases the operating cost. This reduces the cost competitveness of small scale fuel producing companies (Biomass Thermal Market Overview , 2017)

Lack of technology availability in developing countries
- Most of the developing countries do not have technology / equipment to produce biomass based energy and chemicals. They have to bear additional cost to import the technologies / equipment. This hampers the growth of biomass based fuels or chemicals industry (Eubia Newsletter, 2020).

FIGURE 13.2 Key challenges for biomass and biobased feedstocks. Biomass Thermal Market Overview by BTEC Biomass Thermal Energy Council (2017), Eubia Newsletter – (September 2020).

As shown in Figure 13.3, the total supply of biomass for fuels was 55.5 EJ in 2018. Solid biomass sources such as wood chips, wood pellets and traditional biomass sources accounted for 47.6 EJ, which is more than 85% of the total supply. The second largest segment was liquid biofuels, which contributed for 3.98 EJ with a share of 5.8%. The third largest segment was municipal waste, which contributed for 1.45 EJ with a share of 2.6%, as shown in Figure 13.3.

Figure 13.4 shows that the global biomass supply for energy increased from 50.5 to 55.5 EJ with a CAGR of 1.2% from 2010 to 2018. Biogases achieved the highest CAGR of 6.1% from 2010 to 2018. The supply of biogases was 0.85 EJ in 2010, and it reached 1.36 EJ in 2018. Liquid biofuels achieved the second highest CAGR of 5.8% from 2010 to 2018. The supply of liquid biofuels was 2.53 EJ in 2010, and it reached 3.98 EJ in 2018. Industrial waste achieved the third highest CAGR of 4.9% from 2010 to 2018. The supply of industrial waste was 0.77 EJ in 2010, and it reached 1.33 EJ in 2018. Rapid industrialisation was a major contributor to the growth.

Regionally, the total supply of biomass for the fuels was 55.5 EJ in 2018. As shown in Figure 13.5, the largest region was Asia, which contributed for 19.7 EJ with a share of 37%. The second largest region was Africa, which contributed for 15.9 EJ with a share of 29%. In Africa, 100% contribution was made by solid biomass segment. The third largest region was the Americas, which contributed for 11.2 EJ with a share of 14.9%. Solid biomass was the major contributor in all the regions.

Below section gives a detailed analysis of biomass supply by source of biomass types:

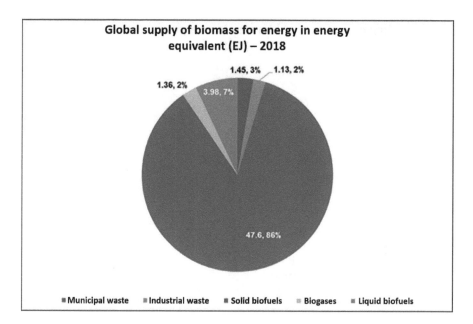

FIGURE 13.3   Global supply of biomass in energy equivalent (EJ) – 2018 (WBA Global Bioenergy Statistics, 2020).

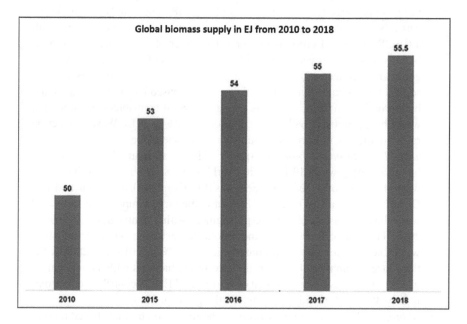

FIGURE 13.4   Global biomass supply in EJ from 2010 to 2018 (WBA Global Bioenergy Statistics, 2020).

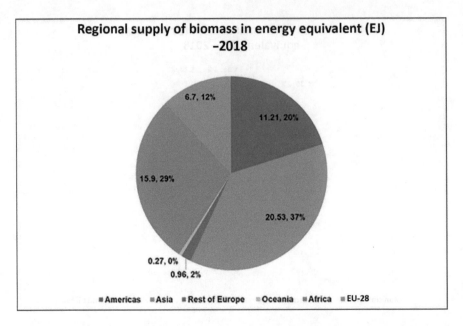

FIGURE 13.5 Regional supply of biomass in energy equivalent (EJ) (WBA Global Bioenergy Statistics, 2020).

Municipal waste-based biomass supply: The total municipal waste-based biomass supply was 1.45 EJ in the world in 2018. The supply of municipal waste-based biomass was 1.18 EJ in 2010, and it reached 1.45 EJ in 2018 with a CAGR of 2.6%. Europe is the largest supplier of municipal waste-based biomass in the energy segment with a contribution of 60% in 2018 owing to good waste collection rates in European countries. The Americas was the second largest contributor of municipal waste-based biomass with a contribution of 21% in 2018. Asia was the third largest contributor of municipal waste-based biomass with a contribution of 14% in 2018. Waste collection rates are highest in European countries across the globe.

Industrial waste-based biomass supply: The total industrial waste-based biomass supply was 1.13 EJ in the world in 2018. The supply of industrial waste-based biomass for solid fuels was 0.77 EJ in 2010, and it reached 1.13 EJ in 2018 with a CAGR of 4.9%. Asia is the largest supplier of industrial waste-based biomass in the energy segment with a contribution of 56% in 2018. The rest of Europe was the second largest contributor of industrial waste-based biomass with a contribution of 23% in 2018. EU-28 was the third largest supplier of industrial waste-based biomass with a contribution of 17% in 2018. Together, the rest of Europe and EU-28 contributed approximately 40% of the total supply across the globe in 2018.

Solid biofuels based on biomass supply: The total solid biofuels based on biomass supply was 47.6 EJ in the world in 2018. The supply of biomass for solid fuels was 45.1 EJ in 2010, and it reached 47.6 EJ in 2018 with a CAGR of 0.7%. Asia was the largest supplier of solid biomass in the energy segment with a contribution of 39% in 2018. Africa was the second largest contributor of solid biomass with a contribution of 33% in 2018. The Americas was the third largest contributor of solid biomass with a contribution of 17% in 2018. Together, the rest of Europe and EU-28 contributed approximately 10% of the total solid biomass supply across the globe in 2018.

Biogases based on biomass supply: The total biomass-based biogases supply was 1.36 EJ in the world in 2018. The supply of biomass-based biogases was 0.85 EJ in 2010, and it reached 1.36 EJ in 2018 with a CAGR of 6.1%. EU-28 was the largest supplier of biomass-based biogases in the energy segment with a contribution of 51% in 2018. Asia was the second largest supplier of the biomass-based biogases in the energy segment with a contribution of 32% in 2018.

Liquid biofuels based on biomass supply: The total liquid biofuels based on biomass supply was 3.98 EJ in the world in 2018. The supply of liquid biofuels based on biomass was 2.53 EJ in 2010, and it reached 3.98 EJ in 2018 with a CAGR of 5.8%. The Americas was the largest supplier of biomass-based liquid fuels in the energy segment with a contribution of 68% in 2018. EU-28 was the second largest contributor of biomass-based liquid fuels with a contribution of 20% in 2018. Asia was the third largest contributor of biomass-based liquid fuels with a contribution of 11% in 2018.

13.5 MARKET ANALYSIS OF BIOMASS-BASED FUEL PRODUCTION

Wood pellets production: The total production of wood pellet was 38.9 million tonnes in 2019. EU-28 was the largest producer of wood pellets in 2019 with 17.8 million tonnes, which is a share of 46% across the globe in 2019. The Americas was the second largest producer in 2019 with the production of 12.3 million tonnes, which is a share of 32%. The third largest region was Asia which produced 5 million tonnes of wood pellets with a share of 13% in 2019.

From 2015 to 2019, the production of wood pellets in the world has grown with a CAGR of 9.2%. The production of wood pellets in Asia has grown with a CAGR of 25.2%, which is the highest among all regions. The production of wood pellet in Africa has also witnessed an increased growth with a CAGR of 23.6%.

Wood-based fuel production: The total production of wood fuel was 1945 million m³ in 2019. Asia was the largest producer of wood fuel in 2019 with 719 million m³, which is a share of 37% across the globe in 2019. Africa was the second largest producer in 2019 with the production of 700 million m³, which is a share of 36% in 2019. The third largest region was the Americas which produced 340 million m³ with a share of 17% in 2019. From 2015 to 2019, the production of wood fuel in the Americas has grown with a CAGR of 2.6%. The growth in the Americas is the highest among all the regions. The production of wood fuel in Asia has declined with a CAGR of -0.5%.

Wood charcoal production: The total production of wood charcoal was 53.1 million tonnes in 2019. Africa was the largest producer of wood charcoal in 2019 with 34.2 million tonnes, which is 64% share across the globe in 2019. The Americas was the second largest producer in 2019 with the production of 9.15 million tonnes and a share of 17.2% in 2019. The third largest region was Asia which produced 9 million tonnes of wood charcoal with a market share of 17% in 2019. From 2015 to 2019, globally, the wood charcoal production has grown with a CAGR of 0.9%. From 2015 to 2019, the production of wood pellets in EU-28 has grown with a CAGR of 2.9%. The growth in EU-28 is the highest among all the regions. The production of wood charcoal in Africa has also witnessed an exponential growth with a CAGR of 1.6%.

Liquid biofuels: The total production of liquid biofuels was 159.8 billion litres in 2018. As shown in Table 13.1, the Americas was the largest producer of liquid biofuels with 120.5 billion litres production capacity and holding 76% of the market share across the globe. Europe was the second largest producer in year 2018 with a production capacity of 21.7 billion litres and 13.6% market share. The third largest region was Asia which produced 17.1 billion litres of biofuels with 17% market share in year 2018.

The total production of liquid biofuels was 159.8 billion litres in 2018. Bioethanol holds the largest share in liquid biofuels in 2018 with 98.5 billion litres and 62% market share. Biodiesel holds the second largest share in liquid biofuels in 2018 with 41.8 billion litres and 26% share. From 2010 to 2018, the production of biofuels in the world has grown with a CAGR of 5.7%. The production of biodiesel has grown with a CAGR of 10%. The growth in biodiesel is the highest among all products. From 2010 to 2018, the production of bioethanol has grown with a CAGR of 4.8%.

Biogas production: The total production of biogas was 59.4 billion m^3 in 2018. EU-28 was the largest producer of biogas with 30.3 billion m^3 and holding 51% share across the globe. Asia was the second largest producer with 32.5% market share and a production capacity of 19.3 billion m^3. The third largest region was the Americas which produced 8.3 billion m^3 of biogas with 14% share in 2018. The production of biogas was 37.1 billion m^3 in 2010, and it reached 59.4 billion m^3 in 2018 with a CAGR of 6.0%.

TABLE 13.1

Liquid Biofuel Production (Billion Litres) – 2018

Biofuel Production (Billion Litres)	Bioethanol	Biodiesel	Other Biofuels	Total
Africa	0.09	0	0	0.09
The Americas	86	16.4	18.2	120.6
Asia	6.9	9.7	0.6	17.2
Europe	5.22	15.7	0.8	21.7
Oceania	0.24	0	0	0.24

Source: WBA Global Bioenergy Statistics (2020).

13.6 GLOBAL MARKET ANALYSIS OF BIOMATERIALS

13.6.1 BIO-BASED POLYMERS – SUPPLY–DEMAND ANALYSIS

Polymers can be divided into two types based on the source of raw material: (a) petroleum based and (b) biomass based. There are many reasons due to which biomass-based polymers are replacing the petroleum-based polymers. The key reasons are the renewable raw materials of bio-based polymers having low carbon emissions in the production process, as well as biodegradability. Bio-based polymers are environment-friendly and provide an apt solution for pollution reduction.

The following are the type of biopolymers: aliphatic polycarbonates (APCs), casein polymers, cellulose acetate (CA), epoxy resins, ethylene propylene diene monomer (EPDM) rubber, polyamides (PA), polybutylene adipate terephthalate (PBAT) and polybutylene succinate (PBS), and the following are the type of copolymers: polyethylene (PE), polyethylene furanoate (PEF), polyethylene terephthalate (PET), polyhydroxyalkanoates (PHAs), polylactic acid (PLA), polypropylene (PP), polytrimethylene terephthalate (PTT), polyurethanes (PURs) and starch-containing polymer compounds.

As shown in Figure 13.6, the global capacity of biopolymers was 4300 kilotonnes in 2019 and it is expected to reach 4900 kilotonnes in 2024 with a CAGR of 2.6% (Figure 13.6). The total production of biopolymers was 3800 kilotonnes in the year 2019, and the operating rates were 88% in the year 2019. Bio-based cellulose acetate and epoxy resin held the major share in year 2019, and this trend is expected to continue till the year 2024.

Textile segment accounted for 22% total consumption volume of biopolymers in 2019 and is expected to contribute 20% of the total consumption volume of biopolymers in 2024.

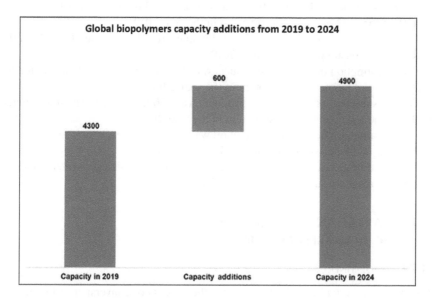

FIGURE 13.6 Global biopolymers capacity additions from 2019 to 2024 (Market and Trends Reports (Service of Nova Institute) http://www.bio-based.eu/reports).

Automotive and transportation segment accounted for 15% total consumption volume of biopolymers in 2019 and is expected to contribute 15% of the total consumption volume of biopolymers in 2024.

Building and construction segment accounted for 13% total consumption volume of biopolymers in 2019 and is expected to contribute 14% of the total consumption volume of biopolymers in 2024.

Consumer goods segment accounted for 13% total consumption volume of biopolymers in 2019 and is expected to contribute 15% of the total consumption volume of biopolymers in 2024.

Flexible packaging segment accounted for 13% total consumption volume of biopolymers in 2019 and is expected to contribute 13% of the total consumption volume of biopolymers in 2024.

Rigid packaging segment accounted for 11% total consumption volume of biopolymers in 2019 and is expected to contribute 10% of the total consumption volume of biopolymers in 2024.

Electrical and electronics segment accounted for 5% total consumption volume of biopolymers in 2019 and is expected to contribute 5% of the total consumption volume of biopolymers in 2024.

13.6.2 Analysis of Biomass Consumption to Produce Biopolymers in 2019

Total tonnage of biomass feedstock consumed to produce 3800 kilotonnes of biopolymers was 5000 kilotonnes in the year 2019 which is distributed below (Figure 13.7).

The consumption of biogenic by-products was 2350 kilotonnes, which accounted for 47% of the total biomass consumption in the year 2019.

The consumption of starch was 1000 kilotonnes, which accounted for 20% of the total biomass consumption in the year 2019.

The consumption of sugars by-products was 850 kilotonnes, which accounted for 17% of the total biomass consumption in the year 2019.

The consumption of cellulose by-products was 450 kilotonnes, which accounted for 9% of the total biomass consumption in the year 2019.

The consumption of non-edible plant oil was 350 kilotonnes, which accounted for 7% of the total biomass consumption in the year 2019.

The consumption of edible plant oil was 50 kilotonnes, which accounted for 1% of the total biomass consumption in the year 2019.

13.7 MARKET AND INDUSTRY ANALYSIS OF BIO-BASED CHEMICALS

13.7.1 Biorefinery

Biorefinery is the facility which integrates the process of conversion of biomass to produce fuels, powers and various chemicals (Energy Today, 2018). One of the key reasons of integrations in biorefineries and petroleum-based refineries is to reduce

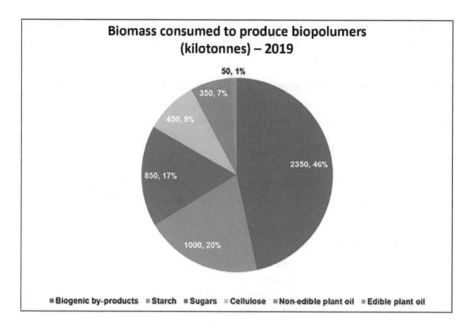

Biomass consumed to produce biopolumers (kilotonnes) – 2019

50, 1%
350, 7%
450, 9%
850, 17%
2350, 46%
1000, 20%

■ Biogenic by-products ■ Starch ■ Sugars ■ Cellulose ■ Non-edible plant oil ■ Edible plant oil

FIGURE 13.7 Biomass consumed to produce biopolymers (Kilo tonnes)-2019 (Source: Market and Trends Reports (Service of Nova Institute) http://www.bio-based.eu/reports)

the cost of production of the end products, and more importantly, biorefineries provide renewable and sustainable solutions towards clean energy. Cost reduction for fuels is based on the economies of scale, while backward integration (raw materials produced in the same plant) is the key factor in value-added chemicals. Recovered utilities also impact the production cost of chemicals.

The classification of biorefineries is based on four basic ideas: (a) platforms, (b) products, (c) feedstock and (d) processes.

The key platforms in biorefinery are as follows.

13.7.1.1 Syngas Platform

Syngas is a mixture of carbon monoxide (CO) and hydrogen (H_2), and it is produced by gasification. Gasification is heating at a temperature of more than 430°C in the presence of oxygen or air (The Concept of Biorefinery by Salman Zafar, 2020).

13.7.1.2 Biogas Platform

The major components of biogas are CH_4 and H_2. The key process used to prepare biogas is anaerobic digestion of biomass (The Concept of Biorefinery by Salman Zafar, 2020).

13.7.1.3 C6 and C6/C5 Sugar Platform

They are prepared from biomass hydrolysis of various types of sugar-based content products such as sucrose, starch, hemicellulose and cellulose (The Concept of Biorefinery by Salman Zafar, 2020). The key process to convert these sugars into building blocks is fermentation. The key products from fermentation are succinic acid, itaconic acid, adipic acid, 3-hydroxypropionic acid/aldehyde, isoprene, glutamic acid and aspartic acid.

13.7.1.4 Selective Dehydration, Hydrogenation and Oxidation Processes
Sorbitol, furfural, glucaric acid, hydroxymethylfurfural (HMF) and levulinic acid are the key products produced from these processes.

13.7.1.5 Organic Solutions Platform
This platform involves processing of fresh wet biomass. Dewatering is the basic process to separate organic solutions and press cake. Organic solution contains compounds such as carbohydrates, amino acids and organic acids. Press cake contains fibrous materials.

13.7.1.6 Lignin-Based Platform
This platform converts lignocellulosic biomass into various chemicals and other products. This platform gives vast opportunities in the area of bio-based chemicals such as syngas derivatives, base chemicals (BTX, etc.), phenol and its derivatives, oxidised products (vanillin, cyclohexanol, etc.), and other products (carbon fibres, resins, etc.) (The Concept of Biorefinery by Salman Zafar, 2020).

13.7.2 BIO-BASED CHEMICALS INDUSTRY
Chemicals and petrochemicals industry defines products based on carbon chains, e.g. C1, C2, C3, C4, etc. Here, the potential of bio-based chemicals is identified based on their carbon chain length.

13.7.2.1 C1-Containing Compounds
Bio-methanol: Major companies manufacturing bio-methanol are OCI (BioMCN), SODRA, Carbon Recycling International and Shell's waste-to-chemicals facility in Rotterdam (W2C Project) (The Concept of Biorefinery by Salman Zafar, 2020, and Bioenergy International, 2020). Some of the key developments/activities done by various companies in the area of bio-methanol are depicted below:

Bio-methanol Key Activities	W2C Project is the Shell's waste-to-chemicals facility in Rotterdam. It is expected that this facility will use 360,000 tonnes of waste to produce 220,000 tonnes of bio-methanol (Amanda Jasi, https://www.thechemicalengineer.com/news/shell-joins-w2c-rotterdam-project/).
	BASF is producing bio-methanol according to biomass balance approach. The trade name of BASF's bio-methanol is EU-REDcert-methanol (Refuelling the future – https://www.sodra.com/en/global/Bioproducts/biomethanol/).
	The bio-methanol capacity of Södra is 5250 tonnes/year. Södra uses pulp to produce the bio-methanol (Ed de Jong et al., 2020).
	Veolia and Metsä have collaborated with each other to produce bio-methanol with the plant capacity of 12,000 tonnes/year. The raw methanol from pulp will be used to produce bio-methanol. It is expected that Veolia's bio-methanol plant will start producing in 2023 (Klaus-Peter Rieser, 2018).

Bio-based formaldehyde: BASF is conducting research in producing bio-based formaldehyde (Bioenergy International, 2020).

Syngas: The total capacity of bio-based syngas is 760,000 tonnes/year (The Concept of Biorefinery by Salman Zafar, 2020).

13.7.2.2 C2-Containing Compounds

Bio-ethylene: Braskem produces bio-ethylene which uses bioethanol as feedstock. Sugarcane or cellulosic-based biomass is used to produce bioethanol. Braskem uses bio-ethylene to produce bio-based polyethylene. The capacity of bio-based polyethylene is 200,000 tonnes/annum (The Chemical Engineers News, 2019).

Bio-based ethylene glycols: Petro-based monoethylene glycol is the key raw material for producing polyester and PET. The global volume of PET-based products is humongous, and they have become threat to the environment. They have become one of the main sources of land-based pollution. This is the key reason which led the companies to look for sustainable alternatives. Bio-based monoethylene glycol is one of the sustainable alternatives to petro-based monoethylene glycol. Currently, there are three companies which are involved in the production of bio-based monoethylene glycols. India Glycols Limited, India, and Greencol Taiwan Corporation, Taiwan, have commercialised the production of bio-based monoethylene glycol, while the collaboration of Haldor Topsoe and Braskem from Lyngby, Denmark has recently opened a demo unit.

Following are key developments/activities done by various companies in the area of bio-based monoethylene glycol:

Bio-based Ethylene Glycols	India Glycols Limited is the commercial-scale manufacturer of bio-based ethylene glycols. Their plant has a capacity of 175,000 tonnes/year. They are using molasses-based bioethanol as a feedstock to produce bio-based ethylene glycols (I'm Green™ Polyethylene – http://plasticoverde.braskem.com.br/site.aspx/Im-greenTM-Polyethylene).
	Braskem and Haldor Topsoe have collaborated to make bio-based MEG technology development. MOSAIK™ is the technology they are using to develop bio-based MEG which uses the sugar to produce bio-based MEG. They have also opened a demonstration unit in Lyngby, Denmark (News Article, 2020).
	Greencol Taiwan Corporation is the joint venture between Toyota Tsusho Corporation (TTC) and Taipei-based China Man-made Fiber Corporation. Currently, the company manufactures biomass-based monoethylene glycol. The plant is located in Kaohsiung Taiwan, and the installed capacity is 100,000 tonnes/year. The company sources bio-based ethanol from Brazil and converts it into bio-based ethylene; it produces bio-based ethylene glycol from bio-based ethylene (News Article, 2020).

Bio-based acetic acid: Godavari Biorefineries, Mumbai, India, is the only producer of bio-based acetic acid with a capacity of 7000 tonnes/year (The Concept of Biorefinery by Salman Zafar, 2020). They are using molasses-based bioethanol as a feedstock to produce bio-based acetic acid (https://www.somaiya.com/about-us).

13.7.2.3 C3-Containing Compounds

Biopropane: Neste has started the world's first bio-based propane plant on a commercial scale. This facility has a total capacity of 40,000 tonnes/year (Neste Corporation Press Release, March 19, 2018).

Bio-based propylene: Most of the bio-based propylene produced is converted into bio-based polypropylene, bio-based propylene glycol and other derivatives. SABIC is one of the producers of bio-based polypropylene and has undergone two major collaborations/agreements, which are vital steps towards the commercialisation of bio-based polypropylene.

Beiersdorf will be using the SABIC's bio-based polypropylene for packaging products. Beiersdorf will introduce this bio-based polypropylene packaging from the year 2021 (SABIC News, 2020).

UPM will be using SABIC's bio-based polypropylene to make 100% renewable biocomposite by using the residues from paper and pulp production (SABIC News, 2020).

Mitsui Chemicals has done demonstration tests for the development of bio-based polypropylene. It is expected that Mitsui shall start commercial production of bio-based polypropylene by the year 2024 (Mitsui Chemical Inc. News Release, 2019).

13.7.2.4   Bio-based Propylene Glycol

ADM has a manufacturing plant in Decatur, Illinois, with an installed capacity of 100,000 tonnes/year (The Global Bio-Based Polymer Market, 2019). Olean has its manufacturing plant of propylene glycol in Ertvelde, Belgium. This plant has a capacity of 20,000 tonnes/year (The Global Bio-Based Polymer Market, 2019).

Glycerol: Glycerol is one of the most important and high-value bio-based C3 compounds. The total capacity across the globe is 1,500,000 tonnes/year (The Concept of Biorefinery by Salman Zafar, 2020).

Bio-based epichlorohydrin: Epichlorohydrin is one of the important building blocks for bio-based epoxy resins. The global capacity of epichlorohydrin is 540,000 tonnes/annum.

Bio-based lactic acid: The most important derivative of lactic acid is polylactic acid.

Corbion is the market leader in terms of lactic acid capacity of 75,000 tonnes/year (The Concept of Biorefinery by Salman Zafar, 2020). Corbion is building a new plant in Thailand with a capacity of 125,000 tonnes/year (Biomass Magazine News, 2018). NatureWorks is the market leader of polylactic acid with an installed capacity of 150,000 metric tonnes/year. The plant is located in Blair, Nebraska, the USA (Energy Today, 2018).

Bio-based acrylic acid: The key companies involved in the development of bio-based acrylic acid are Cargill/Novozymes, ADM/LC Chemicals, Perstorp and Arkema (The Concept of Biorefinery by Salman Zafar, 2020).

13.7.2.5   C4-Containing Compounds

Biobutanol: Green Biologics has a biobutanol plant in Little Falls, Minnesota (the USA) (ICIS News, 2012). Celtic Renewables is starting its demonstration plant in Caledon Green, Grangemouth, Scotland (Business Standard News, 2016).

Bio-based 1,4-butanediol: The total global capacity is around 30,000 tonnes/year, and the key producers are Genomatica, Novamont, Dupont Tate & Lyle and Godavari Biorefineries Ltd (The Concept of Biorefinery by Salman Zafar, 2020).

Bio-based ethyl acetate: Prairie Catalytic is planning to produce bio-based ethyl acetate at its production plant in Nebraska (Business Standard News, 2016a). The company did not disclose the capacity, but ICIS supply and demand database quotes this at approximately 50,000 tonnes/year (Article by Amanda Jasi, Joint Venture & Partnerships News (2020); Petrochemicals News, 2019).

Bio-based succinic acid: Succinity GmbH had a plant capacity of 30,000 tonnes/year; it has gone bankrupt in 2019 (ICIS News, 2019).

Myriant has a plant capacity of 13,500 tonnes/year. The plant is located in Louisiana (the USA) (ICIS News, 2019). Stepan acquired this plant this year, and it plans to make bio-based surfactants in this plant with a capacity of 20,000 tonnes/year (Chemical & Engineering News (48)).

Roquette has a plant capacity of 10,500 tonnes/year. The plant is located in Cassano, Italy (ICIS News, 2019).

13.7.2.6   C5-Containing Compounds

1,5-Pentanediamine: Cathay Industrial Biotech, CJ CheilJedang and Daesang are the key players which are involved in the development and commercialisation (John Wain et al., 2019).

Levulinic acid: GF Biochemicals has an installed capacity of 10,000 tonnes/year (GF Biochemicals Products). GF Biochemicals claims that it is the only company that produces levulinic acid at commercial scale directly from biomass (Report on Evolution of the Bio-based Chemicals Market: Growth and Commercialization Strategies, 2020).

Xylitol: The total capacity is 190,000 tonnes/year. Danisco/Lenzing and Fortress are the key players (https://tuprints.ulb.tu-darmstadt.de/17599/1/Dissertation%20Mrani%20IWAR%202021.pdf).

Furfural: The total capacity is 360,000 tonnes/year. TransFurans Chemicals, Pennakem and Silvateam are the key players (https://tuprints.ulb.tu-darmstadt.de/-17599/1/Dissertation%20Mrani%20IWAR%202021.pdf).

Itaconic acid: The total capacity is 90,000 tonnes/year. Qingdao Kehai, Zhejiang Guoguang, Jinan Huaming Biochemistry are the key players (https://tuprints.ulb.tu-darmstadt.de/17599/1/Dissertation%20Mrani%20IWAR%202021.pdf).

13.7.2.7   C6-Containing Compounds

Lysine: The total capacity is 1,100,000 tonnes/year. Global Biotech, Evonik/RusBiotech, BBCA and Ajinomoto are the key players (https://tuprints.ulb.tu-darmstadt.de/17599/1/Dissertation%20Mrani%20IWAR%202021.pdf).

Sorbitol: The total capacity is 1,800,000 tonnes/year. Roquette, Cargill, ADM and Ingredion are the key players (https://tuprints.ulb.tu-darmstadt.de/17599/1/-Dissertation%20Mrani%20IWAR%202021.pdf).

Citric acid: The total capacity is 2,000,000 tonnes/year. Cargill, DSM, BBCA and Ensign are the key players (https://tuprints.ulb.tu-darmstadt.de/17599/1/-Dissertation%20Mrani%20IWAR%202021.pdf).

13.7.2.8   Others

Bio-based adipic acid: It is technically feasible, but it lacks in cost-competitiveness. Genomatica is the key player in this product (John Wain et al., 2019).

13.8　CONCLUSIONS

The use of biomass in these industries is highly aligned with the United Nations Sustainable Development Goals such as affordable and clean energy, climate action, life below water, good health and well-being.

Currently, bioenergy is the largest renewable energy source globally and accounts for more than two-third of the renewable energy mix. In the overall energy scenario, bioenergy accounts for 13%–14% of the total energy consumption. The total supply of biomass for fuels was 55.5 EJ in 2018. Historically, the demand of biomass for fuels has witnessed a CAGR of 1.2% from 2010 to 2018. Environmental sustainability, rapid industrialisation and urbanisation are the key drivers for the historical growth. In addition to the historical factors, government policies and technological advancements (such as process improvement, waste to energy, etc.) of biomass-based energy processes are expected drive the future growth.

The global capacity of biopolymers was 4300 kilotonnes in 2019, and it is expected to reach 4900 kilotonnes in 2024 with a CAGR of 2.6%. The whole world is struggling with the waste generated from petrochemical-based plastics. Plastic products take around 450–600 years for complete degradation (Hannah Ritchie, 2018). Approximately 8.5 million tonnes of non-biodegradable plastics go into the ocean heavily impacting the marine life. Bio-based polymers degrade readily, which helps in reducing the pollution and is less harmful to marine life. Lots of initiatives have been taken by big companies to make the whole supply chain bio-based, which includes packaging also, and this will boost the growth in the consumption of biopolymers in primary and secondary packaging.

Chemicals and petrochemicals industry defines products based on carbon chains, e.g. C1, C2, C3, C4, etc. Market analysis of bio-based chemicals shows the transition of chemical companies moving from chemicals to biochemicals. Biorefineries are using various platforms such as syngas, biogas, C6/C5 sugars, organic solutions, selective dehydration, hydrogenation and lignin-based platforms to produce biochemicals.

Industries using fossil-based chemicals are focusing on sustainable products/-ingredients to reduce greenhouse gas intensity, water consumption and energy consumption throughout the life cycle of the end product.

In addition to environmental benefits, an increase in awareness from customer, ingredient transparency of the specialty chemicals suppliers, regulations and cost-efficient production process will play a crucial role in increasing the demand of downstream derivatives and specialty chemicals based on bio-based chemicals. Most of the large petrochemicals/chemical firms have commercialised the processes for bio-based chemicals.

REFERENCES

About Us- Godavari Biorefineries Ltd, https://www.somaiya.com/about-us
Article by Amanda Jasi, https://www.thechemicalengineer.com/news/shell-joins-w2c-rotterdam-project/
Bioenergy International (2020), Veolia and Metsä eye biomethanol plant at Äänekoski bioproduct mill, https://bioenergyinternational.com/biofuels-oils/veolia-and-metsa-eye-biomethanol-plant-at-aanekoski-bioproduct-mill

Biomass for heating & cooling Vision Document – Executive Summary (July 2010), http://-www.eurosfaire.prd.fr/7pc/doc/1305645731_rhc_vision_biomass_summary.pdf

Biomass Magazine News (2018), ADM partners with PNNL to produce biobased propylene glycol. http://biomassmagazine.com/articles/14968/adm-partners-with-pnnl-to-produce-biobased-propylene-glycol

Biomass Thermal Market Overview by BTEC Biomass Thermal Energy Council (2017), http://www.biomassthermal.org/wp-content/uploads/2017/12/Fact-Sheet-2.pdf

Business Standard News (2016a), Researchers in Germany plan to produce formaldehyde from carbon dioxide, https://www.business-standard.com/content/b2b-chemicals/researchers-in-germany-plan-to-produce-formaldehyde-from-carbon-dioxide-116093000698_1.html

Business Standard News (2016b), Green Biologics starts shipments of bio-based n-butanol & acetone, https://www.business-standard.com/content/b2b-plastics-polymers/green-biologics-starts-shipments-of-bio-based-n-butanol-acetone-116121400452_1.html

Biomass Wastes (2009) https://www.altenergymag.com/article/2009/08/biomass-wastes/530/ Caledon Green Facility – Production Facility, https://www.celtic-renewables.com/production-facility/

Chemical & Engineering News-Stepan Makes a Pair of Acquisitions (February 6, 2021), volume 99, issue 5, https://cen.acs.org/business/biobased-chemicals/Stepan-makes-pair-acquisitions/99/i5

Felix Colmorgen, Cosette Khawaja, Dominik Rutz (2020), *Handbook on Regional and Local Bio-Based Economies*. Published by WIP Renewable Energies, Munich, Germany, 1st edition, ISBN: 978-3-936338-61-4, https://be-rural.eu/wp-content/uploads/2020/03/BE-Rural_D2.5_Handbook.pdf

Energy Today (2018), *Barriers to Renewable Energy Technologies Development*, Publication of The American Energy Society Published by De Gruyter DOI: dx.doi.org/10.1515/-energytoday -2018–2302, Published Online 2018-01-25, https://www.energytoday.net/economics-policy/barriers-renewable-energy-technologies-development/

Eubia Newsletter – (September 2020), Challenges related to biomass, European Biomass Industry Association, https://www.eubia.org/cms/wiki-biomass/biomass-resources/challenges-related-to-biomass/

GFBiochemicals Products, http://www.gfbiochemicals.com/company/

Goal 7: Affordable and clean energy Website Content, https://www.undp.org/content/undp/en/-home/sustainable-development-goals/goal-7-affordable-and-clean-energy.html

Greencol Taiwan Corporation Bio-MEG Plant, Taiwan News (2021), https://www.lzbxgmjj.com/key/greenon%E5%8F%B0%E6%B9%BE.html

I'm Green™ Polyethylene, http://plasticoverde.braskem.com.br/site.aspx/Im-greenTM-Polyethylene

ICIS News (2012), *Oleon Officially Starts Up New Bio-Based MPG Unit in Belgium*, Author: Heidi Finch, https://www.icis.com/explore/resources/news/2012/06/28/9573676/oleon-officially-starts-up-new-bio-based-mpg-unit-in-belgium/

ICIS News (2019), *US Prairie Catalytic's Bioethanol-to-Etac Plant Start-Up Delayed to Late Q1*, Author: Adam Yanelli, https://www.icis.com/explore/resources/news/2019/01/10/-10305062/us-prairie-catalytics-bioethanol-to-etac-plant-start-up-delayed-to-late-q1/

India Glycols Limited Product News, MEG / DEG / TEG, https://www.indiaglycols.com/-product_groups/monoethylene_glycol.htm

Jorge Islas, Fabio Manzini, Omar Masera, Viridiana Vargas (2019), Chapter Four - Solid Biomass to Heat and Power, Editor(s): Carmen Lago, Natalia Caldés, Yolanda Lechón, *The Role of Bioenergy in the Bioeconomy*, Academic Press, pp. 145–177, ISBN 9780128130568.

Joint Venture & Partnerships News (2020), Sabic Supplies Bio-PP for UPM Wood Bio-composites, https://bioplasticsnews.com/2020/06/05/sabic-trucircle-bio-pp-upm-wood-bio-composites/

Ed de Jong, Avantium, Geoff Bell, Microbiogen (2020), Bio-Based Chemicals a 2020 Update-IEA Bioenergy: Task 42: 2020: 01, ISBN 978-1-910154-69-4 Published by IEA Bioenergy, https://www.ieabioenergy.com/wp-content/uploads/2020/02/Bio-based-chemicals-a-2020-update-final-200213.pdf

Market and Trends Reports (Service of Nova Institute), http://www.bio-based.eu/reports

Mitusi Chemical Inc. News Release (September 26, 2019), Mitusi Chemicals Groups Bio-Propylene Adopted for project commissioned by Ministry of Environment, https://jp.mitsuichemicals.com/sites/default/files/media/document/2019/190926e.pdf

Neste Corporation Press Release (March 19, 2018), Neste delivers first batch of 100% renewable propane to European market, https://www.neste.com/releases-and-news/renewable-solutions/neste-delivers-first-batch-100-renewable-propane-european-market

News Article (2020). Braskem and Haldor Topsoe achieve first production of bio-based MEG, https://www.fuelsandlubes.com/flo-article/braskem-and-haldor-topsoe-achieve-first-production-of-bio-based-meg/

Petrochemicals News by Josh Pedrick (June 11, 2019). US-based Prairie Catalytic producing on-spec ethyl acetate: Source, https://www.spglobal.com/platts/en/market-insights/latest-news/petrochemicals/061119-us-based-prairie-catalytic-producing-on-spec-ethyl-acetate-source

PetroChemical News (2020), Thailand: Corbion boosting lactic acid capacity to meet demand for PLA polymers, https://www.gupta-verlag.com/news/industry/23749/thailand-corbion-boosting-lactic-acid-capacity-to-meet-demand-for-pla-polymers

Press Release (August 25, 2017), Trash to treasure: The benefits of waste-to-energy technologies by Ronald Walli, https://www.anl.gov/article/trash-to-treasure-the-benefits-of-wastetoenergy-technologies#:~:text=according%20to%20Lee.-, %E2%80%9CBy%20using%20waste%20to%20produce%20energy%2C%20we%20can%20avoid%20emissions, trillion%20pounds%20of%20waste%2C%20according

Refuelling the Future, https://www.sodra.com/en/global/Bioproducts/biomethanol/

Report on Evolution of the Bio-based Chemicals Market: Growth and Commercialization Strategies (2020), https://www.aranca.com/knowledge-library/articles/business-research/evolution-of-the-bio-based-chemicals-market-growth-and-commercialization-strategies

Klaus-Peter Rieser (2018), Trade News (November 12, 2018), BASF produces methanol according to the biomass balance approach, https://www.basf.com/global/en/media/news-releases/2018/11/p-18-370.html

Hannah Ritchie (2018), FAQs on Plastics, https://ourworldindata.org/faq-on-plastics#:~:text=Estimated%20decomposition%20times%20for%20plastics, take%20an%20estimated%20450%20years.

Integration of life cycle assessment and life cycle costing in a descriptive multi-criteria decision analysis for Purpose of evaluating innovative technology concepts, https://tuprints.ulb.tu-darmstadt.de/17599/1/Dissertation%20Mrani%20IWAR%202021.pdf

SABIC News (2020), SABIC Collaborates with BEIERSDORF to Implement Sustainable Cosmetics Packaging Using Certified Renewable Polypropylene, https://www.sabic.com/en/news/25625-beiersdorf-to-implement-sustainable-cosmetics-packaging-made-with-sabic-s-certified-renewable-pp

Tackling Increasing Plastic Waste (2019), Plastic waste at the Thilafushi waste disposal site, Maldives - by Mohamed Abdulraheem, https://datatopics.worldbank.org/what-a-waste/tackling_increasing_plastic_waste.html

The Chemical Engineers News (2019), Shell joins W2C Rotterdam project. https://www.thechemicalengineer.com/news/shell-joins-w2c-rotterdam-project/

The Concept of Biorefinery by Salman Zafar, (December 27, 2020), https://www.bioenergyconsult.com/biorefinery/#:~:text=A%20biorefinery%20is%20a%20facility, fuels%20and%20products%20from%20petroleum

The Global Bio-Based Polymer Market (2019), A revised view on a turbulent and grow-
    ing market, https://news.bio-based.eu/the-global-bio-based-polymer-market-2019-a-
    revised-view-on-a-turbulent-and-growing-market/
The Guardian News Article from Environment Section (2020), https://www.theguardian.
    com/environment/2020/oct/06/more-than-14m-tonnes-of-plastic-believed-to-be-at-the-
    bottom-of-the-ocean
Trends in Solid Waste Management, The World Bank Group Web Article (2020), https://
    datatopics.worldbank.org/what-a-waste/trends_in_solid_waste_management.html
UN Environment Program, https://www.unenvironment.org/interactive/beat-plastic-pollution/
    #:~:text=Today%2C%20we%20produce%20about%20300,of%20the%20entire%20
    human%20population
John Wain, Keith Waldron, Graham Moates, Peter Metcalfe (2019), Viability of bio-based
    chemicals from food waste. D6.8 report investigating fuels and chemicals from mixed,
    post-consumer food waste and the selection of bacterial strains for growth on food waste
    https://ec.europa.eu/research/participants/documents/downloadPublic?documentIds=0
    80166e5ce33970d&appId=PPGMS
Waste-to-Energy from Municipal Solid Wastes Report (2019), https://www.energy.gov/sites/
    prod/files/2019/08/f66/BETO--Waste-to-Energy-Report-August--2019.pdf
WBA Fact Sheet, Global Biomass Potential towards 2035 (2012), https://worldbioenergy.org/
    uploads/Factsheet_Biomass%20potential.pdf
WBA Global Bioenergy Statistics (2020), World Bioenergy Association, https://worldbioenergy.
    org/global-bioenergy-statistics

Index

A

Abd-Elrsoul, R.M.M.A. 163
acetic acid 107, 249–251, 363
acetogenesis 226
acetone–butanol–ethanol (ABE) fermentation
  310–311
Achten, W.M.J. 87
acidogenesis 226
acid pretreatment 99, 296–297
acid treatment 154
acrylic acid 364
acyl-acyl carrier protein (ACP) 255–256
acyl-acyl carrier protein reductase (AAR) 255–256
advanced biorefineries 213
agricultural biomass 351–352
agro-industrial residues 163
Akhtar, J. 111
Alba, G. 111
alcohols 26
aliphatic alcohols 26
alkaline extraction method 14
alkaline pretreatment 297
alkali pretreatment 100
alkanes/alkenes, production of 256–258
Alvira, P. 167
amber acid *see* succinic acid (SA)
ammonia fibre explosion technique 102–103, 298
ammonia recycle percolation (ARP) 77
Amsterdam-based Shell Research Laboratory 114
amylopectin 21–23
amylose 20, 22
anti-infective ointment 49
arabidopsis 201
arabinoglucuronoxylan (AGX) 185
arabinoxylan 12
Aro, N. 160
Aspen Plus 78
A-type crystalline starch 23
automotive, and transportation segment, of
  polymers 360

B

bacteria, in biodiesel production 273
bacterial cellulase systems 157–159
bacterial enzymes 44
Bakhiet, S.E.A. 163
Ballesteros, M. 167
batch fermentation 338
beer column 304

benzene, toluene, and xylene (BTX) 225
BiGG model 133
Biller, P. 112
bioalcohols 253, 315–316
  isobutanol 254–255
  n-butanol 253–254
biobutanol 310, 351, 364; *see also* second-
  generation biofuels
  ABE/IBE fermentation 310–311, 312
  amelioration 311
  consolidated bioprocessing 311
  downstream processing 313
    hybrid technologies 315
    pervaporation 314–315
  opportunities and challenges 315
biochemical biorefineries 213
biochemical processes 94
BioCyc database 132
biodiesel 267
bioenergy *see individual entries*
bioethanol 292, 351; *see also* second-generation
  biofuels
  downstream processing 304–308
  fermentative microorganisms 302–303
  opportunities and challenges 309
  research and development 308–309
  thermophilic fermentative microorganisms
    303–304
Bioethanol Plant-Gate Price Assessment Model
  (BPAM) 77, 78
bio-ethylene 363
biofuels 65–66, 292
  alkanes/alkenes production 256–258
  bioalcohols 253
    isobutanol 254–255
    n-butanol 253–254
  first-generation 242–244
  hydrocarbons 255–256
  renewable 292
  second-generation 244–245
    resistance engineered against inhibitors
      245–251
  third-generation 251–253
biogas 226–227
  production 358
biogenic by-products, consumption of 360
biological pretreatment methods 98–99, 299
biomass
  definition 2
  liquefaction 94
  thermochemical conversion of 94

biomass-derived products 352–353, 354
bio-methanol 350, 362
BioModels 133
Bio-PET 79
biopolymers 359
biopropane 363
bio-propane 351
biorefinery 94, 360–361
    biogas platform 361
    C6 and C6/ C5 sugar platform 361–362
    lignin-based platform 362
    organic solutions platform 362
    oxidation processes 362
    selective dehydration and hydrogenation 362
    syngas platform 361
bio-refinery 79
BIOSUCCINIUM® 221
biosyngas 228, 350
BLAST 135
Bligh & Dyer and Folch methods 276
BNICE 141–142
Boocock, D.G.B. 111
Borrion, A.L. 85, 88
BRENDA 133
Brundtland Commission Report 66
B type starch 23
Bugg, T.D.H. 44
building, and construction segment,
            of polymers 360
Bumpus, J.A. 44
Bundesforschungsanstalt für Forst- und
            Holzwirtschaft (BFH) 113
butanedioic acid see succinic acid (SA)
1,4-butanediol (1,4-BDO) 332, 351, 364
2,3-butanediol (2,3-BDO) 222, 331–332, 334–335
    downstream processing 339–340
    feedstock 335, 336
    fermentation process 336, 338
    potential microorganism 335, 337
    substrate concentration effects 338

C

caffeoyl alcohol 197
carbohydrate-active enzymes (CAZymes) 154
carbon catabolite repression (CCR) 160–162
carbon sequestration capability 67
carbon sources 138, 162–163
Caruso, T. 107
CarveMe 140
cassava 244
catalase-peroxidases 46
catalysescellulose synthase complexes (CSCs) 181
catechol tannins see condensed tannins
catechyl (C) lignin 195
cell-free cellulase enzymes 159
cellobiose 301

cellulases 154–156
    applications 167–168
    bacterial cellulase systems 157–159
    enzymes hydrolysis 300
    fungus 156–157
    industrial success stories, in India 168–169
    influencing factors
        carbon sources 162–163
        nitrogen sources 163
        pH and temperature 164
    production
        carbon catabolite repression 160–162
        transcriptional activators and
            repressors 160
cellulolytic enzymes 154
cellulose 32
    biosynthesis 9, 181
        crystallinity 183
        degrading enzymes 184
        disrupting native cellulose pathway
            182–183
        SuSy activity 181–182
    crystalline polymorphs 6
    forms 8
    with inter- and intra-hydrogen bonds 8
    lignocellulosic biomass 96, 97
    microfibrils 6
    in planta modification of 180
    polymeric characteristics 9
    polysaccharide 6
    secondary cell wall 6
    synthesis 8
    X-ray diffraction analysis 9
cellulose microfibrils (CMs) 180
cellulose synthase complexes (CSCs) 181
cellulose synthases (CESAs) 181, 182
cellulosic ethanol production 78
cellulosomes 157–159
Chemical Engineering Index 75
chemical pretreatment methods 99–101, 296–297
    acid 99
    alkali 100
    ionic liquids 101
    organosolv 100–101
chemicals industry
    adipic acid 365
    bio-based propylene glycol 364
    C1-containing compounds 362–363
    C2-containing compounds 363
    C3-containing compounds 363–364
    C4-containing compounds 364–365
    C5-containing compounds 365
    C6-containing compounds 365
Cheng, J. 98
Ciamician, G. 37
cinnamyl-alcohol dehydrogenase (CAD) 196
citric acid 351, 365

Clostridium acetobutylicum 253
clustered regularly interspaced short palindromic repeats (CRISPR/Cas9) 144
^{13}C-metabolic flux analysis (^{13}C-MFA) 145
commercially available biomass 351–352
Comparative ReConstruction (CoReCo) 140
compressed biogas (CBG) 227
compressed natural gas (CNG) 227
condensed tannins 26
coniferyl alcohol 50
consolidated bioprocessing (CBP)
    biobutanol 311
    enzyme production 270
Constraint-Based Reconstruction and Analysis-Python (COBRApy) 139
Constraint-Based Reconstruction and Analysis Toolbox (COBRA Toolbox) 134, 138–139
consumer goods segment, of polymers 360
contactors 304
conventional biorefineries 213
corbion 364
cradle-to-gate system boundary 82, 86
Cre1/CreA 161–162
Crestini, C. 34
crude glycerol 333
crystalline cellulose 183
C-type starch 23
Curvelo, A.A.S. 100
cyanobacteria 251

D

Datar, R. 102
DBT-IOCL, Faridabad 169
δ-aminolevulinic acid (DALA) 223
de Graaff, L.H. 161
degree of polymerization (DP) 9
Dehghanzad, M. 79
depolymerization 37–43, 105, 118
Dermibas, A. 110
DESHARKY 141
Deshpande, S.K. 164
diols
    applications 333–334
    biological production
        1,4-butanediol 332
        2,3-butanediol 331–332
        2, 4-pentanediol 332–333
        1,2-propanediol 330–331
        1,3-propanediol 329–330
    2,3-butanediol production 334–335
        downstream processing 339–340
        feedstock 335, 336
        fermentation process 336, 338
        potential microorganism 335, 337
        substrate concentration effects 338

compounds 221–222
    importance and production 327–329
diphenolic acid 223
direct liquefaction 95
disaccharides 162
DNA assembly tools 142
Donev, E. 180
downstream processing
    bioethanol 304–308
    2,3-butanediol production 339–340
    of butanol
        ABE fermentation process 313
        conventional distillation process 313
        hybrid technologies 315
        liquid–liquid extraction 313
        pervaporation 314–315
    of ethanol
        adsorption 308
        alternative recovery techniques 305
        distillation 304–305
        gas stripping 307
        pervaporation 305–307
        solvent extraction 308
        vacuum fermentation 307–308
downstream saccharification 162–164
draft metabolic model 133–135
dried distillers grains with solubles (DDGS) 164
drop-in fuels 229–231
drug delivery system 49
dry washing method 279

E

edible oil-derived biodiesel 268
edible plant oil, consumption of 360
efficiency 215
Eggeman, T. 77
electrical, and electronics segment, of polymers 360
electricity 87
electron beam irradiation 295
El-Hadi, A.A. 164
Emanuel, E. 77
endocellulase 159, 165, 184
energy return on investment (EROI) 244
enzymatic hydrolysis 167, 299
    factors affect 300–301
    lignocellulosic biomass 300
enzyme-based pretreatment 299
enzyme-based saccharification
    cellulases 154–156
        applications 167–168
        bacterial cellulase systems 157–159
        fungus 156–157
        industrial success stories, in India 168–169
        influencing factors 162–164
    enzyme cocktails 165–167
    laboratory and industrial scales 164–165

enzyme-related factors 301
ethyl acetate 351, 365
ethyl alcohol 26
ethylene glycols 363
exocellulase 165
The Expert Protein Analysis System (ExPASy) 133

F

fatty acid ethyl esters (FAEEs) 276, 277
FBA *see* flux balance analysis (FBA)
FDCA *see* 2,5-furandicarboxylic acid (FDCA)
feasibility analyses 67
fed-batch enzymatic hydrolysis 77
Feedstock Cost Estimation Model (FCEM) 77, 78
Feng, S. 110
fermentation process 336, 338
ferulic acid esterase (FAE) 190
F5H-deficient *fah1* 198–199
filamentous fungi 273
first-generation biofuels 242–244
Fischer–Tropsch (F-T) process 94
flexibility 216
flexible packaging segment, of polymers 360
flux balance analysis (FBA)
    function 136
    with genetic algorithm 141
    of metabolic models 137
    optimizations 135–137
    OptKnock 140–141
flux distribution 129
flux scanning based on enforced objective flux
        (FSEOF) method 141
forest biomass 351
formic acid 106
fossil fuels, of extraction and combustion
        241–242
Freeman, G.G. 156
Freund, A. 329
Fujishima, A. 37
functional unit (FU) 83, 86
fungal species 44
2,5-furandicarboxylic acid (FDCA) 223–224
furfural 351, 365
furfural/5-hydroxymethylfurfural 245–249
futile cycles 131, 135

G

galactoglucomannan (GGM) 193
galactomannan 13
Galanopoulou, A.P. 159
gamma irradiation 295
gap-filling process 134
gapped metabolites 134
gapped reactions/pathways 134
gasification 94, 95

gasoline 315–316
Gautam, S.P. 163, 164
gene-protein-reaction (GPR) 129
genome annotation 130, 131
genome-scale metabolic model (GSMM) 129
    experimental data 137–138
    integration 145
    metabolic engineering
        minimization of metabolic adjustments 141
        OptGene 141
        OptKnock 140–141
        OptPipe 141
        OptReg 141
        regulatory on/off minimization 141
        RELATive CHange 141
        synthetic biology applications 141–142
    reconstruction of 130–133
genomic library 250
Glebes, T.Y. 248
global biomass 350, 353
    biogases based on biomass supply 357
    for energy 354
    for fuels 354
    industrial waste-based biomass supply 356
    liquid biofuels based on biomass supply 357
    municipal waste-based biomass supply 356
    solid biofuels based on biomass supply
        356–357
    supply by source of 354–356
glucomannan 13
glucose 181
glucuronoxylan 12
glycerol 364
Gough, S. 304
Grassmann, W. 156
greenhouse gas emissions 85
Greenhouse gases, Regulated Emissions,
            and Energy use in Transportation
            (GREET) model 85
GSMM *see* genome-scale metabolic model
            (GSMM)
gymnosperm xylan 185

H

Hakkinen, M. 160
Hari Krishna, S. 304
Harmsen, P. 100
heme peroxidases 46
hemicelluloses 32, 128
    chemical structures 10, 12
    components 10, 11
    forms 10
    hydrolysis of 99
    insoluble characteristics 10
    lignin 14, 17–20
    lignocellulosic biomass 96, 97

mannan 13
  structural characteristics 10
  xylan 10, 12–13
  xylogalactan 13–14
  xyloglucan 14, 15–17
herbaceous biomass 3–4
heterogeneous catalysts 112
heterogeneous photocatalysts 37
high-energy advanced biofuels; *see also* biofuels
  alkane/alkene production 256–258
  bioalcohols 253–255
    isobutanol 254–255
    n-butanol 253–254
  hydrocarbons 255–256
H-lignin 198
5-HMF *see* 5-hydroxymethylfurfural (5-HMF)
Hochschule für Angewandte Wissenschaften
      Hamburg 114
homomannan 13
homoxylan-type hemicellulose 12
Honda, K. 37
Hongzhang, C. 101
HTL *see* hydrothermal liquefaction (HTL)
Hwang, H.J. 332
Hybrid ISPR 315
hydrocarbons 255–256
hydrodeoxygenation (HDO) 229–230
hydrolysis 226
hydrolyzable tannins 24, 25
hydrothermal liquefaction (HTL) 95, 104
  challenges 114–115
  industrial applications 113–114
  operating parameters effect 109
    biomass- to-water mass ratio 111–112
    catalyst 112–113
    feedstock type and particle size 109, 110
    heating rate and thermal gradient 110–111
    influence 109, 110
    pressure 112
    residence time 111
    temperature 110
  reaction mechanism 105
    for degradation of carbohydrates 105–108
    for degradation of lignin 108–109
    plausible reaction pathway 105, 106
  sub- and supercritical water properties 104–105
5-hydroxymethylfurfural (5-HMF) 107–108
2-hydroxypropanoic acid *see* lactic acid (LA)

I

Ilmen, M. 161, 163
India Glycols Ltd. 168–169
International Energy Report 2
ionic liquid (IL) 79
  membranes 314–315
  pre-treatment 101

irradiation pretreatment 295–296
isobutanol 254–255
isoprene 224–225
itaconic acid 351, 365
Itoh, H. 99

J

Jatropha-based biodiesel production 87
Jeon, E. 331
Jia, W. 332
Jin, F. 106, 107
Ji, X.J. 335

K

Kabel, M.A. 167
Kabyemela, B.M. 112
Kadam, K.L. 85
Kahn, A. 159
Karagöz, S. 109, 111
Kazi, F.K. 78
Klein-Marcuschamer, D. 77, 79
Kong, L. 107
KORRIGAN1 (KOR1) 183
Kumar, M. 44
Kuster, B.F.M. 107
Kyoto Encyclopedia of Genes and Genomes
      (KEGG) database 132

L

laccases 46–47
lactic acid (LA) 106–107, 219–220, 364
lactic acid bacteria (LAB) 219
laminarin 252
land use 67
Lark, N. 304
Lawrence Berkeley National
      Laboratory 113
LCA *see* life cycle analysis (LCA)
Lee, J.B. 50
levulinic acid 108, 222–223, 351, 365
Li, C. 101
life cycle analysis (LCA) 66
  of biofuel 244
  case studies 85–87
  challenges 87–89
  concept 81
  criteria and indicators 66
  goal and scope 80–83
  impact assessment 84–85
  interpretation 85
  inventory 83–84
  methodology 80
  product-related assessment 68
  research gaps 87–89

lignin 33, 35–36, 128
    in biological agents
        activity 47–48
        drug carrier 49
        microbicidal agent 50
        polyphenolic nature 47
        theranostic agent 50–50
        therapeutic properties 47
    biosynthesis 14
    definition 14
    disrupting cross-linkages 199–200
    functional groups 18
    labile monolignols 200
    lignocellulosic biomass 97
    photodynamic chemotherapy effect 52
    in planta modification of 195–197
    plant cell wall composition 17–20
    reaction mechanism 108–109
    reducing effect, in plant morphology
        197–198
    structure and its components 33–35, 36
    substantial research 36
    valorization 35–36
        consumption 36
        enzymatic degradation 39, 44–47
        photocatalytic degradation 37–39, 40–43
        properties 37
        substantial research 36
        utilization 37
    virucidal and bactericidal effects 50
lignin biosynthesis-related transcription factor 1
        (LTF1) 198
lignin–carbohydrate complexes (LCs) 50
lignin-degrading (LD) enzumes 44, 46
lignin-modifying (LM) enzymes 44, 46
lignin peroxidases (LiP) 46
lignocellulosic biomass (LB) 32–33, 128, 212, 214
    agro-based economies 94
    pre-treatment methods 97–98
        biological 98–99
        chemical 99–101
        physical 98
        physicochemical 101–104
    structure and composition 95–97
lignocellulosic biorefineries 212
    advantages 213
    chemical production 218–219
    classification 212–213
    commercialization 217–218
    diol compounds 221–222
    fractionation technologies and importance
        214–215, 216
    2,5-furandicarboxylic acid 223–224
    isoprene 224–225
    lactic acid 219–220
    levulinic acid 222–223
    lignin-derived chemicals 225–226

succinic acid 220–221
sustainable technology 215–217
valorization
    biogas 226–227
    drop-in fuels 229–231
    synthesis gas 227–229
Li, L. 100
Lio, J. 164
LiP see lignin peroxidases (LiP)
lipid extraction methods 275–276
liquefaction 94, 230; see also hydrothermal
        liquefaction (HTL)
    direct 95
    Fischer–Tropsch (F-T) process 94
    hydrothermal 95
    indirect 94
    thermodynamic 95
liquid ammonia 163
liquid biofuels 358
liquid hot water pretreatment 102, 298
liquid petroleum fuels 128
Liu, A. 109
Liu, H.M. 111
Liying, L. 101
lower-grade biomass 95
low molecular weight organic compounds 23
    extractives 23–26
    fluid content 26–27
    inorganic constituents 26
low-temperature conversion (LTC) process 114
Lu, K.-M. 315
lysine 351, 365
lytic polysaccharide monooxygenases
        (LPMOs) 159

M

macroalgae 252, 253
Magnusson, L. 99
Mandade, P. 86
Mandels, G.R. 164
manganese-dependent peroxidase (MnP) 46
mannan
    biosynthesis 193–194
        acetylation in plants 194
        degrading enzymes 194–195
    hemicellulose 13
    in planta modification of 193
    structural features 193
mannan O-acetyltransferases (MOATs) 194
mannan synthesis-related (MSR) proteins 194
market analysis
    of biomass-based fuel production 357–358
    of biomaterials 359–360
mass balance analysis 131
maturity 68, 69–70, 216
McGinnis, G.D. 101

McKone, T.E. 88
mechanical pretreatment 295
Meng, W. 332
MERLIN 140
metabolic engineering modelling
    alkanes/alkenes production 256–258
    draft model 133–135
    experimental data 137–138
    genome-scale metabolic model 130–133
    integrative analyses 145
    minimization of metabolic adjustments 141
    OptGene 141
    optimization criteria 135–137
    OptKnock 140–141
    OptPipe 141
    OptReg 141
    pathway-centric approach 130
    regulatory on/off minimization 141
    RELATive CHange 141
    synthetic biology applications
        biofuel molecules production 142, 143
        BNICE 141–142
        DNA assembly tools 142
        enzyme discovery 145
        improving tolerance to inhibitors 144
        MAPPS 142
        naringenin biosynthesis pathway 142
    traditional approach 130
metabolic modelling
    COBRApy 139
    COBRA Toolbox 138–139
    pathway tools 139
    ScrumPy 139
    sybil 139
Metabolic network Analysis and Pathway
        Prediction Server (MAPPS) 142
MetaCyc database 132
methanogenesis 226
methyl glucuronic acid level 187–188
microalgae 252
    biodiesel production 271–273
microbes-based pretreatment 299
microbial industrial organisms 253
microcrystalline cellulose (MCC) 49
microwave irradiation 103, 295–296
milling 98
mineral acids 112
minimization of metabolic adjustments
        (MOMA) 141
Minowa, T. 112
MnP see manganese-dependent
        peroxidase (MnP)
ModelSEED 133, 139–140
monolignols 195
motor spirit (MS) see gasoline
Mrudula, S. 163
multifeedstock biorefinery 218

municipal solid waste (MSW) 95–96
Murali, G. 86
Murugammal, R. 163

N

Nakayama, S. 311
naringenin biosynthesis pathway 142
n-butanol 253–254
Negro, M.J. 167
net energy ratio (NER) 86
Nimz, H. 34
Ninomiya, K. 101
nitrogen sources 163
non-edible plant oil, consumption of 360

O

O-acetyl glucuronoxylans (AcGXs) 185
Obiaga, T.I. 34
oligosaccharides 162
Oliva, J.M. 304
OptGene 141
OptKnock 140–141
OptPipe 141
OptReg 141
organic acids 112
organic solvent pretreatment 297
organosolv pre-treatment 100–101
oxidation 38
oxygen oxidoreductases 46–47

P

PACT see photodynamic antimicrobial
        chemotherapy (PACT)
Park, Y.K. 78
pathway genome database (PGDB) 132
p-Coumaroyl-CoA 196
pectin
    bioengineering 191–193
    degrading enzymes 191
    in planta modification of 190–191
pectin acetylesterase (PAEI) 191
1,5-pentanediamine 365
2,4-pentanediol 332–333
Penttila, M. 161
Pereira, R. 100
peroxidases (POD) 44, 46–47
pervaporation (PV) technique 305–307, 314–315
phenylalanine (Phe) 195–196
photocatalysis 37–38
photodynamic antimicrobial chemotherapy
        (PACT) 51
photodynamic therapy 50–51
photosynthetic microorganisms 128
physical pretreatment methods 98, 295–296

physicochemical pretreatment methods 101–102,
        297–298
    ammonia fibre explosion 102–103
    liquid hot water 102
    microwave irradiation 103
    steam explosion 102
    ultrasound irradiation 103
    wet oxidation 103–104
Pittsburgh Energy Research Centre (PERC) 113
plant biomass
    application 2
    attractive inherent fractional composition 2
    classification 3–4
    low molecular weight constituents 23
        extractives 23–26
        fluid content 26–27
        inorganic constituents 26
        plant cell wall composition and architecture
            4–6, 7
            cellulose 6, 8–9
            hemicellulose 10–14
            lignin 14–19
            starch 19–23
        sources 2
plant cell wall
    cellulose 6, 8–9
    components 180
    development 4, 5
    hemicellulose 10–14
    lignin 14–19
    polysaccharides 5
    primary wall and middle lamella 4
    secondary 5–6
    starch 19–23
    types 4
polyamines 246
polybutylene terephthalate (PBT) 333
polydimethylsiloxane (PDMS) 314
polyethylene furanoate (PEF) 223, 224
polyethylene terephthalate (PET) 223
polymers 359
polysaccharides 5
porphyrin-loaded lignin nanoparticles 51
potassium hydroxide (KOH) 100
Prairie Catalytic 365
PRAJ Industries, Pune 169
Prasanna, H.N. 164
pretreatment methods 97–98
    advantages 153–154
    biological 98–99
    chemical 99–101
        acid 99
        alkali 100
        ionic liquids 101
        organosolv 100–101
    for fractionation of LBM 216
    physical 98
    physicochemical 101–102

    ammonia fibre explosion 102–103
    liquid hot water 102
    microwave irradiation 103
    steam explosion 102
    ultrasound irradiation 103
    wet oxidation 103–104
second-generation biofuels 294–295
    biological 299
    chemical 296–297
    physical 295–296
    physicochemical 297–298
primary biomass 4
product-related assessment 68
profitability 217
profitability analysis 71
1,2-propanediol (1,2-PDO) 330–331
1,3-propanediol (1,3-PDO) 222, 329–330
propylene 364
protopectinases 191
pseudo-tannins 26
python, metabolic modelling in 139

Q

quantum dots (QDs) 39

R

Rajasekhar Reddy, B. 164
Ramanjaneyulu, G. 164
Rashid, G.M.M. 44
reactive oxygen species (ROS) 38
recalcitrance 294
Reconstruction, Analysis and Visualization of
        Metabolic Networks (RAVEN) 140
reduced epidermal fluorescence8 (ref8) 199
Reese, E.T. 156
regulatory on/off minimization (ROOM) 141
RELATive CHange (RELATCH) 141
renewable biofuels 292
renewable energy sources 94
RetroPath 142
rigid packaging segment, of polymers 360
robustness 217
Rughani, J. 101

S

saccharides
    biodiesel production 271
        bacteria 273
        filamentous fungi 273
        microalgae 271–273
        microorganisms co-cultivation 274
        properties 279
        purification 278–279
        yeast 273
    chemical constitution of 268–269

fatty acid profiles 279
lipid extraction methods 275–276
physicochemical properties 268–269
source of 270–271
sugar conversion 274–275
transesterification 276–278
Saccharomyces cerevisiae 254, 294
Sant'Ana da Silva, A. 101
Sassner, P.G. 77
Schmidt, A.S. 104
Scown, C.D. 85
ScrumPy 139
Searchinger, T. 88
secondary biomass 4
second-generation biofuels 244–251, 292
    bioalcohols 315–316
    biobutanol 310
        ABE/IBE fermentation 310–311, 312
        amelioration 311
        consolidated bioprocessing 311
        downstream processing 313–315
        opportunities and challenges 315
    bioethanol 292
        downstream processing 304–308
        fermentative microorganisms 302–303
        opportunities and challenges 309
        research and development 308–309
        thermophilic fermentative
            microorganisms 303–304
    conversion of biomass 293–294
    enzymatic hydrolysis 299
        factors affect 300–301
        lignocellulosic biomass 300
    feedstocks and composition 293
    gasoline 315–316
    pretreatment strategies 294–295
        biological 299
        chemical 296–297
        physical 295–296
        physicochemical 297–298
Shastri, Y. 86
simultaneous saccharification and
    co-fermentation (SSCF) 86–87
simultaneous saccharification and fermentation
    (SSF) 77, 270
sinapyl alcohol 50
Singh, A. 85
Singh, R. 111, 112
single-cell oils (SCOs) 268
S-lignin 199
Soam, S. 86, 87
sodium hydroxide (NaOH) 100
solar energy 37
solid biomass 2
sorbitol 351, 365
Spatari, S. 87, 88
Sreekumar, A. 87
Srokol, Z. 106

starch
    amylopectin 21–23
    amylose 20, 22
    consumption of 360
    molecular arrangement 19, 21
    synthesis 19
steam explosion 102, 297–298
Sternberg, D. 164
sterols 26
substrate-related factors 301
succinic acid (SA) 220–221, 351, 365
sucrose synthase (SuSy) catalyses 181
Sugano, M. 110
sugars by-products, consumption of 360
sulfur 316
Sun, F.F. 167
Sun, Y. 98
superoxide dismutases 47
SuperPro Designer® 79
supply chain optimization 217–28
sustainability 66
sustainable technology
    efficiency 215
    flexibility 216
    maturity 216
    profitability 217
    robustness 217
Swanson, R.M. 78
sybil package 139
synthesis gas (Syngas) 227–229
Systems Biology Markup Language
    (SBML) 139

T

Takagi 303
tannins 24
TEA *see* techno-economic analysis (TEA)
techno-economic analysis (TEA) 66
    accuracy 74
    case studies 77–79
    challenges 79–80
    commodity materials 76
    constant, reliable supply of utilities and raw
        materials 75
    criteria and indicators 66
    financial assumptions 75–76
        capital costs 75
        operating costs 75–76
    goal and scope 70–71
    interpretation 76–77
    inventory 71–74
    methodology 68–69
    non-optimized model 67
    phases 68–69
    research gaps 79–80
    stable, reliable equipment operation
        assumption 75

techno-economic analysis (TEA) (*cont.*)
    steady state assumption 74–75
    technology maturity 69–70
technology readiness levels (TRLs) 70
technology readiness standards 70
terpenes 26
tertiary biomass 4
tetrahydrofuran (THF) 333
textile segment, of polymers 359
thermal processes 94
thermochemical processes 230
thermodynamically infeasible cycles 131, 135
thermodynamic liquefaction 95
third-generation biofuels 251–253
Thrane, C. 161
Tran, H.T.M. 311
Trichoderma reesei QM6a 156
trimesters 26
tyrosine ammonia lyase (TAL) enzyme 196

U

ultrasound irradiation 103
ultrasound pretreatment 296
Umwelttechnik Stefan Bothur 114
Universal Protein (UniProt) database 133
Uracil-Specific Excision Reagent (USER)
        cloning 144

V

valorization
    of lignin 35–36
        consumption 36
        enzymatic degradation 39, 44–47
        photocatalytic degradation 37–39, 40–43
        properties 37
        substantial research 36
        utilization 37
    of lignocellulosic biorefineries
        biogas 226–227
        drop-in fuels 229–231
        synthesis gas 227–229
van der Veen, P. 161
vascular-related NAC-domain 6 (VND6) 197–198
virgin biomass 33
Viswanathan, M.B. 79
von Bertalanffy 1.0 algorithm 131

W

Wang, C. 111
Wang, H. 37
Wang, M. 85
Wang, T. 164
Ward, H.M. 156

waste biomass 33
waste-derived biodiesel 267
waste streams 351–352
wax starch 21
Wayman, M. 34
Weil, J. 102
wet oxidation 103–104
whole-crop biorefineries (WCBRs) 212–213
Wiloso, E.I. 88
Wong, D.W.S. 46
wood-based fuel production 357
wood charcoal production 358
wood pellets production 357
Woodward 156
woody biomass 3
Wu, J. 332
Wu, L. 39

X

Xiang, Z. 39
XlnR 160
xylan 10, 12–13
    biosynthesis 185–187
        downregulating expression 187
        hydrolytic enzymes 189–190
        modifying polysaccharide
            acetyltransferase expression 188–189
        reduce end sequence 187–188
    in planta modification of 184–185
    structural feature 185–190
xylanases 300
xylan synthesis complex (XSC) 187
xylitol 351, 365
xylogalactan 13–14
xyloglucan hemicellulose 14, 15–17
xylosyl arabinosyltransferase (XAT) 187
Xyr1 160

Y

yeast 273
Yim, H. 332
Yun, J. 106
Yu, Z. 111

Z

zero waste technology 218
Zhang, B. 108, 110, 111
Zhang, G. 38
Zhao, L. 78
Zheng, M. 100
Zhou, Z. 100
Zhu, Z. 110
Zou, G. 162